P9-AQI-325

# SELECTED TABLES IN MATHEMATICAL STATISTICS

## Volume VII

## THE PRODUCT OF TWO NORMALLY DISTRIBUTED RANDOM VARIABLES

*by*


**WILLIAM Q. MEEKER, JR., LARRY W. CORNWELL, and LEO A. AROIAN**


**Edited by the Institute of Mathematical Statistics**

*Coeditors*

**W. J. Kennedy**
*Iowa State University*

*and*

**R. E. Odeh**
*University of Victoria*

*Managing Editor*

**J. M. Davenport**
*Texas Tech University*


**AMERICAN MATHEMATICAL SOCIETY**

**PROVIDENCE, RHODE ISLAND**

**This volume was prepared with the aid of:**

J. E. Gentle, International Mathematical & Statistical Libraries, Inc.

K. Hinklemann, Virginia Polytechnic Institute and State University

R. L. Iman, Sandia Laboratories

D. B. Owen, Southern Methodist University

S. Pearson, Bell Laboratories

R. H. Wampler, National Bureau of Standards

1980 *Mathematics Subject Classification*
Primary 62Q05; Secondary 62E15, 60E05.

International Standard Serial Number 0094-8837
International Standard Book Number 0-8218-1907-0
Library of Congress Card Number 74-6283

# PREFACE

This volume of mathematical tables has been prepared under the aegis of the Institute of Mathematical Statistics. The Institute of Mathematical Statistics is a professional society for mathematically oriented statisticians. The purpose of the Institute is to encourage the development, dissemination, and application of mathematical statistics. The Committee on Mathematical Tables of the Institute of Mathematical Statistics is responsible for preparing and editing this series of tables. The Institute of Mathematical Statistics has entered into an agreement with the American Mathematical Society to jointly publish this series of volumes. At the time of this writing, submissions for future volumes are being solicited. No set number of volumes has been established for this series. The editors will consider publishing as many volumes as are necessary to disseminate meritorious material.

Potential authors should consider the following rules when submitting material.

1. The manuscript must be prepared by the author in a form acceptable for photo-offset. This includes both the tables and introductory material. The author should assume that nothing will be set in type although the editors reserve the right to make editorial changes. The tables must be produced from computer output.

2. While there are no fixed upper and lower limits on the length of tables, authors should be aware that the purpose of this series is to provide an outlet for tables of high quality and utility which are too long to be accepted by a technical journal but too short for separate publication in book form.

3. The author must, wherever applicable, include in his introduction the following:

(a) He should give the formula used in the calculation, and the computational procedure (or algorithm) used to generate his tables. Generally speaking, FORTRAN or ALGOL programs will not be included but the description of the algorithm used should be complete enough that such programs can be easily prepared.

(b) A recommendation for interpolation in the tables should be given. The author should give the number of figures of accuracy which can be obtained with linear (and higher degree) interpolation.

(c) Adequate references must be given.

(d) The author should give the accuracy of the table and his method of rounding.

(e) In considering possible formats for his tables, the author should attempt to give as much information as possible in as little space as possible. Generally speaking, critical values of a distribution convey more information than the distribution itself, but each case must be judged on its own merits. The text portion of the tables (including column headings, titles, etc.) must be proportional to the size 5–1/4″ by 8–1/4″. Tables may be printed proportional to the size 8–1/4″ by 5–1/4″ (i.e., turned sideways on the page) when absolutely necessary; but this should be avoided and every attempt made to orient the tables in a vertical manner.

(f) The table should adequately cover the entire function. Asymptotic results should be given and tabulated if informative.

(g) An example or examples of the use of the tables should be included.

4. The author should submit as accurate a tabulation as he can. The table will be checked before publication, and any excess of errors will be considered grounds for rejection. The manuscript introduction will be subjected to refereeing and an inadequate introduction may also lead to rejection.

5. Authors having tables they wish to submit should send two copies to:

Dr. Robert E. Odeh, Coeditor
Department of Mathematics
University of Victoria
Victoria, B. C., Canada V8W 2Y2

At the same time, a third copy should be sent to:

Dr. William J. Kennedy, Coeditor
117 Snedecor Hall
Statistical Laboratory
Iowa State University
Ames, Iowa 50011

Additional copies may be required, as needed for the editorial process. After the editorial process is complete, a camera-ready copy must be prepared for the publisher.

Authors should check several current issues of *The Institute of Mathematical Statistics Bulletin* and *The AMSTAT News* for any up-to-date announcements about submissions to this series.

## ACKNOWLEDGMENTS

The tables included in the present volume were checked at the University of Victoria. Dr. R. E. Odeh arranged for, and directed this checking with the assistance of Mr. Bruce Wilson. The editors and the Institute of Mathematical Statistics wish to express their great appreciation for this invaluable assistance. So many other people have contributed to the instigation and preparation of this volume that it would be impossible to record their names here. To all these people, who will remain anonymous, the editors and the Institute also wish to express their thanks.

*To*: **KAREN, SARA, and ARMINÉ**

## Contents of VOLUMES I, II, III, IV, V, and VI of this Series

# TABLE OF CONTENTS

# THE PRODUCT OF TWO NORMALLY DISTRIBUTED RANDOM VARIABLES

William Q. Meeker, Jr., Iowa State University

Larry W. Cornwell, Western Illinois University

Leo A. Aroian, Union College and University

ABSTRACT

Tables for the fractiles of the distribution of the product of two normal random variables are presented. The numerical methods used to compute and check the tables are described and some of the applications of this distribution are reviewed. Interpolation in the tables is discussed and some examples are given.

## 1. INTRODUCTION

The product of two normally distributed random variables occurs frequently in applications, for example, in the fields of physical, engineering, and social sciences, and also in biometry, sampling, auditing, and other business applications. Some specific applications are outlined in Section 6. The important properties of this distribution were first investigated by Craig (1936). Some further mathematical results are given by Aroian et al. (1978). Aroian (1947) showed that under certain conditions the distribution is asymptotically normal. For situations when these conditions are not met, the tabulations of the fractiles, presented here can be used.

Tables of the distribution of the product of two <u>independent</u> normally distributed random variables were originally given by Aroian (1959). The present tables contain fractiles of the distribution of the

---

Received by the editors January 1980 and in revised form May 1980.

product of two possibly dependent normally distributed random variables.

The important properties of this distribution are reviewed in Section 2 and numerical methods for computing the distribution function are given in Section 3. The tables of the distribution function and the fractiles are explained in Sections 4 and 5 respectively. Interpolation in the tables is explained in Section 7 and some examples of interpolation are given in Section 8.

## 2. NOTATION AND PROPERTIES OF THE DISTRIBUTION

Let $X_1$ and $X_2$ be two normally distributed random variables with means $\mu_i$, variances $\sigma_i^2$, i = 1, 2, and correlation coefficient $\rho$. Define $Z = X_1X_2/(\sigma_1\sigma_2)$, $\delta_1 = \mu_1/\sigma_1$, and $\delta_2 = \mu_2/\sigma_2$. For the random variable Z, Craig (1936) derived the distribution function, the moment generating function, the cumulants, mean, variance and measures of skewness and kurtosis. Briefly, his results are as follows.

The mean and variance of Z are

$$\begin{aligned} \mu_Z &= \delta_1\delta_2 + \rho \\ \sigma_Z^2 &= \delta_1^2 + \delta_2^2 + 2\delta_1\delta_2\rho + 1 + \rho^2. \end{aligned} \tag{2.1}$$

Measures of skewness and kurtosis are

$$\xi_3 = \mu_3/\sigma_Z^3 = \{6[\rho(\delta_1^2 + \delta_2^2) + \delta_1\delta_2(1 + \rho^2)] + 2\rho(3 + \rho^2)\}/\sigma_Z^3$$

$$\begin{aligned} \xi_4 = \mu_4/\sigma_Z^4 = [&12(\delta_1^2 + \delta_2^2)(1 + 3\rho^2) + 24\delta_1\delta_2\rho(3 + \rho^2) \\ &+ 6(1 + 6\rho^2 + \rho^4) + 3\sigma_Z^4]/\sigma_Z^4 \end{aligned}$$

where $\mu_3$ and $\mu_4$ are the third and fourth central moments, respectively, of Z.

Let $F_Z(z) = F_Z(z;\delta_1,\delta_2,\rho) = P(Z \leq z)$ be the cumulative distribution function of Z. Evident relationships for the distribution are

$$\begin{aligned} F_Z(z;\delta_1,\delta_2,\rho) &= F_Z(z;\delta_2,\delta_1,\rho) \\ F_Z(-z;-\delta_1,\delta_2,-\rho) &= 1 - F_Z(z;\delta_1,\delta_2,\rho). \end{aligned} \tag{2.2}$$

Also, there is a singularity in the pdf $f_Z(z)$ at $z = 0$. The characteristic function of Z is

$$\varphi_Z(t) = [\exp(-N/D)]/E, \text{ where}$$

$$N = (\delta_1^2 + \delta_2^2 - 2\rho\delta_1\delta_2)t^2 - 2\delta_1\delta_2 it$$

$$D = 2[1 - (1 + \rho)it][1 + (1 - \rho)it]$$

$$E = (D/2)^{1/2}$$

For further information, see Craig (1936) and Aroian, Taneja, and Cornwell (1978).

Aroian et al. (1978) discuss several special cases for the distribution of Z. For $\rho = 1$ and $\delta_1 = \delta_2 = 0$, the distribution is chisquare with 1 degree of freedom. When $\rho = 1$ and $\delta_1 = \delta_2 = \delta$, the distribution is noncentral chisquare with 1 degree of freedom and noncentrality parameter $\lambda = \delta^2/2$. A further result for $\rho = 1$ $(-1)$ is that the distribution of $Z - \mu_Z$ depends only on $\delta_1 + \delta_2$ $(\delta_1 - \delta_2)$ and is thus related to the noncentral chisquare distribution in the obvious way. This can be seen in the tables as well as by manipulation of the characteristic function of Z.

## 3. NUMERICAL METHODS

The distribution function and fractiles were initially computed as described in Sections 4 and 5 using the algorithm of Cornwell, et al. (1978). This was done on the CDC Cyber 170 model 730 using single precision (60 bit words). The entire fractile table was checked with the Iowa State University ITEL AS/6 system using double precision (64 bit words) with a more accurate algorithm due to Meeker and Miller (1979). This algorithm is described here.

Assume that $X_1$ and $X_2$ follow a bivariate normal distribution with means $\delta_1$ and $\delta_2$ respectively, unit variances and correlation $\rho$. Then $Z = X_1X_2$ follows the distribution of the product of two normally distributed random variables with parameters $\delta_1$, $\delta_2$ and $\rho$.

The distribution of Z can be obtained by integrating

$$F_Z(z) = P(Z \leq z) = \iint_{x_1x_2<z} g(x_1,x_2;\delta_1,\delta_2,\rho)dx_1dx_2 \tag{3.1}$$

where

$$g(x_1,x_2;\delta_1,\delta_2,\rho) = [2\pi(1-\rho^2)^{1/2}]^{-1}\exp(-Q/2)$$

and

$$Q = \frac{1}{1-\rho^2}\{(x_1-\delta_1)^2 - 2\rho(x_1-\delta_1)(x_2-\delta_2) + (x_2-\delta_2)^2\} .$$

A numerical algorithm for computing this integral is developed here for $z \geq 0$. For $z < 0$, the relationship in (2.2) can be used.

The probability (volume) represented by (3.1) is evaluated by dividing the $x_1x_2$ plane into 8 parts and integrating over each of these parts, thus avoiding the problem of singularities and improving computational efficiency (see Figure 1). The task is somewhat simplified because the integrands of the interated integrals can be expressed as functions of the standard normal cumulative distribution and density functions which can be efficiently computed to near machine accuracy.

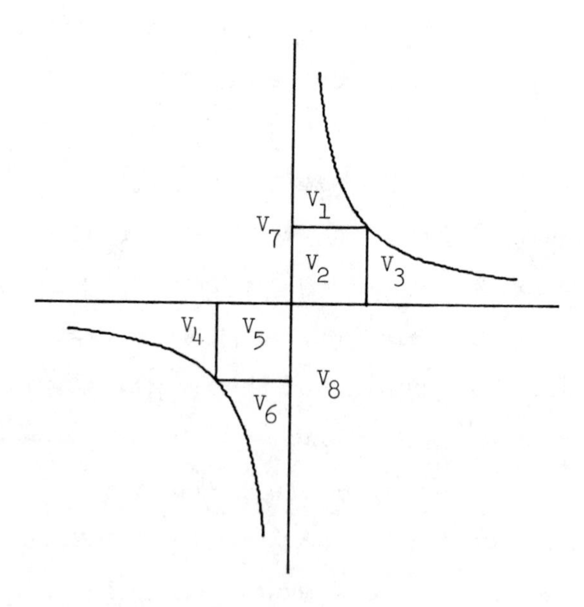


Figure 1. Regions of Integration

For $z \geq 0$, the probability in (3.1) is then computed as the sum of the following integrals.

$$
\begin{aligned}
V_1 &= \int_{\sqrt{z}}^{\infty} G_E(x_2, z; \delta_2, \delta_1, \rho)\,dx_2 \\
V_2 &= \int_{0}^{\sqrt{z}} G_R(x_1, 0, \sqrt{z}; \delta_1, \delta_2, \rho)\,dx_1 \\
V_3 &= \int_{\sqrt{z}}^{\infty} G_E(x_1, z; \delta_1, \delta_2, \rho)\,dx_1 \\
V_4 &= -\int_{-\infty}^{-\sqrt{z}} G_E(x_1, z; \delta_1, \delta_2, \rho)\,dx_1 \\
V_5 &= \int_{-\sqrt{z}}^{0} G_R(x_1, -\sqrt{z}, 0; \delta_1, \delta_2, \rho)\,dx_1 \\
V_6 &= -\int_{-\infty}^{-\sqrt{z}} G_E(x_2, z; \delta_2, \delta_1, \rho)\,dx_2 \\
V_7 &= \int_{-\infty}^{0} G_R(x_1, 0, \infty; \delta_1, \delta_2, \rho)\,dx_1 \\
V_8 &= \int_{0}^{\infty} G_R(x_1, -\infty, 0; \delta_1, \delta_2, \rho)\,dx_1
\end{aligned}
\tag{3.2}
$$

where

$$
\begin{aligned}
&G_E(y, z; \delta_y, \delta_x, \rho) = \\
&\qquad \{\Phi[R_1(z/y - \delta_x - \rho(y-\delta_y))] - \Phi[R_1(-\delta_x - \rho(y-\delta_y))]\}\varphi(y-\delta_y) \\
&G_R(y, x_L, x_u; \delta_y, \delta_x, \rho) = \\
&\qquad \{\Phi[R_1(x_u - \delta_x - \rho(y-\delta_y))] - \Phi(R_1(x_L - \delta_x - \rho(y-\delta_y))]\}\varphi(y-\delta_y)
\end{aligned}
\tag{3.3}
$$

and $\Phi(\cdot)$ and $\varphi(\cdot)$ are the standard normal cumulative distribution and density functions respectively and $R_1 = (1-\rho^2)^{-1/2}$.

The normal cumulative distribution functions were computed with the IBM FORTRAN supplied double precision complementary error function (DERFC). Integration was done with the IMSL (1975) subroutine DCADRE for cautious, adaptive Romberg integration, with a specified absolute error tolerance of $10^{-10}$. Several values of $\delta_i + \gamma$ were used to replace $\infty$ in the limits of integration. It was determined that $\delta_i + \gamma = 8$ yields accuracy of at least 10 decimal places in $F_Z(z)$.

When $\rho = 1$, the distribution is degenerate and the following scheme can be used. Again, let $X_i \sim N(\delta_i,1)$, i=1,2 and now, without loss of generality, let $\rho = 1$. Then $X_1 - \delta_1 = X_2 - \delta_2$ or $X_2 = X_1 + a$ where $a = \delta_2 - \delta_1$. The distribution of $Z = X_1X_2 = X_1^2 + aX_1$ is

$$F_Z(z;\delta_1,\delta_2,1) = P(Z < z) =$$

$$P(X_1^2 + aX_1 - z < 0) = \begin{cases} \Phi(r_U-\delta_1) - \Phi(r_L-\delta_1) & z > -a^2/4 \\ 0 & z \leq -a^2/4 \end{cases}$$

where

$$r_L = [-a - (a^2 + 4z)^{1/2}]/2$$

$$r_U = [-a + (a^2 + 4z)^{1/2}]/2.$$

For $\rho = -1$, use (2.2).

## 4. THE STANDARDIZED DISTRIBUTION FUNCTION

Let $W = (Z-\mu_Z)/\sigma_Z$, where $\mu_Z$ and $\sigma_Z$ are defined in (2.1). As shown by Aroian (1947), W is asymptotically N(0,1) for either $\delta_1$ or $\delta_2$ or both large in absolute value, except when $\rho = 1$ and $\delta_1 = -\delta_2$ or when $\rho = -1$ and $\delta_1 = \delta_2$ in which case the asymptotic distribution is related to the noncentral chisquare distribution, as described in Section 2. The standardized distribution function $F_W(w) = P(W \leq w)$ has been computed for values of $\delta_1 \leq \delta_2$ with $\delta_1,\delta_2 = 0(.4)4,6,12$, $-6 \leq w \leq 6$ and $\rho = 0.00, \pm 0.20, \pm 0.40, \pm 0.60, \pm 0.80, \pm 0.90, \pm 0.95$, and $\pm 1.00$. The interval in w is 0.2 from -3.2 to 3.2 and 0.4 outside this interval. A computer tape of these tabulations can be obtained from the authors.

## 5. TABLES OF THE FRACTILES $w_\alpha$

The present tables contain the fractiles of the standardized distribution function $F_W(w)$. These are solutions of the equations $F_W(w) = \alpha$ and are tabled here for 39 values of $\alpha$ between 0.0005 and 0.9995 for the same values of $\delta_1$, $\delta_2$, and $\rho$ used in the tabulation of $F_W(w)$ described in Section 4. The choice of $\delta_1 \leq \delta_2 = 4, 6, 12, \infty$ was made to facilitate harmonic interpolation but this must be used judiciously.

The fractiles were computed using the algorithm of Cornwell et al. (1978) and an inverse iterative interpolation scheme. Iterations were stopped when a value of $w_\alpha$ was found such that $|F_W(w_\alpha)-\alpha| < .5\text{x}10^{-7}$ or $|W_{\alpha_i} - W_{\alpha_{i-1}}| < .5\text{x}10^{-6}$ where $W_{\alpha_i}$ is the $i^{th}$ iterate of $W_\alpha$.

The fractiles were checked with the algorithm described in Section 3 using the following procedure. For particular values of $\delta_1$, $\delta_2$, $\rho$, and $\alpha$, the fractile estimate of $w_\alpha$ was rounded to exactly five places after the decimal and then incremented (if $F_W(w_\alpha) < \alpha$) or decremented (if $F_W(w_\alpha) > \alpha$) by 0.00001 to see if $|F_W(w_\alpha) - \alpha|$ could be reduced.

The asymptotic normal distribution fractiles (for $\delta_1 = \infty$ or $\delta_2 = \infty$ or $\delta_1 = \delta_2 = \infty$) are given for all values of $\rho$ except -1. When $\rho = -1$ and as $\delta = \delta_1 = \delta_2$ increases, the asymptotic distribution is the same as that for $\rho = -1$, $\delta_1 = \delta_2 = 0$. When $\delta_1$ is finite and fixed and $\delta_2$ increases (or vice versa) the asymptotic distribution is again normal.

## 6. APPLICATIONS

Craig (1936) mentions two inquiries concerning the distribution of a product, one from an investigator in business statistics and another from a psychologist. The third author's own interest arose from his use of sampling methods in auditing while at the Metropolitan Life Insurance Company, particularly for efficient verification of policy reserves. Problems such as the following are typical.

Let $A = \sum_{i=1}^{T} v_i$ be a population total which needs to be estimated. The mean of the population $\mu = \sum_{i=1}^{T} v_i/T$ is unknown. Also, it is often too costly to determine T exactly. However, both $\mu$ and T can be estimated economically by sampling. Estimates $\hat{\mu}$ and $\hat{T}$ and their estimated standard errors $s_{\hat{\mu}}$ and

$s_{\hat{T}}$ are thus obtained. If the estimates $\hat{\mu}$ and $\hat{T}$ are independent, $\rho = 0$; otherwise $\rho$ must be determined or estimated. It is also assumed that $\hat{T}$ and $\hat{\mu}$ approximately follow a normal distribution. In practice, these usually are not restrictive assumptions. In some situations, T is estimated with different sample units than those used to estimate $\mu$, often because T is more readily estimated than $\mu$.

Under these assumptions, the estimated total

$$\hat{A} = \hat{\mu}\hat{T}$$

follows the distribution of the product tabled here with estimated standard error

$$s_{\hat{A}} = (\hat{T}^2 s_{\hat{\mu}}^2 + \hat{\mu}^2 s_{\hat{T}}^2 + 2\hat{T} s_{\hat{T}} \hat{\mu} s_{\hat{\mu}} \hat{\rho} + s_{\hat{\mu}}^2 s_{\hat{T}}^2 (1 + \hat{\rho}^2))^{1/2}$$

For the distribution of $\hat{A}$, $\delta_\mu = \mu/\sigma_{\hat{\mu}}$ and $\delta_T = T/\sigma_{\hat{T}}$. If either $\delta_\mu$ or $\delta_T$ is large, one can take advantage of the asymptotic normality of $\hat{A}$ (Aroian (1947)); in any case, the tables presented here can be used to find the fractiles of $\hat{A}$ and, for example, to find approximate confidence intervals for the true value of A in the obvious manner. Inspection of the tables gives an indication of how large the $\delta$'s need to be to use the asymptotic distribution. It is seen, for example, that at least one of the $\delta$'s must be large to use the approximation in the tails of the distribution.

The distribution of the product is a function of $\delta_1$, $\delta_2$ and $\rho$. Use of the distribution in inferential applications assumes that these quantities and the $\sigma$'s are known. In practice, the $\delta$'s and/or the $\sigma$'s are unknown and sample estimates must be used, yielding approximate results.

An example of the above situation is the estimation of the total accounts receivable (A) for a large company, the total number of accounts (T) being unknown. First, a random sample is taken from a known number of files (or offices) in order to estimate the total number of accounts. Then a random sample of accounts is taken to estimate $\mu$, the average account size. The estimate of the total accounts receivable is then the product of these two estimates. Similar problems arise in studies of wildlife population, and in the estimation of total production, crop yield, etc. For example, Cordeiro and Rathie (1979) use the distribution to estimate total soybean and wheat yields for a particular area.

Sampson and Bruening (1971) have applied the distribution of the product to the uniformity specification of drugs in tablet form. Let tablet weights be designated by y and drug substance by p, then the problem is to estimate $\mu_{yp}$ by $\hat{\mu}_{yp} = \bar{y}\,\bar{p}$ where $\bar{y}$ and $\bar{p}$ are sample estimates, which they assume are independent. They emphasize that y and p need not be distributed normally since for moderately large samples $\bar{y}$ and $\bar{p}$ will be.

The distribution of the product is useful in economic theory for supply and demand situations involving prices and quantities. In reliability problems the product distribution has been used to find confidence intervals in a series system to estimate $p_1p_2$, where $p_i$ is the reliability of system i and the individual estimates are approximately normally distributed. Better methods are given by Mann (1972). Other applications are given by Broadbent (1956) and Barnard (1962) who use the lognormal to approximate the distribution of the product.

Donohue (1964) notes uses of the product in the cyclic firing rate of cannons and in the distribution of measurement errors. His report includes a large list of references as do the reports of Springer et al. (1964, 1966). Both discuss some more general problems concerning products and quotients. Aroian used the distribution of the product in some undocumented random noise problems at Hughes Aircraft Company, and Deutsch (1962) shows how the distribution of the product can be applied to the study of multiplicative random processes which occur in certain physical systems. We have been told that the distribution of the product is useful in problems associated with reliability assessment of nuclear reactors.

In quite a different example, Devlin et al. (1974) use the distribution of the product in their researches on robust estimation and outlier detection for correlation coefficients. For the case, $\delta_1 = \delta_2 = 0$, they provide the fractiles of the distribution. When $\{y_1, y_2\} \sim N(\mu, \Sigma)$ the influence function $z(\rho) = \tanh^{-1}\rho$ has the distribution of the product of two independent standard normal variables. For further information on this application see Devlin et al. (1975).

Important applications of the distribution of the product occur in business. Some are discussed by Ferrara et al. (1972), and Hayya et al. (1972). In the first paper the authors discuss the Jaedicke-Robichek cost-volume-profit analysis under conditions of uncertainty. The total profit, Z, is given by

$$Z = Q(P-V)-F$$

where

$Q$ = unit sales

$P$ = price per unit

$V$ = variable cost per unit

$F$ = fixed cost.

To study the distribution of profit, Jaedicke-Robichek assumes that all components of profit are normally distributed and independent of each other and that Z, the resulting profit is also normally distributed, which is, of course, not necessarily true. Ferrara et al. (1972) analyze the question correctly. Under the assumption that the estimates $\hat{Q}$, $\hat{P}$, and $\hat{V}$ are distributed normally, $\hat{Q}(\hat{P}-\hat{V})$ is the product of two normally distributed random variables. They also cover the relevant literature of the product distribution and provide guidelines for which values of $\delta_1$ and $\delta_2$ normality of profit can be assumed. For a lengthier discussion, see Ferrara et al. (1972).

Hayya et al. (1972) consider a probabilistic budget for a company when it is desired to predict the quantities

$$Z_1 = a + bXY$$

$$Z_2 = a + bX + cX^2$$

$$Z_3 = a + bXY + cX^2$$

where X, the unit sales volume and Y, the unit manufacturing cost, are two normally distributed random variables. The quantities a, b and c are known constants, not necessarily the same in each equation. Here $Z_1$ is the total manufacturing cost, $Z_2$ is managed administrative cost and $Z_3$ is net income. The distribution of the product is used to evaluate $Z_1$ and $Z_3$.

## 7. INTERPOLATION

One can interpolate in the table to 1) determine $F_W(w;\ \delta_1,\ \delta_2,\ \rho)$ for specified values of w, $\delta_1$, $\delta_2$ and $\rho$ or 2) to determine $w_\alpha$ for specified values of $\alpha$, $\delta_1$, $\delta_2$ and $\rho$.

We suggest the use of Aitken's method of interpolation (see page xi of Abramowitz and Stegun (1964) for a brief but informative description of this method which has proven to be very versatile). It is easily applied to both direct and inverse interpolation whether intervals are equal or unequal. It can be used for more than one variable by simply repeating the methods for a single variable.

A few general suggestions are made concerning interpolation in the tables. For values of w corresponding to values of z close to 0, one should use $z^{1/2}$ as the argument, as illustrated in Example 1 of the next section. Also, interpolation should not be attempted over values of w corresponding to z = 0. This is because the density $f_Z(z)$ is infinite at z = 0. In the lower tail of the distribution ($F_W(w) < 0.10$), interpolation using log $[F_W(w)]$ is more accurate; in the upper tail ($F_W(w) > 0.90$) one should use log $[1 - F_W(w)]$.

If $\delta_1$ or $\delta_2$ is greater than 4, harmonic interpolation using reciprocals of the $\delta$'s is advised, and is illustrated in Example 5 of the next section. However, care must be exercised in such cases. This is because the approach to normality if either $\delta_1$ of $\delta_2$ or both become large is peculiar since $F_W(w;\delta_1,\delta_2,\rho) = F_W(w;\delta_2,\delta_1,\rho)$. For example, with $\rho = 0$ the fractiles for $F_W(w) = 0.001$ and $\delta_1 = 12$ increase steadily from -3.15672 for $\delta_2 = 0$ to -2.83707 for $\delta_2 = 12$, but decrease to -3.09024 as $\delta_2$ ranges from 12 to $\infty$. This also happens for other $\alpha \leq 0.01$. For $0.1 \leq \alpha \leq 0.5$, the values of w behave somewhat differently. Similar results hold for $\alpha > 0.50$. Some general comments on the accuracy of interpolations are made in the examples in the next section.

## 8. EXAMPLES OF INTERPOLATION

This section contains 6 examples of interpolation and inverse interpolation in the tables.

Example 1. Find $F_W(w)$ for $\delta_1 = \delta_2 = \rho = 0$ and $w = 0.075$ not tabled. Because this corresponds to a value of z close to 0 ($z = w$ for this case), the square root transformation is used on the argument; $w^{1/2} = 0.273861$. The tableau for Aitken's method is shown for this example.

| w | $w^{1/2}$ | $F_W(w)$ | | |
|---|---|---|---|---|
| 0.00000 | 0.00000 | 0.5 | | |
| 0.03519 | 0.187590 | 0.55 | 0.57299 | |
| 0.08873 | 0.297876 | 0.60 | 0.59194 | 0.58781 |
| 0.15917 | 0.3998961 | 0.65 | 0.60297 | 0.58843 |

This gives $F(0.075) \cong 0.58843$. This compares with the actual value which is 0.58852. Three point interpolation using $w^{1/2}$ yields 0.58781 and linear interpolation with $w^{1/2}$ gives 0.58911. Without the square root transformation, linear interpolation yields 0.58718. Accuracy is somewhat limited here because of the singularity at $z = 0$.

Example 2. Find $F_W(w)$ for $w = 0.45$ with $\rho = 0$, $\delta_1 = 1.2$, and $\delta_2 = 1.8$. Bivariate 4-point interpolation is used. Using $\alpha = .65, .70, .75$ and $.80$, interpolate 4 values for $w_\alpha$ corresponding to $\delta_1 = 1.2$ and $\delta_2 = 1.2, 1.6, 2.0$, and 2.4. One further 4-point interpolation among the results of the first ones, for $\delta_2 = 1.8$, yields $F_W(0.45) \cong 0.72792$. Linear bivariate interpolation yields 0.72718. The correct value is 0.72792.

Example 3. Find $F_W(w)$ for $w = .8$ for $\rho = .4$, $\delta_1 = 1.6$ and $\delta_2 = 12$. Four-point inverse interpolation in the table yields 0.79323. Linear interpolation in the table yields 0.79281 and the correct value is 0.79324. If interpolation is not for values of w corresponding to z near 0 when $\delta_1$ and $\delta_2$ are small, 4 to

5 significant figures are readily obtained from 4-point interpolations in the tables; under the same conditions, linear interpolation generally yields results that are correct to within 2 or 3 units in the third decimal place.

Example 4. Find $w_{.001}$ with $\rho = 0$, $\delta_1 = 1.42$ and $\delta_2 = 2.67$. Using $\delta_1 = 1.2, 1.6$, and $\delta_2 = 2.4, 2.8$, bivariate 2-point interpolation yields $w_\alpha \cong -2.84345$. The correct fractile is -2.84261.

Example 5. Find $w_{.001}$ with $\rho = 0$ and $\delta_1 = \delta_2 = 12$ assuming the value is not tabled. Tabulated values of $w_\alpha$ for $\delta_1 = \delta_2 = 6$ and $\delta_1 = 12$, $\delta_2 = \infty$ (which is the same as $\delta_1 = \delta_2 = \infty$) are used. Harmonic interpolation using the reciprocals of the $\delta_i$'s (1/6 and 0) was used to interpolate for $1/\delta = 1/12$ giving $w_\alpha = -2.84017$. The exact answer given in Table 2 is -2.83707. If one uses harmonic interpolation with $1/\delta_1 = 1/12$, $1/\delta_2 = 1/6$ and $1/\delta_1 = 1/12$, $1/\delta_2 = 0$, $w_{.001} \cong -2.97613$ is obtained. The inaccuracy is because with $\delta_1 = 12$, the values of $w_\alpha$ tend to increase with $\delta_2$ from 0 to 12, but decrease as w goes from 12 to $\infty$.

Example 6. $X_1$ and $X_2$ follow a bivariate normal distribution with $\mu_1 = 3.0$, $\mu_2 = 8.0$, standard deviations $\sigma_1 = 1.5$, $\sigma_2 = 2.0$, and correlation $\rho = 0.2$. It is desired to find $P(X_1X_2 < 20)$. Using the notation and standardization of Section 2, $Z = X_1X_2/(\sigma_1\sigma_2)$, $\delta_1 = 2$, $\delta_2 = 4$, $\mu_Z = 8.2$, $\sigma_Z = 4.06435$ and $P(X_1X_2 < 20) = P(W < (20-8.2)/4.06435) = P(W < -.33309)$. Interpolation using 4 points $F_W(w)$ yields $P(X_1X_2 < 20) \cong 0.40590$, which is correct to the number of places shown.

Example 7. For $\delta_1 = \delta_2 = 6$ and $\rho = .5$, find $w_\alpha$ for $\alpha = .0005$, .2, and .5. Four point interpolation from values for $\rho = .2, .4, .6, .8$ yields -2.57031, -0.85815, and -0.07166 respectively for $w_\alpha$. The correct value for $w_{.0005}$ is -2.57033 and the others are correct to the number of places shown. Linear (2-point) interpolation yields -2.57091, -0.85813, and -0.07161. Except when $\rho$'s near -1 or 1, linear interpolation for untabled values of $\rho$ provides good accuracy.

ACKNOWLEDGMENTS

The authors would like to express their appreciation to the IMS Committee on Tables, to the co-editors, Dr. W. J. Kennedy and Dr. R. E. Odeh, and to the managing editor, Dr. J. M. Davenport, all of whom made very helpful suggestions. Special thanks are due to Dr. Odeh for his encouragement and for suggesting a double integration scheme for computing the distribution. Dr. Robert W. Miller helped develop the algorithm for $F_Z(z)$ in Section 3, and Ms. Sharon Govekar helped develop the inverse interpolation scheme that was used to invert the distribution function. The authors would also like to thank Ms. Linda Wheeler for carefully preparing the manuscript and tables for publication.

The project was partially supported by the Office of Naval Research, Project NR 042-302 and a Faculty Research Grant from Union College. Computer time was provided by the Mid-Illinois Computer Cooperative and Iowa State University. We also acknowledge the Institute of Administration and Management of Union College and University, the Departments of Mathematics and Quantative and Information Sciences at Western Illinois University, and the Statistical Laboratory at Iowa State University for encouragement and support.

REFERENCES

1. Abramowitz, M. and Stegun, I. A. (ed.) (1964). Handbook of Mathematical Functions, Applied Mathematics Series No. 55, Washington, D.C.: U.S. Government Printing Office.

2. Aroian, L. A. (1947). The probability function of a product of two normally distributed variables, Ann. Math. Statist. 18, 265-271.

3. Aroian, L. A. (1959). Tables and Percentage Points of the Distribution Function of a Product, Hughes Aircraft, Los Angeles, Calif. (out of print).

4. Aroian, L. A., Taneja, V. S., and Cornwell, L. W. (1978). Mathematical forms of the distribution of the product of two normal variables. Communications in Statistics A, Theory and Methods, Vol. A7, No. 2, 165-172.

5. Barnard, G. A. (1962). Distribution of products of independent variables. Technometrics 4, 277-278.

6. Broadbent, T. A. A. (1956). Lognormal approximations to products and quotients, Biometrika 43, 404-417.

7. Cordeiro, J. A. and Rathie, P. N. (1979). The exact distribution of the distribution of two normal variates. (Abstract 79t-165) Institute of Mathematical Statistics Bulletin 8, 334.

8. Cornwell, L. W., Aroian, L. A., and Taneja, V. S. (1978). Numerical evaluation of the distribution of the product of two normal variables. Journal of Statistical Computation and Simulation, Vol. 7, No. 2, 85-92.

9. Craig, C. C. (1936). On the frequency function of xy, Ann. Math. Statist. 7, 1-15.

10. Deutsch, R. (1962). Nonlinear Transformations of Random Processes, Prentice-Hall, Englewood Cliffs, N.J.

11. Devlin, S. J., Gnanadesikan, R., and Kettenring, J. P. (1974). Robust estimation and outlier detection with correlation coefficients, Internal report, Bell Laboratories, Murray Hill, N.J. 07974.

12. Devlin, S. J., Gnanadesikan, R., and Kettenring, J. P. (1975). Robust estimation and outlier detection with correlation coefficients, Biometrika 62, 531-546.

13. Donohue, J. D. (1964). Products and quotients of random variables and applications, Defense Documentation Center, Cameron Station, Alexandria, Virginia 22314.

14. Ferrara, W. L., Hayya, J. C. and Nachman, D. A., (1972). Normalcy of profit in the Jaedicke-Robichek Model, The Accounting Review 18, 299-307.

15. Hayya, J. C. and Ferrara, W. L. (1972). On normal approximations of the frequency functions of standard forms where the main variables are normally distributed, Management Science 19, 173-186.

16. IMSL (1975). International Mathematical and Statistical Libraries Reference Manual, Edition 5, Houston Texas.

17. Mann, N. R. (1972). A survey and comparison of methods for determining confidence bounds on system reliability from subsystems data, Proceedings of 1972 NATO Conference on Reliability Testing and Reliability Evaluation, California State University, Northridge, Calif.

18. Meeker, W. Q. and Miller, R. W. (1979). An improved algorithm for computing the distribution of the product of two normally distributed random variables, Department of Statistics, Iowa State University.

19. Sampson, F. B. and Bruening, H. L. (1971). Some statistical aspects of pharmaceutical content uniformity, Journal of Quality Technology 3, 170-178.

20. Springer, M. D. and Thompson, W. E. (1964). The distribution of the products of independent random variables, GM Defense Research Laboratories, Santa Barbara, Calif.

21. Springer, M. D. and Thompson, W. E. (1966). The distribution of the products of independent random variables, SIAM J. Appl. Math. 14, 511-526.

FRACTILES $W_\alpha$ RHO = -1.00

| DELTAS | | α:0.0005 | 0.0010 | 0.0025 | 0.0050 | 0.0075 |
|---|---|---|---|---|---|---|
| 0.0 | 0.0 | -7.85996 | -6.94914 | -5.75627 | -4.86450 | -4.34810 |
| 0.4 | 0.0 | -7.84737 | -6.93996 | -5.75087 | -4.86138 | -4.34608 |
| | 0.4 | -7.85996 | -6.94914 | -5.75627 | -4.86450 | -4.34810 |
| 0.8 | 0.0 | -7.72198 | -6.84509 | -5.69148 | -4.82472 | -4.32091 |
| | 0.4 | -7.84737 | -6.93996 | -5.75087 | -4.86138 | -4.34608 |
| | 0.8 | -7.85996 | -6.94914 | -5.75627 | -4.86450 | -4.34810 |
| 1.2 | 0.0 | -7.43597 | -6.61944 | -5.53969 | -4.72331 | -4.24633 |
| | 0.4 | -7.72198 | -6.84509 | -5.69148 | -4.82472 | -4.32091 |
| | 0.8 | -7.84737 | -6.93996 | -5.75087 | -4.86138 | -4.34608 |
| | 1.2 | -7.85996 | -6.94914 | -5.75627 | -4.86450 | -4.34810 |
| 1.6 | 0.0 | -7.06875 | -6.32058 | -5.32709 | -4.57201 | -4.12887 |
| | 0.4 | -7.43597 | -6.61944 | -5.53969 | -4.72331 | -4.24633 |
| | 0.8 | -7.72198 | -6.84509 | -5.69148 | -4.82472 | -4.32091 |
| | 1.2 | -7.84737 | -6.93996 | -5.75087 | -4.86138 | -4.34608 |
| | 1.6 | -7.85996 | -6.94914 | -5.75627 | -4.86450 | -4.34810 |
| 2.0 | 0.0 | -6.69891 | -6.01367 | -5.10076 | -4.40407 | -3.99380 |
| | 0.4 | -7.06875 | -6.32058 | -5.32709 | -4.57201 | -4.12887 |
| | 0.8 | -7.43597 | -6.61944 | -5.53969 | -4.72331 | -4.24633 |
| | 1.2 | -7.72198 | -6.84509 | -5.69148 | -4.82472 | -4.32091 |
| | 1.6 | -7.84737 | -6.93996 | -5.75087 | -4.86138 | -4.34608 |
| | 2.0 | -7.85996 | -6.94914 | -5.75627 | -4.86450 | -4.34810 |
| 2.4 | 0.0 | -6.36285 | -5.73151 | -4.88801 | -4.24208 | -3.86061 |
| | 0.4 | -6.69891 | -6.01367 | -5.10076 | -4.40407 | -3.99380 |
| | 0.8 | -7.06875 | -6.32058 | -5.32709 | -4.57201 | -4.12887 |
| | 1.2 | -7.43597 | -6.61944 | -5.53969 | -4.72331 | -4.24633 |
| | 1.6 | -7.72198 | -6.84509 | -5.69148 | -4.82472 | -4.32091 |
| | 2.0 | -7.84737 | -6.93996 | -5.75087 | -4.86138 | -4.34608 |
| | 2.4 | -7.85996 | -6.94914 | -5.75627 | -4.86450 | -4.34810 |
| 2.8 | 0.0 | -6.07006 | -5.48386 | -4.69867 | -4.09555 | -3.73848 |
| | 0.4 | -6.36285 | -5.73151 | -4.88801 | -4.24208 | -3.86061 |
| | 0.8 | -6.69891 | -6.01367 | -5.10076 | -4.40407 | -3.99380 |
| | 1.2 | -7.06875 | -6.32058 | -5.32709 | -4.57201 | -4.12887 |
| | 1.6 | -7.43597 | -6.61944 | -5.53969 | -4.72331 | -4.24633 |
| | 2.0 | -7.72198 | -6.84509 | -5.69148 | -4.82472 | -4.32091 |
| | 2.4 | -7.84737 | -6.93996 | -5.75087 | -4.86138 | -4.34608 |
| | 2.8 | -7.85996 | -6.94914 | -5.75627 | -4.86450 | -4.34810 |
| 3.2 | 0.0 | -5.81873 | -5.27023 | -4.53384 | -3.96663 | -3.63008 |
| | 0.4 | -6.07006 | -5.48386 | -4.69867 | -4.09555 | -3.73848 |
| | 0.8 | -6.36285 | -5.73151 | -4.88801 | -4.24208 | -3.86061 |
| | 1.2 | -6.69891 | -6.01367 | -5.10076 | -4.40407 | -3.99380 |
| | 1.6 | -7.06875 | -6.32058 | -5.32709 | -4.57201 | -4.12887 |
| | 2.0 | -7.43597 | -6.61944 | -5.53969 | -4.72331 | -4.24633 |
| | 2.4 | -7.72198 | -6.84509 | -5.69148 | -4.82472 | -4.32091 |
| | 2.8 | -7.84737 | -6.93996 | -5.75087 | -4.86138 | -4.34608 |
| | 3.2 | -7.85996 | -6.94914 | -5.75627 | -4.86450 | -4.34810 |

FRACTILES $W_\alpha$ RHO = -1.00

| DELTAS | | α:0.0005 | 0.0010 | 0.0025 | 0.0050 | 0.0075 |
|---|---|---|---|---|---|---|
| 3.6 | 0.0 | -5.60354 | -5.08669 | -4.39131 | -3.85434 | -3.53508 |
| | 0.4 | -5.81873 | -5.27023 | -4.53384 | -3.96663 | -3.63008 |
| | 0.8 | -6.07006 | -5.48386 | -4.69867 | -4.09555 | -3.73848 |
| | 1.2 | -6.36285 | -5.73151 | -4.88801 | -4.24208 | -3.86061 |
| | 1.6 | -6.69891 | -6.01367 | -5.10076 | -4.40407 | -3.99380 |
| | 2.0 | -7.06875 | -6.32058 | -5.32709 | -4.57201 | -4.12887 |
| | 2.4 | -7.43597 | -6.61944 | -5.53969 | -4.72331 | -4.24633 |
| | 2.8 | -7.72198 | -6.84509 | -5.69148 | -4.82472 | -4.32091 |
| | 3.2 | -7.84737 | -6.93996 | -5.75087 | -4.86138 | -4.34608 |
| | 3.6 | -7.85996 | -6.94914 | -5.75627 | -4.86450 | -4.34810 |
| 4.0 | 0.0 | -5.41872 | -4.92864 | -4.26800 | -3.75667 | -3.45209 |
| | 0.4 | -5.60354 | -5.08669 | -4.39131 | -3.85434 | -3.53508 |
| | 0.8 | -5.81873 | -5.27023 | -4.53384 | -3.96663 | -3.63008 |
| | 1.2 | -6.07006 | -5.48386 | -4.69867 | -4.09555 | -3.73848 |
| | 1.6 | -6.36285 | -5.73151 | -4.88801 | -4.24208 | -3.86061 |
| | 2.0 | -6.69891 | -6.01367 | -5.10076 | -4.40407 | -3.99380 |
| | 2.4 | -7.06875 | -6.32058 | -5.32709 | -4.57201 | -4.12887 |
| | 2.8 | -7.43597 | -6.61944 | -5.53969 | -4.72331 | -4.24633 |
| | 3.2 | -7.72198 | -6.84509 | -5.69148 | -4.82472 | -4.32091 |
| | 3.6 | -7.84737 | -6.93996 | -5.75087 | -4.86138 | -4.34608 |
| | 4.0 | -7.85996 | -6.94914 | -5.75627 | -4.86450 | -4.34810 |
| 6.0 | 0.0 | -4.79701 | -4.39473 | -3.84816 | -3.42123 | -3.16506 |
| | 0.4 | -4.89187 | -4.47640 | -3.91267 | -3.47303 | -3.20956 |
| | 0.8 | -4.99887 | -4.56844 | -3.98525 | -3.53120 | -3.25946 |
| | 1.2 | -5.12032 | -4.67279 | -4.06739 | -3.59690 | -3.31573 |
| | 1.6 | -5.25909 | -4.79188 | -4.16090 | -3.67150 | -3.37949 |
| | 2.0 | -5.41872 | -4.92864 | -4.26800 | -3.75667 | -3.45209 |
| | 2.4 | -5.60354 | -5.08669 | -4.39131 | -3.85434 | -3.53508 |
| | 2.8 | -5.81873 | -5.27023 | -4.53384 | -3.96663 | -3.63008 |
| | 3.2 | -6.07006 | -5.48386 | -4.69867 | -4.09555 | -3.73848 |
| | 3.6 | -6.36285 | -5.73151 | -4.88801 | -4.24208 | -3.86061 |
| | 4.0 | -6.69891 | -6.01367 | -5.10076 | -4.40407 | -3.99380 |
| | 6.0 | -7.85996 | -6.94914 | -5.75627 | -4.86450 | -4.34810 |
| 12.0 | 0.0 | -4.08125 | -3.77656 | -3.35709 | -3.02447 | -2.82255 |
| | 0.4 | -4.10732 | -3.79913 | -3.37510 | -3.03909 | -2.83522 |
| | 0.8 | -4.13515 | -3.82323 | -3.39432 | -3.05469 | -2.84873 |
| | 1.2 | -4.16493 | -3.84900 | -3.41487 | -3.07136 | -2.86316 |
| | 1.6 | -4.19686 | -3.87663 | -3.43689 | -3.08922 | -2.87862 |
| | 2.0 | -4.23118 | -3.90632 | -3.46054 | -3.10839 | -2.89522 |
| | 2.4 | -4.26817 | -3.93830 | -3.48602 | -3.12903 | -2.91307 |
| | 2.8 | -4.30814 | -3.97287 | -3.51353 | -3.15130 | -2.93234 |
| | 3.2 | -4.35146 | -4.01031 | -3.54332 | -3.17542 | -2.95318 |
| | 3.6 | -4.39857 | -4.05102 | -3.57569 | -3.20159 | -2.97579 |
| | 4.0 | -4.44998 | -4.09543 | -3.61098 | -3.23011 | -3.00042 |
| | 6.0 | -4.79701 | -4.39473 | -3.84816 | -3.42123 | -3.16506 |
| | 12.0 | -7.85996 | -6.94914 | -5.75627 | -4.86450 | -4.34810 |

FRACTILES $W_\alpha$ RHO = -1.00

| DELTAS | | $\alpha$:0.0100 | 0.0175 | 0.0250 | 0.0375 | 0.0500 |
|---|---|---|---|---|---|---|
| 0.0 | 0.0 | -3.98447 | -3.28488 | -2.84532 | -2.35294 | -2.00921 |
| 0.4 | 0.0 | -3.98314 | -3.28461 | -2.84556 | -2.35360 | -2.01008 |
| | 0.4 | -3.98447 | -3.28488 | -2.84532 | -2.35294 | -2.00921 |
| 0.8 | 0.0 | -3.96526 | -3.27874 | -2.84576 | -2.35914 | -2.01834 |
| | 0.4 | -3.98314 | -3.28461 | -2.84556 | -2.35360 | -2.01008 |
| | 0.8 | -3.98447 | -3.28488 | -2.84532 | -2.35294 | -2.00921 |
| 1.2 | 0.0 | -3.90835 | -3.25255 | -2.83628 | -2.36552 | -2.03372 |
| | 0.4 | -3.96526 | -3.27874 | -2.84576 | -2.35914 | -2.01834 |
| | 0.8 | -3.98314 | -3.28461 | -2.84556 | -2.35360 | -2.01008 |
| | 1.2 | -3.98447 | -3.28488 | -2.84532 | -2.35294 | -2.00921 |
| 1.6 | 0.0 | -3.81383 | -3.19961 | -2.80731 | -2.36084 | -2.04400 |
| | 0.4 | -3.90835 | -3.25255 | -2.83628 | -2.36552 | -2.03372 |
| | 0.8 | -3.96526 | -3.27874 | -2.84576 | -2.35914 | -2.01834 |
| | 1.2 | -3.98314 | -3.28461 | -2.84556 | -2.35360 | -2.01008 |
| | 1.6 | -3.98447 | -3.28488 | -2.84532 | -2.35294 | -2.00921 |
| 2.0 | 0.0 | -3.70138 | -3.12912 | -2.76187 | -2.34183 | -2.04209 |
| | 0.4 | -3.81383 | -3.19961 | -2.80731 | -2.36084 | -2.04400 |
| | 0.8 | -3.90835 | -3.25255 | -2.83628 | -2.36552 | -2.03372 |
| | 1.2 | -3.96526 | -3.27874 | -2.84576 | -2.35914 | -2.01834 |
| | 1.6 | -3.98314 | -3.28461 | -2.84556 | -2.35360 | -2.01008 |
| | 2.0 | -3.98447 | -3.28488 | -2.84532 | -2.35294 | -2.00921 |
| 2.4 | 0.0 | -3.58815 | -3.05338 | -2.70888 | -2.31334 | -2.02990 |
| | 0.4 | -3.70138 | -3.12912 | -2.76187 | -2.34183 | -2.04209 |
| | 0.8 | -3.81383 | -3.19961 | -2.80731 | -2.36084 | -2.04400 |
| | 1.2 | -3.90835 | -3.25255 | -2.83628 | -2.36552 | -2.03372 |
| | 1.6 | -3.96526 | -3.27874 | -2.84576 | -2.35914 | -2.01834 |
| | 2.0 | -3.98314 | -3.28461 | -2.84556 | -2.35360 | -2.01008 |
| | 2.4 | -3.98447 | -3.28488 | -2.84532 | -2.35294 | -2.00921 |
| 2.8 | 0.0 | -3.48299 | -2.98024 | -2.65534 | -2.28110 | -2.01200 |
| | 0.4 | -3.58815 | -3.05338 | -2.70888 | -2.31334 | -2.02990 |
| | 0.8 | -3.70138 | -3.12912 | -2.76187 | -2.34183 | -2.04209 |
| | 1.2 | -3.81383 | -3.19961 | -2.80731 | -2.36084 | -2.04400 |
| | 1.6 | -3.90835 | -3.25255 | -2.83628 | -2.36552 | -2.03372 |
| | 2.0 | -3.96526 | -3.27874 | -2.84576 | -2.35914 | -2.01834 |
| | 2.4 | -3.98314 | -3.28461 | -2.84556 | -2.35360 | -2.01008 |
| | 2.8 | -3.98447 | -3.28488 | -2.84532 | -2.35294 | -2.00921 |
| 3.2 | 0.0 | -3.38887 | -2.91317 | -2.60488 | -2.24879 | -1.99199 |
| | 0.4 | -3.48299 | -2.98024 | -2.65534 | -2.28110 | -2.01200 |
| | 0.8 | -3.58815 | -3.05338 | -2.70888 | -2.31334 | -2.02990 |
| | 1.2 | -3.70138 | -3.12912 | -2.76187 | -2.34183 | -2.04209 |
| | 1.6 | -3.81383 | -3.19961 | -2.80731 | -2.36084 | -2.04400 |
| | 2.0 | -3.90835 | -3.25255 | -2.83628 | -2.36552 | -2.03372 |
| | 2.4 | -3.96526 | -3.27874 | -2.84576 | -2.35914 | -2.01834 |
| | 2.8 | -3.98314 | -3.28461 | -2.84556 | -2.35360 | -2.01008 |
| | 3.2 | -3.98447 | -3.28488 | -2.84532 | -2.35294 | -2.00921 |

FRACTILES $w_\alpha$ RHO = -1.00

| DELTAS | | $\alpha$: 0.0100 | 0.0175 | 0.0250 | 0.0375 | 0.0500 |
|---|---|---|---|---|---|---|
| 3.6 | 0.0 | -3.30593 | -2.85310 | -2.55889 | -2.21824 | -1.97192 |
| | 0.4 | -3.38887 | -2.91317 | -2.60488 | -2.24879 | -1.99199 |
| | 0.8 | -3.48299 | -2.98024 | -2.65534 | -2.28110 | -2.01200 |
| | 1.2 | -3.58815 | -3.05338 | -2.70888 | -2.31334 | -2.02990 |
| | 1.6 | -3.70138 | -3.12912 | -2.76187 | -2.34183 | -2.04209 |
| | 2.0 | -3.81383 | -3.19961 | -2.80731 | -2.36084 | -2.04400 |
| | 2.4 | -3.90835 | -3.25255 | -2.83628 | -2.36552 | -2.03372 |
| | 2.8 | -3.96526 | -3.27874 | -2.84576 | -2.35914 | -2.01834 |
| | 3.2 | -3.98314 | -3.28461 | -2.84556 | -2.35360 | -2.01008 |
| | 3.6 | -3.98447 | -3.28488 | -2.84532 | -2.35294 | -2.00921 |
| 4.0 | 0.0 | -3.23320 | -2.79981 | -2.51761 | -2.19012 | -1.95278 |
| | 0.4 | -3.30593 | -2.85310 | -2.55889 | -2.21824 | -1.97192 |
| | 0.8 | -3.38887 | -2.91317 | -2.60488 | -2.24879 | -1.99199 |
| | 1.2 | -3.48299 | -2.98024 | -2.65534 | -2.28110 | -2.01200 |
| | 1.6 | -3.58815 | -3.05338 | -2.70888 | -2.31334 | -2.02990 |
| | 2.0 | -3.70138 | -3.12912 | -2.76187 | -2.34183 | -2.04209 |
| | 2.4 | -3.81383 | -3.19961 | -2.80731 | -2.36084 | -2.04400 |
| | 2.8 | -3.90835 | -3.25255 | -2.83628 | -2.36552 | -2.03372 |
| | 3.2 | -3.96526 | -3.27874 | -2.84576 | -2.35914 | -2.01834 |
| | 3.6 | -3.98314 | -3.28461 | -2.84556 | -2.35360 | -2.01008 |
| | 4.0 | -3.98447 | -3.28488 | -2.84532 | -2.35294 | -2.00921 |
| 6.0 | 0.0 | -2.98000 | -2.61101 | -2.36863 | -2.08501 | -1.87766 |
| | 0.4 | -3.01939 | -2.64066 | -2.39226 | -2.10198 | -1.89008 |
| | 0.8 | -3.06351 | -2.67377 | -2.41855 | -2.12075 | -1.90370 |
| | 1.2 | -3.11318 | -2.71089 | -2.44790 | -2.14154 | -1.91863 |
| | 1.6 | -3.16937 | -2.75266 | -2.48076 | -2.16460 | -1.93499 |
| | 2.0 | -3.23320 | -2.79981 | -2.51761 | -2.19012 | -1.95278 |
| | 2.4 | -3.30593 | -2.85310 | -2.55889 | -2.21824 | -1.97192 |
| | 2.8 | -3.38887 | -2.91317 | -2.60488 | -2.24879 | -1.99199 |
| | 3.2 | -3.48299 | -2.98024 | -2.65534 | -2.28110 | -2.01200 |
| | 3.6 | -3.58815 | -3.05338 | -2.70888 | -2.31334 | -2.02990 |
| | 4.0 | -3.70138 | -3.12912 | -2.76187 | -2.34183 | -2.04209 |
| | 6.0 | -3.98447 | -3.28488 | -2.84532 | -2.35294 | -2.00921 |
| 12.0 | 0.0 | -2.67549 | -2.37899 | -2.18165 | -1.94782 | -1.77470 |
| | 0.4 | -2.68679 | -2.38768 | -2.18871 | -1.95308 | -1.77871 |
| | 0.8 | -2.69884 | -2.39693 | -2.19623 | -1.95867 | -1.78298 |
| | 1.2 | -2.71171 | -2.40681 | -2.20424 | -1.96462 | -1.78751 |
| | 1.6 | -2.72549 | -2.41738 | -2.21282 | -1.97098 | -1.79235 |
| | 2.0 | -2.74027 | -2.42871 | -2.22200 | -1.97779 | -1.79752 |
| | 2.4 | -2.75617 | -2.44089 | -2.23186 | -1.98509 | -1.80306 |
| | 2.8 | -2.77333 | -2.45401 | -2.24248 | -1.99293 | -1.80899 |
| | 3.2 | -2.79188 | -2.46819 | -2.25394 | -2.00138 | -1.81537 |
| | 3.6 | -2.81200 | -2.48355 | -2.26634 | -2.01051 | -1.82225 |
| | 4.0 | -2.83390 | -2.50024 | -2.27980 | -2.02040 | -1.82968 |
| | 6.0 | -2.98000 | -2.61101 | -2.36863 | -2.08501 | -1.87766 |
| | 12.0 | -3.98447 | -3.28488 | -2.84532 | -2.35294 | -2.00921 |

FRACTILES $W_\alpha$ RHO = -1.00

| DELTAS | | $\alpha$: 0.0750 | 0.1000 | 0.1500 | 0.2000 | 0.2500 |
|---|---|---|---|---|---|---|
| 0.0 | 0.0 | -1.53446 | -1.20600 | -0.75820 | -0.45423 | -0.22861 |
| 0.4 | 0.0 | -1.53546 | -1.20701 | -0.75909 | -0.45496 | -0.22917 |
| | 0.4 | -1.53446 | -1.20600 | -0.75820 | -0.45423 | -0.22861 |
| 0.8 | 0.0 | -1.54604 | -1.21814 | -0.76946 | -0.46375 | -0.23621 |
| | 0.4 | -1.53546 | -1.20701 | -0.75909 | -0.45496 | -0.22917 |
| | 0.8 | -1.53446 | -1.20600 | -0.75820 | -0.45423 | -0.22861 |
| 1.2 | 0.0 | -1.57053 | -1.24628 | -0.79839 | -0.48995 | -0.25835 |
| | 0.4 | -1.54604 | -1.21814 | -0.76946 | -0.46375 | -0.23621 |
| | 0.8 | -1.53546 | -1.20701 | -0.75909 | -0.45496 | -0.22917 |
| | 1.2 | -1.53446 | -1.20600 | -0.75820 | -0.45423 | -0.22861 |
| 1.6 | 0.0 | -1.59794 | -1.28248 | -0.84108 | -0.53218 | -0.29681 |
| | 0.4 | -1.57053 | -1.24628 | -0.79839 | -0.48995 | -0.25835 |
| | 0.8 | -1.54604 | -1.21814 | -0.76946 | -0.46375 | -0.23621 |
| | 1.2 | -1.53546 | -1.20701 | -0.75909 | -0.45496 | -0.22917 |
| | 1.6 | -1.53446 | -1.20600 | -0.75820 | -0.45423 | -0.22861 |
| 2.0 | 0.0 | -1.61719 | -1.31407 | -0.88496 | -0.57993 | -0.34387 |
| | 0.4 | -1.59794 | -1.28248 | -0.84108 | -0.53218 | -0.29681 |
| | 0.8 | -1.57053 | -1.24628 | -0.79839 | -0.48995 | -0.25835 |
| | 1.2 | -1.54604 | -1.21814 | -0.76946 | -0.46375 | -0.23621 |
| | 1.6 | -1.53546 | -1.20701 | -0.75909 | -0.45496 | -0.22917 |
| | 2.0 | -1.53446 | -1.20600 | -0.75820 | -0.45423 | -0.22861 |
| 2.4 | 0.0 | -1.62597 | -1.33589 | -0.92160 | -0.62349 | -0.38987 |
| | 0.4 | -1.61719 | -1.31407 | -0.88496 | -0.57993 | -0.34387 |
| | 0.8 | -1.59794 | -1.28248 | -0.84108 | -0.53218 | -0.29681 |
| | 1.2 | -1.57053 | -1.24628 | -0.79839 | -0.48995 | -0.25835 |
| | 1.6 | -1.54604 | -1.21814 | -0.76946 | -0.46375 | -0.23621 |
| | 2.0 | -1.53546 | -1.20701 | -0.75909 | -0.45496 | -0.22917 |
| | 2.4 | -1.53446 | -1.20600 | -0.75820 | -0.45423 | -0.22861 |
| 2.8 | 0.0 | -1.62690 | -1.34892 | -0.94920 | -0.65895 | -0.42936 |
| | 0.4 | -1.62597 | -1.33589 | -0.92160 | -0.62349 | -0.38987 |
| | 0.8 | -1.61719 | -1.31407 | -0.88496 | -0.57993 | -0.34387 |
| | 1.2 | -1.59794 | -1.28248 | -0.84108 | -0.53218 | -0.29681 |
| | 1.6 | -1.57053 | -1.24628 | -0.79839 | -0.48995 | -0.25835 |
| | 2.0 | -1.54604 | -1.21814 | -0.76946 | -0.46375 | -0.23621 |
| | 2.4 | -1.53546 | -1.20701 | -0.75909 | -0.45496 | -0.22917 |
| | 2.8 | -1.53446 | -1.20600 | -0.75820 | -0.45423 | -0.22861 |
| 3.2 | 0.0 | -1.62318 | -1.35583 | -0.96926 | -0.68656 | -0.46135 |
| | 0.4 | -1.62690 | -1.34892 | -0.94920 | -0.65895 | -0.42936 |
| | 0.8 | -1.62597 | -1.33589 | -0.92160 | -0.62349 | -0.38987 |
| | 1.2 | -1.61719 | -1.31407 | -0.88496 | -0.57993 | -0.34387 |
| | 1.6 | -1.59794 | -1.28248 | -0.84108 | -0.53218 | -0.29681 |
| | 2.0 | -1.57053 | -1.24628 | -0.79839 | -0.48995 | -0.25835 |
| | 2.4 | -1.54604 | -1.21814 | -0.76946 | -0.46375 | -0.23621 |
| | 2.8 | -1.53546 | -1.20701 | -0.75909 | -0.45496 | -0.22917 |
| | 3.2 | -1.53446 | -1.20600 | -0.75820 | -0.45423 | -0.22861 |

FRACTILES $W_{\alpha}$ RHO = -1.00

| DELTAS | | $\alpha$:0.0750 | 0.1000 | 0.1500 | 0.2000 | 0.2500 |
|---|---|---|---|---|---|---|
| 3.6 | 0.0 | -1.61708 | -1.35890 | -0.98386 | -0.70796 | -0.48690 |
| | 0.4 | -1.62318 | -1.35583 | -0.96926 | -0.68656 | -0.46135 |
| | 0.8 | -1.62690 | -1.34892 | -0.94920 | -0.65895 | -0.42936 |
| | 1.2 | -1.62597 | -1.33589 | -0.92160 | -0.62349 | -0.38987 |
| | 1.6 | -1.61719 | -1.31407 | -0.88496 | -0.57993 | -0.34387 |
| | 2.0 | -1.59794 | -1.28248 | -0.84108 | -0.53218 | -0.29681 |
| | 2.4 | -1.57053 | -1.24628 | -0.79839 | -0.48995 | -0.25835 |
| | 2.8 | -1.54604 | -1.21814 | -0.76946 | -0.46375 | -0.23621 |
| | 3.2 | -1.53546 | -1.20701 | -0.75909 | -0.45496 | -0.22917 |
| | 3.6 | -1.53446 | -1.20600 | -0.75820 | -0.45423 | -0.22861 |
| 4.0 | 0.0 | -1.60994 | -1.35967 | -0.99465 | -0.72474 | -0.50745 |
| | 0.4 | -1.61708 | -1.35890 | -0.98386 | -0.70796 | -0.48690 |
| | 0.8 | -1.62318 | -1.35583 | -0.96926 | -0.68656 | -0.46135 |
| | 1.2 | -1.62690 | -1.34892 | -0.94920 | -0.65895 | -0.42936 |
| | 1.6 | -1.62597 | -1.33589 | -0.92160 | -0.62349 | -0.38987 |
| | 2.0 | -1.61719 | -1.31407 | -0.88496 | -0.57993 | -0.34387 |
| | 2.4 | -1.59794 | -1.28248 | -0.84108 | -0.53218 | -0.29681 |
| | 2.8 | -1.57053 | -1.24628 | -0.79839 | -0.48995 | -0.25835 |
| | 3.2 | -1.54604 | -1.21814 | -0.76946 | -0.46375 | -0.23621 |
| | 3.6 | -1.53546 | -1.20701 | -0.75909 | -0.45496 | -0.22917 |
| | 4.0 | -1.53446 | -1.20600 | -0.75820 | -0.45423 | -0.22861 |
| 6.0 | 0.0 | -1.57508 | -1.35158 | -1.02083 | -0.77186 | -0.56808 |
| | 0.4 | -1.58136 | -1.35376 | -1.01773 | -0.76550 | -0.55959 |
| | 0.8 | -1.58805 | -1.35584 | -1.01387 | -0.75800 | -0.54970 |
| | 1.2 | -1.59512 | -1.35768 | -1.00901 | -0.74902 | -0.53807 |
| | 1.6 | -1.60249 | -1.35907 | -1.00277 | -0.73814 | -0.52420 |
| | 2.0 | -1.60994 | -1.35967 | -0.99465 | -0.72474 | -0.50745 |
| | 2.4 | -1.61708 | -1.35890 | -0.98386 | -0.70796 | -0.48690 |
| | 2.8 | -1.62318 | -1.35583 | -0.96926 | -0.68656 | -0.46135 |
| | 3.2 | -1.62690 | -1.34892 | -0.94920 | -0.65895 | -0.42936 |
| | 3.6 | -1.62597 | -1.33589 | -0.92160 | -0.62349 | -0.38987 |
| | 4.0 | -1.61719 | -1.31407 | -0.88496 | -0.57993 | -0.34387 |
| | 6.0 | -1.53446 | -1.20600 | -0.75820 | -0.45423 | -0.22861 |
| 12.0 | 0.0 | -1.51838 | -1.32591 | -1.03545 | -0.81170 | -0.62474 |
| | 0.4 | -1.52071 | -1.32710 | -1.03516 | -0.81048 | -0.62289 |
| | 0.8 | -1.52317 | -1.32836 | -1.03484 | -0.80915 | -0.62089 |
| | 1.2 | -1.52579 | -1.32968 | -1.03447 | -0.80772 | -0.61874 |
| | 1.6 | -1.52856 | -1.33107 | -1.03405 | -0.80616 | -0.61641 |
| | 2.0 | -1.53152 | -1.33253 | -1.03357 | -0.80445 | -0.61388 |
| | 2.4 | -1.53466 | -1.33407 | -1.03301 | -0.80258 | -0.61112 |
| | 2.8 | -1.53802 | -1.33569 | -1.03237 | -0.80051 | -0.60810 |
| | 3.2 | -1.54160 | -1.33739 | -1.03163 | -0.79823 | -0.60479 |
| | 3.6 | -1.54543 | -1.33918 | -1.03076 | -0.79570 | -0.60114 |
| | 4.0 | -1.54954 | -1.34106 | -1.02974 | -0.79287 | -0.59710 |
| | 6.0 | -1.57508 | -1.35158 | -1.02083 | -0.77186 | -0.56808 |
| | 12.0 | -1.53446 | -1.20600 | -0.75820 | -0.45423 | -0.22861 |

FRACTILES $W_\alpha$ RHO = -1.00

| DELTAS | | $\alpha$:0.3000 | 0.3500 | 0.4000 | 0.4500 | 0.5000 |
|---|---|---|---|---|---|---|
| 0.0 | 0.0 | -0.05246 | 0.08948 | 0.20624 | 0.30359 | 0.38542 |
| 0.4 | 0.0 | -0.05287 | 0.08921 | 0.20610 | 0.30357 | 0.38549 |
| | 0.4 | -0.05246 | 0.08948 | 0.20624 | 0.30359 | 0.38542 |
| 0.8 | 0.0 | -0.05816 | 0.08558 | 0.20400 | 0.30286 | 0.38605 |
| | 0.4 | -0.05287 | 0.08921 | 0.20610 | 0.30357 | 0.38549 |
| | 0.8 | -0.05246 | 0.08948 | 0.20624 | 0.30359 | 0.38542 |
| 1.2 | 0.0 | -0.07577 | 0.07258 | 0.19550 | 0.29863 | 0.38580 |
| | 0.4 | -0.05816 | 0.08558 | 0.20400 | 0.30286 | 0.38605 |
| | 0.8 | -0.05287 | 0.08921 | 0.20610 | 0.30357 | 0.38549 |
| | 1.2 | -0.05246 | 0.08948 | 0.20624 | 0.30359 | 0.38542 |
| 1.6 | 0.0 | -0.10870 | 0.04614 | 0.17602 | 0.28627 | 0.38049 |
| | 0.4 | -0.07577 | 0.07258 | 0.19550 | 0.29863 | 0.38580 |
| | 0.8 | -0.05816 | 0.08558 | 0.20400 | 0.30286 | 0.38605 |
| | 1.2 | -0.05287 | 0.08921 | 0.20610 | 0.30357 | 0.38549 |
| | 1.6 | -0.05246 | 0.08948 | 0.20624 | 0.30359 | 0.38542 |
| 2.0 | 0.0 | -0.15220 | 0.00817 | 0.14498 | 0.26315 | 0.36594 |
| | 0.4 | -0.10870 | 0.04614 | 0.17602 | 0.28627 | 0.38049 |
| | 0.8 | -0.07577 | 0.07258 | 0.19550 | 0.29863 | 0.38580 |
| | 1.2 | -0.05816 | 0.08558 | 0.20400 | 0.30286 | 0.38605 |
| | 1.6 | -0.05287 | 0.08921 | 0.20610 | 0.30357 | 0.38549 |
| | 2.0 | -0.05246 | 0.08948 | 0.20624 | 0.30359 | 0.38542 |
| 2.4 | 0.0 | -0.19760 | -0.03436 | 0.10717 | 0.23162 | 0.34206 |
| | 0.4 | -0.15220 | 0.00817 | 0.14498 | 0.26315 | 0.36594 |
| | 0.8 | -0.10870 | 0.04614 | 0.17602 | 0.28627 | 0.38049 |
| | 1.2 | -0.07577 | 0.07258 | 0.19550 | 0.29863 | 0.38580 |
| | 1.6 | -0.05816 | 0.08558 | 0.20400 | 0.30286 | 0.38605 |
| | 2.0 | -0.05287 | 0.08921 | 0.20610 | 0.30357 | 0.38549 |
| | 2.4 | -0.05246 | 0.08948 | 0.20624 | 0.30359 | 0.38542 |
| 2.8 | 0.0 | -0.23852 | -0.07469 | 0.06913 | 0.19742 | 0.31317 |
| | 0.4 | -0.19760 | -0.03436 | 0.10717 | 0.23162 | 0.34206 |
| | 0.8 | -0.15220 | 0.00817 | 0.14498 | 0.26315 | 0.36594 |
| | 1.2 | -0.10870 | 0.04614 | 0.17602 | 0.28627 | 0.38049 |
| | 1.6 | -0.07577 | 0.07258 | 0.19550 | 0.29863 | 0.38580 |
| | 2.0 | -0.05816 | 0.08558 | 0.20400 | 0.30286 | 0.38605 |
| | 2.4 | -0.05287 | 0.08921 | 0.20610 | 0.30357 | 0.38549 |
| | 2.8 | -0.05246 | 0.08948 | 0.20624 | 0.30359 | 0.38542 |
| 3.2 | 0.0 | -0.27276 | -0.10956 | 0.03500 | 0.16528 | 0.28426 |
| | 0.4 | -0.23852 | -0.07469 | 0.06913 | 0.19742 | 0.31317 |
| | 0.8 | -0.19760 | -0.03436 | 0.10717 | 0.23162 | 0.34206 |
| | 1.2 | -0.15220 | 0.00817 | 0.14498 | 0.26315 | 0.36594 |
| | 1.6 | -0.10870 | 0.04614 | 0.17602 | 0.28627 | 0.38049 |
| | 2.0 | -0.07577 | 0.07258 | 0.19550 | 0.29863 | 0.38580 |
| | 2.4 | -0.05816 | 0.08558 | 0.20400 | 0.30286 | 0.38605 |
| | 2.8 | -0.05287 | 0.08921 | 0.20610 | 0.30357 | 0.38549 |
| | 3.2 | -0.05246 | 0.08948 | 0.20624 | 0.30359 | 0.38542 |

FRACTILES $W_\alpha$ RHO = -1.00

| DELTAS | | $\alpha$:0.3000 | 0.3500 | 0.4000 | 0.4500 | 0.5000 |
|---|---|---|---|---|---|---|
| 3.6 | 0.0 | -0.30071 | -0.13859 | 0.00598 | 0.13726 | 0.25817 |
| | 0.4 | -0.27276 | -0.10956 | 0.03500 | 0.16528 | 0.28426 |
| | 0.8 | -0.23852 | -0.07469 | 0.06913 | 0.19742 | 0.31317 |
| | 1.2 | -0.19760 | -0.03436 | 0.10717 | 0.23162 | 0.34206 |
| | 1.6 | -0.15220 | 0.00817 | 0.14498 | 0.26315 | 0.36594 |
| | 2.0 | -0.10870 | 0.04614 | 0.17602 | 0.28627 | 0.38049 |
| | 2.4 | -0.07577 | 0.07258 | 0.19550 | 0.29863 | 0.38580 |
| | 2.8 | -0.05816 | 0.08558 | 0.20400 | 0.30286 | 0.38605 |
| | 3.2 | -0.05287 | 0.08921 | 0.20610 | 0.30357 | 0.38549 |
| | 3.6 | -0.05246 | 0.08948 | 0.20624 | 0.30359 | 0.38542 |
| 4.0 | 0.0 | -0.32353 | -0.16259 | -0.01831 | 0.11346 | 0.23563 |
| | 0.4 | -0.30071 | -0.13859 | 0.00598 | 0.13726 | 0.25817 |
| | 0.8 | -0.27276 | -0.10956 | 0.03500 | 0.16528 | 0.28426 |
| | 1.2 | -0.23852 | -0.07469 | 0.06913 | 0.19742 | 0.31317 |
| | 1.6 | -0.19760 | -0.03436 | 0.10717 | 0.23162 | 0.34206 |
| | 2.0 | -0.15220 | 0.00817 | 0.14498 | 0.26315 | 0.36594 |
| | 2.4 | -0.10870 | 0.04614 | 0.17602 | 0.28627 | 0.38049 |
| | 2.8 | -0.07577 | 0.07258 | 0.19550 | 0.29863 | 0.38580 |
| | 3.2 | -0.05816 | 0.08558 | 0.20400 | 0.30286 | 0.38605 |
| | 3.6 | -0.05287 | 0.08921 | 0.20610 | 0.30357 | 0.38549 |
| | 4.0 | -0.05246 | 0.08948 | 0.20624 | 0.30359 | 0.38542 |
| 6.0 | 0.0 | -0.39280 | -0.23691 | -0.09478 | 0.03735 | 0.16222 |
| | 0.4 | -0.38291 | -0.22616 | -0.08361 | 0.04857 | 0.17314 |
| | 0.8 | -0.37148 | -0.21380 | -0.07081 | 0.06138 | 0.18557 |
| | 1.2 | -0.35814 | -0.19944 | -0.05601 | 0.07614 | 0.19984 |
| | 1.6 | -0.34238 | -0.18259 | -0.03872 | 0.09331 | 0.21636 |
| | 2.0 | -0.32353 | -0.16259 | -0.01831 | 0.11346 | 0.23563 |
| | 2.4 | -0.30071 | -0.13859 | 0.00598 | 0.13726 | 0.25817 |
| | 2.8 | -0.27276 | -0.10956 | 0.03500 | 0.16528 | 0.28426 |
| | 3.2 | -0.23852 | -0.07469 | 0.06913 | 0.19742 | 0.31317 |
| | 3.6 | -0.19760 | -0.03436 | 0.10717 | 0.23162 | 0.34206 |
| | 4.0 | -0.15220 | 0.00817 | 0.14498 | 0.26315 | 0.36594 |
| | 6.0 | -0.05246 | 0.08948 | 0.20624 | 0.30359 | 0.38542 |
| 12.0 | 0.0 | -0.46079 | -0.31220 | -0.17416 | -0.04334 | 0.08276 |
| | 0.4 | -0.45851 | -0.30962 | -0.17140 | -0.04052 | 0.08557 |
| | 0.8 | -0.45605 | -0.30685 | -0.16845 | -0.03749 | 0.08858 |
| | 1.2 | -0.45340 | -0.30388 | -0.16529 | -0.03424 | 0.09181 |
| | 1.6 | -0.45054 | -0.30068 | -0.16188 | -0.03074 | 0.09528 |
| | 2.0 | -0.44745 | -0.29721 | -0.15819 | -0.02697 | 0.09901 |
| | 2.4 | -0.44409 | -0.29345 | -0.15420 | -0.02289 | 0.10305 |
| | 2.8 | -0.44042 | -0.28936 | -0.14987 | -0.01847 | 0.10743 |
| | 3.2 | -0.43641 | -0.28490 | -0.14514 | -0.01364 | 0.11220 |
| | 3.6 | -0.43201 | -0.28001 | -0.13997 | -0.00838 | 0.11740 |
| | 4.0 | -0.42715 | -0.27462 | -0.13429 | -0.00259 | 0.12309 |
| | 6.0 | -0.39280 | -0.23691 | -0.09478 | 0.03735 | 0.16222 |
| | 12.0 | -0.05246 | 0.08948 | 0.20624 | 0.30359 | 0.38542 |

FRACTILES $W_\alpha$ RHO = -1.00

| DELTAS | | $\alpha$: 0.5500 | 0.6000 | 0.6500 | 0.7000 | 0.7500 |
|---|---|---|---|---|---|---|
| 0.0 | 0.0 | 0.45445 | 0.51266 | 0.56151 | 0.60212 | 0.63531 |
| 0.4 | 0.0 | 0.45461 | 0.51290 | 0.56182 | 0.60249 | 0.63573 |
| | 0.4 | 0.45445 | 0.51266 | 0.56151 | 0.60212 | 0.63531 |
| 0.8 | 0.0 | 0.45629 | 0.51557 | 0.56536 | 0.60677 | 0.64064 |
| | 0.4 | 0.45461 | 0.51290 | 0.56182 | 0.60249 | 0.63573 |
| | 0.8 | 0.45445 | 0.51266 | 0.56151 | 0.60212 | 0.63531 |
| 1.2 | 0.0 | 0.45970 | 0.52229 | 0.57502 | 0.61898 | 0.65502 |
| | 0.4 | 0.45629 | 0.51557 | 0.56536 | 0.60677 | 0.64064 |
| | 0.8 | 0.45461 | 0.51290 | 0.56182 | 0.60249 | 0.63573 |
| | 1.2 | 0.45445 | 0.51266 | 0.56151 | 0.60212 | 0.63531 |
| 1.6 | 0.0 | 0.46119 | 0.53018 | 0.58882 | 0.63810 | 0.67876 |
| | 0.4 | 0.45970 | 0.52229 | 0.57502 | 0.61898 | 0.65502 |
| | 0.8 | 0.45629 | 0.51557 | 0.56536 | 0.60677 | 0.64064 |
| | 1.2 | 0.45461 | 0.51290 | 0.56182 | 0.60249 | 0.63573 |
| | 1.6 | 0.45445 | 0.51266 | 0.56151 | 0.60212 | 0.63531 |
| 2.0 | 0.0 | 0.45558 | 0.53363 | 0.60115 | 0.65888 | 0.70730 |
| | 0.4 | 0.46119 | 0.53018 | 0.58882 | 0.63810 | 0.67876 |
| | 0.8 | 0.45970 | 0.52229 | 0.57502 | 0.61898 | 0.65502 |
| | 1.2 | 0.45629 | 0.51557 | 0.56536 | 0.60677 | 0.64064 |
| | 1.6 | 0.45461 | 0.51290 | 0.56182 | 0.60249 | 0.63573 |
| | 2.0 | 0.45445 | 0.51266 | 0.56151 | 0.60212 | 0.63531 |
| 2.4 | 0.0 | 0.44052 | 0.52835 | 0.60636 | 0.67497 | 0.73421 |
| | 0.4 | 0.45558 | 0.53363 | 0.60115 | 0.65888 | 0.70730 |
| | 0.8 | 0.46119 | 0.53018 | 0.58882 | 0.63810 | 0.67876 |
| | 1.2 | 0.45970 | 0.52229 | 0.57502 | 0.61898 | 0.65502 |
| | 1.6 | 0.45629 | 0.51557 | 0.56536 | 0.60677 | 0.64064 |
| | 2.0 | 0.45461 | 0.51290 | 0.56182 | 0.60249 | 0.63573 |
| | 2.4 | 0.45445 | 0.51266 | 0.56151 | 0.60212 | 0.63531 |
| 2.8 | 0.0 | 0.41842 | 0.51453 | 0.60230 | 0.68209 | 0.75372 |
| | 0.4 | 0.44052 | 0.52835 | 0.60636 | 0.67497 | 0.73421 |
| | 0.8 | 0.45558 | 0.53363 | 0.60115 | 0.65888 | 0.70730 |
| | 1.2 | 0.46119 | 0.53018 | 0.58882 | 0.63810 | 0.67876 |
| | 1.6 | 0.45970 | 0.52229 | 0.57502 | 0.61898 | 0.65502 |
| | 2.0 | 0.45629 | 0.51557 | 0.56536 | 0.60677 | 0.64064 |
| | 2.4 | 0.45461 | 0.51290 | 0.56182 | 0.60249 | 0.63573 |
| | 2.8 | 0.45445 | 0.51266 | 0.56151 | 0.60212 | 0.63531 |
| 3.2 | 0.0 | 0.39403 | 0.49605 | 0.59134 | 0.68047 | 0.76354 |
| | 0.4 | 0.41842 | 0.51453 | 0.60230 | 0.68209 | 0.75372 |
| | 0.8 | 0.44052 | 0.52835 | 0.60636 | 0.67497 | 0.73421 |
| | 1.2 | 0.45558 | 0.53363 | 0.60115 | 0.65888 | 0.70730 |
| | 1.6 | 0.46119 | 0.53018 | 0.58882 | 0.63810 | 0.67876 |
| | 2.0 | 0.45970 | 0.52229 | 0.57502 | 0.61898 | 0.65502 |
| | 2.4 | 0.45629 | 0.51557 | 0.56536 | 0.60677 | 0.64064 |
| | 2.8 | 0.45461 | 0.51290 | 0.56182 | 0.60249 | 0.63573 |
| | 3.2 | 0.45445 | 0.51266 | 0.56151 | 0.60212 | 0.63531 |

FRACTILES $W_\alpha$ RHO = -1.00

| DELTAS | | $\alpha$: 0.5500 | 0.6000 | 0.6500 | 0.7000 | 0.7500 |
|---|---|---|---|---|---|---|
| 3.6 | 0.0 | 0.37086 | 0.47691 | 0.57752 | 0.67356 | 0.76561 |
| | 0.4 | 0.39403 | 0.49605 | 0.59134 | 0.68047 | 0.76354 |
| | 0.8 | 0.41842 | 0.51453 | 0.60230 | 0.68209 | 0.75372 |
| | 1.2 | 0.44052 | 0.52835 | 0.60636 | 0.67497 | 0.73421 |
| | 1.6 | 0.45558 | 0.53363 | 0.60115 | 0.65888 | 0.70730 |
| | 2.0 | 0.46119 | 0.53018 | 0.58882 | 0.63810 | 0.67876 |
| | 2.4 | 0.45970 | 0.52229 | 0.57502 | 0.61898 | 0.65502 |
| | 2.8 | 0.45629 | 0.51557 | 0.56536 | 0.60677 | 0.64064 |
| | 3.2 | 0.45461 | 0.51290 | 0.56182 | 0.60249 | 0.63573 |
| | 3.6 | 0.45445 | 0.51266 | 0.56151 | 0.60212 | 0.63531 |
| 4.0 | 0.0 | 0.35034 | 0.45924 | 0.56368 | 0.66479 | 0.76352 |
| | 0.4 | 0.37086 | 0.47691 | 0.57752 | 0.67356 | 0.76561 |
| | 0.8 | 0.39403 | 0.49605 | 0.59134 | 0.68047 | 0.76354 |
| | 1.2 | 0.41842 | 0.51453 | 0.60230 | 0.68209 | 0.75372 |
| | 1.6 | 0.44052 | 0.52835 | 0.60636 | 0.67497 | 0.73421 |
| | 2.0 | 0.45558 | 0.53363 | 0.60115 | 0.65888 | 0.70730 |
| | 2.4 | 0.46119 | 0.53018 | 0.58882 | 0.63810 | 0.67876 |
| | 2.8 | 0.45970 | 0.52229 | 0.57502 | 0.61898 | 0.65502 |
| | 3.2 | 0.45629 | 0.51557 | 0.56536 | 0.60677 | 0.64064 |
| | 3.6 | 0.45461 | 0.51290 | 0.56182 | 0.60249 | 0.63573 |
| | 4.0 | 0.45445 | 0.51266 | 0.56151 | 0.60212 | 0.63531 |
| 6.0 | 0.0 | 0.28197 | 0.39840 | 0.51318 | 0.62803 | 0.74492 |
| | 0.4 | 0.29224 | 0.40766 | 0.52102 | 0.63396 | 0.74833 |
| | 0.8 | 0.30389 | 0.41812 | 0.52983 | 0.64055 | 0.75199 |
| | 1.2 | 0.31722 | 0.43003 | 0.53977 | 0.64789 | 0.75588 |
| | 1.6 | 0.33257 | 0.44364 | 0.55102 | 0.65601 | 0.75987 |
| | 2.0 | 0.35034 | 0.45924 | 0.56368 | 0.66479 | 0.76352 |
| | 2.4 | 0.37086 | 0.47691 | 0.57752 | 0.67356 | 0.76561 |
| | 2.8 | 0.39403 | 0.49605 | 0.59134 | 0.68047 | 0.76354 |
| | 3.2 | 0.41842 | 0.51453 | 0.60230 | 0.68209 | 0.75372 |
| | 3.6 | 0.44052 | 0.52835 | 0.60636 | 0.67497 | 0.73421 |
| | 4.0 | 0.45558 | 0.53363 | 0.60115 | 0.65888 | 0.70730 |
| | 6.0 | 0.45445 | 0.51266 | 0.56151 | 0.60212 | 0.63531 |
| 12.0 | 0.0 | 0.20625 | 0.32905 | 0.45315 | 0.58080 | 0.71496 |
| | 0.4 | 0.20896 | 0.33157 | 0.45536 | 0.58259 | 0.71618 |
| | 0.8 | 0.21185 | 0.33425 | 0.45772 | 0.58449 | 0.71746 |
| | 1.2 | 0.21496 | 0.33712 | 0.46024 | 0.58652 | 0.71882 |
| | 1.6 | 0.21829 | 0.34020 | 0.46294 | 0.58869 | 0.72027 |
| | 2.0 | 0.22187 | 0.34351 | 0.46584 | 0.59102 | 0.72181 |
| | 2.4 | 0.22575 | 0.34708 | 0.46896 | 0.59352 | 0.72346 |
| | 2.8 | 0.22994 | 0.35094 | 0.47233 | 0.59620 | 0.72522 |
| | 3.2 | 0.23449 | 0.35513 | 0.47598 | 0.59910 | 0.72710 |
| | 3.6 | 0.23946 | 0.35969 | 0.47994 | 0.60224 | 0.72912 |
| | 4.0 | 0.24489 | 0.36467 | 0.48425 | 0.60564 | 0.73128 |
| | 6.0 | 0.28197 | 0.39840 | 0.51318 | 0.62803 | 0.74492 |
| | 12.0 | 0.45445 | 0.51266 | 0.56151 | 0.60212 | 0.63531 |

FRACTILES $W_{\alpha}$ RHO = −1.00

| DELTAS | | $\alpha$:0.8000 | 0.8500 | 0.9000 | 0.9250 | 0.9500 |
|---|---|---|---|---|---|---|
| 0.0 | 0.0 | 0.66172 | 0.68182 | 0.69594 | 0.70084 | 0.70433 |
| 0.4 | 0.0 | 0.66218 | 0.68230 | 0.69645 | 0.70135 | 0.70485 |
| | 0.4 | 0.66172 | 0.68182 | 0.69594 | 0.70084 | 0.70433 |
| 0.8 | 0.0 | 0.66759 | 0.68810 | 0.70253 | 0.70753 | 0.71109 |
| | 0.4 | 0.66218 | 0.68230 | 0.69645 | 0.70135 | 0.70485 |
| | 0.8 | 0.66172 | 0.68182 | 0.69594 | 0.70084 | 0.70433 |
| 1.2 | 0.0 | 0.68375 | 0.70565 | 0.72107 | 0.72642 | 0.73022 |
| | 0.4 | 0.66759 | 0.68810 | 0.70253 | 0.70753 | 0.71109 |
| | 0.8 | 0.66218 | 0.68230 | 0.69645 | 0.70135 | 0.70485 |
| | 1.2 | 0.66172 | 0.68182 | 0.69594 | 0.70084 | 0.70433 |
| 1.6 | 0.0 | 0.71137 | 0.73635 | 0.75400 | 0.76014 | 0.76451 |
| | 0.4 | 0.68375 | 0.70565 | 0.72107 | 0.72642 | 0.73022 |
| | 0.8 | 0.66759 | 0.68810 | 0.70253 | 0.70753 | 0.71109 |
| | 1.2 | 0.66218 | 0.68230 | 0.69645 | 0.70135 | 0.70485 |
| | 1.6 | 0.66172 | 0.68182 | 0.69594 | 0.70084 | 0.70433 |
| 2.0 | 0.0 | 0.74670 | 0.77726 | 0.79906 | 0.80669 | 0.81214 |
| | 0.4 | 0.71137 | 0.73635 | 0.75400 | 0.76014 | 0.76451 |
| | 0.8 | 0.68375 | 0.70565 | 0.72107 | 0.72642 | 0.73022 |
| | 1.2 | 0.66759 | 0.68810 | 0.70253 | 0.70753 | 0.71109 |
| | 1.6 | 0.66218 | 0.68230 | 0.69645 | 0.70135 | 0.70485 |
| | 2.0 | 0.66172 | 0.68182 | 0.69594 | 0.70084 | 0.70433 |
| 2.4 | 0.0 | 0.78385 | 0.82342 | 0.85233 | 0.86259 | 0.86997 |
| | 0.4 | 0.74670 | 0.77726 | 0.79906 | 0.80669 | 0.81214 |
| | 0.8 | 0.71137 | 0.73635 | 0.75400 | 0.76014 | 0.76451 |
| | 1.2 | 0.68375 | 0.70565 | 0.72107 | 0.72642 | 0.73022 |
| | 1.6 | 0.66759 | 0.68810 | 0.70253 | 0.70753 | 0.71109 |
| | 2.0 | 0.66218 | 0.68230 | 0.69645 | 0.70135 | 0.70485 |
| | 2.4 | 0.66172 | 0.68182 | 0.69594 | 0.70084 | 0.70433 |
| 2.8 | 0.0 | 0.81648 | 0.86898 | 0.90923 | 0.92400 | 0.93480 |
| | 0.4 | 0.78385 | 0.82342 | 0.85233 | 0.86259 | 0.86997 |
| | 0.8 | 0.74670 | 0.77726 | 0.79906 | 0.80669 | 0.81214 |
| | 1.2 | 0.71137 | 0.73635 | 0.75400 | 0.76014 | 0.76451 |
| | 1.6 | 0.68375 | 0.70565 | 0.72107 | 0.72642 | 0.73022 |
| | 2.0 | 0.66759 | 0.68810 | 0.70253 | 0.70753 | 0.71109 |
| | 2.4 | 0.66218 | 0.68230 | 0.69645 | 0.70135 | 0.70485 |
| | 2.8 | 0.66172 | 0.68182 | 0.69594 | 0.70084 | 0.70433 |
| 3.2 | 0.0 | 0.83997 | 0.90811 | 0.96453 | 0.98662 | 1.00340 |
| | 0.4 | 0.81648 | 0.86898 | 0.90923 | 0.92400 | 0.93480 |
| | 0.8 | 0.78385 | 0.82342 | 0.85233 | 0.86259 | 0.86997 |
| | 1.2 | 0.74670 | 0.77726 | 0.79906 | 0.80669 | 0.81214 |
| | 1.6 | 0.71137 | 0.73635 | 0.75400 | 0.76014 | 0.76451 |
| | 2.0 | 0.68375 | 0.70565 | 0.72107 | 0.72642 | 0.73022 |
| | 2.4 | 0.66759 | 0.68810 | 0.70253 | 0.70753 | 0.71109 |
| | 2.8 | 0.66218 | 0.68230 | 0.69645 | 0.70135 | 0.70485 |
| | 3.2 | 0.66172 | 0.68182 | 0.69594 | 0.70084 | 0.70433 |

FRACTILES $w_\alpha$ RHO = -1.00

| DELTAS | | $\alpha$:0.8000 | 0.8500 | 0.9000 | 0.9250 | 0.9500 |
|---|---|---|---|---|---|---|
| 3.6 | 0.0 | 0.85377 | 0.93720 | 1.01286 | 1.04542 | 1.07200 |
| | 0.4 | 0.83997 | 0.90811 | 0.96453 | 0.98662 | 1.00340 |
| | 0.8 | 0.81648 | 0.86898 | 0.90923 | 0.92400 | 0.93480 |
| | 1.2 | 0.78385 | 0.82342 | 0.85233 | 0.86259 | 0.86997 |
| | 1.6 | 0.74670 | 0.77726 | 0.79906 | 0.80669 | 0.81214 |
| | 2.0 | 0.71137 | 0.73635 | 0.75400 | 0.76014 | 0.76451 |
| | 2.4 | 0.68375 | 0.70565 | 0.72107 | 0.72642 | 0.73022 |
| | 2.8 | 0.66759 | 0.68810 | 0.70253 | 0.70753 | 0.71109 |
| | 3.2 | 0.66218 | 0.68230 | 0.69645 | 0.70135 | 0.70485 |
| | 3.6 | 0.66172 | 0.68182 | 0.69594 | 0.70084 | 0.70433 |
| 4.0 | 0.0 | 0.86070 | 0.95677 | 1.05084 | 1.09555 | 1.13590 |
| | 0.4 | 0.85377 | 0.93720 | 1.01286 | 1.04542 | 1.07200 |
| | 0.8 | 0.83997 | 0.90811 | 0.96453 | 0.98662 | 1.00340 |
| | 1.2 | 0.81648 | 0.86898 | 0.90923 | 0.92400 | 0.93480 |
| | 1.6 | 0.78385 | 0.82342 | 0.85233 | 0.86259 | 0.86997 |
| | 2.0 | 0.74670 | 0.77726 | 0.79906 | 0.80669 | 0.81214 |
| | 2.4 | 0.71137 | 0.73635 | 0.75400 | 0.76014 | 0.76451 |
| | 2.8 | 0.68375 | 0.70565 | 0.72107 | 0.72642 | 0.73022 |
| | 3.2 | 0.66759 | 0.68810 | 0.70253 | 0.70753 | 0.71109 |
| | 3.6 | 0.66218 | 0.68230 | 0.69645 | 0.70135 | 0.70485 |
| | 4.0 | 0.66172 | 0.68182 | 0.69594 | 0.70084 | 0.70433 |
| 6.0 | 0.0 | 0.86649 | 0.99675 | 1.14316 | 1.22719 | 1.32428 |
| | 0.4 | 0.86650 | 0.99203 | 1.13130 | 1.21002 | 1.29935 |
| | 0.8 | 0.86623 | 0.98630 | 1.11731 | 1.18984 | 1.27006 |
| | 1.2 | 0.86552 | 0.97916 | 1.10038 | 1.16552 | 1.23468 |
| | 1.6 | 0.86397 | 0.96984 | 1.07907 | 1.13503 | 1.19060 |
| | 2.0 | 0.86070 | 0.95677 | 1.05084 | 1.09555 | 1.13590 |
| | 2.4 | 0.85377 | 0.93720 | 1.01286 | 1.04542 | 1.07200 |
| | 2.8 | 0.83997 | 0.90811 | 0.96453 | 0.98662 | 1.00340 |
| | 3.2 | 0.81648 | 0.86898 | 0.90923 | 0.92400 | 0.93480 |
| | 3.6 | 0.78385 | 0.82342 | 0.85233 | 0.86259 | 0.86997 |
| | 4.0 | 0.74670 | 0.77726 | 0.79906 | 0.80669 | 0.81214 |
| | 6.0 | 0.66172 | 0.68182 | 0.69594 | 0.70084 | 0.70433 |
| 12.0 | 0.0 | 0.85998 | 1.02317 | 1.21958 | 1.34090 | 1.49240 |
| | 0.4 | 0.86039 | 1.02247 | 1.21716 | 1.33720 | 1.48682 |
| | 0.8 | 0.86083 | 1.02170 | 1.21455 | 1.33321 | 1.48081 |
| | 1.2 | 0.86128 | 1.02085 | 1.21173 | 1.32890 | 1.47435 |
| | 1.6 | 0.86174 | 1.01991 | 1.20866 | 1.32424 | 1.46735 |
| | 2.0 | 0.86221 | 1.01888 | 1.20532 | 1.31918 | 1.45977 |
| | 2.4 | 0.86269 | 1.01772 | 1.20167 | 1.31366 | 1.45153 |
| | 2.8 | 0.86319 | 1.01643 | 1.19766 | 1.30762 | 1.44252 |
| | 3.2 | 0.86368 | 1.01498 | 1.19324 | 1.30099 | 1.43266 |
| | 3.6 | 0.86418 | 1.01334 | 1.18835 | 1.29368 | 1.42180 |
| | 4.0 | 0.86467 | 1.01148 | 1.18291 | 1.28557 | 1.40980 |
| | 6.0 | 0.86649 | 0.99675 | 1.14316 | 1.22719 | 1.32428 |
| | 12.0 | 0.66172 | 0.68182 | 0.69594 | 0.70084 | 0.70433 |

FRACTILES $W_\alpha$ RHO = -1.00

| DELTAS | | $\alpha$:0.9625 | 0.9750 | 0.9825 | 0.9900 | 0.9925 |
|---|---|---|---|---|---|---|
| 0.0 | 0.0 | 0.70554 | 0.70641 | 0.70677 | 0.70700 | 0.70704 |
| 0.4 | 0.0 | 0.70606 | 0.70693 | 0.70729 | 0.70752 | 0.70757 |
| | 0.4 | 0.70554 | 0.70641 | 0.70677 | 0.70700 | 0.70704 |
| 0.8 | 0.0 | 0.71233 | 0.71322 | 0.71358 | 0.71382 | 0.71387 |
| | 0.4 | 0.70606 | 0.70693 | 0.70729 | 0.70752 | 0.70757 |
| | 0.8 | 0.70554 | 0.70641 | 0.70677 | 0.70700 | 0.70704 |
| 1.2 | 0.0 | 0.73155 | 0.73250 | 0.73289 | 0.73314 | 0.73319 |
| | 0.4 | 0.71233 | 0.71322 | 0.71358 | 0.71382 | 0.71387 |
| | 0.8 | 0.70606 | 0.70693 | 0.70729 | 0.70752 | 0.70757 |
| | 1.2 | 0.70554 | 0.70641 | 0.70677 | 0.70700 | 0.70704 |
| 1.6 | 0.0 | 0.76604 | 0.76713 | 0.76757 | 0.76786 | 0.76792 |
| | 0.4 | 0.73155 | 0.73250 | 0.73289 | 0.73314 | 0.73319 |
| | 0.8 | 0.71233 | 0.71322 | 0.71358 | 0.71382 | 0.71387 |
| | 1.2 | 0.70606 | 0.70693 | 0.70729 | 0.70752 | 0.70757 |
| | 1.6 | 0.70554 | 0.70641 | 0.70677 | 0.70700 | 0.70704 |
| 2.0 | 0.0 | 0.81405 | 0.81541 | 0.81596 | 0.81632 | 0.81640 |
| | 0.4 | 0.76604 | 0.76713 | 0.76757 | 0.76786 | 0.76792 |
| | 0.8 | 0.73155 | 0.73250 | 0.73289 | 0.73314 | 0.73319 |
| | 1.2 | 0.71233 | 0.71322 | 0.71358 | 0.71382 | 0.71387 |
| | 1.6 | 0.70606 | 0.70693 | 0.70729 | 0.70752 | 0.70757 |
| | 2.0 | 0.70554 | 0.70641 | 0.70677 | 0.70700 | 0.70704 |
| 2.4 | 0.0 | 0.87257 | 0.87442 | 0.87518 | 0.87567 | 0.87578 |
| | 0.4 | 0.81405 | 0.81541 | 0.81596 | 0.81632 | 0.81640 |
| | 0.8 | 0.76604 | 0.76713 | 0.76757 | 0.76786 | 0.76792 |
| | 1.2 | 0.73155 | 0.73250 | 0.73289 | 0.73314 | 0.73319 |
| | 1.6 | 0.71233 | 0.71322 | 0.71358 | 0.71382 | 0.71387 |
| | 2.0 | 0.70606 | 0.70693 | 0.70729 | 0.70752 | 0.70757 |
| | 2.4 | 0.70554 | 0.70641 | 0.70677 | 0.70700 | 0.70704 |
| 2.8 | 0.0 | 0.93864 | 0.94140 | 0.94253 | 0.94326 | 0.94341 |
| | 0.4 | 0.87257 | 0.87442 | 0.87518 | 0.87567 | 0.87578 |
| | 0.8 | 0.81405 | 0.81541 | 0.81596 | 0.81632 | 0.81640 |
| | 1.2 | 0.76604 | 0.76713 | 0.76757 | 0.76786 | 0.76792 |
| | 1.6 | 0.73155 | 0.73250 | 0.73289 | 0.73314 | 0.73319 |
| | 2.0 | 0.71233 | 0.71322 | 0.71358 | 0.71382 | 0.71387 |
| | 2.4 | 0.70606 | 0.70693 | 0.70729 | 0.70752 | 0.70757 |
| | 2.8 | 0.70554 | 0.70641 | 0.70677 | 0.70700 | 0.70704 |
| 3.2 | 0.0 | 1.00951 | 1.01395 | 1.01579 | 1.01698 | 1.01723 |
| | 0.4 | 0.93864 | 0.94140 | 0.94253 | 0.94326 | 0.94341 |
| | 0.8 | 0.87257 | 0.87442 | 0.87518 | 0.87567 | 0.87578 |
| | 1.2 | 0.81405 | 0.81541 | 0.81596 | 0.81632 | 0.81640 |
| | 1.6 | 0.76604 | 0.76713 | 0.76757 | 0.76786 | 0.76792 |
| | 2.0 | 0.73155 | 0.73250 | 0.73289 | 0.73314 | 0.73319 |
| | 2.4 | 0.71233 | 0.71322 | 0.71358 | 0.71382 | 0.71387 |
| | 2.8 | 0.70606 | 0.70693 | 0.70729 | 0.70752 | 0.70757 |
| | 3.2 | 0.70554 | 0.70641 | 0.70677 | 0.70700 | 0.70704 |

FRACTILES $W_\alpha$ RHO = -1.00

| DELTAS | | $\alpha$:0.9625 | 0.9750 | 0.9825 | 0.9900 | 0.9925 |
|---|---|---|---|---|---|---|
| 3.6 | 0.0 | 1.08221 | 1.08986 | 1.09308 | 1.09519 | 1.09564 |
| | 0.4 | 1.00951 | 1.01395 | 1.01579 | 1.01698 | 1.01723 |
| | 0.8 | 0.93864 | 0.94140 | 0.94253 | 0.94326 | 0.94341 |
| | 1.2 | 0.87257 | 0.87442 | 0.87518 | 0.87567 | 0.87578 |
| | 1.6 | 0.81405 | 0.81541 | 0.81596 | 0.81632 | 0.81640 |
| | 2.0 | 0.76604 | 0.76713 | 0.76757 | 0.76786 | 0.76792 |
| | 2.4 | 0.73155 | 0.73250 | 0.73289 | 0.73314 | 0.73319 |
| | 2.8 | 0.71233 | 0.71322 | 0.71358 | 0.71382 | 0.71387 |
| | 3.2 | 0.70606 | 0.70693 | 0.70729 | 0.70752 | 0.70757 |
| | 3.6 | 0.70554 | 0.70641 | 0.70677 | 0.70700 | 0.70704 |
| 4.0 | 0.0 | 1.15291 | 1.16650 | 1.17248 | 1.17651 | 1.17738 |
| | 0.4 | 1.08221 | 1.08986 | 1.09308 | 1.09519 | 1.09564 |
| | 0.8 | 1.00951 | 1.01395 | 1.01579 | 1.01698 | 1.01723 |
| | 1.2 | 0.93864 | 0.94140 | 0.94253 | 0.94326 | 0.94341 |
| | 1.6 | 0.87257 | 0.87442 | 0.87518 | 0.87567 | 0.87578 |
| | 2.0 | 0.81405 | 0.81541 | 0.81596 | 0.81632 | 0.81640 |
| | 2.4 | 0.76604 | 0.76713 | 0.76757 | 0.76786 | 0.76792 |
| | 2.8 | 0.73155 | 0.73250 | 0.73289 | 0.73314 | 0.73319 |
| | 3.2 | 0.71233 | 0.71322 | 0.71358 | 0.71382 | 0.71387 |
| | 3.6 | 0.70606 | 0.70693 | 0.70729 | 0.70752 | 0.70757 |
| | 4.0 | 0.70554 | 0.70641 | 0.70677 | 0.70700 | 0.70704 |
| 6.0 | 0.0 | 1.38089 | 1.44659 | 1.49291 | 1.54764 | 1.56840 |
| | 0.4 | 1.35027 | 1.40767 | 1.44640 | 1.48862 | 1.50288 |
| | 0.8 | 1.31422 | 1.36164 | 1.39124 | 1.41948 | 1.42753 |
| | 1.2 | 1.27057 | 1.30609 | 1.32572 | 1.34165 | 1.34555 |
| | 1.6 | 1.21686 | 1.24011 | 1.25135 | 1.25941 | 1.26122 |
| | 2.0 | 1.15291 | 1.16650 | 1.17248 | 1.17651 | 1.17738 |
| | 2.4 | 1.08221 | 1.08986 | 1.09308 | 1.09519 | 1.09564 |
| | 2.8 | 1.00951 | 1.01395 | 1.01579 | 1.01698 | 1.01723 |
| | 3.2 | 0.93864 | 0.94140 | 0.94253 | 0.94326 | 0.94341 |
| | 3.6 | 0.87257 | 0.87442 | 0.87518 | 0.87567 | 0.87578 |
| | 4.0 | 0.81405 | 0.81541 | 0.81596 | 0.81632 | 0.81640 |
| | 6.0 | 0.70554 | 0.70641 | 0.70677 | 0.70700 | 0.70704 |
| 12.0 | 0.0 | 1.58863 | 1.71133 | 1.80874 | 1.94523 | 2.00877 |
| | 0.4 | 1.58168 | 1.70241 | 1.79805 | 1.93171 | 1.99378 |
| | 0.8 | 1.57421 | 1.69282 | 1.78657 | 1.91721 | 1.97770 |
| | 1.2 | 1.56616 | 1.68250 | 1.77421 | 1.90161 | 1.96041 |
| | 1.6 | 1.55747 | 1.67136 | 1.76089 | 1.88478 | 1.94177 |
| | 2.0 | 1.54806 | 1.65931 | 1.74646 | 1.86658 | 1.92161 |
| | 2.4 | 1.53782 | 1.64621 | 1.73081 | 1.84684 | 1.89974 |
| | 2.8 | 1.52666 | 1.63194 | 1.71375 | 1.82535 | 1.87595 |
| | 3.2 | 1.51444 | 1.61633 | 1.69511 | 1.80188 | 1.84995 |
| | 3.6 | 1.50100 | 1.59919 | 1.67465 | 1.77613 | 1.82145 |
| | 4.0 | 1.48616 | 1.58028 | 1.65210 | 1.74776 | 1.79007 |
| | 6.0 | 1.38089 | 1.44659 | 1.49291 | 1.54764 | 1.56840 |
| | 12.0 | 0.70554 | 0.70641 | 0.70677 | 0.70700 | 0.70704 |

FRACTILES $W_\alpha$ RHO = -1.00

| DELTAS | | $\alpha$: 0.9950 | 0.9975 | 0.9990 | 0.9995 |
|---|---|---|---|---|---|
| 0.0 | 0.0 | 0.70708 | 0.70710 | 0.70711 | 0.70711 |
| 0.4 | 0.0 | 0.70760 | 0.70762 | 0.70763 | 0.70763 |
| | 0.4 | 0.70708 | 0.70710 | 0.70711 | 0.70711 |
| 0.8 | 0.0 | 0.71390 | 0.71392 | 0.71393 | 0.71393 |
| | 0.4 | 0.70760 | 0.70762 | 0.70763 | 0.70763 |
| | 0.8 | 0.70708 | 0.70710 | 0.70711 | 0.70711 |
| 1.2 | 0.0 | 0.73323 | 0.73325 | 0.73326 | 0.73327 |
| | 0.4 | 0.71390 | 0.71392 | 0.71393 | 0.71393 |
| | 0.8 | 0.70760 | 0.70762 | 0.70763 | 0.70763 |
| | 1.2 | 0.70708 | 0.70710 | 0.70711 | 0.70711 |
| 1.6 | 0.0 | 0.76797 | 0.76799 | 0.76800 | 0.76800 |
| | 0.4 | 0.73323 | 0.73325 | 0.73326 | 0.73327 |
| | 0.8 | 0.71390 | 0.71392 | 0.71393 | 0.71393 |
| | 1.2 | 0.70760 | 0.70762 | 0.70763 | 0.70763 |
| | 1.6 | 0.70708 | 0.70710 | 0.70711 | 0.70711 |
| 2.0 | 0.0 | 0.81645 | 0.81649 | 0.81649 | 0.81650 |
| | 0.4 | 0.76797 | 0.76799 | 0.76800 | 0.76800 |
| | 0.8 | 0.73323 | 0.73325 | 0.73326 | 0.73327 |
| | 1.2 | 0.71390 | 0.71392 | 0.71393 | 0.71393 |
| | 1.6 | 0.70760 | 0.70762 | 0.70763 | 0.70763 |
| | 2.0 | 0.70708 | 0.70710 | 0.70711 | 0.70711 |
| 2.4 | 0.0 | 0.87585 | 0.87589 | 0.87590 | 0.87591 |
| | 0.4 | 0.81645 | 0.81649 | 0.81649 | 0.81650 |
| | 0.8 | 0.76797 | 0.76799 | 0.76800 | 0.76800 |
| | 1.2 | 0.73323 | 0.73325 | 0.73326 | 0.73327 |
| | 1.6 | 0.71390 | 0.71392 | 0.71393 | 0.71393 |
| | 2.0 | 0.70760 | 0.70762 | 0.70763 | 0.70763 |
| | 2.4 | 0.70708 | 0.70710 | 0.70711 | 0.70711 |
| 2.8 | 0.0 | 0.94352 | 0.94359 | 0.94361 | 0.94361 |
| | 0.4 | 0.87585 | 0.87589 | 0.87590 | 0.87591 |
| | 0.8 | 0.81645 | 0.81649 | 0.81649 | 0.81650 |
| | 1.2 | 0.76797 | 0.76799 | 0.76800 | 0.76800 |
| | 1.6 | 0.73323 | 0.73325 | 0.73326 | 0.73327 |
| | 2.0 | 0.71390 | 0.71392 | 0.71393 | 0.71393 |
| | 2.4 | 0.70760 | 0.70762 | 0.70763 | 0.70763 |
| | 2.8 | 0.70708 | 0.70710 | 0.70711 | 0.70711 |
| 3.2 | 0.0 | 1.01741 | 1.01752 | 1.01755 | 1.01756 |
| | 0.4 | 0.94352 | 0.94359 | 0.94361 | 0.94361 |
| | 0.8 | 0.87585 | 0.87589 | 0.87590 | 0.87591 |
| | 1.2 | 0.81645 | 0.81649 | 0.81649 | 0.81650 |
| | 1.6 | 0.76797 | 0.76799 | 0.76800 | 0.76800 |
| | 2.0 | 0.73323 | 0.73325 | 0.73326 | 0.73327 |
| | 2.4 | 0.71390 | 0.71392 | 0.71393 | 0.71393 |
| | 2.8 | 0.70760 | 0.70762 | 0.70763 | 0.70763 |
| | 3.2 | 0.70708 | 0.70710 | 0.70711 | 0.70711 |

FRACTILES $W_{\alpha}$ RHO = -1.00

| DELTAS | | α: 0.9950 | 0.9975 | 0.9990 | 0.9995 |
|---|---|---|---|---|---|
| 3.6 | 0.0 | 1.09597 | 1.09616 | 1.09621 | 1.09622 |
| | 0.4 | 1.01741 | 1.01752 | 1.01755 | 1.01756 |
| | 0.8 | 0.94352 | 0.94359 | 0.94361 | 0.94361 |
| | 1.2 | 0.87585 | 0.87589 | 0.87590 | 0.87591 |
| | 1.6 | 0.81645 | 0.81649 | 0.81649 | 0.81650 |
| | 2.0 | 0.76797 | 0.76799 | 0.76800 | 0.76800 |
| | 2.4 | 0.73323 | 0.73325 | 0.73326 | 0.73327 |
| | 2.8 | 0.71390 | 0.71392 | 0.71393 | 0.71393 |
| | 3.2 | 0.70760 | 0.70762 | 0.70763 | 0.70763 |
| | 3.6 | 0.70708 | 0.70710 | 0.70711 | 0.70711 |
| 4.0 | 0.0 | 1.17801 | 1.17838 | 1.17849 | 1.17851 |
| | 0.4 | 1.09597 | 1.09616 | 1.09621 | 1.09622 |
| | 0.8 | 1.01741 | 1.01752 | 1.01755 | 1.01756 |
| | 1.2 | 0.94352 | 0.94359 | 0.94361 | 0.94361 |
| | 1.6 | 0.87585 | 0.87589 | 0.87590 | 0.87591 |
| | 2.0 | 0.81645 | 0.81649 | 0.81649 | 0.81650 |
| | 2.4 | 0.76797 | 0.76799 | 0.76800 | 0.76800 |
| | 2.8 | 0.73323 | 0.73325 | 0.73326 | 0.73327 |
| | 3.2 | 0.71390 | 0.71392 | 0.71393 | 0.71393 |
| | 3.6 | 0.70760 | 0.70762 | 0.70763 | 0.70763 |
| | 4.0 | 0.70708 | 0.70710 | 0.70711 | 0.70711 |
| 6.0 | 0.0 | 1.59030 | 1.61136 | 1.62022 | 1.62170 |
| | 0.4 | 1.51614 | 1.52643 | 1.52984 | 1.53035 |
| | 0.8 | 1.43409 | 1.43846 | 1.43975 | 1.43994 |
| | 1.2 | 1.34848 | 1.35030 | 1.35082 | 1.35089 |
| | 1.6 | 1.26254 | 1.26334 | 1.26356 | 1.26360 |
| | 2.0 | 1.17801 | 1.17838 | 1.17849 | 1.17851 |
| | 2.4 | 1.09597 | 1.09616 | 1.09621 | 1.09622 |
| | 2.8 | 1.01741 | 1.01752 | 1.01755 | 1.01756 |
| | 3.2 | 0.94352 | 0.94359 | 0.94361 | 0.94361 |
| | 3.6 | 0.87585 | 0.87589 | 0.87590 | 0.87591 |
| | 4.0 | 0.81645 | 0.81649 | 0.81649 | 0.81650 |
| | 6.0 | 0.70708 | 0.70710 | 0.70711 | 0.70711 |
| 12.0 | 0.0 | 2.09178 | 2.21839 | 2.36143 | 2.45458 |
| | 0.4 | 2.07470 | 2.19771 | 2.33591 | 2.42536 |
| | 0.8 | 2.05638 | 2.17552 | 2.30855 | 2.39406 |
| | 1.2 | 2.03669 | 2.15168 | 2.27915 | 2.36042 |
| | 1.6 | 2.01546 | 2.12598 | 2.24748 | 2.32418 |
| | 2.0 | 1.99251 | 2.09821 | 2.21326 | 2.28503 |
| | 2.4 | 1.96763 | 2.06811 | 2.17617 | 2.24262 |
| | 2.8 | 1.94055 | 2.03536 | 2.13585 | 2.19651 |
| | 3.2 | 1.91098 | 1.99962 | 2.09185 | 2.14622 |
| | 3.6 | 1.87857 | 1.96046 | 2.04367 | 2.09113 |
| | 4.0 | 1.84289 | 1.91737 | 1.99065 | 2.03048 |
| | 6.0 | 1.59030 | 1.61136 | 1.62022 | 1.62170 |
| | 12.0 | 0.70708 | 0.70710 | 0.70711 | 0.70711 |

FRACTILES $W_\alpha$ RHO = −0.95

| DELTAS | | $\alpha$: 0.0005 | 0.0010 | 0.0025 | 0.0050 | 0.0075 |
|---|---|---|---|---|---|---|
| 0.0 | 0.0 | -7.85763 | -6.94710 | -5.75463 | -4.86316 | -4.34693 |
| 0.4 | 0.0 | -7.83772 | -6.93164 | -5.74425 | -4.85602 | -4.34145 |
| | 0.4 | -7.82791 | -6.92122 | -5.73378 | -4.84608 | -4.33204 |
| 0.8 | 0.0 | -7.69292 | -6.82027 | -5.67209 | -4.80928 | -4.30772 |
| | 0.4 | -7.78334 | -6.88430 | -5.70610 | -4.82477 | -4.31420 |
| | 0.8 | -7.74103 | -6.84555 | -5.67283 | -4.79618 | -4.28856 |
| 1.2 | 0.0 | -7.38598 | -6.57659 | -5.50610 | -4.69654 | -4.22346 |
| | 0.4 | -7.62858 | -6.76409 | -5.62665 | -4.77195 | -4.27512 |
| | 0.8 | -7.67849 | -6.79292 | -5.63249 | -4.76449 | -4.26168 |
| | 1.2 | -7.60332 | -6.72565 | -5.57631 | -4.71720 | -4.21978 |
| 1.6 | 0.0 | -7.00445 | -6.26511 | -5.28318 | -4.53670 | -4.09853 |
| | 0.4 | -7.32420 | -6.52245 | -5.46206 | -4.66017 | -4.19157 |
| | 0.8 | -7.52538 | -6.67398 | -5.55382 | -4.71214 | -4.22290 |
| | 1.2 | -7.52990 | -6.66353 | -5.52832 | -4.67926 | -4.18745 |
| | 1.6 | -7.42402 | -6.56961 | -5.45081 | -4.61461 | -4.13051 |
| 2.0 | 0.0 | -6.62736 | -5.95165 | -5.05123 | -4.36391 | -3.95907 |
| | 0.4 | -6.95062 | -6.21779 | -5.24449 | -4.50458 | -4.07027 |
| | 0.8 | -7.23486 | -6.44417 | -5.39843 | -4.60762 | -4.14552 |
| | 1.2 | -7.38872 | -6.55472 | -5.45748 | -4.63309 | -4.15392 |
| | 1.6 | -7.34647 | -6.50388 | -5.39989 | -4.57429 | -4.09613 |
| | 2.0 | -7.21376 | -6.38673 | -5.30388 | -4.49466 | -4.02625 |
| 2.4 | 0.0 | -6.28885 | -5.66712 | -4.83629 | -4.19988 | -3.82395 |
| | 0.4 | -6.58218 | -5.91183 | -5.01856 | -4.33669 | -3.93506 |
| | 0.8 | -6.87772 | -6.15371 | -5.19213 | -4.46113 | -4.03204 |
| | 1.2 | -7.12153 | -6.34491 | -5.31778 | -4.54108 | -4.08724 |
| | 1.6 | -7.22516 | -6.41203 | -5.34232 | -4.53867 | -4.07162 |
| | 2.0 | -7.13773 | -6.32231 | -5.25402 | -4.45522 | -3.99266 |
| | 2.4 | -6.98314 | -6.18628 | -5.14306 | -4.36360 | -3.91249 |
| 2.8 | 0.0 | -5.99635 | -5.41957 | -4.64681 | -4.05306 | -3.70145 |
| | 0.4 | -6.25126 | -5.63395 | -4.80900 | -4.17709 | -3.80381 |
| | 0.8 | -6.52369 | -5.86030 | -4.97630 | -4.30149 | -3.90401 |
| | 1.2 | -6.78796 | -6.07483 | -5.12771 | -4.40770 | -3.98507 |
| | 1.6 | -6.98840 | -6.22834 | -5.22314 | -4.46306 | -4.01895 |
| | 2.0 | -7.04158 | -6.25197 | -5.21327 | -4.43301 | -3.97960 |
| | 2.4 | -6.91289 | -6.12688 | -5.09723 | -4.32747 | -3.88180 |
| | 2.8 | -6.74167 | -5.97661 | -4.97513 | -4.22703 | -3.79417 |
| 3.2 | 0.0 | -5.74676 | -5.20735 | -4.48297 | -3.92483 | -3.59356 |
| | 0.4 | -5.96497 | -5.39185 | -4.62397 | -4.03395 | -3.68455 |
| | 0.8 | -6.20418 | -5.59242 | -4.77485 | -4.14857 | -3.77862 |
| | 1.2 | -6.45325 | -5.79827 | -4.92545 | -4.25915 | -3.86669 |
| | 1.6 | -6.68386 | -5.98340 | -5.05308 | -4.34585 | -3.93073 |
| | 2.0 | -6.83986 | -6.09835 | -5.11770 | -4.37623 | -3.94302 |
| | 2.4 | -6.84464 | -6.08038 | -5.07510 | -4.32006 | -3.88137 |
| | 2.8 | -6.68005 | -5.92467 | -4.93529 | -4.19580 | -3.76775 |
| | 3.2 | -6.49727 | -5.76460 | -4.80569 | -4.08957 | -3.67534 |

FRACTILES $W_{\alpha}$ RHO = -0.95

| DELTAS | | α:0.0005 | 0.0010 | 0.0025 | 0.0050 | 0.0075 |
|---|---|---|---|---|---|---|
| 3.6 | 0.0 | -5.53401 | -5.02586 | -4.34200 | -3.81373 | -3.49955 |
| | 0.4 | -5.72034 | -5.18400 | -4.46371 | -3.90871 | -3.57929 |
| | 0.8 | -5.92666 | -5.35801 | -4.59610 | -4.01065 | -3.66394 |
| | 1.2 | -6.14848 | -5.54328 | -4.73447 | -4.11488 | -3.74885 |
| | 1.6 | -6.37241 | -5.72710 | -4.86713 | -4.21064 | -3.82394 |
| | 2.0 | -6.56812 | -5.88178 | -4.97021 | -4.27723 | -3.87048 |
| | 2.4 | -6.68023 | -5.95872 | -5.00458 | -4.28319 | -3.86174 |
| | 2.8 | -6.64031 | -5.90248 | -4.93208 | -4.20335 | -3.78002 |
| | 3.2 | -6.44583 | -5.72148 | -4.77291 | -4.06410 | -3.65393 |
| | 3.6 | -6.25604 | -5.55561 | -4.63907 | -3.95480 | -3.55913 |
| 4.0 | 0.0 | -5.35191 | -4.87015 | -4.22050 | -3.71750 | -3.41778 |
| | 0.4 | -5.51155 | -5.00600 | -4.32561 | -3.80000 | -3.48739 |
| | 0.8 | -5.68874 | -5.15607 | -4.44068 | -3.88943 | -3.56223 |
| | 1.2 | -5.88197 | -5.31855 | -4.56362 | -3.98350 | -3.63993 |
| | 1.6 | -6.08511 | -5.48740 | -4.68857 | -4.07660 | -3.71506 |
| | 2.0 | -6.28280 | -5.64823 | -4.80257 | -4.15698 | -3.77668 |
| | 2.4 | -6.44341 | -5.77234 | -4.88104 | -4.20348 | -3.80578 |
| | 2.8 | -6.51350 | -5.81299 | -4.88665 | -4.18634 | -3.77725 |
| | 3.2 | -6.43363 | -5.72270 | -4.78780 | -4.08587 | -3.67820 |
| | 3.6 | -6.21536 | -5.52179 | -4.61371 | -3.93535 | -3.54294 |
| | 4.0 | -6.02238 | -5.35346 | -4.47836 | -3.82528 | -3.44778 |
| 6.0 | 0.0 | -4.74309 | -4.34743 | -3.80963 | -3.38935 | -3.13707 |
| | 0.4 | -4.82419 | -4.41695 | -3.86415 | -3.43281 | -3.17420 |
| | 0.8 | -4.91276 | -4.49272 | -3.92334 | -3.47980 | -3.21422 |
| | 1.2 | -5.00926 | -4.57505 | -3.98736 | -3.53036 | -3.25711 |
| | 1.6 | -5.11391 | -4.66401 | -4.05609 | -3.58427 | -3.30258 |
| | 2.0 | -5.22633 | -4.75912 | -4.12894 | -3.64085 | -3.34993 |
| | 2.4 | -5.34514 | -4.85895 | -4.20443 | -3.69863 | -3.39771 |
| | 2.8 | -5.46708 | -4.96033 | -4.27999 | -3.75483 | -3.44325 |
| | 3.2 | -5.58569 | -5.05721 | -4.34891 | -3.80444 | -3.48187 |
| | 3.6 | -5.68940 | -5.13892 | -4.40303 | -3.83909 | -3.50585 |
| | 4.0 | -5.75906 | -5.18806 | -4.42697 | -3.84583 | -3.50345 |
| | 6.0 | -5.03645 | -4.50525 | -3.81162 | -3.29572 | -2.99880 |
| 12.0 | 0.0 | -4.04858 | -3.74791 | -3.33377 | -3.00518 | -2.80562 |
| | 0.4 | -4.07047 | -3.76680 | -3.34875 | -3.01729 | -2.81607 |
| | 0.8 | -4.09342 | -3.78659 | -3.36445 | -3.02995 | -2.82698 |
| | 1.2 | -4.11750 | -3.80734 | -3.38087 | -3.04318 | -2.83839 |
| | 1.6 | -4.14276 | -3.82908 | -3.39806 | -3.05701 | -2.85029 |
| | 2.0 | -4.16923 | -3.85185 | -3.41603 | -3.07144 | -2.86270 |
| | 2.4 | -4.19695 | -3.87567 | -3.43479 | -3.08648 | -2.87562 |
| | 2.8 | -4.22596 | -3.90055 | -3.45435 | -3.10213 | -2.88903 |
| | 3.2 | -4.25624 | -3.92650 | -3.47469 | -3.11835 | -2.90290 |
| | 3.6 | -4.28779 | -3.95347 | -3.49577 | -3.13510 | -2.91720 |
| | 4.0 | -4.32054 | -3.98141 | -3.51751 | -3.15232 | -2.93184 |
| | 6.0 | -4.49483 | -4.12830 | -3.62944 | -3.23897 | -3.00427 |
| | 12.0 | -3.67309 | -3.37766 | -2.99181 | -2.69983 | -2.52702 |
| INFINITY | | -3.29051 | -3.09024 | -2.80703 | -2.57583 | -2.43238 |

FRACTILES $W_\alpha$ RHO = -0.95

| DELTAS | | $\alpha$: 0.0100 | 0.0175 | 0.0250 | 0.0375 | 0.0500 |
|---|---|---|---|---|---|---|
| 0.0 | 0.0 | -3.98343 | -3.28407 | -2.84466 | -2.35245 | -2.00884 |
| 0.4 | 0.0 | -3.97901 | -3.28145 | -2.84300 | -2.35172 | -2.00866 |
| | 0.4 | -3.97009 | -3.27373 | -2.83621 | -2.34614 | -2.00405 |
| 0.8 | 0.0 | -3.95362 | -3.27003 | -2.83886 | -2.35421 | -2.01476 |
| | 0.4 | -3.95459 | -3.26252 | -2.82754 | -2.34018 | -1.99991 |
| | 0.8 | -3.93114 | -3.24355 | -2.81158 | -2.32780 | -1.99015 |
| 1.2 | 0.0 | -3.88820 | -3.23756 | -2.82447 | -2.35721 | -2.02780 |
| | 0.4 | -3.92436 | -3.24727 | -2.82022 | -2.34025 | -2.00412 |
| | 0.8 | -3.90755 | -3.22607 | -2.79781 | -2.31805 | -1.98315 |
| | 1.2 | -3.86956 | -3.19590 | -2.77276 | -2.29896 | -1.96837 |
| 1.6 | 0.0 | -3.78698 | -3.17942 | -2.79128 | -2.34943 | -2.03578 |
| | 0.4 | -3.85950 | -3.21506 | -2.80593 | -2.34317 | -2.01697 |
| | 0.8 | -3.87752 | -3.21086 | -2.79044 | -2.31799 | -1.98720 |
| | 1.2 | -3.84110 | -3.17468 | -2.75596 | -2.28701 | -1.95975 |
| | 1.6 | -3.78970 | -3.13424 | -2.72264 | -2.26191 | -1.94057 |
| 2.0 | 0.0 | -3.67048 | -3.10560 | -2.74298 | -2.32813 | -2.03200 |
| | 0.4 | -3.76146 | -3.15926 | -2.77455 | -2.33662 | -2.02576 |
| | 0.8 | -3.81806 | -3.18261 | -2.77921 | -2.32299 | -2.00145 |
| | 1.2 | -3.81568 | -3.16288 | -2.75126 | -2.28881 | -1.96511 |
| | 1.6 | -3.75942 | -3.11166 | -2.70477 | -2.24923 | -1.93148 |
| | 2.0 | -3.69653 | -3.06253 | -2.66454 | -2.21923 | -1.90884 |
| 2.4 | 0.0 | -3.55540 | -3.02819 | -2.68846 | -2.29828 | -2.01859 |
| | 0.4 | -3.64874 | -3.08833 | -2.72857 | -2.31699 | -2.02321 |
| | 0.8 | -3.72696 | -3.13202 | -2.75197 | -2.31935 | -2.01229 |
| | 1.2 | -3.76564 | -3.14161 | -2.74551 | -2.29760 | -1.98199 |
| | 1.6 | -3.74195 | -3.10579 | -2.70475 | -2.25433 | -1.93918 |
| | 2.0 | -3.66697 | -3.04057 | -2.64723 | -2.20706 | -1.90022 |
| | 2.4 | -3.59501 | -2.98470 | -2.60175 | -2.17352 | -1.87527 |
| 2.8 | 0.0 | -3.44982 | -2.95454 | -2.63436 | -2.26544 | -2.00008 |
| | 0.4 | -3.53715 | -3.01363 | -2.67628 | -2.28881 | -2.01106 |
| | 0.8 | -3.62066 | -3.06603 | -2.70998 | -2.30265 | -2.01190 |
| | 1.2 | -3.68458 | -3.09861 | -2.72430 | -2.29825 | -1.99587 |
| | 1.6 | -3.70427 | -3.09370 | -2.70620 | -2.26810 | -1.95948 |
| | 2.0 | -3.65961 | -3.04221 | -2.65312 | -2.21629 | -1.91082 |
| | 2.4 | -3.56807 | -2.96481 | -2.58619 | -2.16273 | -1.86779 |
| | 2.8 | -3.48958 | -2.90429 | -2.53724 | -2.12710 | -1.84177 |
| 3.2 | 0.0 | -3.35610 | -2.88765 | -2.58395 | -2.23304 | -1.97988 |
| | 0.4 | -3.43448 | -2.94228 | -2.62408 | -2.25743 | -1.99368 |
| | 0.8 | -3.51432 | -2.99544 | -2.66106 | -2.27701 | -2.00169 |
| | 1.2 | -3.58690 | -3.03926 | -2.68769 | -2.28548 | -1.99839 |
| | 1.6 | -3.63557 | -3.06003 | -2.69240 | -2.27399 | -1.97706 |
| | 2.0 | -3.63607 | -3.04059 | -2.66273 | -2.23564 | -1.93487 |
| | 2.4 | -3.57180 | -2.97466 | -2.59848 | -2.17634 | -1.88136 |
| | 2.8 | -3.46648 | -2.88741 | -2.52417 | -2.11823 | -1.83584 |
| | 3.2 | -3.38393 | -2.82420 | -2.47343 | -2.08190 | -1.80998 |

FRACTILES $W_\alpha$ RHO = -0.95

| DELTAS | | α:0.0100 | 0.0175 | 0.0250 | 0.0375 | 0.0500 |
|---|---|---|---|---|---|---|
| 3.6 | 0.0 | -3.27400 | -2.82814 | -2.53836 | -2.20270 | -1.95989 |
| | 0.4 | -3.34314 | -2.87728 | -2.57524 | -2.22624 | -1.97444 |
| | 0.8 | -3.41579 | -2.92735 | -2.61157 | -2.24768 | -1.98591 |
| | 1.2 | -3.48737 | -2.97398 | -2.64313 | -2.26311 | -1.99068 |
| | 1.6 | -3.54827 | -3.00865 | -2.66224 | -2.26593 | -1.98305 |
| | 2.0 | -3.58128 | -3.01737 | -2.65719 | -2.24730 | -1.95648 |
| | 2.4 | -3.56315 | -2.98395 | -2.61651 | -2.20133 | -1.90908 |
| | 2.8 | -3.48134 | -2.90537 | -2.54270 | -2.13598 | -1.85204 |
| | 3.2 | -3.36531 | -2.81079 | -2.46321 | -2.07520 | -1.80575 |
| | 3.6 | -3.28086 | -2.74667 | -2.41223 | -2.03949 | -1.78128 |
| 4.0 | 0.0 | -3.20233 | -2.77563 | -2.49766 | -2.17497 | -1.94100 |
| | 0.4 | -3.26297 | -2.81931 | -2.53093 | -2.19689 | -1.95524 |
| | 0.8 | -3.32767 | -2.86490 | -2.56486 | -2.21814 | -1.96796 |
| | 1.2 | -3.39403 | -2.90999 | -2.59703 | -2.23638 | -1.97691 |
| | 1.6 | -3.45678 | -2.94965 | -2.62282 | -2.24741 | -1.97826 |
| | 2.0 | -3.50557 | -2.97489 | -2.63420 | -2.24444 | -1.96625 |
| | 2.4 | -3.52302 | -2.97170 | -2.61959 | -2.21893 | -1.93470 |
| | 2.8 | -3.48743 | -2.92535 | -2.56886 | -2.16621 | -1.88295 |
| | 3.2 | -3.39061 | -2.83625 | -2.48738 | -2.09646 | -1.82392 |
| | 3.6 | -3.26690 | -2.73685 | -2.40492 | -2.03497 | -1.77872 |
| | 4.0 | -3.18241 | -2.67332 | -2.35502 | -2.00106 | -1.75664 |
| 6.0 | 0.0 | -2.95477 | -2.59112 | -2.35215 | -2.07237 | -1.86772 |
| | 0.4 | -2.98748 | -2.61547 | -2.37133 | -2.08589 | -1.87739 |
| | 0.8 | -3.02265 | -2.64144 | -2.39165 | -2.10001 | -1.88731 |
| | 1.2 | -3.06020 | -2.66889 | -2.41291 | -2.11452 | -1.89725 |
| | 1.6 | -3.09981 | -2.69745 | -2.43472 | -2.12899 | -1.90677 |
| | 2.0 | -3.14076 | -2.72638 | -2.45634 | -2.14270 | -1.91519 |
| | 2.4 | -3.18164 | -2.75433 | -2.47647 | -2.15445 | -1.92139 |
| | 2.8 | -3.21985 | -2.77893 | -2.49292 | -2.16227 | -1.92356 |
| | 3.2 | -3.25096 | -2.79625 | -2.50213 | -2.16304 | -1.91897 |
| | 3.6 | -3.26773 | -2.80007 | -2.49857 | -2.15214 | -1.90370 |
| | 4.0 | -3.25934 | -2.78147 | -2.47468 | -2.12374 | -1.87334 |
| | 6.0 | -2.79098 | -2.39542 | -2.15045 | -1.87893 | -1.68997 |
| 12.0 | 0.0 | -2.66022 | -2.36696 | -2.17167 | -1.94014 | -1.76863 |
| | 0.4 | -2.66952 | -2.37405 | -2.17739 | -1.94436 | -1.77183 |
| | 0.8 | -2.67922 | -2.38144 | -2.18335 | -1.94874 | -1.77513 |
| | 1.2 | -2.68935 | -2.38914 | -2.18955 | -1.95329 | -1.77854 |
| | 1.6 | -2.69991 | -2.39715 | -2.19598 | -1.95799 | -1.78205 |
| | 2.0 | -2.71091 | -2.40547 | -2.20265 | -1.96284 | -1.78566 |
| | 2.4 | -2.72234 | -2.41408 | -2.20953 | -1.96782 | -1.78935 |
| | 2.8 | -2.73419 | -2.42298 | -2.21662 | -1.97292 | -1.79310 |
| | 3.2 | -2.74643 | -2.43214 | -2.22388 | -1.97811 | -1.79688 |
| | 3.6 | -2.75901 | -2.44150 | -2.23126 | -1.98333 | -1.80065 |
| | 4.0 | -2.77187 | -2.45099 | -2.23870 | -1.98854 | -1.80436 |
| | 6.0 | -2.83450 | -2.49537 | -2.27210 | -2.01020 | -1.81825 |
| | 12.0 | -2.40272 | -2.15481 | -1.99091 | -1.79685 | -1.65276 |
| INFINITY | | -2.32635 | -2.10836 | -1.95996 | -1.78046 | -1.64485 |

FRACTILES $W_\alpha$ RHO = -0.95

| DELTAS | | $\alpha$:0.0750 | 0.1000 | 0.1500 | 0.2000 | 0.2500 |
|---|---|---|---|---|---|---|
| 0.0 | 0.0 | -1.53425 | -1.20592 | -0.75828 | -0.45444 | -0.22893 |
| 0.4 | 0.0 | -1.53470 | -1.20671 | -0.75945 | -0.45580 | -0.23041 |
| | 0.4 | -1.53160 | -1.20478 | -0.75932 | -0.45708 | -0.23289 |
| 0.8 | 0.0 | -1.54431 | -1.21767 | -0.77073 | -0.46626 | -0.23971 |
| | 0.4 | -1.52986 | -1.20465 | -0.76137 | -0.46067 | -0.23773 |
| | 0.8 | -1.52394 | -1.20157 | -0.76250 | -0.46500 | -0.24476 |
| 1.2 | 0.0 | -1.56784 | -1.24577 | -0.80081 | -0.49436 | -0.26432 |
| | 0.4 | -1.53835 | -1.21506 | -0.77292 | -0.47202 | -0.24845 |
| | 0.8 | -1.52067 | -1.20087 | -0.76535 | -0.47044 | -0.25237 |
| | 1.2 | -1.51213 | -1.19687 | -0.76804 | -0.47824 | -0.26461 |
| 1.6 | 0.0 | -1.59407 | -1.28156 | -0.84412 | -0.53786 | -0.30448 |
| | 0.4 | -1.56156 | -1.24277 | -0.80252 | -0.49960 | -0.27250 |
| | 0.8 | -1.52897 | -1.21107 | -0.77673 | -0.48166 | -0.26304 |
| | 1.2 | -1.50807 | -1.19598 | -0.77163 | -0.48519 | -0.27457 |
| | 1.6 | -1.49739 | -1.19150 | -0.77632 | -0.49712 | -0.29342 |
| 2.0 | 0.0 | -1.61207 | -1.31236 | -0.88787 | -0.58594 | -0.35218 |
| | 0.4 | -1.58802 | -1.27838 | -0.84507 | -0.54189 | -0.31107 |
| | 0.8 | -1.55264 | -1.23859 | -0.80523 | -0.50749 | -0.28479 |
| | 1.2 | -1.51693 | -1.20623 | -0.78235 | -0.49528 | -0.28379 |
| | 1.6 | -1.49324 | -1.19081 | -0.78062 | -0.50547 | -0.30560 |
| | 2.0 | -1.48117 | -1.18648 | -0.78801 | -0.52275 | -0.33111 |
| 2.4 | 0.0 | -1.61985 | -1.33338 | -0.92398 | -0.62919 | -0.39801 |
| | 0.4 | -1.60661 | -1.30928 | -0.88823 | -0.58883 | -0.35714 |
| | 0.8 | -1.57995 | -1.27419 | -0.84651 | -0.54755 | -0.32029 |
| | 1.2 | -1.54160 | -1.23362 | -0.80910 | -0.51810 | -0.30139 |
| | 1.6 | -1.50312 | -1.20116 | -0.79013 | -0.51337 | -0.31168 |
| | 2.0 | -1.47747 | -1.18627 | -0.79310 | -0.53216 | -0.34339 |
| | 2.4 | -1.46491 | -1.18291 | -0.80422 | -0.55426 | -0.37194 |
| 2.8 | 0.0 | -1.62017 | -1.34579 | -0.95100 | -0.66409 | -0.43698 |
| | 0.4 | -1.61508 | -1.33059 | -0.92401 | -0.63126 | -0.40173 |
| | 0.8 | -1.59961 | -1.30539 | -0.88880 | -0.59269 | -0.36372 |
| | 1.2 | -1.57020 | -1.26925 | -0.84857 | -0.55490 | -0.33221 |
| | 1.6 | -1.52906 | -1.22827 | -0.81434 | -0.53172 | -0.32294 |
| | 2.0 | -1.48851 | -1.19655 | -0.80067 | -0.53657 | -0.34494 |
| | 2.4 | -1.46206 | -1.18344 | -0.80992 | -0.56331 | -0.38247 |
| | 2.8 | -1.45006 | -1.18211 | -0.82472 | -0.58773 | -0.41114 |
| 3.2 | 0.0 | -1.61614 | -1.35231 | -0.97057 | -0.69115 | -0.46836 |
| | 0.4 | -1.61607 | -1.34334 | -0.95088 | -0.66565 | -0.43985 |
| | 0.8 | -1.60917 | -1.32715 | -0.92412 | -0.63395 | -0.40649 |
| | 1.2 | -1.59131 | -1.30084 | -0.88964 | -0.59755 | -0.37194 |
| | 1.6 | -1.55917 | -1.26384 | -0.85136 | -0.56403 | -0.34711 |
| | 2.0 | -1.51566 | -1.22302 | -0.82131 | -0.54884 | -0.34894 |
| | 2.4 | -1.47407 | -1.19319 | -0.81463 | -0.56357 | -0.37960 |
| | 2.8 | -1.44828 | -1.18343 | -0.83028 | -0.59533 | -0.41935 |
| | 3.2 | -1.43797 | -1.18487 | -0.84734 | -0.61993 | -0.44665 |

FRACTILES $W_\alpha$ RHO = −0.95

| DELTAS | | $\alpha$:0.0750 | 0.1000 | 0.1500 | 0.2000 | 0.2500 |
|---|---|---|---|---|---|---|
| 3.6 | 0.0 | −1.60994 | −1.35517 | −0.98479 | −0.71205 | −0.49333 |
| | 0.4 | −1.61263 | −1.35018 | −0.97039 | −0.69236 | −0.47064 |
| | 0.8 | −1.61111 | −1.34038 | −0.95078 | −0.66760 | −0.44341 |
| | 1.2 | −1.60226 | −1.32317 | −0.92434 | −0.63724 | −0.41225 |
| | 1.6 | −1.58196 | −1.29580 | −0.89086 | −0.60345 | −0.38185 |
| | 2.0 | −1.54728 | −1.25825 | −0.85508 | −0.57519 | −0.36518 |
| | 2.4 | −1.50206 | −1.21837 | −0.83049 | −0.56921 | −0.37715 |
| | 2.8 | −1.46073 | −1.19192 | −0.83165 | −0.59180 | −0.41294 |
| | 3.2 | −1.43729 | −1.18672 | −0.85209 | −0.62576 | −0.45263 |
| | 3.6 | −1.42949 | −1.19076 | −0.86989 | −0.64928 | −0.47787 |
| 4.0 | 0.0 | −1.60284 | −1.35584 | −0.99530 | −0.72843 | −0.51336 |
| | 0.4 | −1.60693 | −1.35332 | −0.98460 | −0.71302 | −0.49519 |
| | 0.8 | −1.60846 | −1.34766 | −0.97021 | −0.69383 | −0.47341 |
| | 1.2 | −1.60538 | −1.33700 | −0.95073 | −0.66994 | −0.44763 |
| | 1.6 | −1.59451 | −1.31878 | −0.92474 | −0.64114 | −0.41903 |
| | 2.0 | −1.57184 | −1.29049 | −0.89253 | −0.61044 | −0.39359 |
| | 2.4 | −1.53495 | −1.25280 | −0.85996 | −0.58858 | −0.38572 |
| | 2.8 | −1.48888 | −1.21486 | −0.84206 | −0.59150 | −0.40543 |
| | 3.2 | −1.44938 | −1.19327 | −0.85034 | −0.61924 | −0.44363 |
| | 3.6 | −1.42968 | −1.19271 | −0.87355 | −0.65342 | −0.48196 |
| | 4.0 | −1.42454 | −1.19859 | −0.89099 | −0.67525 | −0.50491 |
| 6.0 | 0.0 | −1.56891 | −1.34802 | −1.02084 | −0.77427 | −0.57223 |
| | 0.4 | −1.57343 | −1.34915 | −1.01767 | −0.76852 | −0.56485 |
| | 0.8 | −1.57776 | −1.34981 | −1.01372 | −0.76184 | −0.55649 |
| | 1.2 | −1.58164 | −1.34975 | −1.00875 | −0.75402 | −0.54696 |
| | 1.6 | −1.58466 | −1.34857 | −1.00243 | −0.74480 | −0.53608 |
| | 2.0 | −1.58619 | −1.34569 | −0.99430 | −0.73384 | −0.52364 |
| | 2.4 | −1.58523 | −1.34028 | −0.98375 | −0.72077 | −0.50952 |
| | 2.8 | −1.58031 | −1.33110 | −0.97002 | −0.70528 | −0.49389 |
| | 3.2 | −1.56927 | −1.31650 | −0.95242 | −0.68769 | −0.47831 |
| | 3.6 | −1.54940 | −1.29483 | −0.93144 | −0.67089 | −0.46791 |
| | 4.0 | −1.51872 | −1.26642 | −0.91182 | −0.66256 | −0.47045 |
| | 6.0 | −1.42607 | −1.23779 | −0.96487 | −0.76045 | −0.59160 |
| 12.0 | 0.0 | −1.51455 | −1.32362 | −1.03525 | −0.81288 | −0.62691 |
| | 0.4 | −1.51635 | −1.32450 | −1.03493 | −0.81181 | −0.62534 |
| | 0.8 | −1.51820 | −1.32538 | −1.03456 | −0.81066 | −0.62368 |
| | 1.2 | −1.52009 | −1.32626 | −1.03413 | −0.80943 | −0.62192 |
| | 1.6 | −1.52202 | −1.32712 | −1.03364 | −0.80811 | −0.62005 |
| | 2.0 | −1.52397 | −1.32797 | −1.03308 | −0.80669 | −0.61807 |
| | 2.4 | −1.52593 | −1.32879 | −1.03244 | −0.80516 | −0.61597 |
| | 2.8 | −1.52788 | −1.32955 | −1.03170 | −0.80350 | −0.61373 |
| | 3.2 | −1.52979 | −1.33023 | −1.03083 | −0.80171 | −0.61136 |
| | 3.6 | −1.53163 | −1.33081 | −1.02983 | −0.79978 | −0.60885 |
| | 4.0 | −1.53334 | −1.33124 | −1.02866 | −0.79768 | −0.60619 |
| | 6.0 | −1.53725 | −1.32890 | −1.01903 | −0.78427 | −0.59089 |
| | 12.0 | −1.43793 | −1.27487 | −1.02487 | −0.82821 | −0.66062 |
| INFINITY | | −1.43953 | −1.28155 | −1.03643 | −0.84162 | −0.67449 |

FRACTILES $W_\alpha$ RHO = -0.95

| DELTAS | | $\alpha$: 0.3000 | 0.3500 | 0.4000 | 0.4500 | 0.5000 |
|---|---|---|---|---|---|---|
| 0.0 | 0.0 | -0.05287 | 0.08898 | 0.20565 | 0.30290 | 0.38461 |
| 0.4 | 0.0 | -0.05446 | 0.08728 | 0.20383 | 0.30093 | 0.38244 |
| | 0.4 | -0.05802 | 0.08270 | 0.19822 | 0.29423 | 0.37453 |
| 0.8 | 0.0 | -0.06254 | 0.08035 | 0.19791 | 0.29583 | 0.37793 |
| | 0.4 | -0.06399 | 0.07563 | 0.18999 | 0.28467 | 0.36328 |
| | 0.8 | -0.07349 | 0.06369 | 0.17540 | 0.26676 | 0.34064 |
| 1.2 | 0.0 | -0.08308 | 0.06401 | 0.18565 | 0.28736 | 0.37284 |
| | 0.4 | -0.07398 | 0.06627 | 0.18105 | 0.27575 | 0.35349 |
| | 0.8 | -0.08313 | 0.05187 | 0.16076 | 0.24859 | 0.32073 |
| | 1.2 | -0.09975 | 0.03018 | 0.13389 | 0.21902 | 0.29191 |
| 1.6 | 0.0 | -0.11801 | 0.03535 | 0.16378 | 0.27245 | 0.36475 |
| | 0.4 | -0.09395 | 0.05050 | 0.16931 | 0.26760 | 0.34831 |
| | 0.8 | -0.09323 | 0.04205 | 0.15076 | 0.23888 | 0.31241 |
| | 1.2 | -0.11301 | 0.01365 | 0.11563 | 0.20106 | 0.27578 |
| | 1.6 | -0.13856 | -0.01642 | 0.08431 | 0.17121 | 0.24926 |
| 2.0 | 0.0 | -0.16233 | -0.00351 | 0.13186 | 0.24855 | 0.34964 |
| | 0.4 | -0.12693 | 0.02417 | 0.15017 | 0.25592 | 0.34419 |
| | 0.8 | -0.11039 | 0.02960 | 0.14299 | 0.23557 | 0.31305 |
| | 1.2 | -0.12144 | 0.00617 | 0.10948 | 0.19651 | 0.27295 |
| | 1.6 | -0.15359 | -0.03253 | 0.06866 | 0.15704 | 0.23722 |
| | 2.0 | -0.18394 | -0.06425 | 0.03809 | 0.12929 | 0.21335 |
| 2.4 | 0.0 | -0.20766 | -0.04598 | 0.09419 | 0.21739 | 0.32651 |
| | 0.4 | -0.16912 | -0.01206 | 0.12150 | 0.23611 | 0.33440 |
| | 0.8 | -0.13945 | 0.00821 | 0.13010 | 0.23104 | 0.31568 |
| | 1.2 | -0.13317 | 0.00026 | 0.10856 | 0.19945 | 0.27875 |
| | 1.6 | -0.15781 | -0.03503 | 0.06765 | 0.15730 | 0.23853 |
| | 2.0 | -0.19736 | -0.07748 | 0.02589 | 0.11861 | 0.20452 |
| | 2.4 | -0.22828 | -0.10803 | -0.00258 | 0.09329 | 0.18307 |
| 2.8 | 0.0 | -0.24805 | -0.08575 | 0.05681 | 0.18402 | 0.29879 |
| | 0.4 | -0.21279 | -0.05241 | 0.08651 | 0.20835 | 0.31582 |
| | 0.8 | -0.17816 | -0.02351 | 0.10734 | 0.21854 | 0.31319 |
| | 1.2 | -0.15590 | -0.01326 | 0.10375 | 0.20176 | 0.28639 |
| | 1.6 | -0.16218 | -0.03387 | 0.07275 | 0.16504 | 0.24792 |
| | 2.0 | -0.19670 | -0.07527 | 0.02913 | 0.12250 | 0.20877 |
| | 2.4 | -0.23899 | -0.11815 | -0.01169 | 0.08547 | 0.17671 |
| | 2.8 | -0.26852 | -0.14653 | -0.03772 | 0.06256 | 0.15745 |
| 3.2 | 0.0 | -0.28163 | -0.11989 | 0.02351 | 0.15285 | 0.27107 |
| | 0.4 | -0.25202 | -0.09069 | 0.05099 | 0.17733 | 0.29116 |
| | 0.8 | -0.21934 | -0.06062 | 0.07660 | 0.19645 | 0.30138 |
| | 1.2 | -0.18950 | -0.03816 | 0.08895 | 0.19670 | 0.28955 |
| | 1.6 | -0.17664 | -0.03919 | 0.07477 | 0.17241 | 0.25899 |
| | 2.0 | -0.19423 | -0.06839 | 0.03877 | 0.13369 | 0.22062 |
| | 2.4 | -0.23418 | -0.11226 | -0.00528 | 0.09198 | 0.18302 |
| | 2.8 | -0.27654 | -0.15393 | -0.04426 | 0.05702 | 0.15301 |
| | 3.2 | -0.30388 | -0.17983 | -0.06782 | 0.03638 | 0.13574 |

FRACTILES $W_\alpha$ RHO = -0.95

| DELTAS | | $\alpha$:0.3000 | 0.3500 | 0.4000 | 0.4500 | 0.5000 |
|---|---|---|---|---|---|---|
| 3.6 | 0.0 | -0.30891 | -0.14818 | -0.00469 | 0.12575 | 0.24605 |
| | 0.4 | -0.28479 | -0.12380 | 0.01894 | 0.14770 | 0.26534 |
| | 0.8 | -0.25693 | -0.09678 | 0.04378 | 0.16898 | 0.28145 |
| | 1.2 | -0.22728 | -0.07066 | 0.06427 | 0.18151 | 0.28403 |
| | 1.6 | -0.20334 | -0.05609 | 0.06745 | 0.17322 | 0.26612 |
| | 2.0 | -0.20063 | -0.06689 | 0.04595 | 0.14470 | 0.23402 |
| | 2.4 | -0.22631 | -0.10111 | 0.00762 | 0.10556 | 0.19651 |
| | 2.8 | -0.26860 | -0.14535 | -0.03561 | 0.06534 | 0.16072 |
| | 3.2 | -0.30956 | -0.18498 | -0.07231 | 0.03263 | 0.13278 |
| | 3.6 | -0.33445 | -0.20837 | -0.09350 | 0.01412 | 0.11730 |
| 4.0 | 0.0 | -0.33113 | -0.17150 | -0.02822 | 0.10280 | 0.22445 |
| | 0.4 | -0.31150 | -0.15136 | -0.00837 | 0.12164 | 0.24156 |
| | 0.8 | -0.28861 | -0.12851 | 0.01345 | 0.14147 | 0.25839 |
| | 1.2 | -0.26274 | -0.10401 | 0.03517 | 0.15887 | 0.26962 |
| | 1.6 | -0.23664 | -0.08263 | 0.04955 | 0.16436 | 0.26550 |
| | 2.0 | -0.21968 | -0.07638 | 0.04470 | 0.14989 | 0.24395 |
| | 2.4 | -0.22600 | -0.09440 | 0.01859 | 0.11915 | 0.21148 |
| | 2.8 | -0.25678 | -0.13120 | -0.02048 | 0.08051 | 0.17522 |
| | 3.2 | -0.29945 | -0.17458 | -0.06216 | 0.04215 | 0.14137 |
| | 3.6 | -0.33824 | -0.21174 | -0.09638 | 0.01176 | 0.11548 |
| | 4.0 | -0.36067 | -0.23272 | -0.11535 | -0.00481 | 0.10162 |
| 6.0 | 0.0 | -0.39826 | -0.24334 | -0.10192 | 0.02972 | 0.15431 |
| | 0.4 | -0.38987 | -0.23441 | -0.09282 | 0.03868 | 0.16284 |
| | 0.8 | -0.38050 | -0.22454 | -0.08285 | 0.04839 | 0.17196 |
| | 1.2 | -0.37002 | -0.21365 | -0.07201 | 0.05880 | 0.18157 |
| | 1.6 | -0.35830 | -0.20171 | -0.06034 | 0.06976 | 0.19140 |
| | 2.0 | -0.34529 | -0.18880 | -0.04810 | 0.08081 | 0.20072 |
| | 2.4 | -0.33111 | -0.17534 | -0.03609 | 0.09065 | 0.20762 |
| | 2.8 | -0.31654 | -0.16291 | -0.02679 | 0.09595 | 0.20838 |
| | 3.2 | -0.30457 | -0.15576 | -0.02509 | 0.09221 | 0.19977 |
| | 3.6 | -0.30153 | -0.15984 | -0.03527 | 0.07737 | 0.18179 |
| | 4.0 | -0.31294 | -0.17770 | -0.05730 | 0.05312 | 0.15695 |
| | 6.0 | -0.44399 | -0.30991 | -0.18467 | -0.06501 | 0.05153 |
| 12.0 | 0.0 | -0.46368 | -0.31561 | -0.17793 | -0.04735 | 0.07865 |
| | 0.4 | -0.46178 | -0.31349 | -0.17570 | -0.04508 | 0.08088 |
| | 0.8 | -0.45978 | -0.31127 | -0.17336 | -0.04271 | 0.08321 |
| | 1.2 | -0.45766 | -0.30894 | -0.17091 | -0.04023 | 0.08564 |
| | 1.6 | -0.45544 | -0.30649 | -0.16835 | -0.03764 | 0.08816 |
| | 2.0 | -0.45309 | -0.30392 | -0.16567 | -0.03496 | 0.09077 |
| | 2.4 | -0.45062 | -0.30124 | -0.16288 | -0.03217 | 0.09347 |
| | 2.8 | -0.44802 | -0.29842 | -0.15998 | -0.02928 | 0.09624 |
| | 3.2 | -0.44529 | -0.29549 | -0.15698 | -0.02632 | 0.09907 |
| | 3.6 | -0.44243 | -0.29245 | -0.15389 | -0.02329 | 0.10193 |
| | 4.0 | -0.43944 | -0.28931 | -0.15073 | -0.02023 | 0.10479 |
| | 6.0 | -0.42347 | -0.27354 | -0.13583 | -0.00676 | 0.11633 |
| | 12.0 | -0.51083 | -0.37252 | -0.24164 | -0.11529 | 0.00882 |
| INFINITY | | -0.52440 | -0.38532 | -0.25335 | -0.12566 | 0.0 |

FRACTILES $W_\alpha$ RHO = −0.95

| DELTAS | | $\alpha$:0.5500 | 0.6000 | 0.6500 | 0.7000 | 0.7500 |
|---|---|---|---|---|---|---|
| 0.0 | 0.0 | 0.45351 | 0.51155 | 0.56019 | 0.60050 | 0.63325 |
| 0.4 | 0.0 | 0.45107 | 0.50874 | 0.55686 | 0.59641 | 0.62797 |
| | 0.4 | 0.44171 | 0.49751 | 0.54290 | 0.57695 | 0.60575 |
| 0.8 | 0.0 | 0.44684 | 0.50439 | 0.55172 | 0.58904 | 0.61670 |
| | 0.4 | 0.42798 | 0.47944 | 0.52198 | 0.56038 | 0.59773 |
| | 0.8 | 0.40119 | 0.45338 | 0.50102 | 0.54680 | 0.59297 |
| 1.2 | 0.0 | 0.44457 | 0.50405 | 0.55144 | 0.58992 | 0.62458 |
| | 0.4 | 0.41643 | 0.46875 | 0.51487 | 0.55803 | 0.60070 |
| | 0.8 | 0.38252 | 0.43807 | 0.49030 | 0.54146 | 0.59366 |
| | 1.2 | 0.35710 | 0.41777 | 0.47629 | 0.53468 | 0.59502 |
| 1.6 | 0.0 | 0.44291 | 0.50787 | 0.56069 | 0.60526 | 0.64581 |
| | 0.4 | 0.41455 | 0.47068 | 0.52072 | 0.56772 | 0.61415 |
| | 0.8 | 0.37643 | 0.43464 | 0.48977 | 0.54398 | 0.59935 |
| | 1.2 | 0.34378 | 0.40790 | 0.47031 | 0.53300 | 0.59806 |
| | 1.6 | 0.32183 | 0.39137 | 0.45993 | 0.52946 | 0.60218 |
| 2.0 | 0.0 | 0.43704 | 0.51152 | 0.57364 | 0.62627 | 0.67352 |
| | 0.4 | 0.41756 | 0.47980 | 0.53491 | 0.58613 | 0.63616 |
| | 0.8 | 0.38042 | 0.44146 | 0.49898 | 0.55526 | 0.61243 |
| | 1.2 | 0.34268 | 0.40849 | 0.47256 | 0.53686 | 0.60354 |
| | 1.6 | 0.31233 | 0.38473 | 0.45640 | 0.52930 | 0.60571 |
| | 2.0 | 0.29310 | 0.37074 | 0.44821 | 0.52751 | 0.61103 |
| 2.4 | 0.0 | 0.42337 | 0.50882 | 0.58279 | 0.64615 | 0.70210 |
| | 0.4 | 0.41798 | 0.48922 | 0.55155 | 0.60847 | 0.66304 |
| | 0.8 | 0.38868 | 0.45401 | 0.51479 | 0.57355 | 0.63261 |
| | 1.2 | 0.35055 | 0.41783 | 0.48291 | 0.54784 | 0.61480 |
| | 1.6 | 0.31452 | 0.38764 | 0.45992 | 0.53331 | 0.61009 |
| | 2.0 | 0.28633 | 0.36621 | 0.44608 | 0.52797 | 0.61432 |
| | 2.4 | 0.26931 | 0.35408 | 0.43932 | 0.52710 | 0.62001 |
| 2.8 | 0.0 | 0.40303 | 0.49784 | 0.58345 | 0.65942 | 0.72667 |
| | 0.4 | 0.41031 | 0.49274 | 0.56493 | 0.62981 | 0.69068 |
| | 0.8 | 0.39469 | 0.46671 | 0.53261 | 0.59524 | 0.65721 |
| | 1.2 | 0.36205 | 0.43202 | 0.49891 | 0.56494 | 0.63239 |
| | 1.6 | 0.32483 | 0.39831 | 0.47048 | 0.54335 | 0.61919 |
| | 2.0 | 0.29072 | 0.37054 | 0.45019 | 0.53168 | 0.61744 |
| | 2.4 | 0.26453 | 0.35100 | 0.43804 | 0.52776 | 0.62279 |
| | 2.8 | 0.24937 | 0.34032 | 0.43226 | 0.52736 | 0.62837 |
| 3.2 | 0.0 | 0.38019 | 0.48158 | 0.57602 | 0.66349 | 0.74347 |
| | 0.4 | 0.39419 | 0.48721 | 0.57067 | 0.64583 | 0.71527 |
| | 0.8 | 0.39326 | 0.47446 | 0.54783 | 0.61634 | 0.68287 |
| | 1.2 | 0.37161 | 0.44638 | 0.51675 | 0.58524 | 0.65428 |
| | 1.6 | 0.33832 | 0.41323 | 0.48604 | 0.55888 | 0.63408 |
| | 2.0 | 0.30257 | 0.38187 | 0.46053 | 0.54060 | 0.62447 |
| | 2.4 | 0.27040 | 0.35621 | 0.44240 | 0.53105 | 0.62476 |
| | 2.8 | 0.24610 | 0.33829 | 0.43155 | 0.52805 | 0.63058 |
| | 3.2 | 0.23254 | 0.32879 | 0.42648 | 0.52784 | 0.63578 |

FRACTILES $W_{\alpha}$ RHO = -0.95

| DELTAS | | α:0.5500 | 0.6000 | 0.6500 | 0.7000 | 0.7500 |
|---|---|---|---|---|---|---|
| 3.6 | 0.0 | 0.35831 | 0.46407 | 0.56449 | 0.66024 | 0.75135 |
| | 0.4 | 0.37385 | 0.47445 | 0.56767 | 0.65375 | 0.73365 |
| | 0.8 | 0.38274 | 0.47398 | 0.55673 | 0.63326 | 0.70633 |
| | 1.2 | 0.37473 | 0.45662 | 0.53259 | 0.60536 | 0.67758 |
| | 1.6 | 0.35013 | 0.42835 | 0.50338 | 0.57753 | 0.65322 |
| | 2.0 | 0.31727 | 0.39700 | 0.47538 | 0.55452 | 0.63682 |
| | 2.4 | 0.28319 | 0.36780 | 0.45234 | 0.53889 | 0.62999 |
| | 2.8 | 0.25293 | 0.34403 | 0.43598 | 0.53093 | 0.63162 |
| | 3.2 | 0.23042 | 0.32754 | 0.42614 | 0.52848 | 0.63747 |
| | 3.6 | 0.21828 | 0.31904 | 0.42161 | 0.52831 | 0.64217 |
| 4.0 | 0.0 | 0.33885 | 0.44766 | 0.55221 | 0.65362 | 0.75271 |
| | 0.4 | 0.35347 | 0.45888 | 0.55884 | 0.65387 | 0.74424 |
| | 0.8 | 0.36604 | 0.46554 | 0.55765 | 0.64347 | 0.72492 |
| | 1.2 | 0.36929 | 0.45976 | 0.54325 | 0.62224 | 0.69945 |
| | 1.6 | 0.35637 | 0.44004 | 0.51920 | 0.59636 | 0.67410 |
| | 2.0 | 0.33052 | 0.41240 | 0.49197 | 0.57147 | 0.65336 |
| | 2.4 | 0.29858 | 0.38285 | 0.46638 | 0.55129 | 0.64009 |
| | 2.8 | 0.26622 | 0.35564 | 0.44545 | 0.53781 | 0.63537 |
| | 3.2 | 0.23784 | 0.33358 | 0.43056 | 0.53101 | 0.63782 |
| | 3.6 | 0.21702 | 0.31837 | 0.42155 | 0.52888 | 0.64342 |
| | 4.0 | 0.20614 | 0.31073 | 0.41745 | 0.52868 | 0.64757 |
| 6.0 | 0.0 | 0.27399 | 0.39057 | 0.50576 | 0.62133 | 0.73937 |
| | 0.4 | 0.28179 | 0.39734 | 0.51115 | 0.62493 | 0.74067 |
| | 0.8 | 0.29000 | 0.40430 | 0.51647 | 0.62815 | 0.74119 |
| | 1.2 | 0.29845 | 0.41118 | 0.52134 | 0.63044 | 0.74016 |
| | 1.6 | 0.30670 | 0.41737 | 0.52488 | 0.63059 | 0.73605 |
| | 2.0 | 0.31371 | 0.42137 | 0.52515 | 0.62651 | 0.72734 |
| | 2.4 | 0.31694 | 0.42039 | 0.51972 | 0.61688 | 0.71424 |
| | 2.8 | 0.31302 | 0.41214 | 0.50790 | 0.60256 | 0.69872 |
| | 3.2 | 0.30052 | 0.39693 | 0.49125 | 0.58579 | 0.68323 |
| | 3.6 | 0.28086 | 0.37695 | 0.47225 | 0.56902 | 0.67000 |
| | 4.0 | 0.25680 | 0.35488 | 0.45327 | 0.55426 | 0.66067 |
| | 6.0 | 0.16706 | 0.28360 | 0.40329 | 0.52875 | 0.66350 |
| 12.0 | 0.0 | 0.20216 | 0.32511 | 0.44951 | 0.57764 | 0.71252 |
| | 0.4 | 0.20429 | 0.32706 | 0.45119 | 0.57896 | 0.71336 |
| | 0.8 | 0.20650 | 0.32908 | 0.45292 | 0.58031 | 0.71420 |
| | 1.2 | 0.20880 | 0.33116 | 0.45470 | 0.58169 | 0.71504 |
| | 1.6 | 0.21117 | 0.33331 | 0.45653 | 0.58308 | 0.71585 |
| | 2.0 | 0.21362 | 0.33551 | 0.45838 | 0.58447 | 0.71664 |
| | 2.4 | 0.21614 | 0.33775 | 0.46025 | 0.58585 | 0.71738 |
| | 2.8 | 0.21871 | 0.34002 | 0.46211 | 0.58719 | 0.71805 |
| | 3.2 | 0.22130 | 0.34229 | 0.46395 | 0.58846 | 0.71861 |
| | 3.6 | 0.22390 | 0.34452 | 0.46571 | 0.58962 | 0.71902 |
| | 4.0 | 0.22646 | 0.34668 | 0.46735 | 0.59062 | 0.71922 |
| | 6.0 | 0.23559 | 0.35295 | 0.47031 | 0.58980 | 0.71412 |
| | 12.0 | 0.13275 | 0.25853 | 0.38840 | 0.52515 | 0.67265 |
| INFINITY | | 0.12566 | 0.25335 | 0.38532 | 0.52440 | 0.67449 |

FRACTILES $w_{\alpha}$ RHO = -0.95

| DELTAS | | $\alpha$: 0.8000 | 0.8500 | 0.9000 | 0.9250 | 0.9500 |
|---|---|---|---|---|---|---|
| 0.0 | 0.0 | 0.65895 | 0.67781 | 0.68885 | 0.69385 | 0.70323 |
| 0.4 | 0.0 | 0.65155 | 0.66614 | 0.68709 | 0.70234 | 0.72357 |
| | 0.4 | 0.63507 | 0.66773 | 0.70823 | 0.73448 | 0.76911 |
| 0.8 | 0.0 | 0.64369 | 0.67313 | 0.70915 | 0.73234 | 0.76282 |
| | 0.4 | 0.63651 | 0.67973 | 0.73273 | 0.76665 | 0.81090 |
| | 0.8 | 0.64188 | 0.69691 | 0.76461 | 0.80788 | 0.86413 |
| 1.2 | 0.0 | 0.65884 | 0.69576 | 0.73999 | 0.76791 | 0.80403 |
| | 0.4 | 0.64526 | 0.69484 | 0.75529 | 0.79371 | 0.84348 |
| | 0.8 | 0.64932 | 0.71212 | 0.78938 | 0.83867 | 0.90257 |
| | 1.2 | 0.65997 | 0.73371 | 0.82475 | 0.88291 | 0.95832 |
| 1.6 | 0.0 | 0.68567 | 0.72810 | 0.77813 | 0.80929 | 0.84915 |
| | 0.4 | 0.66242 | 0.71582 | 0.78041 | 0.82119 | 0.87368 |
| | 0.8 | 0.65838 | 0.72488 | 0.80642 | 0.85827 | 0.92528 |
| | 1.2 | 0.66829 | 0.74815 | 0.84678 | 0.90974 | 0.99128 |
| | 1.6 | 0.68111 | 0.77125 | 0.88291 | 0.95430 | 1.04682 |
| 2.0 | 0.0 | 0.71910 | 0.76671 | 0.82178 | 0.85560 | 0.89839 |
| | 0.4 | 0.68762 | 0.74392 | 0.81130 | 0.85347 | 0.90741 |
| | 0.8 | 0.67306 | 0.74098 | 0.82377 | 0.87615 | 0.94359 |
| | 1.2 | 0.67540 | 0.75695 | 0.85740 | 0.92137 | 1.00405 |
| | 1.6 | 0.68876 | 0.78368 | 0.90128 | 0.97645 | 1.07379 |
| | 2.0 | 0.70219 | 0.80672 | 0.93655 | 1.01965 | 1.12734 |
| 2.4 | 0.0 | 0.75469 | 0.80818 | 0.86858 | 0.90504 | 0.95063 |
| | 0.4 | 0.71822 | 0.77766 | 0.84776 | 0.89117 | 0.94626 |
| | 0.8 | 0.69462 | 0.76346 | 0.84662 | 0.89889 | 0.96583 |
| | 1.2 | 0.68659 | 0.76765 | 0.86699 | 0.93001 | 1.01118 |
| | 1.6 | 0.69340 | 0.78841 | 0.90587 | 0.98078 | 1.07764 |
| | 2.0 | 0.70863 | 0.81682 | 0.95119 | 1.03717 | 1.14856 |
| | 2.4 | 0.72181 | 0.83887 | 0.98459 | 1.07794 | 1.19896 |
| 2.8 | 0.0 | 0.78845 | 0.84940 | 0.91618 | 0.95566 | 1.00433 |
| | 0.4 | 0.75089 | 0.81444 | 0.88797 | 0.93290 | 0.98939 |
| | 0.8 | 0.72137 | 0.79169 | 0.87561 | 0.92789 | 0.99443 |
| | 1.2 | 0.70409 | 0.78441 | 0.88210 | 0.94370 | 1.02273 |
| | 1.6 | 0.70109 | 0.79408 | 0.90850 | 0.98123 | 1.07500 |
| | 2.0 | 0.71093 | 0.81796 | 0.95062 | 1.03536 | 1.14496 |
| | 2.4 | 0.72693 | 0.84671 | 0.99579 | 1.09128 | 1.21502 |
| | 2.8 | 0.73935 | 0.86726 | 1.02674 | 1.12900 | 1.26161 |
| 3.2 | 0.0 | 0.81689 | 0.88755 | 0.96245 | 1.00564 | 1.05802 |
| | 0.4 | 0.78240 | 0.85156 | 0.92977 | 0.97678 | 1.03522 |
| | 0.8 | 0.75052 | 0.82346 | 0.90917 | 0.96199 | 1.02867 |
| | 1.2 | 0.72681 | 0.80719 | 0.90394 | 0.96449 | 1.04174 |
| | 1.6 | 0.71467 | 0.80555 | 0.91661 | 0.98685 | 1.07707 |
| | 2.0 | 0.71550 | 0.81929 | 0.94741 | 1.02898 | 1.13424 |
| | 2.4 | 0.72727 | 0.84495 | 0.99113 | 1.08461 | 1.20558 |
| | 2.8 | 0.74324 | 0.87310 | 1.03499 | 1.13876 | 1.27330 |
| | 3.2 | 0.75464 | 0.89187 | 1.06322 | 1.17317 | 1.31579 |

FRACTILES $W_\alpha$ RHO = −0.95

| DELTAS | | α: 0.8000 | 0.8500 | 0.9000 | 0.9250 | 0.9500 |
|---|---|---|---|---|---|---|
| 3.6 | 0.0 | 0.83747 | 0.92015 | 1.00544 | 1.05333 | 1.11031 |
| | 0.4 | 0.80984 | 0.88657 | 0.97113 | 1.02097 | 1.08212 |
| | 0.8 | 0.77923 | 0.85630 | 0.94521 | 0.99926 | 1.06685 |
| | 1.2 | 0.75234 | 0.83407 | 0.93115 | 0.99135 | 1.06762 |
| | 1.6 | 0.73354 | 0.82325 | 0.93192 | 1.00018 | 1.08744 |
| | 2.0 | 0.72557 | 0.82612 | 0.94949 | 1.02769 | 1.12825 |
| | 2.4 | 0.72924 | 0.84277 | 0.98325 | 1.07282 | 1.18850 |
| | 2.8 | 0.74207 | 0.86914 | 1.02726 | 1.12847 | 1.25952 |
| | 3.2 | 0.75749 | 0.89605 | 1.06901 | 1.17998 | 1.32387 |
| | 3.6 | 0.76776 | 0.91297 | 1.09448 | 1.21103 | 1.36226 |
| 4.0 | 0.0 | 0.84985 | 0.94545 | 1.04340 | 1.09721 | 1.15995 |
| | 0.4 | 0.83101 | 0.91732 | 1.01020 | 1.06383 | 1.12862 |
| | 0.8 | 0.80494 | 0.88793 | 0.98169 | 1.03779 | 1.10719 |
| | 1.2 | 0.77810 | 0.86271 | 0.96167 | 1.02233 | 1.09857 |
| | 1.6 | 0.75557 | 0.84552 | 0.95325 | 1.02036 | 1.10565 |
| | 2.0 | 0.74090 | 0.83926 | 0.95899 | 1.03444 | 1.13103 |
| | 2.4 | 0.73628 | 0.84568 | 0.98032 | 1.06583 | 1.17590 |
| | 2.8 | 0.74199 | 0.86424 | 1.01584 | 1.11263 | 1.23770 |
| | 3.2 | 0.75521 | 0.89053 | 1.05914 | 1.16716 | 1.30709 |
| | 3.6 | 0.76975 | 0.91580 | 1.09832 | 1.21548 | 1.36747 |
| | 4.0 | 0.77889 | 0.93091 | 1.12111 | 1.24330 | 1.40190 |
| 6.0 | 0.0 | 0.86270 | 0.99572 | 1.14685 | 1.23475 | 1.33808 |
| | 0.4 | 0.86098 | 0.98990 | 1.13497 | 1.21854 | 1.31632 |
| | 0.8 | 0.85796 | 0.98204 | 1.12026 | 1.19960 | 1.29303 |
| | 1.2 | 0.85258 | 0.97104 | 1.10276 | 1.17908 | 1.27035 |
| | 1.6 | 0.84344 | 0.95665 | 1.08407 | 1.15924 | 1.25070 |
| | 2.0 | 0.83039 | 0.94033 | 1.06659 | 1.14252 | 1.23636 |
| | 2.4 | 0.81505 | 0.92447 | 1.05276 | 1.13122 | 1.22944 |
| | 2.8 | 0.79986 | 0.91153 | 1.04481 | 1.12745 | 1.23194 |
| | 3.2 | 0.78721 | 0.90367 | 1.04467 | 1.13304 | 1.24564 |
| | 3.6 | 0.77904 | 0.90256 | 1.05377 | 1.14933 | 1.27182 |
| | 4.0 | 0.77663 | 0.90912 | 1.07270 | 1.17672 | 1.31066 |
| | 6.0 | 0.81297 | 0.98660 | 1.20450 | 1.34475 | 1.52695 |
| 12.0 | 0.0 | 0.85858 | 1.02335 | 1.22232 | 1.34565 | 1.50023 |
| | 0.4 | 0.85877 | 1.02264 | 1.22025 | 1.34260 | 1.49576 |
| | 0.8 | 0.85893 | 1.02185 | 1.21806 | 1.33937 | 1.49107 |
| | 1.2 | 0.85904 | 1.02097 | 1.21572 | 1.33597 | 1.48615 |
| | 1.6 | 0.85911 | 1.02000 | 1.21322 | 1.33237 | 1.48100 |
| | 2.0 | 0.85910 | 1.01891 | 1.21055 | 1.32857 | 1.47562 |
| | 2.4 | 0.85900 | 1.01768 | 1.20770 | 1.32456 | 1.47000 |
| | 2.8 | 0.85879 | 1.01630 | 1.20464 | 1.32033 | 1.46417 |
| | 3.2 | 0.85843 | 1.01473 | 1.20137 | 1.31589 | 1.45815 |
| | 3.6 | 0.85789 | 1.01294 | 1.19787 | 1.31123 | 1.45199 |
| | 4.0 | 0.85711 | 1.01090 | 1.19413 | 1.30639 | 1.44577 |
| | 6.0 | 0.84722 | 0.99578 | 1.17359 | 1.28340 | 1.42119 |
| | 12.0 | 0.83682 | 1.02816 | 1.26894 | 1.42419 | 1.62608 |
| INFINITY | | 0.84162 | 1.03643 | 1.28155 | 1.43953 | 1.64485 |

FRACTILES $w_\alpha$ RHO = -0.95

| DELTAS | | $\alpha$: 0.9625 | 0.9750 | 0.9825 | 0.9900 | 0.9925 |
|---|---|---|---|---|---|---|
| 0.0 | 0.0 | 0.71068 | 0.72186 | 0.73216 | 0.74893 | 0.75778 |
| 0.4 | 0.0 | 0.73839 | 0.75893 | 0.77671 | 0.80414 | 0.81805 |
| | 0.4 | 0.79242 | 0.82390 | 0.85053 | 0.89071 | 0.91074 |
| 0.8 | 0.0 | 0.78328 | 0.81086 | 0.83416 | 0.86928 | 0.88676 |
| | 0.4 | 0.84035 | 0.87977 | 0.91280 | 0.96216 | 0.98657 |
| | 0.8 | 0.90145 | 0.95119 | 0.99270 | 1.05445 | 1.08485 |
| 1.2 | 0.0 | 0.82794 | 0.85980 | 0.88642 | 0.92608 | 0.94565 |
| | 0.4 | 0.87641 | 0.92022 | 0.95673 | 1.01095 | 1.03763 |
| | 0.8 | 0.94484 | 1.00103 | 1.04778 | 1.11708 | 1.15111 |
| | 1.2 | 1.00816 | 1.07436 | 1.12936 | 1.21076 | 1.25066 |
| 1.6 | 0.0 | 0.87529 | 0.90985 | 0.93850 | 0.98085 | 1.00162 |
| | 0.4 | 0.90823 | 0.95398 | 0.99193 | 1.04803 | 1.07551 |
| | 0.8 | 0.96948 | 1.02808 | 1.07669 | 1.14855 | 1.18374 |
| | 1.2 | 1.04511 | 1.11651 | 1.17574 | 1.26327 | 1.30610 |
| | 1.6 | 1.10791 | 1.18891 | 1.25608 | 1.35526 | 1.40377 |
| 2.0 | 0.0 | 0.92621 | 0.96274 | 0.99281 | 1.03697 | 1.05851 |
| | 0.4 | 0.94272 | 0.98925 | 1.02768 | 1.08423 | 1.11183 |
| | 0.8 | 0.98792 | 1.04650 | 1.09497 | 1.16639 | 1.20128 |
| | 1.2 | 1.05853 | 1.13066 | 1.19039 | 1.27848 | 1.32152 |
| | 1.6 | 1.13802 | 1.22311 | 1.29363 | 1.39762 | 1.44844 |
| | 2.0 | 1.19842 | 1.29260 | 1.37064 | 1.48570 | 1.54190 |
| 2.4 | 0.0 | 0.97999 | 1.01825 | 1.04953 | 1.09519 | 1.11734 |
| | 0.4 | 0.98209 | 1.02909 | 1.06771 | 1.12429 | 1.15181 |
| | 0.8 | 1.00965 | 1.06737 | 1.11496 | 1.18486 | 1.21891 |
| | 1.2 | 1.06453 | 1.13499 | 1.19322 | 1.27890 | 1.32068 |
| | 1.6 | 1.14146 | 1.22589 | 1.29575 | 1.39865 | 1.44887 |
| | 2.0 | 1.22204 | 1.31934 | 1.39992 | 1.51864 | 1.57659 |
| | 2.4 | 1.27883 | 1.38461 | 1.47222 | 1.60129 | 1.66428 |
| 2.8 | 0.0 | 1.03534 | 1.07542 | 1.10795 | 1.15510 | 1.17785 |
| | 0.4 | 1.02586 | 1.07342 | 1.11230 | 1.16898 | 1.19643 |
| | 0.8 | 1.03776 | 1.09460 | 1.14130 | 1.20963 | 1.24282 |
| | 1.2 | 1.07448 | 1.14263 | 1.19881 | 1.28124 | 1.32135 |
| | 1.6 | 1.13664 | 1.21804 | 1.28527 | 1.38411 | 1.43228 |
| | 2.0 | 1.21718 | 1.31270 | 1.39172 | 1.50801 | 1.56471 |
| | 2.4 | 1.29666 | 1.40474 | 1.49420 | 1.62594 | 1.69021 |
| | 2.8 | 1.34914 | 1.46504 | 1.56100 | 1.70230 | 1.77122 |
| 3.2 | 0.0 | 1.09096 | 1.13314 | 1.16709 | 1.21592 | 1.23935 |
| | 0.4 | 1.07262 | 1.12106 | 1.16044 | 1.21750 | 1.24502 |
| | 0.8 | 1.07183 | 1.12817 | 1.17425 | 1.24140 | 1.27391 |
| | 1.2 | 1.09211 | 1.15821 | 1.21252 | 1.29196 | 1.33052 |
| | 1.6 | 1.13619 | 1.21407 | 1.27826 | 1.37239 | 1.41818 |
| | 2.0 | 1.20345 | 1.29485 | 1.37033 | 1.48125 | 1.53527 |
| | 2.4 | 1.28530 | 1.39075 | 1.47795 | 1.60623 | 1.66876 |
| | 2.8 | 1.36206 | 1.47957 | 1.57681 | 1.71996 | 1.78975 |
| | 3.2 | 1.40993 | 1.53459 | 1.63777 | 1.78967 | 1.86373 |

FRACTILES $W_\alpha$ RHO = -0.95

| DELTAS | | α: 0.9625 | 0.9750 | 0.9825 | 0.9900 | 0.9925 |
|---|---|---|---|---|---|---|
| 3.6 | 0.0 | 1.14562 | 1.19034 | 1.22601 | 1.27685 | 1.30108 |
| | 0.4 | 1.12085 | 1.17063 | 1.21081 | 1.26867 | 1.29644 |
| | 0.8 | 1.11028 | 1.16664 | 1.21251 | 1.27903 | 1.31111 |
| | 1.2 | 1.11708 | 1.18172 | 1.23463 | 1.31173 | 1.34905 |
| | 1.6 | 1.14440 | 1.21920 | 1.28067 | 1.37057 | 1.41421 |
| | 2.0 | 1.19419 | 1.28108 | 1.35269 | 1.45771 | 1.50876 |
| | 2.4 | 1.26459 | 1.36508 | 1.44807 | 1.56998 | 1.62933 |
| | 2.8 | 1.34590 | 1.46016 | 1.55463 | 1.69357 | 1.76127 |
| | 3.2 | 1.41883 | 1.54453 | 1.64854 | 1.80160 | 1.87621 |
| | 3.6 | 1.46209 | 1.59428 | 1.70368 | 1.86470 | 1.94318 |
| 4.0 | 0.0 | 1.19820 | 1.24606 | 1.28381 | 1.33711 | 1.36232 |
| | 0.4 | 1.16917 | 1.22083 | 1.26220 | 1.32134 | 1.34956 |
| | 0.8 | 1.15139 | 1.20837 | 1.25447 | 1.32094 | 1.35287 |
| | 1.2 | 1.14769 | 1.21156 | 1.26361 | 1.33913 | 1.37557 |
| | 1.6 | 1.16105 | 1.23353 | 1.29289 | 1.37943 | 1.42132 |
| | 2.0 | 1.19415 | 1.27709 | 1.34528 | 1.44501 | 1.49341 |
| | 2.4 | 1.24813 | 1.34333 | 1.42180 | 1.53686 | 1.59279 |
| | 2.8 | 1.31999 | 1.42870 | 1.51848 | 1.65034 | 1.71452 |
| | 3.2 | 1.39935 | 1.52138 | 1.62227 | 1.77064 | 1.84291 |
| | 3.6 | 1.46777 | 1.60056 | 1.71043 | 1.87208 | 1.95085 |
| | 4.0 | 1.50661 | 1.64526 | 1.75999 | 1.92883 | 2.01112 |
| 6.0 | 0.0 | 1.39983 | 1.47431 | 1.53046 | 1.60570 | 1.63977 |
| | 0.4 | 1.37490 | 1.44630 | 1.50093 | 1.57552 | 1.60983 |
| | 0.8 | 1.34974 | 1.41991 | 1.47449 | 1.55025 | 1.58556 |
| | 1.2 | 1.32668 | 1.39742 | 1.45322 | 1.53174 | 1.56871 |
| | 1.6 | 1.30803 | 1.38093 | 1.43910 | 1.52181 | 1.56108 |
| | 2.0 | 1.29593 | 1.37245 | 1.43405 | 1.52239 | 1.56458 |
| | 2.4 | 1.29243 | 1.37397 | 1.44008 | 1.53548 | 1.58128 |
| | 2.8 | 1.29948 | 1.38746 | 1.45917 | 1.56319 | 1.61332 |
| | 3.2 | 1.31888 | 1.41473 | 1.49319 | 1.60746 | 1.66269 |
| | 3.6 | 1.35187 | 1.45702 | 1.54338 | 1.66954 | 1.73066 |
| | 4.0 | 1.39852 | 1.51424 | 1.60953 | 1.74906 | 1.81679 |
| | 6.0 | 1.64731 | 1.80670 | 1.93858 | 2.13258 | 2.22708 |
| 12.0 | 0.0 | 1.59880 | 1.72501 | 1.82569 | 1.96766 | 2.03418 |
| | 0.4 | 1.59332 | 1.71810 | 1.81753 | 1.95757 | 2.02312 |
| | 0.8 | 1.58759 | 1.71090 | 1.80906 | 1.94714 | 2.01171 |
| | 1.2 | 1.58160 | 1.70342 | 1.80029 | 1.93641 | 2.00000 |
| | 1.6 | 1.57536 | 1.69567 | 1.79124 | 1.92542 | 1.98806 |
| | 2.0 | 1.56887 | 1.68767 | 1.78196 | 1.91424 | 1.97597 |
| | 2.4 | 1.56215 | 1.67945 | 1.77251 | 1.90299 | 1.96388 |
| | 2.8 | 1.55524 | 1.67109 | 1.76297 | 1.89181 | 1.95196 |
| | 3.2 | 1.54817 | 1.66268 | 1.75349 | 1.88093 | 1.94049 |
| | 3.6 | 1.54105 | 1.65435 | 1.74427 | 1.87063 | 1.92978 |
| | 4.0 | 1.53398 | 1.64631 | 1.73557 | 1.86128 | 1.92027 |
| | 6.0 | 1.50947 | 1.62337 | 1.71525 | 1.84698 | 1.90981 |
| | 12.0 | 1.75952 | 1.93628 | 2.08253 | 2.29762 | 2.40235 |
| INFINITY | | 1.78046 | 1.95996 | 2.10836 | 2.32635 | 2.43238 |

FRACTILES $w_\alpha$ RHO = -0.95

| DELTAS | | $\alpha$: 0.9950 | 0.9975 | 0.9990 | 0.9995 |
|---|---|---|---|---|---|
| 0.0 | 0.0 | 0.77045 | 0.79255 | 0.82238 | 0.84529 |
| 0.4 | 0.0 | 0.83745 | 0.87016 | 0.91264 | 0.94431 |
| | 0.4 | 0.93836 | 0.98419 | 1.04259 | 1.08545 |
| 0.8 | 0.0 | 0.91087 | 0.95083 | 1.00172 | 1.03904 |
| | 0.4 | 1.02003 | 1.07509 | 1.14453 | 1.19503 |
| | 0.8 | 1.12641 | 1.19447 | 1.27979 | 1.34150 |
| 1.2 | 0.0 | 0.97244 | 1.01646 | 1.07187 | 1.11212 |
| | 0.4 | 1.07406 | 1.13368 | 1.20835 | 1.26231 |
| | 0.8 | 1.19750 | 1.27324 | 1.36775 | 1.43584 |
| | 1.2 | 1.30500 | 1.39352 | 1.50369 | 1.58285 |
| 1.6 | 0.0 | 1.02992 | 1.07611 | 1.13379 | 1.17539 |
| | 0.4 | 1.11293 | 1.17391 | 1.24987 | 1.30452 |
| | 0.8 | 1.23162 | 1.30957 | 1.40649 | 1.47607 |
| | 1.2 | 1.36436 | 1.45911 | 1.57673 | 1.66105 |
| | 1.6 | 1.46970 | 1.57682 | 1.70960 | 1.80465 |
| 2.0 | 0.0 | 1.08775 | 1.13521 | 1.19409 | 1.23631 |
| | 0.4 | 1.14932 | 1.21018 | 1.28562 | 1.33967 |
| | 0.8 | 1.24867 | 1.32560 | 1.42093 | 1.48917 |
| | 1.2 | 1.37999 | 1.47490 | 1.59244 | 1.67653 |
| | 1.6 | 1.51746 | 1.62946 | 1.76808 | 1.86716 |
| | 2.0 | 1.61821 | 1.74197 | 1.89502 | 2.00432 |
| 2.4 | 0.0 | 1.14730 | 1.19569 | 1.25536 | 1.29793 |
| | 0.4 | 1.18908 | 1.24936 | 1.32376 | 1.37685 |
| | 0.8 | 1.26508 | 1.33983 | 1.43214 | 1.49803 |
| | 1.2 | 1.37736 | 1.46920 | 1.58265 | 1.66364 |
| | 1.6 | 1.51702 | 1.62745 | 1.76390 | 1.86128 |
| | 2.0 | 1.65524 | 1.78269 | 1.94013 | 2.05244 |
| | 2.4 | 1.74976 | 1.88823 | 2.05919 | 2.18108 |
| 2.8 | 0.0 | 1.20851 | 1.25778 | 1.31818 | 1.36105 |
| | 0.4 | 1.23351 | 1.29327 | 1.36668 | 1.41887 |
| | 0.8 | 1.28772 | 1.36022 | 1.44944 | 1.51294 |
| | 1.2 | 1.37569 | 1.46354 | 1.57178 | 1.64887 |
| | 1.6 | 1.49756 | 1.60321 | 1.73347 | 1.82629 |
| | 2.0 | 1.64162 | 1.76612 | 1.91970 | 2.02913 |
| | 2.4 | 1.77737 | 1.91850 | 2.09260 | 2.21663 |
| | 2.8 | 1.86471 | 2.01603 | 2.20265 | 2.33555 |
| 3.2 | 0.0 | 1.27079 | 1.32107 | 1.38231 | 1.42557 |
| | 0.4 | 1.28208 | 1.34157 | 1.41430 | 1.46581 |
| | 0.8 | 1.31779 | 1.38842 | 1.47502 | 1.53646 |
| | 1.2 | 1.38266 | 1.46676 | 1.57008 | 1.64349 |
| | 1.6 | 1.48016 | 1.58028 | 1.70346 | 1.79106 |
| | 2.0 | 1.60846 | 1.72679 | 1.87252 | 1.97621 |
| | 2.4 | 1.75352 | 1.89063 | 2.05958 | 2.17983 |
| | 2.8 | 1.88438 | 2.03751 | 2.22621 | 2.36052 |
| | 3.2 | 1.96415 | 2.12662 | 2.32680 | 2.46925 |

FRACTILES $W_\alpha$ RHO = −0.95

| DELTAS | | $\alpha$:0.9950 | 0.9975 | 0.9990 | 0.9995 |
|---|---|---|---|---|---|
| 3.6 | 0.0 | 1.33346 | 1.38495 | 1.44725 | 1.49104 |
| | 0.4 | 1.33371 | 1.39327 | 1.46572 | 1.51681 |
| | 0.8 | 1.35430 | 1.42358 | 1.50819 | 1.56801 |
| | 1.2 | 1.39941 | 1.48042 | 1.57963 | 1.64993 |
| | 1.6 | 1.47319 | 1.56826 | 1.68493 | 1.76773 |
| | 2.0 | 1.57787 | 1.68941 | 1.82652 | 1.92393 |
| | 2.4 | 1.70973 | 1.83963 | 1.99947 | 2.11310 |
| | 2.8 | 1.85302 | 2.00135 | 2.18399 | 2.31387 |
| | 3.2 | 1.97734 | 2.14090 | 2.34233 | 2.48559 |
| | 3.6 | 2.04958 | 2.22164 | 2.43351 | 2.58418 |
| 4.0 | 0.0 | 1.39585 | 1.44883 | 1.51248 | 1.55697 |
| | 0.4 | 1.38730 | 1.44732 | 1.51991 | 1.57088 |
| | 0.8 | 1.39573 | 1.46421 | 1.54747 | 1.60613 |
| | 1.2 | 1.42463 | 1.50330 | 1.59931 | 1.66713 |
| | 1.6 | 1.47785 | 1.56875 | 1.67998 | 1.75873 |
| | 2.0 | 1.55882 | 1.66421 | 1.79345 | 1.88510 |
| | 2.4 | 1.66848 | 1.79061 | 1.94062 | 2.04711 |
| | 2.8 | 1.80144 | 1.94183 | 2.11446 | 2.23710 |
| | 3.2 | 1.94063 | 2.09910 | 2.29383 | 2.43224 |
| | 3.6 | 2.05762 | 2.23022 | 2.44265 | 2.59366 |
| | 4.0 | 2.12264 | 2.30294 | 2.52483 | 2.68255 |
| 6.0 | 0.0 | 1.68371 | 1.75039 | 1.82683 | 1.87835 |
| | 0.4 | 1.65459 | 1.72349 | 1.80373 | 1.85843 |
| | 0.8 | 1.63203 | 1.70435 | 1.78961 | 1.84825 |
| | 1.2 | 1.61770 | 1.69459 | 1.78610 | 1.84949 |
| | 1.6 | 1.61338 | 1.69602 | 1.79510 | 1.86411 |
| | 2.0 | 1.62102 | 1.71065 | 1.81876 | 1.89439 |
| | 2.4 | 1.64273 | 1.74074 | 1.85950 | 1.94287 |
| | 2.8 | 1.68075 | 1.78866 | 1.91988 | 2.01228 |
| | 3.2 | 1.73713 | 1.85657 | 2.00226 | 2.10507 |
| | 3.6 | 1.81318 | 1.94584 | 2.10804 | 2.22270 |
| | 4.0 | 1.90833 | 2.05574 | 2.23629 | 2.36410 |
| | 6.0 | 2.35510 | 2.56187 | 2.81599 | 2.99634 |
| 12.0 | 0.0 | 2.12158 | 2.25622 | 2.41092 | 2.51372 |
| | 0.4 | 2.10917 | 2.24158 | 2.39352 | 2.49442 |
| | 0.8 | 2.09641 | 2.22663 | 2.37592 | 2.47502 |
| | 1.2 | 2.08338 | 2.21148 | 2.35827 | 2.45575 |
| | 1.6 | 2.07015 | 2.19624 | 2.34078 | 2.43688 |
| | 2.0 | 2.05686 | 2.18113 | 2.32374 | 2.41877 |
| | 2.4 | 2.04368 | 2.16638 | 2.30751 | 2.40183 |
| | 2.8 | 2.03085 | 2.15234 | 2.29250 | 2.38655 |
| | 3.2 | 2.01869 | 2.13940 | 2.27923 | 2.37347 |
| | 3.6 | 2.00759 | 2.12804 | 2.26823 | 2.36317 |
| | 4.0 | 1.99802 | 2.11880 | 2.26010 | 2.35626 |
| | 6.0 | 1.99365 | 2.12630 | 2.28523 | 2.39566 |
| | 12.0 | 2.54416 | 2.77305 | 3.05396 | 3.25304 |
| INFINITY | | 2.57583 | 2.80703 | 3.09024 | 3.29051 |

FRACTILES $W_\alpha$ RHO = -0.90

| DELTAS | | $\alpha$: 0.0005 | 0.0010 | 0.0025 | 0.0050 | 0.0075 |
|---|---|---|---|---|---|---|
| 0.0 | 0.0 | -7.85012 | -6.94056 | -5.74936 | -4.85883 | -4.34316 |
| 0.4 | 0.0 | -7.82260 | -6.91855 | -5.73379 | -4.84750 | -4.33405 |
| | 0.4 | -7.78798 | -6.88642 | -5.70570 | -4.82304 | -4.31193 |
| 0.8 | 0.0 | -7.65802 | -6.79035 | -5.64858 | -4.79046 | -4.29156 |
| | 0.4 | -7.71023 | -6.82060 | -5.65479 | -4.78273 | -4.27754 |
| | 0.8 | -7.61118 | -6.73242 | -5.58159 | -4.72135 | -4.22327 |
| 1.2 | 0.0 | -7.33048 | -6.52891 | -5.46857 | -4.66651 | -4.19772 |
| | 0.4 | -7.52689 | -6.67581 | -5.55589 | -4.71424 | -4.22495 |
| | 0.8 | -7.50116 | -6.63845 | -5.50800 | -4.66247 | -4.17271 |
| | 1.2 | -7.34501 | -6.50064 | -5.39500 | -4.56867 | -4.09029 |
| 1.6 | 0.0 | -6.93547 | -6.20553 | -5.23588 | -4.49855 | -4.06568 |
| | 0.4 | -7.20618 | -6.41994 | -5.37989 | -4.59320 | -4.13342 |
| | 0.8 | -7.32465 | -6.49921 | -5.41311 | -4.59695 | -4.12253 |
| | 1.2 | -7.22132 | -6.39480 | -5.31189 | -4.50207 | -4.03306 |
| | 1.6 | -7.02137 | -6.21911 | -5.16875 | -4.38390 | -3.92964 |
| 2.0 | 0.0 | -6.55206 | -5.88629 | -4.99894 | -4.32142 | -3.92227 |
| | 0.4 | -6.82819 | -6.11117 | -5.15869 | -4.43444 | -4.00923 |
| | 0.8 | -7.03295 | -6.26814 | -5.25646 | -4.49127 | -4.04408 |
| | 1.2 | -7.07094 | -6.27783 | -5.23434 | -4.45033 | -3.99465 |
| | 1.6 | -6.89929 | -6.11470 | -5.08686 | -4.31840 | -3.87346 |
| | 2.0 | -6.67075 | -5.91447 | -4.92449 | -4.18498 | -3.75709 |
| 2.4 | 0.0 | -6.21186 | -5.60008 | -4.78234 | -4.15579 | -3.78560 |
| | 0.4 | -6.46255 | -5.80740 | -4.93418 | -4.26743 | -3.87462 |
| | 0.8 | -6.68762 | -5.98761 | -5.05771 | -4.35062 | -3.93550 |
| | 1.2 | -6.82352 | -6.08473 | -5.10750 | -4.36842 | -3.93651 |
| | 1.6 | -6.78620 | -6.02958 | -5.03424 | -4.28653 | -3.85203 |
| | 2.0 | -6.56036 | -5.82025 | -4.85089 | -4.12636 | -3.70700 |
| | 2.4 | -6.31705 | -5.60763 | -4.67921 | -3.98594 | -3.58498 |
| 2.8 | 0.0 | -5.92024 | -5.35315 | -4.59316 | -4.00904 | -3.66304 |
| | 0.4 | -6.13770 | -5.53462 | -4.72848 | -4.11079 | -3.74582 |
| | 0.8 | -6.35006 | -5.70829 | -4.85289 | -4.19972 | -3.81490 |
| | 1.2 | -6.52158 | -5.84174 | -4.93863 | -4.25191 | -3.84874 |
| | 1.6 | -6.59084 | -5.88112 | -4.94239 | -4.23248 | -3.81768 |
| | 2.0 | -6.48833 | -5.77019 | -4.82561 | -4.11620 | -3.70407 |
| | 2.4 | -6.22357 | -5.52814 | -4.61751 | -3.93714 | -3.54349 |
| | 2.8 | -5.97645 | -5.31271 | -4.44434 | -3.79623 | -3.42157 |
| 3.2 | 0.0 | -5.67284 | -5.14273 | -4.43063 | -3.88177 | -3.55591 |
| | 0.4 | -5.85851 | -5.29860 | -4.54821 | -3.97143 | -3.62976 |
| | 0.8 | -6.04723 | -5.45480 | -4.66286 | -4.05600 | -3.69741 |
| | 1.2 | -6.21936 | -5.59321 | -4.75859 | -4.12126 | -3.74576 |
| | 1.6 | -6.33800 | -5.68058 | -4.80724 | -4.14316 | -3.75329 |
| | 2.0 | -6.34656 | -5.66758 | -4.76955 | -4.09051 | -3.69378 |
| | 2.4 | -6.19118 | -5.51180 | -4.61835 | -3.94754 | -3.55796 |
| | 2.8 | -5.90153 | -5.24935 | -4.39561 | -3.75806 | -3.38939 |
| | 3.2 | -5.65840 | -5.03796 | -4.22654 | -3.62129 | -3.27165 |

FRACTILES $w_\alpha$ RHO = −0.90

| DELTAS | | $\alpha$: 0.0005 | 0.0010 | 0.0025 | 0.0050 | 0.0075 |
|---|---|---|---|---|---|---|
| 3.6 | 0.0 | -5.46286 | -4.96360 | -4.29147 | -3.77208 | -3.46308 |
| | 0.4 | -5.62097 | -5.09687 | -4.39280 | -3.85009 | -3.52787 |
| | 0.8 | -5.78488 | -5.23357 | -4.49466 | -3.92666 | -3.59017 |
| | 1.2 | -5.94348 | -5.36331 | -4.58770 | -3.99332 | -3.64208 |
| | 1.6 | -6.07535 | -5.46649 | -4.65489 | -4.03510 | -3.66992 |
| | 2.0 | -6.14423 | -5.51062 | -4.66891 | -4.02888 | -3.65315 |
| | 2.4 | -6.10016 | -5.45244 | -4.59583 | -3.94819 | -3.56988 |
| | 2.8 | -5.90437 | -5.26281 | -4.41930 | -3.78621 | -3.41869 |
| | 3.2 | -5.60146 | -4.99018 | -4.19028 | -3.59330 | -3.24833 |
| | 3.6 | -5.36745 | -4.78730 | -4.02894 | -3.46371 | -3.13748 |
| 4.0 | 0.0 | -5.28374 | -4.81044 | -4.17198 | -3.67744 | -3.38267 |
| | 0.4 | -5.41882 | -4.92465 | -4.25932 | -3.74515 | -3.43923 |
| | 0.8 | -5.56021 | -5.04317 | -4.34854 | -3.81306 | -3.49509 |
| | 1.2 | -5.70133 | -5.15982 | -4.43398 | -3.87598 | -3.54538 |
| | 1.6 | -5.82954 | -5.26289 | -4.50531 | -3.92469 | -3.58156 |
| | 2.0 | -5.92268 | -5.33227 | -4.54520 | -3.94412 | -3.58994 |
| | 2.4 | -5.94655 | -5.33742 | -4.52822 | -3.91292 | -3.55171 |
| | 2.8 | -5.85871 | -5.24193 | -4.42631 | -3.80976 | -3.44969 |
| | 3.2 | -5.63387 | -5.02846 | -4.23268 | -3.63569 | -3.28931 |
| | 3.6 | -5.32672 | -4.75353 | -4.00384 | -3.44476 | -3.12198 |
| | 4.0 | -5.10488 | -4.56185 | -3.85240 | -3.32417 | -3.01969 |
| 6.0 | 0.0 | -4.68848 | -4.29950 | -3.77057 | -3.35701 | -3.10866 |
| | 0.4 | -4.75625 | -4.35727 | -3.81542 | -3.39240 | -3.13867 |
| | 0.8 | -4.82738 | -4.41764 | -3.86196 | -3.42883 | -3.16937 |
| | 1.2 | -4.90099 | -4.47978 | -3.90938 | -3.46555 | -3.20003 |
| | 1.6 | -4.97551 | -4.54217 | -3.95630 | -3.50128 | -3.22946 |
| | 2.0 | -5.04831 | -4.60238 | -4.00053 | -3.53404 | -3.25582 |
| | 2.4 | -5.11521 | -4.65653 | -4.03865 | -3.56080 | -3.27632 |
| | 2.8 | -5.16982 | -4.69880 | -4.06573 | -3.57714 | -3.28689 |
| | 3.2 | -5.20291 | -4.72082 | -4.07424 | -3.57680 | -3.28189 |
| | 3.6 | -5.20193 | -4.71128 | -4.05496 | -3.55161 | -3.25397 |
| | 4.0 | -5.15129 | -4.65635 | -3.99630 | -3.49199 | -3.19474 |
| | 6.0 | -4.17049 | -3.77050 | -3.25160 | -2.86967 | -2.65128 |
| 12.0 | 0.0 | -4.01567 | -3.71905 | -3.31027 | -2.98575 | -2.78855 |
| | 0.4 | -4.03353 | -3.73439 | -3.32235 | -2.99543 | -2.79687 |
| | 0.8 | -4.05184 | -3.75010 | -3.33469 | -3.00530 | -2.80533 |
| | 1.2 | -4.07058 | -3.76613 | -3.34726 | -3.01533 | -2.81391 |
| | 1.6 | -4.08966 | -3.78244 | -3.35999 | -3.02545 | -2.82255 |
| | 2.0 | -4.10901 | -3.79893 | -3.37282 | -3.03560 | -2.83118 |
| | 2.4 | -4.12849 | -3.81549 | -3.38563 | -3.04570 | -2.83974 |
| | 2.8 | -4.14796 | -3.83198 | -3.39831 | -3.05562 | -2.84810 |
| | 3.2 | -4.16720 | -3.84820 | -3.41069 | -3.06522 | -2.85615 |
| | 3.6 | -4.18596 | -3.86391 | -3.42255 | -3.07432 | -2.86371 |
| | 4.0 | -4.20389 | -3.87881 | -3.43363 | -3.08268 | -2.87056 |
| | 6.0 | -4.26232 | -3.92404 | -3.46281 | -3.10099 | -2.88310 |
| | 12.0 | -3.35109 | -3.12912 | -2.82484 | -2.58259 | -2.43438 |
| INFINITY | | -3.29051 | -3.09024 | -2.80703 | -2.57583 | -2.43238 |

FRACTILES $w_\alpha$ RHO = -0.90

| DELTAS | | α: 0.0100 | 0.0175 | 0.0250 | 0.0375 | 0.0500 |
|---|---|---|---|---|---|---|
| 0.0 | 0.0 | -3.98005 | -3.28144 | -2.84251 | -2.35084 | -2.00761 |
| 0.4 | 0.0 | -3.97239 | -3.27634 | -2.83884 | -2.34862 | -2.00630 |
| | 0.4 | -3.95205 | -3.25968 | -2.82469 | -2.33748 | -1.99740 |
| 0.8 | 0.0 | -3.93931 | -3.25924 | -2.83022 | -2.34795 | -2.01014 |
| | 0.4 | -3.92173 | -3.23697 | -2.80661 | -2.32448 | -1.98787 |
| | 0.8 | -3.87259 | -3.19800 | -2.77427 | -2.29978 | -1.96870 |
| 1.2 | 0.0 | -3.86546 | -3.22052 | -2.81095 | -2.34756 | -2.02081 |
| | 0.4 | -3.87950 | -3.21262 | -2.79198 | -2.31921 | -1.98813 |
| | 0.8 | -3.82780 | -3.16412 | -2.74712 | -2.28008 | -1.95416 |
| | 1.2 | -3.75352 | -3.10586 | -2.69917 | -2.24397 | -1.92653 |
| 1.6 | 0.0 | -3.75784 | -3.15740 | -2.77370 | -2.33680 | -2.02655 |
| | 0.4 | -3.80755 | -3.17505 | -2.77343 | -2.31909 | -1.99878 |
| | 0.8 | -3.78761 | -3.14114 | -2.73348 | -2.27544 | -1.95480 |
| | 1.2 | -3.70282 | -3.06752 | -2.66851 | -2.22185 | -1.91034 |
| | 1.6 | -3.60992 | -2.99523 | -2.60944 | -2.17792 | -1.87726 |
| 2.0 | 0.0 | -3.63768 | -3.08053 | -2.72277 | -2.31335 | -2.02102 |
| | 0.4 | -3.70685 | -3.11707 | -2.74020 | -2.31109 | -2.00641 |
| | 0.8 | -3.72715 | -3.11206 | -2.72155 | -2.27987 | -1.96859 |
| | 1.2 | -3.67301 | -3.05230 | -2.66103 | -2.22159 | -1.91417 |
| | 1.6 | -3.56022 | -2.95787 | -2.57975 | -2.15679 | -1.86211 |
| | 2.0 | -3.45601 | -2.87742 | -2.51456 | -2.10908 | -1.82694 |
| 2.4 | 0.0 | -3.52111 | -3.00172 | -2.66694 | -2.28232 | -2.00652 |
| | 0.4 | -3.59455 | -3.04622 | -2.69413 | -2.29120 | -2.00350 |
| | 0.8 | -3.64030 | -3.06453 | -2.69663 | -2.27777 | -1.98041 |
| | 1.2 | -3.63044 | -3.03651 | -2.65951 | -2.23326 | -1.93296 |
| | 1.6 | -3.54540 | -2.95385 | -2.58112 | -2.16279 | -1.87040 |
| | 2.0 | -3.41184 | -2.84454 | -2.48871 | -2.09110 | -1.81450 |
| | 2.4 | -3.30293 | -2.76127 | -2.42193 | -2.04326 | -1.78039 |
| 2.8 | 0.0 | -3.41538 | -2.92778 | -2.61247 | -2.24903 | -1.98752 |
| | 0.4 | -3.48504 | -2.97295 | -2.64284 | -2.26358 | -1.99161 |
| | 0.8 | -3.54052 | -3.00333 | -2.65838 | -2.26362 | -1.98175 |
| | 1.2 | -3.56205 | -3.00290 | -2.64565 | -2.23897 | -1.95032 |
| | 1.6 | -3.52375 | -2.95348 | -2.59161 | -2.18263 | -1.89468 |
| | 2.0 | -3.41328 | -2.85253 | -2.49945 | -2.10352 | -1.82715 |
| | 2.4 | -3.26655 | -2.73459 | -2.40129 | -2.02945 | -1.77145 |
| | 2.8 | -3.15815 | -2.65271 | -2.33653 | -1.98456 | -1.74125 |
| 3.2 | 0.0 | -3.32228 | -2.86125 | -2.56226 | -2.21667 | -1.96724 |
| | 0.4 | -3.38518 | -2.90364 | -2.59221 | -2.23324 | -1.97490 |
| | 0.8 | -3.44119 | -2.93801 | -2.61363 | -2.24092 | -1.97363 |
| | 1.2 | -3.47802 | -2.95382 | -2.61720 | -2.23197 | -1.95692 |
| | 1.6 | -3.47606 | -2.93540 | -2.59000 | -2.19687 | -1.91791 |
| | 2.0 | -3.41271 | -2.86750 | -2.52167 | -2.13104 | -1.85624 |
| | 2.4 | -3.28317 | -2.75355 | -2.42036 | -2.04724 | -1.78735 |
| | 2.8 | -3.13013 | -2.63262 | -2.32140 | -1.97511 | -1.73594 |
| | 3.2 | -3.02598 | -2.55519 | -2.26140 | -1.93568 | -1.71164 |

FRACTILES $W_\alpha$ RHO = -0.90

| DELTAS | | $\alpha$:0.0100 | 0.0175 | 0.0250 | 0.0375 | 0.0500 |
|---|---|---|---|---|---|---|
| 3.6 | 0.0 | -3.24120 | -2.80246 | -2.51719 | -2.18663 | -1.94742 |
| | 0.4 | -3.29683 | -2.84088 | -2.54515 | -2.20330 | -1.95654 |
| | 0.8 | -3.34929 | -2.87498 | -2.56820 | -2.21455 | -1.96002 |
| | 1.2 | -3.39110 | -2.89818 | -2.58038 | -2.21521 | -1.95331 |
| | 1.6 | -3.40954 | -2.89971 | -2.57231 | -2.19764 | -1.93014 |
| | 2.0 | -3.38597 | -2.86495 | -2.53215 | -2.15345 | -1.88482 |
| | 2.4 | -3.30188 | -2.78222 | -2.45276 | -2.08092 | -1.81968 |
| | 2.8 | -3.15956 | -2.66048 | -2.34691 | -1.99648 | -1.75328 |
| | 3.2 | -3.00590 | -2.54128 | -2.25141 | -1.93020 | -1.70929 |
| | 3.6 | -2.90847 | -2.47048 | -2.19820 | -1.89755 | -1.69094 |
| 4.0 | 0.0 | -3.17072 | -2.75083 | -2.47718 | -2.15936 | -1.92883 |
| | 0.4 | -3.21955 | -2.78512 | -2.50263 | -2.17524 | -1.93829 |
| | 0.8 | -3.26710 | -2.81711 | -2.52522 | -2.18775 | -1.94413 |
| | 1.2 | -3.30871 | -2.84264 | -2.54115 | -2.19355 | -1.94335 |
| | 1.6 | -3.33636 | -2.85475 | -2.54422 | -2.18736 | -1.93140 |
| | 2.0 | -3.33740 | -2.84290 | -2.52534 | -2.16193 | -1.90249 |
| | 2.4 | -3.29487 | -2.79406 | -2.47423 | -2.11040 | -1.85247 |
| | 2.8 | -3.19467 | -2.70035 | -2.38720 | -2.03417 | -1.78665 |
| | 3.2 | -3.04521 | -2.57554 | -2.28103 | -1.95298 | -1.72623 |
| | 3.6 | -2.89536 | -2.46196 | -2.19258 | -1.89512 | -1.69053 |
| | 4.0 | -2.80625 | -2.39924 | -2.14717 | -1.86903 | -1.67696 |
| 6.0 | 0.0 | -2.92915 | -2.57093 | -2.33540 | -2.05950 | -1.85759 |
| | 0.4 | -2.95541 | -2.59013 | -2.35028 | -2.06968 | -1.86461 |
| | 0.8 | -2.98214 | -2.60938 | -2.36497 | -2.07945 | -1.87106 |
| | 1.2 | -3.00862 | -2.62802 | -2.37888 | -2.08825 | -1.87648 |
| | 1.6 | -3.03372 | -2.64506 | -2.39108 | -2.09530 | -1.88013 |
| | 2.0 | -3.05569 | -2.65896 | -2.40019 | -2.09940 | -1.88100 |
| | 2.4 | -3.07194 | -2.66748 | -2.40423 | -2.09886 | -1.87761 |
| | 2.8 | -3.07867 | -2.66741 | -2.40039 | -2.09136 | -1.86801 |
| | 3.2 | -3.07066 | -2.65440 | -2.38488 | -2.07384 | -1.84972 |
| | 3.6 | -3.04118 | -2.62300 | -2.35317 | -2.04289 | -1.82025 |
| | 4.0 | -2.98275 | -2.56762 | -2.30107 | -1.99631 | -1.77914 |
| | 6.0 | -2.49845 | -2.20474 | -2.01831 | -1.80455 | -1.64999 |
| 12.0 | 0.0 | -2.64484 | -2.35483 | -2.16160 | -1.93239 | -1.76252 |
| | 0.4 | -2.65220 | -2.36039 | -2.16605 | -1.93563 | -1.76492 |
| | 0.8 | -2.65969 | -2.36602 | -2.17054 | -1.93887 | -1.76732 |
| | 1.2 | -2.66726 | -2.37169 | -2.17504 | -1.94210 | -1.76969 |
| | 1.6 | -2.67486 | -2.37735 | -2.17952 | -1.94529 | -1.77200 |
| | 2.0 | -2.68245 | -2.38297 | -2.18393 | -1.94839 | -1.77423 |
| | 2.4 | -2.68994 | -2.38846 | -2.18821 | -1.95136 | -1.77632 |
| | 2.8 | -2.69723 | -2.39375 | -2.19229 | -1.95414 | -1.77823 |
| | 3.2 | -2.70421 | -2.39873 | -2.19607 | -1.95664 | -1.77989 |
| | 3.6 | -2.71071 | -2.40327 | -2.19945 | -1.95879 | -1.78122 |
| | 4.0 | -2.71654 | -2.40722 | -2.20228 | -1.96045 | -1.78214 |
| | 6.0 | -2.72529 | -2.40941 | -2.20093 | -1.95577 | -1.77561 |
| | 12.0 | -2.32565 | -2.10392 | -1.95408 | -1.77375 | -1.63803 |
| INFINITY | | -2.32635 | -2.10836 | -1.95996 | -1.78046 | -1.64485 |

FRACTILES $w_\alpha$ RHO = -0.90

| DELTAS | | α: 0.0750 | 0.1000 | 0.1500 | 0.2000 | 0.2500 |
|---|---|---|---|---|---|---|
| 0.0 | 0.0 | -1.53356 | -1.20561 | -0.75851 | -0.45507 | -0.22988 |
| 0.4 | 0.0 | -1.53338 | -1.20612 | -0.75989 | -0.45699 | -0.23221 |
| | 0.4 | -1.52777 | -1.20296 | -0.76034 | -0.46016 | -0.23765 |
| 0.8 | 0.0 | -1.54191 | -1.21680 | -0.77196 | -0.46897 | -0.24359 |
| | 0.4 | -1.52298 | -1.20143 | -0.76333 | -0.46640 | -0.24654 |
| | 0.8 | -1.51171 | -1.19588 | -0.76611 | -0.47542 | -0.26078 |
| 1.2 | 0.0 | -1.56444 | -1.24481 | -0.80310 | -0.49886 | -0.27052 |
| | 0.4 | -1.52940 | -1.21107 | -0.77593 | -0.48012 | -0.26069 |
| | 0.8 | -1.50430 | -1.19345 | -0.77073 | -0.48525 | -0.27505 |
| | 1.2 | -1.48876 | -1.18666 | -0.77671 | -0.50105 | -0.29972 |
| 1.6 | 0.0 | -1.58950 | -1.28017 | -0.84696 | -0.54354 | -0.31228 |
| | 0.4 | -1.55154 | -1.23845 | -0.80619 | -0.50899 | -0.28652 |
| | 0.8 | -1.51083 | -1.20304 | -0.78315 | -0.49879 | -0.28922 |
| | 1.2 | -1.48084 | -1.18459 | -0.78309 | -0.51414 | -0.31964 |
| | 1.6 | -1.46326 | -1.17828 | -0.79376 | -0.53931 | -0.35660 |
| 2.0 | 0.0 | -1.60632 | -1.31020 | -0.89057 | -0.59190 | -0.36054 |
| | 0.4 | -1.57727 | -1.27363 | -0.84864 | -0.55133 | -0.32517 |
| | 0.8 | -1.53416 | -1.23028 | -0.81139 | -0.52432 | -0.31064 |
| | 1.2 | -1.48892 | -1.19462 | -0.79448 | -0.52584 | -0.33134 |
| | 1.6 | -1.45650 | -1.17756 | -0.80244 | -0.55557 | -0.37700 |
| | 2.0 | -1.43941 | -1.17399 | -0.82028 | -0.58766 | -0.41560 |
| 2.4 | 0.0 | -1.61319 | -1.33046 | -0.92616 | -0.63481 | -0.40618 |
| | 0.4 | -1.59537 | -1.30399 | -0.89115 | -0.59749 | -0.37024 |
| | 0.8 | -1.56169 | -1.26555 | -0.85145 | -0.56236 | -0.34324 |
| | 1.2 | -1.51418 | -1.22161 | -0.81935 | -0.54548 | -0.34442 |
| | 1.6 | -1.46656 | -1.18785 | -0.81166 | -0.56288 | -0.38208 |
| | 2.0 | -1.43488 | -1.17547 | -0.83059 | -0.60236 | -0.43151 |
| | 2.4 | -1.42106 | -1.17718 | -0.85298 | -0.63426 | -0.46644 |
| 2.8 | 0.0 | -1.61296 | -1.34231 | -0.95261 | -0.66916 | -0.44460 |
| | 0.4 | -1.60371 | -1.32488 | -0.92615 | -0.63884 | -0.41345 |
| | 0.8 | -1.58194 | -1.29652 | -0.89229 | -0.60501 | -0.38309 |
| | 1.2 | -1.54399 | -1.25681 | -0.85582 | -0.57692 | -0.36727 |
| | 1.6 | -1.49353 | -1.21383 | -0.83117 | -0.57373 | -0.38526 |
| | 2.0 | -1.44641 | -1.18507 | -0.83582 | -0.60389 | -0.43042 |
| | 2.4 | -1.41933 | -1.18064 | -0.86222 | -0.64542 | -0.47772 |
| | 2.8 | -1.41087 | -1.18713 | -0.88485 | -0.67408 | -0.50783 |
| 3.2 | 0.0 | -1.60869 | -1.34849 | -0.97172 | -0.69566 | -0.47536 |
| | 0.4 | -1.60486 | -1.33744 | -0.95234 | -0.67219 | -0.45021 |
| | 0.8 | -1.59238 | -1.31831 | -0.92640 | -0.64406 | -0.42269 |
| | 1.2 | -1.56680 | -1.28836 | -0.89424 | -0.61458 | -0.39933 |
| | 1.6 | -1.52543 | -1.24831 | -0.86235 | -0.59583 | -0.39697 |
| | 2.0 | -1.47403 | -1.20845 | -0.84798 | -0.60623 | -0.42603 |
| | 2.4 | -1.43093 | -1.18805 | -0.86313 | -0.64237 | -0.47238 |
| | 2.8 | -1.41134 | -1.19110 | -0.89193 | -0.68188 | -0.51539 |
| | 3.2 | -1.40772 | -1.19988 | -0.91259 | -0.70656 | -0.54081 |

FRACTILES $W_\alpha$ RHO = -0.90

| DELTAS | | $\alpha$: 0.0750 | 0.1000 | 0.1500 | 0.2000 | 0.2500 |
|---|---|---|---|---|---|---|
| 3.6 | 0.0 | -1.60247 | -1.35118 | -0.98558 | -0.71608 | -0.49974 |
| | 0.4 | -1.60179 | -1.34429 | -0.97135 | -0.69802 | -0.47982 |
| | 0.8 | -1.59538 | -1.33182 | -0.95219 | -0.67598 | -0.45711 |
| | 1.2 | -1.57972 | -1.31114 | -0.92709 | -0.65051 | -0.43389 |
| | 1.6 | -1.55078 | -1.28010 | -0.89732 | -0.62652 | -0.41926 |
| | 2.0 | -1.50718 | -1.24095 | -0.87185 | -0.61862 | -0.42871 |
| | 2.4 | -1.45732 | -1.20691 | -0.86832 | -0.63846 | -0.46306 |
| | 2.8 | -1.42150 | -1.19570 | -0.88953 | -0.67585 | -0.50750 |
| | 3.2 | -1.40924 | -1.20339 | -0.91757 | -0.71171 | -0.54563 |
| | 3.6 | -1.40867 | -1.21256 | -0.93558 | -0.73257 | -0.56690 |
| 4.0 | 0.0 | -1.59546 | -1.35180 | -0.99583 | -0.73205 | -0.51925 |
| | 0.4 | -1.59655 | -1.34755 | -0.98519 | -0.71797 | -0.50338 |
| | 0.8 | -1.59383 | -1.33950 | -0.97104 | -0.70090 | -0.48518 |
| | 1.2 | -1.58485 | -1.32567 | -0.95229 | -0.68056 | -0.46524 |
| | 1.6 | -1.56625 | -1.30372 | -0.92837 | -0.65826 | -0.44717 |
| | 2.0 | -1.53463 | -1.27229 | -0.90194 | -0.64107 | -0.44182 |
| | 2.4 | -1.49030 | -1.23569 | -0.88437 | -0.64297 | -0.45936 |
| | 2.8 | -1.44473 | -1.20936 | -0.88945 | -0.66795 | -0.49530 |
| | 3.2 | -1.41731 | -1.20554 | -0.91306 | -0.70401 | -0.53642 |
| | 3.6 | -1.41044 | -1.21530 | -0.93887 | -0.73576 | -0.56978 |
| | 4.0 | -1.41142 | -1.22388 | -0.95420 | -0.75325 | -0.58754 |
| 6.0 | 0.0 | -1.56259 | -1.34436 | -1.02080 | -0.77666 | -0.57638 |
| | 0.4 | -1.56542 | -1.34447 | -1.01756 | -0.77149 | -0.57007 |
| | 0.8 | -1.56755 | -1.34382 | -1.01354 | -0.76561 | -0.56314 |
| | 1.2 | -1.56857 | -1.34208 | -1.00854 | -0.75890 | -0.55558 |
| | 1.6 | -1.56792 | -1.33878 | -1.00227 | -0.75125 | -0.54743 |
| | 2.0 | -1.56479 | -1.33331 | -0.99443 | -0.74260 | -0.53884 |
| | 2.4 | -1.55809 | -1.32487 | -0.98469 | -0.73304 | -0.53033 |
| | 2.8 | -1.54641 | -1.31251 | -0.97289 | -0.72322 | -0.52340 |
| | 3.2 | -1.52820 | -1.29547 | -0.95985 | -0.71547 | -0.52142 |
| | 3.6 | -1.50263 | -1.27452 | -0.94898 | -0.71423 | -0.52825 |
| | 4.0 | -1.47233 | -1.25439 | -0.94580 | -0.72308 | -0.54502 |
| | 6.0 | -1.42486 | -1.25734 | -1.00465 | -0.80843 | -0.64245 |
| 12.0 | 0.0 | -1.51068 | -1.32131 | -1.03505 | -0.81406 | -0.62908 |
| | 0.4 | -1.51198 | -1.32188 | -1.03468 | -0.81314 | -0.62779 |
| | 0.8 | -1.51325 | -1.32241 | -1.03427 | -0.81215 | -0.62644 |
| | 1.2 | -1.51448 | -1.32289 | -1.03380 | -0.81112 | -0.62505 |
| | 1.6 | -1.51564 | -1.32329 | -1.03326 | -0.81003 | -0.62361 |
| | 2.0 | -1.51671 | -1.32361 | -1.03265 | -0.80887 | -0.62213 |
| | 2.4 | -1.51765 | -1.32381 | -1.03195 | -0.80766 | -0.62061 |
| | 2.8 | -1.51843 | -1.32388 | -1.03115 | -0.80638 | -0.61907 |
| | 3.2 | -1.51900 | -1.32377 | -1.03024 | -0.80504 | -0.61751 |
| | 3.6 | -1.51931 | -1.32345 | -1.02920 | -0.80365 | -0.61595 |
| | 4.0 | -1.51929 | -1.32287 | -1.02802 | -0.80220 | -0.61441 |
| | 6.0 | -1.51101 | -1.31404 | -1.01957 | -0.79501 | -0.60885 |
| | 12.0 | -1.43318 | -1.27595 | -1.03239 | -0.83901 | -0.67313 |
| INFINITY | | -1.43953 | -1.28155 | -1.03643 | -0.84162 | -0.67449 |

FRACTILES $W_\alpha$ RHO = -0.90

| DELTAS | | $\alpha$:0.3000 | 0.3500 | 0.4000 | 0.4500 | 0.5000 |
|---|---|---|---|---|---|---|
| 0.0 | 0.0 | -0.05412 | 0.08746 | 0.20387 | 0.30084 | 0.38224 |
| 0.4 | 0.0 | -0.05679 | 0.08446 | 0.20053 | 0.29713 | 0.37811 |
| | 0.4 | -0.06424 | 0.07512 | 0.18933 | 0.28398 | 0.36280 |
| 0.8 | 0.0 | -0.06744 | 0.07451 | 0.19114 | 0.28810 | 0.36914 |
| | 0.4 | -0.07549 | 0.06160 | 0.17342 | 0.26537 | 0.34075 |
| | 0.8 | -0.09451 | 0.03779 | 0.14422 | 0.22837 | 0.29666 |
| 1.2 | 0.0 | -0.09071 | 0.05507 | 0.17541 | 0.27573 | 0.35960 |
| | 0.4 | -0.08991 | 0.04680 | 0.15787 | 0.24811 | 0.31955 |
| | 0.8 | -0.11309 | 0.01401 | 0.11406 | 0.19611 | 0.26746 |
| | 1.2 | -0.14746 | -0.02888 | 0.06843 | 0.15283 | 0.22949 |
| 1.6 | 0.0 | -0.12753 | 0.02432 | 0.15128 | 0.25837 | 0.34877 |
| | 0.4 | -0.11208 | 0.02836 | 0.14275 | 0.23515 | 0.31055 |
| | 0.8 | -0.12822 | -0.00264 | 0.09785 | 0.18231 | 0.25700 |
| | 1.2 | -0.17281 | -0.05590 | 0.04269 | 0.13005 | 0.21058 |
| | 1.6 | -0.21567 | -0.09944 | 0.00174 | 0.09356 | 0.17971 |
| 2.0 | 0.0 | -0.17259 | -0.01538 | 0.11852 | 0.23372 | 0.33304 |
| | 0.4 | -0.14506 | 0.00216 | 0.12389 | 0.22416 | 0.30735 |
| | 0.8 | -0.14507 | -0.01454 | 0.09095 | 0.17990 | 0.25844 |
| | 1.2 | -0.18353 | -0.06478 | 0.03600 | 0.12558 | 0.20824 |
| | 1.6 | -0.23697 | -0.11970 | -0.01644 | 0.07804 | 0.16717 |
| | 2.0 | -0.27687 | -0.15793 | -0.05134 | 0.04748 | 0.14164 |
| 2.4 | 0.0 | -0.21781 | -0.05775 | 0.08104 | 0.20294 | 0.31068 |
| | 0.4 | -0.18595 | -0.03228 | 0.09778 | 0.20809 | 0.30153 |
| | 0.8 | -0.17006 | -0.03061 | 0.08345 | 0.17951 | 0.26359 |
| | 1.2 | -0.19021 | -0.06625 | 0.03845 | 0.13090 | 0.21565 |
| | 1.6 | -0.23996 | -0.12089 | -0.01615 | 0.07955 | 0.16966 |
| | 2.0 | -0.29231 | -0.17207 | -0.06373 | 0.03709 | 0.13339 |
| | 2.4 | -0.32693 | -0.20458 | -0.09308 | 0.01155 | 0.11214 |
| 2.8 | 0.0 | -0.25765 | -0.09693 | 0.04434 | 0.17045 | 0.28420 |
| | 0.4 | -0.22788 | -0.07040 | 0.06579 | 0.18467 | 0.28848 |
| | 0.8 | -0.20377 | -0.05544 | 0.06868 | 0.17408 | 0.26587 |
| | 1.2 | -0.20339 | -0.07097 | 0.04028 | 0.13752 | 0.22567 |
| | 1.6 | -0.23754 | -0.11470 | -0.00752 | 0.08965 | 0.18052 |
| | 2.0 | -0.28945 | -0.16803 | -0.05894 | 0.04230 | 0.13877 |
| | 2.4 | -0.33751 | -0.21406 | -0.10127 | 0.00478 | 0.10687 |
| | 2.8 | -0.36681 | -0.24134 | -0.12581 | -0.01655 | 0.08913 |
| 3.2 | 0.0 | -0.29054 | -0.13030 | 0.01189 | 0.14028 | 0.25772 |
| | 0.4 | -0.26542 | -0.10661 | 0.03289 | 0.15717 | 0.26874 |
| | 0.8 | -0.24065 | -0.08663 | 0.04573 | 0.16053 | 0.26133 |
| | 1.2 | -0.22664 | -0.08457 | 0.03528 | 0.13945 | 0.23291 |
| | 1.6 | -0.24050 | -0.11118 | 0.00051 | 0.10067 | 0.19341 |
| | 2.0 | -0.28084 | -0.15693 | -0.04653 | 0.05520 | 0.15153 |
| | 2.4 | -0.33088 | -0.20678 | -0.09378 | 0.01216 | 0.11388 |
| | 2.8 | -0.37374 | -0.24745 | -0.13100 | -0.02076 | 0.08593 |
| | 3.2 | -0.39828 | -0.27024 | -0.15149 | -0.03857 | 0.07108 |

FRACTILES $W_\alpha$ RHO = −0.90

| DELTAS | | α:0.3000 | 0.3500 | 0.4000 | 0.4500 | 0.5000 |
|---|---|---|---|---|---|---|
| 3.6 | 0.0 | -0.31715 | -0.15783 | -0.01545 | 0.11414 | 0.23381 |
| | 0.4 | -0.29673 | -0.13795 | 0.00296 | 0.13016 | 0.24640 |
| | 0.8 | -0.27494 | -0.11847 | 0.01869 | 0.14042 | 0.24937 |
| | 1.2 | -0.25625 | -0.10666 | 0.02164 | 0.13380 | 0.23411 |
| | 1.6 | -0.25342 | -0.11583 | 0.00240 | 0.10747 | 0.20375 |
| | 2.0 | -0.27655 | -0.14795 | -0.03456 | 0.06890 | 0.16603 |
| | 2.4 | -0.31850 | -0.19284 | -0.07927 | 0.02651 | 0.12752 |
| | 2.8 | -0.36496 | -0.23838 | -0.12206 | -0.01225 | 0.09375 |
| | 3.2 | -0.40260 | -0.27396 | -0.15457 | -0.04101 | 0.06931 |
| | 3.6 | -0.42307 | -0.29296 | -0.17168 | -0.05591 | 0.05685 |
| 4.0 | 0.0 | -0.33876 | -0.18046 | -0.03820 | 0.09206 | 0.21319 |
| | 0.4 | -0.32220 | -0.16404 | -0.02264 | 0.10611 | 0.22503 |
| | 0.8 | -0.30410 | -0.14705 | -0.00768 | 0.11801 | 0.23265 |
| | 1.2 | -0.28612 | -0.13253 | 0.00179 | 0.12098 | 0.22836 |
| | 1.6 | -0.27463 | -0.12934 | -0.00382 | 0.10748 | 0.20878 |
| | 2.0 | -0.28131 | -0.14616 | -0.02795 | 0.07889 | 0.17825 |
| | 2.4 | -0.30942 | -0.18041 | -0.06494 | 0.04169 | 0.14276 |
| | 2.8 | -0.35053 | -0.22302 | -0.10663 | 0.00263 | 0.10758 |
| | 3.2 | -0.39276 | -0.26407 | -0.14501 | -0.03206 | 0.07741 |
| | 3.6 | -0.42555 | -0.29502 | -0.17331 | -0.05712 | 0.05606 |
| | 4.0 | -0.44263 | -0.31089 | -0.18763 | -0.06963 | 0.04556 |
| 6.0 | 0.0 | -0.40372 | -0.24979 | -0.10910 | 0.02205 | 0.14637 |
| | 0.4 | -0.39678 | -0.24260 | -0.10196 | 0.02888 | 0.15264 |
| | 0.8 | -0.38935 | -0.23506 | -0.09464 | 0.03571 | 0.15872 |
| | 1.2 | -0.38150 | -0.22734 | -0.08738 | 0.04221 | 0.16419 |
| | 1.6 | -0.37341 | -0.21974 | -0.08063 | 0.04778 | 0.16826 |
| | 2.0 | -0.36547 | -0.21291 | -0.07532 | 0.05119 | 0.16943 |
| | 2.4 | -0.35865 | -0.20832 | -0.07337 | 0.05024 | 0.16551 |
| | 2.8 | -0.35525 | -0.20874 | -0.07757 | 0.04260 | 0.15503 |
| | 3.2 | -0.35882 | -0.21715 | -0.08986 | 0.02751 | 0.13822 |
| | 3.6 | -0.37180 | -0.23445 | -0.10987 | 0.00617 | 0.11673 |
| | 4.0 | -0.39347 | -0.25881 | -0.13526 | -0.01898 | 0.09285 |
| | 6.0 | -0.49478 | -0.35883 | -0.23043 | -0.10663 | 0.01492 |
| 12.0 | 0.0 | -0.46657 | -0.31902 | -0.18172 | -0.05136 | 0.07453 |
| | 0.4 | -0.46504 | -0.31735 | -0.17998 | -0.04962 | 0.07622 |
| | 0.8 | -0.46347 | -0.31564 | -0.17821 | -0.04786 | 0.07792 |
| | 1.2 | -0.46185 | -0.31390 | -0.17642 | -0.04609 | 0.07962 |
| | 1.6 | -0.46020 | -0.31214 | -0.17463 | -0.04432 | 0.08129 |
| | 2.0 | -0.45853 | -0.31038 | -0.17284 | -0.04259 | 0.08292 |
| | 2.4 | -0.45684 | -0.30862 | -0.17109 | -0.04091 | 0.08448 |
| | 2.8 | -0.45516 | -0.30690 | -0.16940 | -0.03931 | 0.08592 |
| | 3.2 | -0.45351 | -0.30524 | -0.16781 | -0.03784 | 0.08721 |
| | 3.6 | -0.45191 | -0.30368 | -0.16636 | -0.03655 | 0.08830 |
| | 4.0 | -0.45040 | -0.30227 | -0.16511 | -0.03551 | 0.08910 |
| | 6.0 | -0.44664 | -0.30041 | -0.16517 | -0.03747 | 0.08529 |
| | 12.0 | -0.52414 | -0.38603 | -0.25489 | -0.12792 | -0.00285 |
| INFINITY | | -0.52440 | -0.38532 | -0.25335 | -0.12566 | 0.0 |

FRACTILES $W_\alpha$ RHO = -0.90

| DELTAS | | $\alpha$:0.5500 | 0.6000 | 0.6500 | 0.7000 | 0.7500 |
|---|---|---|---|---|---|---|
| 0.0 | 0.0 | 0.45078 | 0.50840 | 0.55651 | 0.59611 | 0.62787 |
| 0.4 | 0.0 | 0.44613 | 0.50310 | 0.55033 | 0.58872 | 0.61861 |
| | 0.4 | 0.42827 | 0.48196 | 0.52439 | 0.55366 | 0.58534 |
| 0.8 | 0.0 | 0.43681 | 0.49281 | 0.53805 | 0.57178 | 0.59835 |
| | 0.4 | 0.40078 | 0.44740 | 0.49135 | 0.53517 | 0.58112 |
| | 0.8 | 0.35660 | 0.41251 | 0.46714 | 0.52267 | 0.58131 |
| 1.2 | 0.0 | 0.42926 | 0.48563 | 0.52864 | 0.56839 | 0.60836 |
| | 0.4 | 0.37911 | 0.43269 | 0.48377 | 0.53482 | 0.58808 |
| | 0.8 | 0.33279 | 0.39514 | 0.45681 | 0.51984 | 0.58651 |
| | 1.2 | 0.30180 | 0.37220 | 0.44273 | 0.51544 | 0.59277 |
| 1.6 | 0.0 | 0.42414 | 0.48470 | 0.53621 | 0.58380 | 0.63087 |
| | 0.4 | 0.37529 | 0.43408 | 0.49012 | 0.54583 | 0.60352 |
| | 0.8 | 0.32600 | 0.39212 | 0.45755 | 0.52434 | 0.59482 |
| | 1.2 | 0.28726 | 0.36238 | 0.43793 | 0.51597 | 0.59905 |
| | 1.6 | 0.26280 | 0.34496 | 0.42816 | 0.51455 | 0.60683 |
| 2.0 | 0.0 | 0.41769 | 0.48838 | 0.54910 | 0.60449 | 0.65825 |
| | 0.4 | 0.37890 | 0.44340 | 0.50424 | 0.56406 | 0.62532 |
| | 0.8 | 0.33067 | 0.39952 | 0.46727 | 0.53604 | 0.60817 |
| | 1.2 | 0.28695 | 0.36397 | 0.44131 | 0.52104 | 0.60572 |
| | 1.6 | 0.25347 | 0.33902 | 0.42579 | 0.51597 | 0.61232 |
| | 2.0 | 0.23349 | 0.32506 | 0.41836 | 0.51565 | 0.61987 |
| 2.4 | 0.0 | 0.40563 | 0.48808 | 0.55973 | 0.62428 | 0.68562 |
| | 0.4 | 0.38203 | 0.45382 | 0.52053 | 0.58512 | 0.65032 |
| | 0.8 | 0.34009 | 0.41220 | 0.48245 | 0.55307 | 0.62652 |
| | 1.2 | 0.29581 | 0.37379 | 0.45166 | 0.53154 | 0.61595 |
| | 1.6 | 0.25674 | 0.34290 | 0.43011 | 0.52055 | 0.61698 |
| | 2.0 | 0.22751 | 0.32145 | 0.41723 | 0.51714 | 0.62418 |
| | 2.4 | 0.21095 | 0.30996 | 0.41123 | 0.51714 | 0.63082 |
| 2.8 | 0.0 | 0.38730 | 0.48036 | 0.56366 | 0.63891 | 0.70935 |
| | 0.4 | 0.37934 | 0.46021 | 0.53442 | 0.60512 | 0.67531 |
| | 0.8 | 0.34844 | 0.42526 | 0.49910 | 0.57244 | 0.64787 |
| | 1.2 | 0.30818 | 0.38769 | 0.46640 | 0.54652 | 0.63057 |
| | 1.6 | 0.26781 | 0.35369 | 0.44020 | 0.52952 | 0.62435 |
| | 2.0 | 0.23283 | 0.32651 | 0.42183 | 0.52106 | 0.62717 |
| | 2.4 | 0.20723 | 0.30785 | 0.41080 | 0.51846 | 0.63403 |
| | 2.8 | 0.19340 | 0.29825 | 0.40577 | 0.51844 | 0.63958 |
| 3.2 | 0.0 | 0.36615 | 0.46676 | 0.55997 | 0.64615 | 0.72694 |
| | 0.4 | 0.36907 | 0.45967 | 0.54275 | 0.62104 | 0.69760 |
| | 0.8 | 0.35167 | 0.43485 | 0.51378 | 0.59115 | 0.66969 |
| | 1.2 | 0.31938 | 0.40173 | 0.48240 | 0.56370 | 0.64824 |
| | 1.6 | 0.28169 | 0.36785 | 0.45402 | 0.54239 | 0.63567 |
| | 2.0 | 0.24495 | 0.33755 | 0.43136 | 0.52864 | 0.63228 |
| | 2.4 | 0.21365 | 0.31348 | 0.41540 | 0.52181 | 0.63582 |
| | 2.8 | 0.19122 | 0.29714 | 0.40576 | 0.51957 | 0.64191 |
| | 3.2 | 0.17961 | 0.28903 | 0.40145 | 0.51942 | 0.64640 |

FRACTILES $W_\alpha$ RHO = −0.90

| DELTAS | | $\alpha$: 0.5500 | 0.6000 | 0.6500 | 0.7000 | 0.7500 |
|---|---|---|---|---|---|---|
| 3.6 | 0.0 | 0.34563 | 0.45109 | 0.55119 | 0.64641 | 0.73741 |
| | 0.4 | 0.35340 | 0.45229 | 0.54425 | 0.63104 | 0.71522 |
| | 0.8 | 0.34794 | 0.43862 | 0.52402 | 0.60679 | 0.68976 |
| | 1.2 | 0.32614 | 0.41288 | 0.49690 | 0.58067 | 0.66688 |
| | 1.6 | 0.29444 | 0.38211 | 0.46901 | 0.55743 | 0.65004 |
| | 2.0 | 0.25948 | 0.35149 | 0.44413 | 0.53965 | 0.64091 |
| | 2.4 | 0.22612 | 0.32436 | 0.42429 | 0.52825 | 0.63928 |
| | 2.8 | 0.19815 | 0.30296 | 0.41025 | 0.52247 | 0.64291 |
| | 3.2 | 0.17851 | 0.28860 | 0.40170 | 0.52038 | 0.64809 |
| | 3.6 | 0.16872 | 0.28170 | 0.39795 | 0.52009 | 0.65166 |
| 4.0 | 0.0 | 0.32729 | 0.43601 | 0.54067 | 0.64230 | 0.74172 |
| | 0.4 | 0.33613 | 0.44077 | 0.53998 | 0.63487 | 0.72718 |
| | 0.8 | 0.33811 | 0.43612 | 0.52860 | 0.61778 | 0.70636 |
| | 1.2 | 0.32686 | 0.41918 | 0.50786 | 0.59540 | 0.68454 |
| | 1.6 | 0.30337 | 0.39396 | 0.48291 | 0.57260 | 0.66575 |
| | 2.0 | 0.27302 | 0.36555 | 0.45803 | 0.55274 | 0.65249 |
| | 2.4 | 0.24075 | 0.33782 | 0.43603 | 0.53771 | 0.64582 |
| | 2.8 | 0.21052 | 0.31346 | 0.41850 | 0.52803 | 0.64524 |
| | 3.2 | 0.18556 | 0.29439 | 0.40601 | 0.52296 | 0.64861 |
| | 3.6 | 0.16834 | 0.28173 | 0.39838 | 0.52092 | 0.65289 |
| | 4.0 | 0.16004 | 0.27583 | 0.39510 | 0.52051 | 0.65570 |
| 6.0 | 0.0 | 0.26598 | 0.38273 | 0.49836 | 0.61469 | 0.73393 |
| | 0.4 | 0.27146 | 0.38718 | 0.50148 | 0.61618 | 0.73338 |
| | 0.8 | 0.27654 | 0.39098 | 0.50372 | 0.61649 | 0.73133 |
| | 1.2 | 0.28069 | 0.39351 | 0.50429 | 0.61472 | 0.72687 |
| | 1.6 | 0.28294 | 0.39360 | 0.50193 | 0.60976 | 0.71937 |
| | 2.0 | 0.28159 | 0.38963 | 0.49544 | 0.60110 | 0.70921 |
| | 2.4 | 0.27490 | 0.38058 | 0.48466 | 0.58942 | 0.69764 |
| | 2.8 | 0.26233 | 0.36679 | 0.47059 | 0.57609 | 0.68619 |
| | 3.2 | 0.24485 | 0.34965 | 0.45478 | 0.56263 | 0.67622 |
| | 3.6 | 0.22425 | 0.33087 | 0.43875 | 0.55032 | 0.66870 |
| | 4.0 | 0.20253 | 0.31213 | 0.42380 | 0.54001 | 0.66407 |
| | 6.0 | 0.13627 | 0.25944 | 0.38670 | 0.52082 | 0.66565 |
| 12.0 | 0.0 | 0.19807 | 0.32118 | 0.44587 | 0.57449 | 0.71010 |
| | 0.4 | 0.19965 | 0.32259 | 0.44706 | 0.57538 | 0.71059 |
| | 0.8 | 0.20123 | 0.32400 | 0.44822 | 0.57623 | 0.71104 |
| | 1.2 | 0.20279 | 0.32537 | 0.44935 | 0.57703 | 0.71142 |
| | 1.6 | 0.20432 | 0.32669 | 0.45040 | 0.57775 | 0.71172 |
| | 2.0 | 0.20578 | 0.32794 | 0.45137 | 0.57837 | 0.71190 |
| | 2.4 | 0.20716 | 0.32908 | 0.45222 | 0.57885 | 0.71195 |
| | 2.8 | 0.20841 | 0.33007 | 0.45290 | 0.57917 | 0.71183 |
| | 3.2 | 0.20947 | 0.33087 | 0.45338 | 0.57928 | 0.71151 |
| | 3.6 | 0.21031 | 0.33141 | 0.45359 | 0.57911 | 0.71093 |
| | 4.0 | 0.21083 | 0.33162 | 0.45346 | 0.57863 | 0.71006 |
| | 6.0 | 0.20527 | 0.32447 | 0.44494 | 0.56906 | 0.69994 |
| | 12.0 | 0.12232 | 0.24965 | 0.38141 | 0.52045 | 0.67072 |
| INFINITY | | 0.12566 | 0.25335 | 0.38532 | 0.52440 | 0.67449 |

FRACTILES $W_\alpha$ RHO = -0.90

| DELTAS | | $\alpha$: 0.8000 | 0.8500 | 0.9000 | 0.9250 | 0.9500 |
|---|---|---|---|---|---|---|
| 0.0 | 0.0 | 0.65205 | 0.66787 | 0.68212 | 0.69581 | 0.71730 |
| 0.4 | 0.0 | 0.63917 | 0.65611 | 0.68800 | 0.71225 | 0.74723 |
| | 0.4 | 0.62240 | 0.66727 | 0.72648 | 0.76642 | 0.82054 |
| 0.8 | 0.0 | 0.63111 | 0.67067 | 0.72265 | 0.75760 | 0.80489 |
| | 0.4 | 0.63175 | 0.69100 | 0.76692 | 0.81703 | 0.88384 |
| | 0.8 | 0.64595 | 0.72130 | 0.81715 | 0.87995 | 0.96309 |
| 1.2 | 0.0 | 0.65122 | 0.70045 | 0.76264 | 0.80335 | 0.85734 |
| | 0.4 | 0.64625 | 0.71356 | 0.79868 | 0.85423 | 0.92760 |
| | 0.8 | 0.65992 | 0.74522 | 0.85318 | 0.92353 | 1.01623 |
| | 1.2 | 0.67818 | 0.77753 | 0.90319 | 0.98494 | 1.09244 |
| 1.6 | 0.0 | 0.68038 | 0.73609 | 0.80499 | 0.84935 | 0.90744 |
| | 0.4 | 0.66600 | 0.73761 | 0.82720 | 0.88514 | 0.96110 |
| | 0.8 | 0.67215 | 0.76162 | 0.87420 | 0.94720 | 1.04298 |
| | 1.2 | 0.69078 | 0.79737 | 0.93189 | 1.01920 | 1.13374 |
| | 1.6 | 0.70898 | 0.82785 | 0.97794 | 1.07534 | 1.20301 |
| 2.0 | 0.0 | 0.71370 | 0.77492 | 0.84923 | 0.89639 | 0.95746 |
| | 0.4 | 0.69096 | 0.76538 | 0.85745 | 0.91648 | 0.99333 |
| | 0.8 | 0.68686 | 0.77734 | 0.89044 | 0.96337 | 1.05862 |
| | 1.2 | 0.69897 | 0.80700 | 0.94282 | 1.03071 | 1.14568 |
| | 1.6 | 0.71896 | 0.84297 | 0.99935 | 1.10068 | 1.23333 |
| | 2.0 | 0.73545 | 0.87003 | 1.03987 | 1.14994 | 1.29399 |
| 2.4 | 0.0 | 0.74752 | 0.81445 | 0.89407 | 0.94387 | 1.00766 |
| | 0.4 | 0.71923 | 0.79639 | 0.89065 | 0.95050 | 1.02786 |
| | 0.8 | 0.70599 | 0.79663 | 0.90902 | 0.98104 | 1.07463 |
| | 1.2 | 0.70847 | 0.81513 | 0.94856 | 1.03454 | 1.14664 |
| | 1.6 | 0.72347 | 0.84700 | 1.00233 | 1.10274 | 1.23391 |
| | 2.0 | 0.74285 | 0.88096 | 1.05508 | 1.16782 | 1.31522 |
| | 2.4 | 0.75705 | 0.90414 | 1.08973 | 1.20992 | 1.36707 |
| 2.8 | 0.0 | 0.77891 | 0.85244 | 0.93803 | 0.99070 | 1.05742 |
| | 0.4 | 0.74834 | 0.82890 | 0.92590 | 0.98683 | 1.06498 |
| | 0.8 | 0.72863 | 0.81984 | 0.93185 | 1.00310 | 1.09516 |
| | 1.2 | 0.72209 | 0.82693 | 0.95721 | 1.04074 | 1.14921 |
| | 1.6 | 0.72865 | 0.84915 | 1.00005 | 1.09726 | 1.22391 |
| | 2.0 | 0.74457 | 0.88091 | 1.05242 | 1.16324 | 1.30790 |
| | 2.4 | 0.76231 | 0.91172 | 1.10010 | 1.22201 | 1.38128 |
| | 2.8 | 0.77421 | 0.93117 | 1.12922 | 1.25743 | 1.42495 |
| 3.2 | 0.0 | 0.80559 | 0.88703 | 0.97969 | 1.03575 | 1.10588 |
| | 0.4 | 0.77595 | 0.86097 | 0.96171 | 1.02425 | 1.10374 |
| | 0.8 | 0.75278 | 0.84554 | 0.95816 | 1.02919 | 1.12039 |
| | 1.2 | 0.73950 | 0.84321 | 0.97107 | 1.05254 | 1.15786 |
| | 1.6 | 0.73768 | 0.85490 | 1.00087 | 1.09451 | 1.21610 |
| | 2.0 | 0.74655 | 0.87880 | 1.04455 | 1.15134 | 1.29043 |
| | 2.4 | 0.76215 | 0.90901 | 1.09380 | 1.21319 | 1.36896 |
| | 2.8 | 0.77784 | 0.93625 | 1.13600 | 1.26523 | 1.43398 |
| | 3.2 | 0.78764 | 0.95237 | 1.16024 | 1.29478 | 1.47047 |

FRACTILES $W_\alpha$ RHO = -0.90

| DELTAS | | $\alpha$: 0.8000 | 0.8500 | 0.9000 | 0.9250 | 0.9500 |
|---|---|---|---|---|---|---|
| 3.6 | 0.0 | 0.82613 | 0.91681 | 1.01788 | 1.07797 | 1.15219 |
| | 0.4 | 0.80019 | 0.89094 | 0.99665 | 1.06142 | 1.14297 |
| | 0.8 | 0.77640 | 0.87191 | 0.98640 | 1.05792 | 1.14910 |
| | 1.2 | 0.75904 | 0.86279 | 0.98951 | 1.06968 | 1.17277 |
| | 1.6 | 0.75061 | 0.86540 | 1.00738 | 1.09799 | 1.21518 |
| | 2.0 | 0.75200 | 0.87997 | 1.03960 | 1.14206 | 1.27513 |
| | 2.4 | 0.76193 | 0.90408 | 1.08238 | 1.19728 | 1.34690 |
| | 2.8 | 0.77651 | 0.93195 | 1.12761 | 1.25401 | 1.41887 |
| | 3.2 | 0.79009 | 0.95565 | 1.16445 | 1.29951 | 1.47581 |
| | 3.6 | 0.79810 | 0.96895 | 1.18456 | 1.32407 | 1.50622 |
| 4.0 | 0.0 | 0.84018 | 0.94088 | 1.05166 | 1.11650 | 1.19557 |
| | 0.4 | 0.81979 | 0.91748 | 1.02946 | 1.09716 | 1.18155 |
| | 0.8 | 0.79780 | 0.89732 | 1.01502 | 1.08778 | 1.17984 |
| | 1.2 | 0.77886 | 0.88395 | 1.01097 | 1.09069 | 1.19259 |
| | 1.6 | 0.76609 | 0.87972 | 1.01915 | 1.10759 | 1.22146 |
| | 2.0 | 0.76127 | 0.88586 | 1.04036 | 1.13908 | 1.26684 |
| | 2.4 | 0.76474 | 0.90199 | 1.07344 | 1.18356 | 1.32658 |
| | 2.8 | 0.77490 | 0.92535 | 1.11420 | 1.23592 | 1.39441 |
| | 3.2 | 0.78813 | 0.95055 | 1.15506 | 1.28718 | 1.45945 |
| | 3.6 | 0.79972 | 0.97097 | 1.18697 | 1.32667 | 1.50899 |
| | 4.0 | 0.80624 | 0.98192 | 1.20364 | 1.34709 | 1.53432 |
| 6.0 | 0.0 | 0.85910 | 0.99503 | 1.15117 | 1.24338 | 1.35410 |
| | 0.4 | 0.85598 | 0.98859 | 1.14046 | 1.23027 | 1.33882 |
| | 0.8 | 0.85104 | 0.98025 | 1.12864 | 1.21712 | 1.32529 |
| | 1.2 | 0.84367 | 0.97018 | 1.11696 | 1.20561 | 1.31532 |
| | 1.6 | 0.83409 | 0.95951 | 1.10711 | 1.19749 | 1.31062 |
| | 2.0 | 0.82343 | 0.94983 | 1.10079 | 1.19438 | 1.31266 |
| | 2.4 | 0.81325 | 0.94275 | 1.09945 | 1.19762 | 1.32267 |
| | 2.8 | 0.80504 | 0.93958 | 1.10419 | 1.20818 | 1.34148 |
| | 3.2 | 0.79993 | 0.94121 | 1.11558 | 1.22648 | 1.36935 |
| | 3.6 | 0.79856 | 0.94789 | 1.13349 | 1.25214 | 1.40558 |
| | 4.0 | 0.80091 | 0.95911 | 1.15678 | 1.28365 | 1.44822 |
| | 6.0 | 0.82712 | 1.01572 | 1.25376 | 1.40770 | 1.60846 |
| 12.0 | 0.0 | 0.85721 | 1.02356 | 1.22508 | 1.35042 | 1.50806 |
| | 0.4 | 0.85720 | 1.02285 | 1.22337 | 1.34800 | 1.50465 |
| | 0.8 | 0.85712 | 1.02207 | 1.22159 | 1.34551 | 1.50119 |
| | 1.2 | 0.85697 | 1.02121 | 1.21974 | 1.34297 | 1.49772 |
| | 1.6 | 0.85672 | 1.02027 | 1.21782 | 1.34039 | 1.49425 |
| | 2.0 | 0.85637 | 1.01922 | 1.21583 | 1.33778 | 1.49085 |
| | 2.4 | 0.85588 | 1.01806 | 1.21379 | 1.33518 | 1.48756 |
| | 2.8 | 0.85524 | 1.01678 | 1.21172 | 1.33263 | 1.48446 |
| | 3.2 | 0.85441 | 1.01538 | 1.20964 | 1.33018 | 1.48166 |
| | 3.6 | 0.85338 | 1.01384 | 1.20760 | 1.32793 | 1.47930 |
| | 4.0 | 0.85211 | 1.01219 | 1.20567 | 1.32596 | 1.47752 |
| | 6.0 | 0.84223 | 1.00390 | 1.20163 | 1.32612 | 1.48487 |
| | 12.0 | 0.83834 | 1.03414 | 1.28115 | 1.44076 | 1.64870 |
| INFINITY | | 0.84162 | 1.03643 | 1.28155 | 1.43953 | 1.64485 |

FRACTILES $W_\alpha$ RHO = -0.90

| DELTAS | | $\alpha$: 0.9625 | 0.9750 | 0.9825 | 0.9900 | 0.9925 |
|---|---|---|---|---|---|---|
| 0.0 | 0.0 | 0.73356 | 0.75747 | 0.77922 | 0.81435 | 0.83279 |
| 0.4 | 0.0 | 0.77229 | 0.80769 | 0.83881 | 0.88749 | 0.91243 |
| | 0.4 | 0.85771 | 0.90872 | 0.95247 | 1.01940 | 1.05311 |
| 0.8 | 0.0 | 0.83733 | 0.88179 | 0.91990 | 0.97815 | 1.00747 |
| | 0.4 | 0.92911 | 0.99054 | 1.04267 | 1.12158 | 1.16099 |
| | 0.8 | 1.01908 | 1.09462 | 1.15837 | 1.25429 | 1.30195 |
| 1.2 | 0.0 | 0.89378 | 0.94310 | 0.98487 | 1.04796 | 1.07942 |
| | 0.4 | 0.97691 | 1.04335 | 1.09936 | 1.18355 | 1.22534 |
| | 0.8 | 1.07838 | 1.16189 | 1.23208 | 1.33722 | 1.38928 |
| | 1.2 | 1.16435 | 1.26077 | 1.34162 | 1.46242 | 1.52208 |
| 1.6 | 0.0 | 0.94623 | 0.99827 | 1.04197 | 1.10744 | 1.13986 |
| | 0.4 | 1.01183 | 1.07981 | 1.13682 | 1.22202 | 1.26414 |
| | 0.8 | 1.10693 | 1.19257 | 1.26431 | 1.37138 | 1.42421 |
| | 1.2 | 1.21019 | 1.31248 | 1.39807 | 1.52564 | 1.58851 |
| | 1.6 | 1.28814 | 1.40193 | 1.49704 | 1.63859 | 1.70826 |
| 2.0 | 0.0 | 0.99788 | 1.05171 | 1.09661 | 1.16338 | 1.19628 |
| | 0.4 | 1.04434 | 1.11238 | 1.16916 | 1.25361 | 1.29519 |
| | 0.8 | 1.12198 | 1.20653 | 1.27713 | 1.38212 | 1.43379 |
| | 1.2 | 1.22222 | 1.32441 | 1.40973 | 1.53658 | 1.59898 |
| | 1.6 | 1.32166 | 1.43957 | 1.53798 | 1.68423 | 1.75612 |
| | 2.0 | 1.38987 | 1.51779 | 1.62450 | 1.78294 | 1.86077 |
| 2.4 | 0.0 | 1.04952 | 1.10487 | 1.15074 | 1.21853 | 1.25175 |
| | 0.4 | 1.07891 | 1.14665 | 1.20293 | 1.28625 | 1.32711 |
| | 0.8 | 1.13662 | 1.21906 | 1.28766 | 1.38934 | 1.43924 |
| | 1.2 | 1.22106 | 1.32017 | 1.40272 | 1.52515 | 1.58525 |
| | 1.6 | 1.32109 | 1.43728 | 1.53411 | 1.67775 | 1.74826 |
| | 2.0 | 1.41324 | 1.54389 | 1.65277 | 1.81426 | 1.89351 |
| | 2.4 | 1.47155 | 1.61078 | 1.72676 | 1.89871 | 1.98305 |
| 2.8 | 0.0 | 1.10081 | 1.15777 | 1.20467 | 1.27353 | 1.30711 |
| | 0.4 | 1.11621 | 1.18386 | 1.23978 | 1.32219 | 1.36245 |
| | 0.8 | 1.15586 | 1.23628 | 1.30296 | 1.40144 | 1.44963 |
| | 1.2 | 1.22097 | 1.31629 | 1.39547 | 1.51260 | 1.56997 |
| | 1.6 | 1.30790 | 1.41961 | 1.51253 | 1.65010 | 1.71752 |
| | 2.0 | 1.40394 | 1.53181 | 1.63824 | 1.79587 | 1.87315 |
| | 2.4 | 1.48709 | 1.62801 | 1.74531 | 1.91906 | 2.00423 |
| | 2.8 | 1.53624 | 1.68442 | 1.80775 | 1.99037 | 2.07986 |
| 3.2 | 0.0 | 1.15105 | 1.20990 | 1.25802 | 1.32823 | 1.36228 |
| | 0.4 | 1.15550 | 1.22346 | 1.27936 | 1.36131 | 1.40120 |
| | 0.8 | 1.18021 | 1.25914 | 1.32433 | 1.42025 | 1.46704 |
| | 1.2 | 1.22727 | 1.31917 | 1.39530 | 1.50757 | 1.56244 |
| | 1.6 | 1.29650 | 1.40321 | 1.49177 | 1.62259 | 1.68659 |
| | 2.0 | 1.38260 | 1.50510 | 1.60690 | 1.75745 | 1.83114 |
| | 2.4 | 1.47231 | 1.60981 | 1.72415 | 1.89333 | 1.97617 |
| | 2.8 | 1.54602 | 1.69513 | 1.81916 | 2.00269 | 2.09258 |
| | 3.2 | 1.58713 | 1.74237 | 1.87148 | 2.06250 | 2.15602 |

FRACTILES $W_\alpha$ RHO = -0.90

| DELTAS | | $\alpha$: 0.9625 | 0.9750 | 0.9825 | 0.9900 | 0.9925 |
|---|---|---|---|---|---|---|
| 3.6 | 0.0 | 1.19949 | 1.26063 | 1.31025 | 1.38217 | 1.41686 |
| | 0.4 | 1.19567 | 1.26444 | 1.32070 | 1.40276 | 1.44253 |
| | 0.8 | 1.20857 | 1.28667 | 1.35092 | 1.44505 | 1.49082 |
| | 1.2 | 1.24042 | 1.32968 | 1.40338 | 1.51172 | 1.56452 |
| | 1.6 | 1.29242 | 1.39465 | 1.47929 | 1.60400 | 1.66489 |
| | 2.0 | 1.36309 | 1.47978 | 1.57657 | 1.71942 | 1.78924 |
| | 2.4 | 1.44601 | 1.57767 | 1.68701 | 1.84856 | 1.92758 |
| | 2.8 | 1.52821 | 1.67359 | 1.79440 | 1.97303 | 2.06043 |
| | 3.2 | 1.59281 | 1.74844 | 1.87781 | 2.06912 | 2.16275 |
| | 3.6 | 1.62710 | 1.78789 | 1.92154 | 2.11916 | 2.21585 |
| 4.0 | 0.0 | 1.24543 | 1.30933 | 1.36080 | 1.43485 | 1.47037 |
| | 0.4 | 1.23564 | 1.30577 | 1.36280 | 1.44552 | 1.48544 |
| | 0.8 | 1.23949 | 1.31746 | 1.38129 | 1.47441 | 1.51953 |
| | 1.2 | 1.25914 | 1.34661 | 1.41856 | 1.52397 | 1.57521 |
| | 1.6 | 1.29623 | 1.39490 | 1.47635 | 1.59605 | 1.65436 |
| | 2.0 | 1.35106 | 1.46253 | 1.55477 | 1.69063 | 1.75692 |
| | 2.4 | 1.42112 | 1.54650 | 1.65045 | 1.80377 | 1.87867 |
| | 2.8 | 1.49936 | 1.63874 | 1.75443 | 1.92526 | 2.00878 |
| | 3.2 | 1.57367 | 1.72548 | 1.85159 | 2.03791 | 2.12903 |
| | 3.6 | 1.62993 | 1.79074 | 1.92437 | 2.12185 | 2.21844 |
| | 4.0 | 1.65854 | 1.82370 | 1.96094 | 2.16373 | 2.26292 |
| 6.0 | 0.0 | 1.42214 | 1.50686 | 1.57297 | 1.66486 | 1.70770 |
| | 0.4 | 1.40619 | 1.49098 | 1.55793 | 1.65214 | 1.69650 |
| | 0.8 | 1.39321 | 1.47963 | 1.54858 | 1.64663 | 1.69317 |
| | 1.2 | 1.38500 | 1.47448 | 1.54652 | 1.64979 | 1.69914 |
| | 1.6 | 1.38316 | 1.47705 | 1.55317 | 1.66303 | 1.71579 |
| | 2.0 | 1.38911 | 1.48868 | 1.56985 | 1.68763 | 1.74443 |
| | 2.4 | 1.40401 | 1.51047 | 1.59765 | 1.72469 | 1.78614 |
| | 2.8 | 1.42863 | 1.54313 | 1.63724 | 1.77483 | 1.84156 |
| | 3.2 | 1.46311 | 1.58670 | 1.68856 | 1.83787 | 1.91044 |
| | 3.6 | 1.50660 | 1.64008 | 1.75032 | 1.91227 | 1.99109 |
| | 4.0 | 1.55681 | 1.70057 | 1.81950 | 1.99449 | 2.07977 |
| | 6.0 | 1.74153 | 1.91826 | 2.06492 | 2.28130 | 2.38696 |
| 12.0 | 0.0 | 1.60895 | 1.73861 | 1.84250 | 1.98981 | 2.05922 |
| | 0.4 | 1.60486 | 1.73358 | 1.83669 | 1.98284 | 2.05169 |
| | 0.8 | 1.60074 | 1.72858 | 1.83095 | 1.97604 | 2.04440 |
| | 1.2 | 1.59664 | 1.72364 | 1.82535 | 1.96951 | 2.03745 |
| | 1.6 | 1.59260 | 1.71886 | 1.81999 | 1.96340 | 2.03102 |
| | 2.0 | 1.58868 | 1.71433 | 1.81500 | 1.95786 | 2.02530 |
| | 2.4 | 1.58498 | 1.71015 | 1.81052 | 1.95312 | 2.02051 |
| | 2.8 | 1.58159 | 1.70649 | 1.80674 | 1.94939 | 2.01691 |
| | 3.2 | 1.57866 | 1.70352 | 1.80389 | 1.94697 | 2.01482 |
| | 3.6 | 1.57635 | 1.70146 | 1.80220 | 1.94614 | 2.01454 |
| | 4.0 | 1.57485 | 1.70055 | 1.80197 | 1.94723 | 2.01641 |
| | 6.0 | 1.58804 | 1.72278 | 1.83277 | 1.99232 | 2.06915 |
| | 12.0 | 1.78635 | 1.96896 | 2.12027 | 2.34312 | 2.45177 |
| INFINITY | | 1.78046 | 1.95996 | 2.10836 | 2.32635 | 2.43238 |

FRACTILES $W_\alpha$ RHO = -0.90

| DELTAS | | α:0.9950 | 0.9975 | 0.9990 | 0.9995 |
|---|---|---|---|---|---|
| 0.0 | 0.0 | 0.85911 | 0.90487 | 0.96646 | 1.01369 |
| 0.4 | 0.0 | 0.94746 | 1.00703 | 1.08520 | 1.14392 |
| | 0.4 | 1.09993 | 1.17837 | 1.27952 | 1.35445 |
| 0.8 | 0.0 | 1.04819 | 1.11638 | 1.20427 | 1.26936 |
| | 0.4 | 1.21540 | 1.30578 | 1.42110 | 1.50579 |
| | 0.8 | 1.36751 | 1.47584 | 1.61314 | 1.71338 |
| 1.2 | 0.0 | 1.12282 | 1.19483 | 1.28660 | 1.35392 |
| | 0.4 | 1.28281 | 1.37771 | 1.49791 | 1.58561 |
| | 0.8 | 1.46068 | 1.57821 | 1.72640 | 1.83410 |
| | 1.2 | 1.60377 | 1.73789 | 1.90641 | 2.02849 |
| 1.6 | 0.0 | 1.18437 | 1.25773 | 1.35044 | 1.41797 |
| | 0.4 | 1.32185 | 1.41672 | 1.53616 | 1.62287 |
| | 0.8 | 1.49653 | 1.61517 | 1.76413 | 1.87198 |
| | 1.2 | 1.67447 | 1.81527 | 1.99164 | 2.11907 |
| | 1.6 | 1.80343 | 1.95906 | 2.15362 | 2.29392 |
| 2.0 | 0.0 | 1.24124 | 1.31495 | 1.40745 | 1.47445 |
| | 0.4 | 1.35200 | 1.44502 | 1.56153 | 1.64575 |
| | 0.8 | 1.50436 | 1.61979 | 1.76418 | 1.86838 |
| | 1.2 | 1.68417 | 1.82338 | 1.99729 | 2.12263 |
| | 1.6 | 1.85422 | 2.01439 | 2.21423 | 2.35806 |
| | 2.0 | 1.96689 | 2.14002 | 2.35571 | 2.51076 |
| 2.4 | 0.0 | 1.29700 | 1.37079 | 1.46285 | 1.52917 |
| | 0.4 | 1.38279 | 1.47362 | 1.58686 | 1.66839 |
| | 0.8 | 1.50725 | 1.61818 | 1.75644 | 1.85590 |
| | 1.2 | 1.66717 | 1.80077 | 1.96720 | 2.08686 |
| | 1.6 | 1.84436 | 2.00103 | 2.19609 | 2.33623 |
| | 2.0 | 2.00150 | 2.17747 | 2.39640 | 2.55356 |
| | 2.4 | 2.09792 | 2.28500 | 2.51751 | 2.68429 |
| 2.8 | 0.0 | 1.35269 | 1.42667 | 1.51842 | 1.58420 |
| | 0.4 | 1.41717 | 1.50610 | 1.61647 | 1.69564 |
| | 0.8 | 1.51517 | 1.62178 | 1.75418 | 1.84915 |
| | 1.2 | 1.64804 | 1.77509 | 1.93292 | 2.04612 |
| | 1.6 | 1.80930 | 1.95869 | 2.14428 | 2.27736 |
| | 2.0 | 1.97835 | 2.14957 | 2.36224 | 2.51470 |
| | 2.4 | 2.12017 | 2.30882 | 2.54304 | 2.71087 |
| | 2.8 | 2.20164 | 2.39971 | 2.64546 | 2.82143 |
| 3.2 | 0.0 | 1.40835 | 1.48276 | 1.57452 | 1.64000 |
| | 0.4 | 1.45527 | 1.54280 | 1.65095 | 1.72822 |
| | 0.8 | 1.53056 | 1.63355 | 1.76100 | 1.85214 |
| | 1.2 | 1.63698 | 1.75799 | 1.90789 | 2.01514 |
| | 1.6 | 1.77360 | 1.91496 | 2.09019 | 2.21560 |
| | 2.0 | 1.93138 | 2.09429 | 2.29629 | 2.44088 |
| | 2.4 | 2.08887 | 2.27208 | 2.49923 | 2.66181 |
| | 2.8 | 2.21485 | 2.41359 | 2.65995 | 2.83621 |
| | 3.2 | 2.28323 | 2.48991 | 2.74600 | 2.92914 |

FRACTILES $W_{\alpha}$ RHO = −0.90

| DELTAS | | $\alpha$: 0.9950 | 0.9975 | 0.9990 | 0.9995 |
|---|---|---|---|---|---|
| 3.6 | 0.0 | 1.46362 | 1.53879 | 1.63094 | 1.69639 |
| | 0.4 | 1.49628 | 1.58296 | 1.68958 | 1.76546 |
| | 0.8 | 1.55281 | 1.65302 | 1.77655 | 1.86462 |
| | 1.2 | 1.63614 | 1.75212 | 1.89535 | 1.99757 |
| | 1.6 | 1.74756 | 1.88159 | 2.04733 | 2.16571 |
| | 2.0 | 1.88410 | 2.03805 | 2.22856 | 2.36470 |
| | 2.4 | 2.03499 | 2.20938 | 2.42530 | 2.57963 |
| | 2.8 | 2.17926 | 2.37225 | 2.61123 | 2.78204 |
| | 3.2 | 2.29004 | 2.49676 | 2.75271 | 2.93563 |
| | 3.6 | 2.34730 | 2.56072 | 2.82487 | 3.01357 |
| 4.0 | 0.0 | 1.51807 | 1.59435 | 1.68730 | 1.75300 |
| | 0.4 | 1.53922 | 1.62561 | 1.73135 | 1.80631 |
| | 0.8 | 1.58049 | 1.67872 | 1.79933 | 1.88504 |
| | 1.2 | 1.64456 | 1.75657 | 1.89445 | 1.99259 |
| | 1.6 | 1.73339 | 1.86127 | 2.01898 | 2.13138 |
| | 2.0 | 1.84686 | 1.99258 | 2.17253 | 2.30088 |
| | 2.4 | 1.98038 | 2.14529 | 2.34913 | 2.49461 |
| | 2.8 | 2.12223 | 2.30630 | 2.53393 | 2.69646 |
| | 3.2 | 2.25285 | 2.45380 | 2.70237 | 2.87987 |
| | 3.6 | 2.34971 | 2.56274 | 2.82625 | 3.01440 |
| | 4.0 | 2.39769 | 2.61638 | 2.88682 | 3.07984 |
| 6.0 | 0.0 | 1.76406 | 1.85174 | 1.95511 | 2.02626 |
| | 0.4 | 1.75527 | 1.84751 | 1.95734 | 2.03348 |
| | 0.8 | 1.75516 | 1.85314 | 1.97071 | 2.05271 |
| | 1.2 | 1.76514 | 1.87004 | 1.99669 | 2.08543 |
| | 1.6 | 1.78660 | 1.89963 | 2.03678 | 2.13323 |
| | 2.0 | 1.82086 | 1.94329 | 2.09244 | 2.19763 |
| | 2.4 | 1.86902 | 2.00216 | 2.16486 | 2.27989 |
| | 2.8 | 1.93171 | 2.07685 | 2.25466 | 2.38062 |
| | 3.2 | 2.00860 | 2.16692 | 2.36127 | 2.49916 |
| | 3.6 | 2.09784 | 2.27025 | 2.48222 | 2.63279 |
| | 4.0 | 2.19534 | 2.38221 | 2.61223 | 2.77577 |
| | 6.0 | 2.53037 | 2.76263 | 3.04905 | 3.25296 |
| 12.0 | 0.0 | 2.15083 | 2.29309 | 2.45860 | 2.57015 |
| | 0.4 | 2.14258 | 2.28372 | 2.44806 | 2.55894 |
| | 0.8 | 2.13464 | 2.27487 | 2.43833 | 2.54879 |
| | 1.2 | 2.12718 | 2.26673 | 2.42967 | 2.54002 |
| | 1.6 | 2.12039 | 2.25954 | 2.42238 | 2.53296 |
| | 2.0 | 2.11449 | 2.25359 | 2.41681 | 2.52797 |
| | 2.4 | 2.10974 | 2.24919 | 2.41332 | 2.52545 |
| | 2.8 | 2.10645 | 2.24667 | 2.41228 | 2.52580 |
| | 3.2 | 2.10493 | 2.24641 | 2.41409 | 2.52942 |
| | 3.6 | 2.10553 | 2.24876 | 2.41914 | 2.53672 |
| | 4.0 | 2.10861 | 2.25412 | 2.42782 | 2.54808 |
| | 6.0 | 2.17238 | 2.33723 | 2.53697 | 2.67705 |
| | 12.0 | 2.59903 | 2.83704 | 3.12970 | 3.33745 |
| INFINITY | | 2.57583 | 2.80703 | 3.09024 | 3.29051 |

FRACTILES $W_\alpha$ RHO = -0.80

| DELTAS | | $\alpha$: 0.0005 | 0.0010 | 0.0025 | 0.0050 | 0.0075 |
|---|---|---|---|---|---|---|
| 0.0 | 0.0 | -7.81624 | -6.91102 | -5.72551 | -4.83925 | -4.32606 |
| 0.4 | 0.0 | -7.77290 | -6.87538 | -5.69913 | -4.81916 | -4.30936 |
| | 0.4 | -7.68116 | -6.79324 | -5.63039 | -4.76112 | -4.25778 |
| 0.8 | 0.0 | -7.56843 | -6.71326 | -5.58760 | -4.74131 | -4.24916 |
| | 0.4 | -7.53434 | -6.66729 | -5.53106 | -4.68113 | -4.18877 |
| | 0.8 | -7.31929 | -6.47780 | -5.37590 | -4.55234 | -4.07556 |
| 1.2 | 0.0 | -7.20162 | -6.41789 | -5.38073 | -4.59584 | -4.13691 |
| | 0.4 | -7.29823 | -6.47705 | -5.39620 | -4.58370 | -4.11125 |
| | 0.8 | -7.12390 | -6.30959 | -5.24262 | -4.44466 | -3.98251 |
| | 1.2 | -6.82792 | -6.05001 | -5.03158 | -4.27063 | -3.83022 |
| 1.6 | 0.0 | -6.78292 | -6.07344 | -5.13061 | -4.41333 | -3.99203 |
| | 0.4 | -6.95184 | -6.19875 | -5.20218 | -4.44805 | -4.00714 |
| | 0.8 | -6.91378 | -6.14126 | -5.12459 | -4.36048 | -3.91626 |
| | 1.2 | -6.63186 | -5.88137 | -4.89823 | -4.16322 | -3.73767 |
| | 1.6 | -6.30110 | -5.59222 | -4.66442 | -3.97150 | -3.57064 |
| 2.0 | 0.0 | -6.38991 | -5.74532 | -4.88583 | -4.22922 | -3.84221 |
| | 0.4 | -6.57106 | -5.88701 | -4.97797 | -4.28638 | -3.88019 |
| | 0.8 | -6.62879 | -5.91560 | -4.97187 | -4.25778 | -3.84033 |
| | 1.2 | -6.47847 | -5.76159 | -4.81831 | -4.10954 | -3.69762 |
| | 1.6 | -6.13132 | -5.44663 | -4.54997 | -3.87993 | -3.49219 |
| | 2.0 | -5.79963 | -5.15758 | -4.31758 | -3.69061 | -3.32816 |
| 2.4 | 0.0 | -6.04880 | -5.45791 | -4.66769 | -4.06186 | -3.70373 |
| | 0.4 | -6.21521 | -5.59127 | -4.75927 | -4.12362 | -3.74896 |
| | 0.8 | -6.31137 | -5.65871 | -4.79135 | -4.13147 | -3.74390 |
| | 1.2 | -6.27179 | -5.60309 | -4.71830 | -4.04890 | -3.65762 |
| | 1.6 | -6.04119 | -5.38093 | -4.51234 | -3.85995 | -3.48095 |
| | 2.0 | -5.66522 | -5.04291 | -4.22828 | -3.61993 | -3.26814 |
| | 2.4 | -5.35201 | -4.77093 | -4.01110 | -3.44445 | -3.11717 |
| 2.8 | 0.0 | -5.76082 | -5.21384 | -4.48043 | -3.91635 | -3.58204 |
| | 0.4 | -5.90515 | -5.33106 | -4.56325 | -3.97454 | -3.62651 |
| | 0.8 | -6.00817 | -5.40886 | -4.60965 | -3.99901 | -3.63906 |
| | 1.2 | -6.02756 | -5.40950 | -4.58810 | -3.96319 | -3.59616 |
| | 1.6 | -5.91165 | -5.28838 | -4.46378 | -3.84001 | -3.47549 |
| | 2.0 | -5.63133 | -5.02502 | -4.22766 | -3.62907 | -3.28152 |
| | 2.4 | -5.25207 | -4.68631 | -3.94611 | -3.39383 | -3.07479 |
| | 2.8 | -4.96619 | -4.43912 | -3.75039 | -3.23738 | -2.94148 |
| 3.2 | 0.0 | -5.51916 | -5.00825 | -4.32153 | -3.79185 | -3.47719 |
| | 0.4 | -5.64182 | -5.10868 | -4.39372 | -3.84378 | -3.51781 |
| | 0.8 | -5.73879 | -5.18429 | -4.44260 | -3.87384 | -3.53757 |
| | 1.2 | -5.78329 | -5.21097 | -4.44767 | -3.86442 | -3.52059 |
| | 1.6 | -5.73887 | -5.15640 | -4.38228 | -3.79334 | -3.44744 |
| | 2.0 | -5.56838 | -4.98906 | -4.22269 | -3.64312 | -3.30452 |
| | 2.4 | -5.26265 | -4.70587 | -3.97394 | -3.42485 | -3.10630 |
| | 2.8 | -4.89578 | -4.38016 | -3.70604 | -3.20369 | -2.91392 |
| | 3.2 | -4.63989 | -4.16005 | -3.53363 | -3.06782 | -2.79977 |

FRACTILES $W_\alpha$ RHO = -0.80

| DELTAS | | $\alpha$:0.0005 | 0.0010 | 0.0025 | 0.0050 | 0.0075 |
|---|---|---|---|---|---|---|
| 3.6 | 0.0 | -5.31574 | -4.83472 | -4.18674 | -3.68561 | -3.38728 |
| | 0.4 | -5.41952 | -4.92015 | -4.24884 | -3.73096 | -3.42328 |
| | 0.8 | -5.50637 | -4.98906 | -4.29526 | -3.76149 | -3.44506 |
| | 1.2 | -5.55893 | -5.02578 | -4.31256 | -3.76554 | -3.44208 |
| | 1.6 | -5.55282 | -5.00845 | -4.28235 | -3.72746 | -3.40031 |
| | 2.0 | -5.45887 | -4.91139 | -4.18376 | -3.63019 | -3.30508 |
| | 2.4 | -5.25277 | -4.71460 | -4.00277 | -3.46462 | -3.15034 |
| | 2.8 | -4.93900 | -4.42679 | -3.75381 | -3.24941 | -2.95711 |
| | 3.2 | -4.59320 | -4.12161 | -3.50565 | -3.04748 | -2.78386 |
| | 3.6 | -4.36678 | -3.92810 | -3.35620 | -2.93213 | -2.68909 |
| 4.0 | 0.0 | -5.14334 | -4.68736 | -4.07183 | -3.59466 | -3.31004 |
| | 0.4 | -5.23135 | -4.76008 | -4.12513 | -3.63399 | -3.34157 |
| | 0.8 | -5.30760 | -4.82126 | -4.16738 | -3.66284 | -3.36302 |
| | 1.2 | -5.36047 | -4.86037 | -4.18952 | -3.67331 | -3.36723 |
| | 1.6 | -5.37355 | -4.86262 | -4.17903 | -3.65464 | -3.34450 |
| | 2.0 | -5.32599 | -4.80947 | -4.12044 | -3.59380 | -3.28328 |
| | 2.4 | -5.19590 | -4.68183 | -3.99857 | -3.47375 | -3.17350 |
| | 2.8 | -4.96924 | -4.46883 | -3.80709 | -3.30701 | -3.01513 |
| | 3.2 | -4.65900 | -4.18650 | -3.56612 | -3.10177 | -2.83318 |
| | 3.6 | -4.33845 | -3.90546 | -3.34071 | -2.92190 | -2.68192 |
| | 4.0 | -4.13958 | -3.73690 | -3.21303 | -2.82636 | -2.60570 |
| 6.0 | 0.0 | -4.57712 | -4.20174 | -3.69083 | -3.29094 | -3.05060 |
| | 0.4 | -4.61962 | -4.23720 | -3.71736 | -3.31104 | -3.06711 |
| | 0.8 | -4.65881 | -4.26942 | -3.74078 | -3.32821 | -3.08081 |
| | 1.2 | -4.69232 | -4.29621 | -3.75922 | -3.34081 | -3.09021 |
| | 1.6 | -4.71693 | -4.31469 | -3.77020 | -3.34669 | -3.09339 |
| | 2.0 | -4.72850 | -4.32110 | -3.77055 | -3.34313 | -3.08788 |
| | 2.4 | -4.72189 | -4.31085 | -3.75639 | -3.32685 | -3.07075 |
| | 2.8 | -4.69109 | -4.27860 | -3.72270 | -3.29410 | -3.03870 |
| | 3.2 | -4.62968 | -4.21867 | -3.66661 | -3.24108 | -2.98841 |
| | 3.6 | -4.53168 | -4.12588 | -3.58227 | -3.16461 | -2.91726 |
| | 4.0 | -4.39285 | -3.99677 | -3.46792 | -3.06333 | -2.82464 |
| | 6.0 | -3.48547 | -3.21123 | -2.85495 | -2.58648 | -2.42773 |
| 12.0 | 0.0 | -3.94910 | -3.66066 | -3.26273 | -2.94642 | -2.75402 |
| | 0.4 | -3.95936 | -3.66932 | -3.26935 | -2.95156 | -2.75834 |
| | 0.8 | -3.96911 | -3.67749 | -3.27552 | -2.95631 | -2.76230 |
| | 1.2 | -3.97817 | -3.68502 | -3.28113 | -2.96056 | -2.76579 |
| | 1.6 | -3.98635 | -3.69174 | -3.28603 | -2.96418 | -2.76871 |
| | 2.0 | -3.99342 | -3.69744 | -3.29004 | -2.96703 | -2.77092 |
| | 2.4 | -3.99913 | -3.70190 | -3.29298 | -2.96894 | -2.77229 |
| | 2.8 | -4.00317 | -3.70484 | -3.29462 | -2.96972 | -2.77264 |
| | 3.2 | -4.00521 | -3.70597 | -3.29471 | -2.96917 | -2.77178 |
| | 3.6 | -4.00486 | -3.70494 | -3.29296 | -2.96705 | -2.76952 |
| | 4.0 | -4.00170 | -3.70138 | -3.28908 | -2.96311 | -2.76562 |
| | 6.0 | -3.92671 | -3.63072 | -3.22557 | -2.90620 | -2.71311 |
| | 12.0 | -3.20622 | -3.01386 | -2.74252 | -2.52113 | -2.38369 |
| INFINITY | | -3.29051 | -3.09024 | -2.80703 | -2.57583 | -2.43238 |

FRACTILES $W_\alpha$ RHO = -0.80

| DELTAS | | $\alpha$:0.0100 | 0.0175 | 0.0250 | 0.0375 | 0.0500 |
|---|---|---|---|---|---|---|
| 0.0 | 0.0 | -3.96470 | -3.26948 | -2.83268 | -2.34342 | -2.00190 |
| 0.4 | 0.0 | -3.95027 | -3.25914 | -2.82474 | -2.33800 | -1.99813 |
| | 0.4 | -3.90338 | -3.22161 | -2.79332 | -2.31365 | -1.97889 |
| 0.8 | 0.0 | -3.90162 | -3.23048 | -2.80699 | -2.33083 | -1.99723 |
| | 0.4 | -3.84200 | -3.17470 | -2.75536 | -2.28563 | -1.95775 |
| | 0.8 | -3.73991 | -3.09435 | -2.68896 | -2.23516 | -1.91865 |
| 1.2 | 0.0 | -3.81155 | -3.17973 | -2.77829 | -2.32388 | -2.00329 |
| | 0.4 | -3.77766 | -3.13355 | -2.72723 | -2.27052 | -1.95070 |
| | 0.8 | -3.65708 | -3.03103 | -2.63780 | -2.19758 | -1.89055 |
| | 1.2 | -3.52026 | -2.92438 | -2.55044 | -2.13220 | -1.84083 |
| 1.6 | 0.0 | -3.69234 | -3.10754 | -2.73362 | -2.30759 | -2.00489 |
| | 0.4 | -3.69456 | -3.08765 | -2.70210 | -2.26580 | -1.95809 |
| | 0.8 | -3.60265 | -2.99731 | -2.61562 | -2.18686 | -1.88684 |
| | 1.2 | -3.43810 | -2.86209 | -2.50059 | -2.09629 | -1.81471 |
| | 1.6 | -3.28863 | -2.74686 | -2.40724 | -2.02793 | -1.76421 |
| 2.0 | 0.0 | -3.56620 | -3.02557 | -2.67821 | -2.28047 | -1.99628 |
| | 0.4 | -3.59124 | -3.02741 | -2.66692 | -2.25625 | -1.96450 |
| | 0.8 | -3.54441 | -2.96995 | -2.60513 | -2.19245 | -1.90157 |
| | 1.2 | -3.40688 | -2.84596 | -2.49253 | -2.09589 | -1.81872 |
| | 1.6 | -3.21936 | -2.69517 | -2.36659 | -1.99972 | -1.74481 |
| | 2.0 | -3.07330 | -2.58422 | -2.27815 | -1.93715 | -1.70098 |
| 2.4 | 0.0 | -3.44777 | -2.94488 | -2.62052 | -2.24763 | -1.98006 |
| | 0.4 | -3.48173 | -2.95830 | -2.62198 | -2.23686 | -1.96169 |
| | 0.8 | -3.46821 | -2.93027 | -2.58637 | -2.19465 | -1.91645 |
| | 1.2 | -3.38030 | -2.84209 | -2.50044 | -2.11422 | -1.84224 |
| | 1.6 | -3.21356 | -2.69800 | -2.37351 | -2.00988 | -1.75634 |
| | 2.0 | -3.02076 | -2.54602 | -2.24901 | -1.91835 | -1.68977 |
| | 2.4 | -2.88725 | -2.44673 | -2.17188 | -1.86726 | -1.65822 |
| 2.8 | 0.0 | -3.34266 | -2.87109 | -2.56592 | -2.21393 | -1.96047 |
| | 0.4 | -3.37774 | -2.88894 | -2.57364 | -2.21112 | -1.95096 |
| | 0.8 | -3.38232 | -2.87940 | -2.55624 | -2.18618 | -1.92178 |
| | 1.2 | -3.33508 | -2.82570 | -2.50011 | -2.12934 | -1.86612 |
| | 1.6 | -3.21718 | -2.71606 | -2.39818 | -2.03916 | -1.78671 |
| | 2.0 | -3.03644 | -2.56438 | -2.26775 | -1.93620 | -1.70606 |
| | 2.4 | -2.85065 | -2.42129 | -2.15362 | -1.85755 | -1.65468 |
| | 2.8 | -2.73389 | -2.33735 | -2.09149 | -1.82107 | -1.63574 |
| 3.2 | 0.0 | -3.25149 | -2.80584 | -2.51660 | -2.18202 | -1.94034 |
| | 0.4 | -3.28436 | -2.82445 | -2.52678 | -2.18338 | -1.93604 |
| | 0.8 | -3.29718 | -2.82481 | -2.52004 | -2.16960 | -1.91807 |
| | 1.2 | -3.27533 | -2.79487 | -2.48613 | -2.13257 | -1.87997 |
| | 1.6 | -3.20141 | -2.72143 | -2.41470 | -2.06555 | -1.81784 |
| | 2.0 | -3.06465 | -2.59956 | -2.30482 | -1.97246 | -1.73941 |
| | 2.4 | -2.88183 | -2.45012 | -2.17964 | -1.87891 | -1.67172 |
| | 2.8 | -2.71065 | -2.32264 | -2.08237 | -1.81805 | -1.63638 |
| | 3.2 | -2.61221 | -2.25583 | -2.03611 | -1.79369 | -1.62531 |

FRACTILES $W_\alpha$ RHO = −0.80

| DELTAS | | $\alpha$: 0.0100 | 0.0175 | 0.0250 | 0.0375 | 0.0500 |
|---|---|---|---|---|---|---|
| 3.6 | 0.0 | −3.17297 | −2.74890 | −2.47295 | −2.15291 | −1.92111 |
| | 0.4 | −3.20255 | −2.76666 | −2.48369 | −2.15630 | −1.91976 |
| | 0.8 | −3.21843 | −2.77186 | −2.48276 | −2.14918 | −1.90886 |
| | 1.2 | −3.21083 | −2.75633 | −2.46305 | −2.12575 | −1.88360 |
| | 1.6 | −3.16694 | −2.70974 | −2.41592 | −2.07946 | −1.83909 |
| | 2.0 | −3.07384 | −2.62281 | −2.33468 | −2.00691 | −1.77465 |
| | 2.4 | −2.92779 | −2.49664 | −2.22385 | −1.91720 | −1.70328 |
| | 2.8 | −2.75140 | −2.35681 | −2.11091 | −1.83896 | −1.65150 |
| | 3.2 | −2.59947 | −2.24928 | −2.03322 | −1.79418 | −1.62750 |
| | 3.6 | −2.51966 | −2.19863 | −1.99986 | −1.77767 | −1.62063 |
| 4.0 | 0.0 | −3.10529 | −2.69937 | −2.43459 | −2.12681 | −1.90334 |
| | 0.4 | −3.13147 | −2.71567 | −2.44503 | −2.13108 | −1.90363 |
| | 0.8 | −3.14792 | −2.72304 | −2.44715 | −2.12786 | −1.89710 |
| | 1.2 | −3.14798 | −2.71588 | −2.43607 | −2.11312 | −1.88039 |
| | 1.6 | −3.12274 | −2.68684 | −2.40549 | −2.08185 | −1.84946 |
| | 2.0 | −3.06175 | −2.62772 | −2.34878 | −2.02937 | −1.80124 |
| | 2.4 | −2.95640 | −2.53307 | −2.26279 | −1.95571 | −1.73863 |
| | 2.8 | −2.80858 | −2.40897 | −2.15692 | −1.87481 | −1.67855 |
| | 3.2 | −2.64455 | −2.28424 | −2.06066 | −1.81281 | −1.64021 |
| | 3.6 | −2.51462 | −2.19734 | −2.00044 | −1.77970 | −1.62324 |
| | 4.0 | −2.45202 | −2.15952 | −1.97618 | −1.76802 | −1.61858 |
| 6.0 | 0.0 | −2.87677 | −2.52958 | −2.30105 | −2.03309 | −1.83676 |
| | 0.4 | −2.89081 | −2.53906 | −2.30781 | −2.03697 | −1.83877 |
| | 0.8 | −2.90215 | −2.54608 | −2.31231 | −2.03885 | −1.83898 |
| | 1.2 | −2.90942 | −2.54950 | −2.31354 | −2.03789 | −1.83670 |
| | 1.6 | −2.91081 | −2.54783 | −2.31023 | −2.03306 | −1.83107 |
| | 2.0 | −2.90409 | −2.53922 | −2.30079 | −2.02311 | −1.82108 |
| | 2.4 | −2.88657 | −2.52149 | −2.28340 | −2.00662 | −1.80562 |
| | 2.8 | −2.85525 | −2.49231 | −2.25614 | −1.98219 | −1.78369 |
| | 3.2 | −2.80721 | −2.44950 | −2.21737 | −1.94887 | −1.75493 |
| | 3.6 | −2.74024 | −2.39182 | −2.16663 | −1.90731 | −1.72092 |
| | 4.0 | −2.65435 | −2.32086 | −2.10679 | −1.86178 | −1.68646 |
| | 6.0 | −2.31346 | −2.08503 | −1.93340 | −1.75304 | −1.61843 |
| 12.0 | 0.0 | −2.61371 | −2.33028 | −2.14122 | −1.91671 | −1.75012 |
| | 0.4 | −2.61746 | −2.33298 | −2.14329 | −1.91810 | −1.75107 |
| | 0.8 | −2.62087 | −2.33539 | −2.14509 | −1.91928 | −1.75183 |
| | 1.2 | −2.62385 | −2.33742 | −2.14658 | −1.92018 | −1.75235 |
| | 1.6 | −2.62629 | −2.33901 | −2.14766 | −1.92075 | −1.75260 |
| | 2.0 | −2.62809 | −2.34005 | −2.14827 | −1.92093 | −1.75252 |
| | 2.4 | −2.62909 | −2.34043 | −2.14832 | −1.92065 | −1.75205 |
| | 2.8 | −2.62916 | −2.34004 | −2.14769 | −1.91983 | −1.75114 |
| | 3.2 | −2.62812 | −2.33873 | −2.14628 | −1.91837 | −1.74971 |
| | 3.6 | −2.62579 | −2.33636 | −2.14397 | −1.91620 | −1.74768 |
| | 4.0 | −2.62196 | −2.33278 | −2.14062 | −1.91320 | −1.74499 |
| | 6.0 | −2.57284 | −2.29090 | −2.10385 | −1.88273 | −1.71932 |
| | 12.0 | −2.28203 | −2.07274 | −1.92999 | −1.75696 | −1.62594 |
| INFINITY | | −2.32635 | −2.10836 | −1.95996 | −1.78046 | −1.64485 |

FRACTILES $w_\alpha$ RHO = -0.80

| DELTAS | | $\alpha$: 0.0750 | 0.1000 | 0.1500 | 0.2000 | 0.2500 |
|---|---|---|---|---|---|---|
| 0.0 | 0.0 | -1.53024 | -1.20397 | -0.75927 | -0.45756 | -0.23377 |
| 0.4 | 0.0 | -1.52861 | -1.20374 | -0.76088 | -0.46039 | -0.23755 |
| | 0.4 | -1.51670 | -1.19713 | -0.76190 | -0.46702 | -0.24871 |
| 0.8 | 0.0 | -1.53477 | -1.21362 | -0.77421 | -0.47500 | -0.25260 |
| | 0.4 | -1.50506 | -1.19211 | -0.76612 | -0.47788 | -0.26495 |
| | 0.8 | -1.48206 | -1.18064 | -0.77120 | -0.49519 | -0.29237 |
| 1.2 | 0.0 | -1.55529 | -1.24133 | -0.80722 | -0.50811 | -0.28365 |
| | 0.4 | -1.50762 | -1.20027 | -0.78053 | -0.49575 | -0.28517 |
| | 0.8 | -1.46717 | -1.17505 | -0.77887 | -0.51276 | -0.31853 |
| | 1.2 | -1.43966 | -1.16353 | -0.79071 | -0.54288 | -0.36715 |
| 1.6 | 0.0 | -1.57815 | -1.27584 | -0.85204 | -0.55489 | -0.32826 |
| | 0.4 | -1.52833 | -1.22744 | -0.81213 | -0.52695 | -0.31410 |
| | 0.8 | -1.47174 | -1.18439 | -0.79354 | -0.53061 | -0.33933 |
| | 1.2 | -1.42729 | -1.16103 | -0.80298 | -0.56878 | -0.40260 |
| | 1.6 | -1.40238 | -1.15497 | -0.82741 | -0.61467 | -0.45603 |
| 2.0 | 0.0 | -1.59281 | -1.30444 | -0.89528 | -0.60364 | -0.37744 |
| | 0.4 | -1.55332 | -1.26220 | -0.85454 | -0.56939 | -0.35277 |
| | 0.8 | -1.49572 | -1.21203 | -0.82178 | -0.55584 | -0.36063 |
| | 1.2 | -1.43608 | -1.17223 | -0.81678 | -0.58417 | -0.41621 |
| | 1.6 | -1.39565 | -1.15806 | -0.84542 | -0.63822 | -0.47981 |
| | 2.0 | -1.37982 | -1.16385 | -0.87666 | -0.67829 | -0.52136 |
| 2.4 | 0.0 | -1.59814 | -1.32333 | -0.92984 | -0.64581 | -0.42257 |
| | 0.4 | -1.57105 | -1.29189 | -0.89597 | -0.61407 | -0.39589 |
| | 0.8 | -1.52455 | -1.24733 | -0.85993 | -0.59026 | -0.38768 |
| | 1.2 | -1.46337 | -1.19932 | -0.83882 | -0.59831 | -0.42271 |
| | 1.6 | -1.40781 | -1.16988 | -0.85472 | -0.64401 | -0.48249 |
| | 2.0 | -1.38017 | -1.17181 | -0.89100 | -0.69358 | -0.53573 |
| | 2.4 | -1.37622 | -1.18420 | -0.91788 | -0.72458 | -0.56688 |
| 2.8 | 0.0 | -1.59710 | -1.33424 | -0.95523 | -0.67904 | -0.45983 |
| | 0.4 | -1.57957 | -1.31230 | -0.92959 | -0.65338 | -0.43642 |
| | 0.8 | -1.54646 | -1.27831 | -0.89831 | -0.62834 | -0.42048 |
| | 1.2 | -1.49563 | -1.23396 | -0.86982 | -0.61962 | -0.43359 |
| | 1.6 | -1.43606 | -1.19335 | -0.86646 | -0.64645 | -0.47874 |
| | 2.0 | -1.39276 | -1.18068 | -0.89400 | -0.69274 | -0.53237 |
| | 2.4 | -1.38013 | -1.19162 | -0.92738 | -0.73386 | -0.57524 |
| | 2.8 | -1.38193 | -1.20396 | -0.94844 | -0.75725 | -0.59850 |
| 3.2 | 0.0 | -1.59259 | -1.33991 | -0.97347 | -0.70444 | -0.48932 |
| | 0.4 | -1.58139 | -1.32476 | -0.95459 | -0.68475 | -0.47053 |
| | 0.8 | -1.55893 | -1.30045 | -0.93032 | -0.66330 | -0.45391 |
| | 1.2 | -1.52146 | -1.26545 | -0.90323 | -0.64735 | -0.45214 |
| | 1.6 | -1.46966 | -1.22457 | -0.88650 | -0.65494 | -0.47868 |
| | 2.0 | -1.41777 | -1.19703 | -0.89733 | -0.68834 | -0.52341 |
| | 2.4 | -1.39018 | -1.19673 | -0.92645 | -0.72960 | -0.56903 |
| | 2.8 | -1.38614 | -1.20950 | -0.95440 | -0.76268 | -0.60314 |
| | 3.2 | -1.38947 | -1.21984 | -0.97050 | -0.78029 | -0.62062 |

FRACTILES $W_\alpha$ RHO = -0.80

| DELTAS | | $\alpha$: 0.0750 | 0.1000 | 0.1500 | 0.2000 | 0.2500 |
|---|---|---|---|---|---|---|
| 3.6 | 0.0 | -1.58649 | -1.34241 | -0.98671 | -0.72393 | -0.51253 |
| | 0.4 | -1.57929 | -1.33181 | -0.97273 | -0.70892 | -0.49784 |
| | 0.8 | -1.56415 | -1.31464 | -0.95457 | -0.69198 | -0.48350 |
| | 1.2 | -1.53781 | -1.28891 | -0.93254 | -0.67585 | -0.47540 |
| | 1.6 | -1.49807 | -1.25482 | -0.91195 | -0.67111 | -0.48602 |
| | 2.0 | -1.44916 | -1.22138 | -0.90761 | -0.68862 | -0.51721 |
| | 2.4 | -1.40943 | -1.20667 | -0.92488 | -0.72188 | -0.55789 |
| | 2.8 | -1.39347 | -1.21202 | -0.95149 | -0.75697 | -0.59590 |
| | 3.2 | -1.39305 | -1.22374 | -0.97414 | -0.78333 | -0.62302 |
| | 3.6 | -1.39630 | -1.23191 | -0.98638 | -0.79668 | -0.63628 |
| 4.0 | 0.0 | -1.57982 | -1.34304 | -0.99650 | -0.73912 | -0.53101 |
| | 0.4 | -1.57517 | -1.33548 | -0.98596 | -0.72754 | -0.51948 |
| | 0.8 | -1.56484 | -1.32321 | -0.97238 | -0.71443 | -0.50788 |
| | 1.2 | -1.54645 | -1.30462 | -0.95548 | -0.70078 | -0.49872 |
| | 1.6 | -1.51755 | -1.27862 | -0.93683 | -0.69133 | -0.49955 |
| | 2.0 | -1.47787 | -1.24776 | -0.92444 | -0.69598 | -0.51733 |
| | 2.4 | -1.43562 | -1.22381 | -0.92865 | -0.71752 | -0.54861 |
| | 2.8 | -1.40757 | -1.21767 | -0.94747 | -0.74799 | -0.58427 |
| | 3.2 | -1.39840 | -1.22477 | -0.97043 | -0.77725 | -0.61568 |
| | 3.6 | -1.39918 | -1.23464 | -0.98855 | -0.79826 | -0.63733 |
| | 4.0 | -1.40189 | -1.24095 | -0.99790 | -0.80848 | -0.64751 |
| 6.0 | 0.0 | -1.54953 | -1.33671 | -1.02053 | -0.78136 | -0.58467 |
| | 0.4 | -1.54917 | -1.33492 | -1.01717 | -0.77730 | -0.58037 |
| | 0.8 | -1.54736 | -1.33196 | -1.01311 | -0.77291 | -0.57609 |
| | 1.2 | -1.54360 | -1.32749 | -1.00821 | -0.76827 | -0.57205 |
| | 1.6 | -1.53731 | -1.32111 | -1.00242 | -0.76353 | -0.56860 |
| | 2.0 | -1.52781 | -1.31243 | -0.99576 | -0.75907 | -0.56642 |
| | 2.4 | -1.51448 | -1.30120 | -0.98857 | -0.75574 | -0.56678 |
| | 2.8 | -1.49696 | -1.28761 | -0.98196 | -0.75521 | -0.57139 |
| | 3.2 | -1.47591 | -1.27310 | -0.97811 | -0.75937 | -0.58136 |
| | 3.6 | -1.45406 | -1.26072 | -0.97915 | -0.76890 | -0.59622 |
| | 4.0 | -1.43571 | -1.25339 | -0.98540 | -0.78259 | -0.61404 |
| | 6.0 | -1.41652 | -1.26220 | -1.02369 | -0.83439 | -0.67185 |
| 12.0 | 0.0 | -1.50285 | -1.31662 | -1.03459 | -0.81642 | -0.63342 |
| | 0.4 | -1.50322 | -1.31663 | -1.03417 | -0.81576 | -0.63264 |
| | 0.8 | -1.50344 | -1.31653 | -1.03370 | -0.81509 | -0.63188 |
| | 1.2 | -1.50350 | -1.31630 | -1.03316 | -0.81442 | -0.63115 |
| | 1.6 | -1.50336 | -1.31594 | -1.03257 | -0.81375 | -0.63047 |
| | 2.0 | -1.50298 | -1.31540 | -1.03191 | -0.81308 | -0.62986 |
| | 2.4 | -1.50232 | -1.31468 | -1.03118 | -0.81243 | -0.62933 |
| | 2.8 | -1.50136 | -1.31374 | -1.03038 | -0.81181 | -0.62890 |
| | 3.2 | -1.50004 | -1.31256 | -1.02951 | -0.81124 | -0.62862 |
| | 3.6 | -1.49831 | -1.31112 | -1.02857 | -0.81074 | -0.62850 |
| | 4.0 | -1.49615 | -1.30939 | -1.02758 | -0.81034 | -0.62860 |
| | 6.0 | -1.47770 | -1.29639 | -1.02262 | -0.81124 | -0.63401 |
| | 12.0 | -1.42704 | -1.27353 | -1.03449 | -0.84371 | -0.67945 |
| INFINITY | | -1.43953 | -1.28155 | -1.03643 | -0.84162 | -0.67449 |

FRACTILES $w_\alpha$ RHO = -0.80

| DELTAS | | $\alpha$:0.3000 | 0.3500 | 0.4000 | 0.4500 | 0.5000 |
|---|---|---|---|---|---|---|
| 0.0 | 0.0 | -0.05921 | 0.08126 | 0.19660 | 0.29250 | 0.37276 |
| 0.4 | 0.0 | -0.06378 | 0.07597 | 0.19062 | 0.28580 | 0.36529 |
| | 0.4 | -0.07889 | 0.05726 | 0.16842 | 0.26006 | 0.33576 |
| 0.8 | 0.0 | -0.07895 | 0.06075 | 0.17524 | 0.27005 | 0.34883 |
| | 0.4 | -0.09984 | 0.03184 | 0.13845 | 0.22501 | 0.29430 |
| | 0.8 | -0.13646 | -0.01399 | 0.08181 | 0.15295 | 0.21864 |
| 1.2 | 0.0 | -0.10705 | 0.03589 | 0.15351 | 0.25105 | 0.33182 |
| | 0.4 | -0.12208 | 0.00741 | 0.11107 | 0.19221 | 0.25462 |
| | 0.8 | -0.17140 | -0.06125 | 0.02835 | 0.10837 | 0.18363 |
| | 1.2 | -0.23610 | -0.12831 | -0.03312 | 0.05506 | 0.13965 |
| 1.6 | 0.0 | -0.14720 | 0.00146 | 0.12543 | 0.22942 | 0.31614 |
| | 0.4 | -0.14808 | -0.01580 | 0.08887 | 0.17151 | 0.24414 |
| | 0.8 | -0.19720 | -0.08593 | 0.00868 | 0.09409 | 0.17455 |
| | 1.2 | -0.27220 | -0.16127 | -0.06154 | 0.03172 | 0.12160 |
| | 1.6 | -0.32544 | -0.21082 | -0.10573 | -0.00619 | 0.09053 |
| 2.0 | 0.0 | -0.19354 | -0.03968 | 0.09123 | 0.20343 | 0.29911 |
| | 0.4 | -0.18092 | -0.04172 | 0.07033 | 0.16256 | 0.24326 |
| | 0.8 | -0.21313 | -0.09548 | 0.00475 | 0.09480 | 0.17907 |
| | 1.2 | -0.28266 | -0.16847 | -0.06574 | 0.03021 | 0.12245 |
| | 1.6 | -0.34735 | -0.23002 | -0.12187 | -0.01916 | 0.08081 |
| | 2.0 | -0.38715 | -0.26653 | -0.15425 | -0.04688 | 0.05811 |
| 2.4 | 0.0 | -0.23838 | -0.08171 | 0.05422 | 0.17347 | 0.27828 |
| | 0.4 | -0.21920 | -0.07258 | 0.04959 | 0.15238 | 0.24203 |
| | 0.8 | -0.23046 | -0.10379 | 0.00362 | 0.09918 | 0.18763 |
| | 1.2 | -0.28327 | -0.16479 | -0.05896 | 0.03917 | 0.13290 |
| | 1.6 | -0.34752 | -0.22823 | -0.11855 | -0.01466 | 0.08616 |
| | 2.0 | -0.39993 | -0.27745 | -0.16321 | -0.05386 | 0.05310 |
| | 2.4 | -0.42946 | -0.30444 | -0.18715 | -0.07438 | 0.03626 |
| 2.8 | 0.0 | -0.27703 | -0.11960 | 0.01899 | 0.14283 | 0.25443 |
| | 0.4 | -0.25768 | -0.10615 | 0.02410 | 0.13676 | 0.23585 |
| | 0.8 | -0.25421 | -0.11771 | -0.00177 | 0.10058 | 0.19430 |
| | 1.2 | -0.28583 | -0.16129 | -0.05116 | 0.04994 | 0.14565 |
| | 1.6 | -0.33978 | -0.21794 | -0.10673 | -0.00205 | 0.09899 |
| | 2.0 | -0.39488 | -0.27129 | -0.15637 | -0.04667 | 0.06037 |
| | 2.4 | -0.43664 | -0.31038 | -0.19184 | -0.07785 | 0.03400 |
| | 2.8 | -0.45864 | -0.33051 | -0.20972 | -0.09322 | 0.02134 |
| 3.2 | 0.0 | -0.30849 | -0.15139 | -0.01167 | 0.11475 | 0.23058 |
| | 0.4 | -0.29189 | -0.13817 | -0.00321 | 0.11660 | 0.22398 |
| | 0.8 | -0.28223 | -0.13782 | -0.01359 | 0.09620 | 0.19612 |
| | 1.2 | -0.29538 | -0.16348 | -0.04773 | 0.05755 | 0.15628 |
| | 1.6 | -0.33388 | -0.20815 | -0.09444 | 0.01171 | 0.11343 |
| | 2.0 | -0.38324 | -0.25817 | -0.14259 | -0.03285 | 0.07373 |
| | 2.4 | -0.42929 | -0.30242 | -0.18364 | -0.06972 | 0.04181 |
| | 2.8 | -0.46244 | -0.33348 | -0.21188 | -0.09462 | 0.02067 |
| | 3.2 | -0.47899 | -0.34865 | -0.22540 | -0.10628 | 0.01102 |

FRACTILES $W_\alpha$ RHO = -0.80

| DELTAS | | $\alpha$:0.3000 | 0.3500 | 0.4000 | 0.4500 | 0.5000 |
|---|---|---|---|---|---|---|
| 3.6 | 0.0 | -0.33373 | -0.17735 | -0.03725 | 0.09061 | 0.20901 |
| | 0.4 | -0.32030 | -0.16598 | -0.02875 | 0.09523 | 0.20852 |
| | 0.8 | -0.30982 | -0.16079 | -0.03029 | 0.08619 | 0.19243 |
| | 1.2 | -0.31146 | -0.17243 | -0.05049 | 0.05985 | 0.16256 |
| | 1.6 | -0.33426 | -0.20341 | -0.08605 | 0.02261 | 0.12593 |
| | 2.0 | -0.37295 | -0.24536 | -0.12837 | -0.01804 | 0.08847 |
| | 2.4 | -0.41629 | -0.28855 | -0.16959 | -0.05600 | 0.05477 |
| | 2.8 | -0.45438 | -0.32507 | -0.20347 | -0.08646 | 0.02834 |
| | 3.2 | -0.48077 | -0.34985 | -0.22607 | -0.10647 | 0.01127 |
| | 3.6 | -0.49335 | -0.36142 | -0.23641 | -0.11543 | 0.00382 |
| 4.0 | 0.0 | -0.35409 | -0.19854 | -0.05839 | 0.07033 | 0.19040 |
| | 0.4 | -0.34334 | -0.18916 | -0.05090 | 0.07536 | 0.19228 |
| | 0.8 | -0.33405 | -0.18297 | -0.04872 | 0.07268 | 0.18431 |
| | 1.2 | -0.33064 | -0.18621 | -0.05871 | 0.05676 | 0.16392 |
| | 1.6 | -0.34128 | -0.20491 | -0.08315 | 0.02889 | 0.13470 |
| | 2.0 | -0.36789 | -0.23673 | -0.11737 | -0.00559 | 0.10162 |
| | 2.4 | -0.40406 | -0.27466 | -0.15494 | -0.04127 | 0.06903 |
| | 2.8 | -0.44140 | -0.31156 | -0.19001 | -0.07350 | 0.04043 |
| | 3.2 | -0.47281 | -0.34169 | -0.21800 | -0.09874 | 0.01847 |
| | 3.6 | -0.49393 | -0.36159 | -0.23623 | -0.11494 | 0.00458 |
| | 4.0 | -0.50362 | -0.37053 | -0.24424 | -0.12190 | -0.00124 |
| 6.0 | 0.0 | -0.41469 | -0.26276 | -0.12354 | 0.00662 | 0.13040 |
| | 0.4 | -0.41046 | -0.25882 | -0.12005 | 0.00951 | 0.13255 |
| | 0.8 | -0.40656 | -0.25549 | -0.11744 | 0.01126 | 0.13333 |
| | 1.2 | -0.40332 | -0.25322 | -0.11627 | 0.01123 | 0.13201 |
| | 1.6 | -0.40131 | -0.25275 | -0.11741 | 0.00844 | 0.12760 |
| | 2.0 | -0.40146 | -0.25521 | -0.12209 | 0.00174 | 0.11918 |
| | 2.4 | -0.40523 | -0.26200 | -0.13141 | -0.00953 | 0.10657 |
| | 2.8 | -0.41400 | -0.27394 | -0.14559 | -0.02508 | 0.09047 |
| | 3.2 | -0.42801 | -0.29059 | -0.16370 | -0.04370 | 0.07217 |
| | 3.6 | -0.44608 | -0.31035 | -0.18405 | -0.06375 | 0.05316 |
| | 4.0 | -0.46603 | -0.33111 | -0.20465 | -0.08346 | 0.03493 |
| | 6.0 | -0.52562 | -0.38979 | -0.26054 | -0.13510 | -0.01123 |
| 12.0 | 0.0 | -0.47237 | -0.32587 | -0.18930 | -0.05940 | 0.06629 |
| | 0.4 | -0.47152 | -0.32502 | -0.18847 | -0.05863 | 0.06699 |
| | 0.8 | -0.47073 | -0.32424 | -0.18773 | -0.05796 | 0.06756 |
| | 1.2 | -0.47000 | -0.32355 | -0.18710 | -0.05742 | 0.06799 |
| | 1.6 | -0.46936 | -0.32297 | -0.18662 | -0.05705 | 0.06825 |
| | 2.0 | -0.46883 | -0.32255 | -0.18632 | -0.05687 | 0.06829 |
| | 2.4 | -0.46844 | -0.32230 | -0.18622 | -0.05693 | 0.06807 |
| | 2.8 | -0.46821 | -0.32227 | -0.18638 | -0.05727 | 0.06756 |
| | 3.2 | -0.46819 | -0.32250 | -0.18684 | -0.05794 | 0.06670 |
| | 3.6 | -0.46842 | -0.32303 | -0.18765 | -0.05899 | 0.06545 |
| | 4.0 | -0.46894 | -0.32393 | -0.18886 | -0.06047 | 0.06375 |
| | 6.0 | -0.47790 | -0.33566 | -0.20272 | -0.07586 | 0.04738 |
| | 12.0 | -0.53147 | -0.39392 | -0.26303 | -0.13604 | -0.01072 |
| INFINITY | | -0.52440 | -0.38532 | -0.25335 | -0.12566 | 0.0 |

FRACTILES $w_\alpha$ RHO = −0.80

| DELTAS | | α:0.5500 | 0.6000 | 0.6500 | 0.7000 | 0.7500 |
|---|---|---|---|---|---|---|
| 0.0 | 0.0 | 0.44005 | 0.49621 | 0.54254 | 0.57982 | 0.60823 |
| 0.4 | 0.0 | 0.43170 | 0.48680 | 0.53177 | 0.56715 | 0.59220 |
| | 0.4 | 0.39778 | 0.44733 | 0.48380 | 0.51120 | 0.55158 |
| 0.8 | 0.0 | 0.41399 | 0.46698 | 0.50819 | 0.53495 | 0.56784 |
| | 0.4 | 0.34525 | 0.38910 | 0.43908 | 0.49377 | 0.55498 |
| | 0.8 | 0.28269 | 0.34722 | 0.41413 | 0.48548 | 0.56393 |
| 1.2 | 0.0 | 0.39769 | 0.44832 | 0.48739 | 0.53271 | 0.58318 |
| | 0.4 | 0.31469 | 0.37417 | 0.43513 | 0.49956 | 0.56996 |
| | 0.8 | 0.25709 | 0.33093 | 0.40715 | 0.48793 | 0.57616 |
| | 1.2 | 0.22310 | 0.30744 | 0.39471 | 0.48725 | 0.58821 |
| 1.6 | 0.0 | 0.38576 | 0.44132 | 0.49572 | 0.55106 | 0.60990 |
| | 0.4 | 0.31196 | 0.37810 | 0.44491 | 0.51459 | 0.58972 |
| | 0.8 | 0.25289 | 0.33129 | 0.41179 | 0.49663 | 0.58871 |
| | 1.2 | 0.21043 | 0.30022 | 0.39304 | 0.49129 | 0.59822 |
| | 1.6 | 0.18664 | 0.28414 | 0.38513 | 0.49216 | 0.60867 |
| 2.0 | 0.0 | 0.37775 | 0.44585 | 0.50963 | 0.57275 | 0.63830 |
| | 0.4 | 0.31769 | 0.38929 | 0.46065 | 0.53413 | 0.61244 |
| | 0.8 | 0.26051 | 0.34142 | 0.42390 | 0.51023 | 0.60331 |
| | 1.2 | 0.21335 | 0.30495 | 0.39931 | 0.49885 | 0.60678 |
| | 1.6 | 0.18018 | 0.28097 | 0.38530 | 0.49575 | 0.61582 |
| | 2.0 | 0.16283 | 0.26929 | 0.37967 | 0.49663 | 0.62385 |
| 2.4 | 0.0 | 0.36881 | 0.44828 | 0.52153 | 0.59253 | 0.66477 |
| | 0.4 | 0.32370 | 0.40114 | 0.47722 | 0.55454 | 0.63593 |
| | 0.8 | 0.27222 | 0.35546 | 0.43957 | 0.52689 | 0.62030 |
| | 1.2 | 0.22470 | 0.31670 | 0.41097 | 0.50993 | 0.61671 |
| | 1.6 | 0.18613 | 0.28724 | 0.39164 | 0.50186 | 0.62136 |
| | 2.0 | 0.15979 | 0.26822 | 0.38057 | 0.49952 | 0.62878 |
| | 2.4 | 0.14687 | 0.25948 | 0.37630 | 0.50008 | 0.63465 |
| 2.8 | 0.0 | 0.35478 | 0.44505 | 0.52815 | 0.60763 | 0.68712 |
| | 0.4 | 0.32561 | 0.40973 | 0.49130 | 0.57311 | 0.65815 |
| | 0.8 | 0.28293 | 0.36921 | 0.45554 | 0.54433 | 0.63850 |
| | 1.2 | 0.23864 | 0.33114 | 0.42530 | 0.52353 | 0.62892 |
| | 1.6 | 0.19866 | 0.29903 | 0.40223 | 0.51076 | 0.62799 |
| | 2.0 | 0.16687 | 0.27486 | 0.38652 | 0.50448 | 0.63238 |
| | 2.4 | 0.14579 | 0.25954 | 0.37749 | 0.50239 | 0.63806 |
| | 2.8 | 0.13602 | 0.25287 | 0.37414 | 0.50264 | 0.64228 |
| 3.2 | 0.0 | 0.33747 | 0.43636 | 0.52868 | 0.61689 | 0.70420 |
| | 0.4 | 0.32185 | 0.41321 | 0.50101 | 0.58809 | 0.67755 |
| | 0.8 | 0.28977 | 0.38005 | 0.46947 | 0.56059 | 0.65636 |
| | 1.2 | 0.25134 | 0.34514 | 0.43990 | 0.53807 | 0.64269 |
| | 1.6 | 0.21312 | 0.31294 | 0.41502 | 0.52185 | 0.63670 |
| | 2.0 | 0.17935 | 0.28604 | 0.39598 | 0.51175 | 0.63689 |
| | 2.4 | 0.15305 | 0.26602 | 0.38294 | 0.50652 | 0.64051 |
| | 2.8 | 0.13605 | 0.25356 | 0.37544 | 0.50452 | 0.64470 |
| | 3.2 | 0.12856 | 0.24838 | 0.37276 | 0.50456 | 0.64772 |

FRACTILES $W_\alpha$ RHO = -0.80

| DELTAS | | α:0.5500 | 0.6000 | 0.6500 | 0.7000 | 0.7500 |
|---|---|---|---|---|---|---|
| 3.6 | 0.0 | 0.31992 | 0.42468 | 0.52438 | 0.62057 | 0.71574 |
| | 0.4 | 0.31320 | 0.41146 | 0.50573 | 0.59863 | 0.69320 |
| | 0.8 | 0.29164 | 0.38665 | 0.48000 | 0.57430 | 0.67256 |
| | 1.2 | 0.26067 | 0.35671 | 0.45299 | 0.55200 | 0.65679 |
| | 1.6 | 0.22647 | 0.32648 | 0.42814 | 0.53394 | 0.64708 |
| | 2.0 | 0.19346 | 0.29902 | 0.40731 | 0.52090 | 0.64319 |
| | 2.4 | 0.16486 | 0.27632 | 0.39134 | 0.51259 | 0.64371 |
| | 2.8 | 0.14303 | 0.25962 | 0.38037 | 0.50805 | 0.64648 |
| | 3.2 | 0.12921 | 0.24939 | 0.37408 | 0.50613 | 0.64950 |
| | 3.6 | 0.12339 | 0.24532 | 0.37191 | 0.50602 | 0.65168 |
| 4.0 | 0.0 | 0.30392 | 0.41250 | 0.51744 | 0.62010 | 0.72241 |
| | 0.4 | 0.30183 | 0.40578 | 0.50599 | 0.60472 | 0.70477 |
| | 0.8 | 0.28886 | 0.38881 | 0.48659 | 0.58473 | 0.68622 |
| | 1.2 | 0.26576 | 0.36482 | 0.46346 | 0.56420 | 0.67009 |
| | 1.6 | 0.23698 | 0.33806 | 0.44018 | 0.54583 | 0.65817 |
| | 2.0 | 0.20669 | 0.31177 | 0.41904 | 0.53103 | 0.65108 |
| | 2.4 | 0.17817 | 0.28822 | 0.40138 | 0.52026 | 0.64840 |
| | 2.8 | 0.15391 | 0.26897 | 0.38783 | 0.51323 | 0.64888 |
| | 3.2 | 0.13569 | 0.25495 | 0.37851 | 0.50920 | 0.65089 |
| | 3.6 | 0.12437 | 0.24648 | 0.37319 | 0.50737 | 0.65303 |
| | 4.0 | 0.11980 | 0.24325 | 0.37141 | 0.50719 | 0.65462 |
| 6.0 | 0.0 | 0.24992 | 0.36705 | 0.48362 | 0.60158 | 0.72338 |
| | 0.4 | 0.25121 | 0.36736 | 0.48281 | 0.59953 | 0.71998 |
| | 0.8 | 0.25091 | 0.36589 | 0.48008 | 0.59552 | 0.71474 |
| | 1.2 | 0.24825 | 0.36187 | 0.47478 | 0.58913 | 0.70767 |
| | 1.6 | 0.24231 | 0.35463 | 0.46660 | 0.58051 | 0.69931 |
| | 2.0 | 0.23258 | 0.34409 | 0.45586 | 0.57031 | 0.69053 |
| | 2.4 | 0.21931 | 0.33085 | 0.44340 | 0.55943 | 0.68219 |
| | 2.8 | 0.20343 | 0.31593 | 0.43019 | 0.54876 | 0.67498 |
| | 3.2 | 0.18618 | 0.30045 | 0.41718 | 0.53898 | 0.66933 |
| | 3.6 | 0.16886 | 0.28544 | 0.40511 | 0.53054 | 0.66535 |
| | 4.0 | 0.15267 | 0.27180 | 0.39456 | 0.52368 | 0.66291 |
| | 6.0 | 0.11309 | 0.23991 | 0.37153 | 0.51088 | 0.66204 |
| 12.0 | 0.0 | 0.18988 | 0.31331 | 0.43863 | 0.56823 | 0.70529 |
| | 0.4 | 0.19047 | 0.31377 | 0.43893 | 0.56836 | 0.70520 |
| | 0.8 | 0.19093 | 0.31410 | 0.43911 | 0.56835 | 0.70500 |
| | 1.2 | 0.19124 | 0.31426 | 0.43912 | 0.56820 | 0.70466 |
| | 1.6 | 0.19136 | 0.31424 | 0.43894 | 0.56786 | 0.70417 |
| | 2.0 | 0.19125 | 0.31399 | 0.43855 | 0.56733 | 0.70350 |
| | 2.4 | 0.19089 | 0.31348 | 0.43790 | 0.56657 | 0.70265 |
| | 2.8 | 0.19021 | 0.31267 | 0.43697 | 0.56554 | 0.70159 |
| | 3.2 | 0.18919 | 0.31151 | 0.43571 | 0.56424 | 0.70031 |
| | 3.6 | 0.18777 | 0.30997 | 0.43411 | 0.56264 | 0.69881 |
| | 4.0 | 0.18591 | 0.30801 | 0.43212 | 0.56072 | 0.69708 |
| | 6.0 | 0.16914 | 0.29145 | 0.41647 | 0.54681 | 0.68598 |
| | 12.0 | 0.11494 | 0.24299 | 0.37572 | 0.51602 | 0.66793 |
| INFINITY | | 0.12566 | 0.25335 | 0.38532 | 0.52440 | 0.67449 |

FRACTILES $W_\alpha$ RHO = -0.80

| DELTAS | | $\alpha$:0.8000 | 0.8500 | 0.9000 | 0.9250 | 0.9500 |
|---|---|---|---|---|---|---|
| 0.0 | 0.0 | 0.62576 | 0.64758 | 0.68798 | 0.72001 | 0.76809 |
| 0.4 | 0.0 | 0.61109 | 0.64559 | 0.70195 | 0.74476 | 0.80733 |
| | 0.4 | 0.60346 | 0.67073 | 0.76434 | 0.82965 | 0.92020 |
| 0.8 | 0.0 | 0.61224 | 0.67040 | 0.75157 | 0.80822 | 0.88675 |
| | 0.4 | 0.62594 | 0.71259 | 0.82803 | 0.90636 | 1.01292 |
| | 0.8 | 0.65352 | 0.76137 | 0.90293 | 0.99786 | 1.12580 |
| 1.2 | 0.0 | 0.64123 | 0.71160 | 0.80474 | 0.86768 | 0.95307 |
| | 0.4 | 0.64993 | 0.74582 | 0.87122 | 0.95512 | 1.06800 |
| | 0.8 | 0.67616 | 0.79556 | 0.95081 | 1.05411 | 1.19241 |
| | 1.2 | 0.70238 | 0.83826 | 1.01415 | 1.13069 | 1.28611 |
| 1.6 | 0.0 | 0.67542 | 0.75278 | 0.85266 | 0.91894 | 1.00766 |
| | 0.4 | 0.67401 | 0.77381 | 0.90260 | 0.98788 | 1.10165 |
| | 0.8 | 0.69241 | 0.81538 | 0.97402 | 1.07889 | 1.21852 |
| | 1.2 | 0.71879 | 0.86178 | 1.04605 | 1.16766 | 1.32928 |
| | 1.6 | 0.73999 | 0.89556 | 1.09566 | 1.22744 | 1.40225 |
| 2.0 | 0.0 | 0.70976 | 0.79249 | 0.89728 | 0.96583 | 1.05658 |
| | 0.4 | 0.69929 | 0.80097 | 0.93067 | 1.01576 | 1.12846 |
| | 0.8 | 0.70742 | 0.83001 | 0.98698 | 1.09010 | 1.22673 |
| | 1.2 | 0.72802 | 0.87119 | 1.05481 | 1.17550 | 1.33536 |
| | 1.6 | 0.75092 | 0.91062 | 1.11550 | 1.25009 | 1.42823 |
| | 2.0 | 0.76699 | 0.93613 | 1.15292 | 1.29518 | 1.48325 |
| 2.4 | 0.0 | 0.74203 | 0.82988 | 0.93924 | 1.00983 | 1.10237 |
| | 0.4 | 0.72516 | 0.82845 | 0.95872 | 1.04342 | 1.15483 |
| | 0.8 | 0.72401 | 0.84523 | 0.99923 | 1.09977 | 1.23232 |
| | 1.2 | 0.73608 | 0.87636 | 1.05533 | 1.17244 | 1.32701 |
| | 1.6 | 0.75544 | 0.91346 | 1.11548 | 1.24782 | 1.42254 |
| | 2.0 | 0.77403 | 0.94542 | 1.16467 | 1.30831 | 1.49790 |
| | 2.4 | 0.78590 | 0.96435 | 1.19251 | 1.34188 | 1.53891 |
| 2.8 | 0.0 | 0.77067 | 0.86404 | 0.97831 | 1.05115 | 1.14570 |
| | 0.4 | 0.75027 | 0.85569 | 0.98710 | 1.07177 | 1.18240 |
| | 0.8 | 0.74221 | 0.86245 | 1.01395 | 1.11222 | 1.24112 |
| | 1.2 | 0.74607 | 0.88296 | 1.05659 | 1.16967 | 1.31834 |
| | 1.6 | 0.75902 | 0.91287 | 1.10874 | 1.23662 | 1.40499 |
| | 2.0 | 0.77580 | 0.94462 | 1.16001 | 1.30080 | 1.48628 |
| | 2.4 | 0.79041 | 0.96994 | 1.19917 | 1.34905 | 1.54651 |
| | 2.8 | 0.79912 | 0.98394 | 1.21987 | 1.37406 | 1.57709 |
| 3.2 | 0.0 | 0.79467 | 0.89423 | 1.01409 | 1.08953 | 1.18657 |
| | 0.4 | 0.77334 | 0.88169 | 1.01515 | 1.10037 | 1.21094 |
| | 0.8 | 0.76090 | 0.88109 | 1.03121 | 1.12792 | 1.25410 |
| | 1.2 | 0.75826 | 0.89247 | 1.06158 | 1.17116 | 1.31467 |
| | 1.6 | 0.76451 | 0.91388 | 1.10317 | 1.22627 | 1.38787 |
| | 2.0 | 0.77678 | 0.94093 | 1.14967 | 1.28574 | 1.46460 |
| | 2.4 | 0.79071 | 0.96736 | 1.19243 | 1.33931 | 1.53253 |
| | 2.8 | 0.80202 | 0.98724 | 1.22341 | 1.37761 | 1.58046 |
| | 3.2 | 0.80844 | 0.99767 | 1.23891 | 1.39636 | 1.60343 |

FRACTILES $W_\alpha$ RHO = -0.80

| DELTAS | | $\alpha$: 0.8000 | 0.8500 | 0.9000 | 0.9250 | 0.9500 |
|---|---|---|---|---|---|---|
| 3.6 | 0.0 | 0.81359 | 0.91996 | 1.04617 | 1.12467 | 1.22471 |
| | 0.4 | 0.79343 | 0.90555 | 1.04206 | 1.12843 | 1.23971 |
| | 0.8 | 0.77888 | 0.90005 | 1.05004 | 1.14599 | 1.27049 |
| | 1.2 | 0.77176 | 0.90438 | 1.07033 | 1.17728 | 1.31674 |
| | 1.6 | 0.77234 | 0.91800 | 1.10160 | 1.22051 | 1.37607 |
| | 2.0 | 0.77940 | 0.93864 | 1.14032 | 1.27138 | 1.44321 |
| | 2.4 | 0.79031 | 0.96227 | 1.18076 | 1.32302 | 1.50980 |
| | 2.8 | 0.80161 | 0.98396 | 1.21604 | 1.36732 | 1.56610 |
| | 3.2 | 0.81034 | 0.99957 | 1.24058 | 1.39776 | 1.60431 |
| | 3.6 | 0.81511 | 1.00741 | 1.25230 | 1.41198 | 1.62176 |
| 4.0 | 0.0 | 0.82757 | 0.94111 | 1.07432 | 1.15631 | 1.25989 |
| | 0.4 | 0.81000 | 0.92664 | 1.06712 | 1.15523 | 1.26795 |
| | 0.8 | 0.79518 | 0.91832 | 1.06937 | 1.16531 | 1.28912 |
| | 1.2 | 0.78547 | 0.91764 | 1.08184 | 1.18705 | 1.32364 |
| | 1.6 | 0.78187 | 0.92492 | 1.10420 | 1.21978 | 1.37047 |
| | 2.0 | 0.78423 | 0.93922 | 1.13465 | 1.26119 | 1.42664 |
| | 2.4 | 0.79122 | 0.95822 | 1.16968 | 1.30699 | 1.48689 |
| | 2.8 | 0.80057 | 0.97846 | 1.20432 | 1.35126 | 1.54402 |
| | 3.2 | 0.80964 | 0.99615 | 1.23332 | 1.38779 | 1.59057 |
| | 3.6 | 0.81638 | 1.00845 | 1.25287 | 1.41212 | 1.62121 |
| | 4.0 | 0.81997 | 1.01441 | 1.26184 | 1.42302 | 1.63461 |
| 6.0 | 0.0 | 0.85247 | 0.99463 | 1.16171 | 1.26319 | 1.38885 |
| | 0.4 | 0.84767 | 0.98858 | 1.15521 | 1.25725 | 1.38475 |
| | 0.8 | 0.84149 | 0.98214 | 1.15003 | 1.25387 | 1.38477 |
| | 1.2 | 0.83442 | 0.97620 | 1.14727 | 1.25411 | 1.38987 |
| | 1.6 | 0.82730 | 0.97176 | 1.14787 | 1.25880 | 1.40074 |
| | 2.0 | 0.82106 | 0.96964 | 1.15243 | 1.26843 | 1.41767 |
| | 2.4 | 0.81643 | 0.97037 | 1.16123 | 1.28306 | 1.44053 |
| | 2.8 | 0.81388 | 0.97412 | 1.17405 | 1.30230 | 1.46865 |
| | 3.2 | 0.81349 | 0.98064 | 1.19024 | 1.32520 | 1.50075 |
| | 3.6 | 0.81504 | 0.98928 | 1.20862 | 1.35027 | 1.53494 |
| | 4.0 | 0.81799 | 0.99905 | 1.22765 | 1.37562 | 1.56884 |
| | 6.0 | 0.83133 | 1.03001 | 1.28215 | 1.44595 | 1.66041 |
| 12.0 | 0.0 | 0.85453 | 1.02405 | 1.23067 | 1.36000 | 1.52369 |
| | 0.4 | 0.85419 | 1.02340 | 1.22966 | 1.35877 | 1.52222 |
| | 0.8 | 0.85375 | 1.02271 | 1.22868 | 1.35764 | 1.52097 |
| | 1.2 | 0.85322 | 1.02197 | 1.22776 | 1.35666 | 1.51999 |
| | 1.6 | 0.85257 | 1.02119 | 1.22691 | 1.35585 | 1.51936 |
| | 2.0 | 0.85180 | 1.02037 | 1.22618 | 1.35528 | 1.51914 |
| | 2.4 | 0.85091 | 1.01953 | 1.22560 | 1.35499 | 1.51942 |
| | 2.8 | 0.84989 | 1.01869 | 1.22521 | 1.35506 | 1.52029 |
| | 3.2 | 0.84875 | 1.01786 | 1.22507 | 1.35555 | 1.52184 |
| | 3.6 | 0.84749 | 1.01709 | 1.22524 | 1.35655 | 1.52415 |
| | 4.0 | 0.84614 | 1.01641 | 1.22579 | 1.35812 | 1.52733 |
| | 6.0 | 0.83931 | 1.01609 | 1.23597 | 1.37638 | 1.55760 |
| | 12.0 | 0.83769 | 1.03639 | 1.28766 | 1.45036 | 1.66271 |
| INFINITY | | 0.84162 | 1.03643 | 1.28155 | 1.43953 | 1.64485 |

FRACTILES $w_\alpha$ RHO = −0.80

| DELTAS | | $\alpha$:0.9625 | 0.9750 | 0.9825 | 0.9900 | 0.9925 |
|---|---|---|---|---|---|---|
| 0.0 | 0.0 | 0.80372 | 0.85549 | 0.90223 | 0.97726 | 1.01649 |
| 0.4 | 0.0 | 0.85274 | 0.91766 | 0.97535 | 1.06656 | 1.11365 |
| | 0.4 | 0.98349 | 1.07151 | 1.14791 | 1.26613 | 1.32621 |
| 0.8 | 0.0 | 0.94162 | 1.01790 | 1.08411 | 1.18651 | 1.23854 |
| | 0.4 | 1.08631 | 1.18717 | 1.27380 | 1.40645 | 1.47331 |
| | 0.8 | 1.21320 | 1.33252 | 1.43434 | 1.58923 | 1.66687 |
| 1.2 | 0.0 | 1.01175 | 1.09229 | 1.16137 | 1.26703 | 1.32024 |
| | 0.4 | 1.14503 | 1.25011 | 1.33971 | 1.47590 | 1.54414 |
| | 0.8 | 1.28635 | 1.41398 | 1.52237 | 1.68642 | 1.76833 |
| | 1.2 | 1.39132 | 1.53379 | 1.65441 | 1.83634 | 1.92691 |
| 1.6 | 0.0 | 1.06796 | 1.14997 | 1.21974 | 1.32556 | 1.37849 |
| | 0.4 | 1.17875 | 1.28328 | 1.37192 | 1.50588 | 1.57268 |
| | 0.8 | 1.31293 | 1.44066 | 1.54871 | 1.71157 | 1.79260 |
| | 1.2 | 1.43834 | 1.58562 | 1.70997 | 1.89697 | 1.98983 |
| | 1.6 | 1.51998 | 1.67869 | 1.81243 | 2.01316 | 2.11266 |
| 2.0 | 0.0 | 1.11771 | 1.20024 | 1.26997 | 1.37501 | 1.42726 |
| | 0.4 | 1.20436 | 1.30674 | 1.39313 | 1.52304 | 1.58756 |
| | 0.8 | 1.31870 | 1.44268 | 1.54718 | 1.70410 | 1.78194 |
| | 1.2 | 1.44290 | 1.58774 | 1.70971 | 1.89263 | 1.98326 |
| | 1.6 | 1.54797 | 1.70909 | 1.84462 | 2.04762 | 2.14807 |
| | 2.0 | 1.60951 | 1.77923 | 1.92181 | 2.13510 | 2.24052 |
| 2.4 | 0.0 | 1.16419 | 1.24713 | 1.31676 | 1.42100 | 1.47261 |
| | 0.4 | 1.22943 | 1.32958 | 1.41370 | 1.53961 | 1.60191 |
| | 0.8 | 1.32117 | 1.44051 | 1.54076 | 1.69077 | 1.76497 |
| | 1.2 | 1.43067 | 1.56992 | 1.68688 | 1.86182 | 1.94829 |
| | 1.6 | 1.53972 | 1.69710 | 1.82923 | 2.02674 | 2.12432 |
| | 2.0 | 1.62501 | 1.79563 | 1.93879 | 2.15263 | 2.25819 |
| | 2.4 | 1.67089 | 1.84793 | 1.99635 | 2.21785 | 2.32710 |
| 2.8 | 0.0 | 1.20840 | 1.29196 | 1.36171 | 1.46554 | 1.51670 |
| | 0.4 | 1.25605 | 1.35446 | 1.43676 | 1.55939 | 1.61986 |
| | 0.8 | 1.32715 | 1.44231 | 1.53870 | 1.68246 | 1.75335 |
| | 1.2 | 1.41774 | 1.55090 | 1.66244 | 1.82883 | 1.91090 |
| | 1.6 | 1.51764 | 1.66863 | 1.79514 | 1.98387 | 2.07694 |
| | 2.0 | 1.61042 | 1.77681 | 1.91622 | 2.12413 | 2.22663 |
| | 2.4 | 1.67864 | 1.85571 | 2.00401 | 2.22507 | 2.33401 |
| | 2.8 | 1.71288 | 1.89476 | 2.04699 | 2.27377 | 2.38545 |
| 3.2 | 0.0 | 1.25041 | 1.33498 | 1.40518 | 1.50908 | 1.56005 |
| | 0.4 | 1.28412 | 1.38146 | 1.46251 | 1.58275 | 1.64182 |
| | 0.8 | 1.33796 | 1.44978 | 1.54308 | 1.68172 | 1.74990 |
| | 1.2 | 1.41027 | 1.53800 | 1.64470 | 1.80343 | 1.88155 |
| | 1.6 | 1.49571 | 1.63993 | 1.76052 | 1.94001 | 2.02837 |
| | 2.0 | 1.58407 | 1.74395 | 1.87769 | 2.07680 | 2.17483 |
| | 2.4 | 1.66165 | 1.83447 | 1.97904 | 2.19428 | 2.30023 |
| | 2.8 | 1.71601 | 1.89743 | 2.04915 | 2.27498 | 2.38611 |
| | 3.2 | 1.74175 | 1.92680 | 2.08149 | 2.31162 | 2.42480 |

FRACTILES $w_\alpha$ RHO = -0.80

| DELTAS | | $\alpha$: 0.9625 | 0.9750 | 0.9825 | 0.9900 | 0.9925 |
|---|---|---|---|---|---|---|
| 3.6 | 0.0 | 1.29004 | 1.37606 | 1.44705 | 1.55155 | 1.60258 |
| | 0.4 | 1.31294 | 1.40988 | 1.49023 | 1.60893 | 1.66704 |
| | 0.8 | 1.35286 | 1.46229 | 1.55326 | 1.68799 | 1.75406 |
| | 1.2 | 1.40932 | 1.53263 | 1.63536 | 1.78776 | 1.86258 |
| | 1.6 | 1.47960 | 1.61773 | 1.73296 | 1.90410 | 1.98819 |
| | 2.0 | 1.55774 | 1.71073 | 1.83847 | 2.02831 | 2.12164 |
| | 2.4 | 1.63442 | 1.80099 | 1.94013 | 2.14701 | 2.24872 |
| | 2.8 | 1.69877 | 1.87616 | 2.02438 | 2.24475 | 2.35309 |
| | 3.2 | 1.74219 | 1.92652 | 2.08052 | 2.30945 | 2.42198 |
| | 3.6 | 1.76175 | 1.94885 | 2.10511 | 2.33731 | 2.45140 |
| 4.0 | 0.0 | 1.32704 | 1.41493 | 1.48707 | 1.59267 | 1.64403 |
| | 0.4 | 1.34170 | 1.43887 | 1.51905 | 1.63699 | 1.69452 |
| | 0.8 | 1.37064 | 1.47854 | 1.56792 | 1.69981 | 1.76431 |
| | 1.2 | 1.41398 | 1.53394 | 1.63358 | 1.78097 | 1.85316 |
| | 1.6 | 1.47044 | 1.60351 | 1.71426 | 1.87833 | 1.95880 |
| | 2.0 | 1.53666 | 1.68332 | 1.80555 | 1.98685 | 2.07584 |
| | 2.4 | 1.60669 | 1.76657 | 1.89994 | 2.09790 | 2.19511 |
| | 2.8 | 1.67251 | 1.84409 | 1.98729 | 2.19994 | 2.30439 |
| | 3.2 | 1.72579 | 1.90642 | 2.05720 | 2.28115 | 2.39115 |
| | 3.6 | 1.76066 | 1.94694 | 2.10243 | 2.33334 | 2.44674 |
| | 4.0 | 1.77570 | 1.96412 | 2.12136 | 2.35478 | 2.46939 |
| 6.0 | 0.0 | 1.46854 | 1.57066 | 1.65268 | 1.76999 | 1.82594 |
| | 0.4 | 1.46631 | 1.57164 | 1.65686 | 1.77967 | 1.83858 |
| | 0.8 | 1.46918 | 1.57891 | 1.66825 | 1.79776 | 1.86018 |
| | 1.2 | 1.47802 | 1.59324 | 1.68752 | 1.82485 | 1.89129 |
| | 1.6 | 1.49341 | 1.61509 | 1.71506 | 1.86125 | 1.93219 |
| | 2.0 | 1.51555 | 1.64454 | 1.75086 | 1.90682 | 1.98268 |
| | 2.4 | 1.54418 | 1.68117 | 1.79439 | 1.96086 | 2.04199 |
| | 2.8 | 1.57847 | 1.72394 | 1.84441 | 2.02191 | 2.10853 |
| | 3.2 | 1.61691 | 1.77106 | 1.89893 | 2.08761 | 2.17980 |
| | 3.6 | 1.65735 | 1.82002 | 1.95512 | 2.15470 | 2.25231 |
| | 4.0 | 1.69709 | 1.86769 | 2.00952 | 2.21922 | 2.32184 |
| | 6.0 | 1.80304 | 1.99305 | 2.15121 | 2.38532 | 2.49996 |
| 12.0 | 0.0 | 1.62914 | 1.76556 | 1.87568 | 2.03325 | 2.10815 |
| | 0.4 | 1.62755 | 1.76387 | 1.87397 | 2.03163 | 2.10663 |
| | 0.8 | 1.62627 | 1.76262 | 1.87283 | 2.03080 | 2.10602 |
| | 1.2 | 1.62537 | 1.76192 | 1.87239 | 2.03091 | 2.10649 |
| | 1.6 | 1.62494 | 1.76187 | 1.87276 | 2.03211 | 2.10818 |
| | 2.0 | 1.62505 | 1.76257 | 1.87408 | 2.03456 | 2.11128 |
| | 2.4 | 1.62583 | 1.76416 | 1.87648 | 2.03841 | 2.11594 |
| | 2.8 | 1.62736 | 1.76675 | 1.88010 | 2.04380 | 2.12231 |
| | 3.2 | 1.62976 | 1.77046 | 1.88507 | 2.05088 | 2.13052 |
| | 3.6 | 1.63312 | 1.77540 | 1.89148 | 2.05974 | 2.14069 |
| | 4.0 | 1.63753 | 1.78165 | 1.89944 | 2.07047 | 2.15289 |
| | 6.0 | 1.67661 | 1.83343 | 1.96255 | 2.15153 | 2.24320 |
| | 12.0 | 1.80351 | 1.99057 | 2.14580 | 2.37481 | 2.48662 |
| INFINITY | | 1.78046 | 1.95996 | 2.10836 | 2.32635 | 2.43238 |

FRACTILES $W_\alpha$ RHO = -0.80

| DELTAS | | $\alpha$:0.9950 | 0.9975 | 0.9990 | 0.9995 |
|---|---|---|---|---|---|
| 0.0 | 0.0 | 1.07239 | 1.16933 | 1.29952 | 1.39921 |
| 0.4 | 0.0 | 1.18018 | 1.29413 | 1.44489 | 1.55890 |
| | 0.4 | 1.41018 | 1.55202 | 1.73671 | 1.87465 |
| 0.8 | 0.0 | 1.31125 | 1.43405 | 1.59391 | 1.71327 |
| | 0.4 | 1.56621 | 1.72191 | 1.92271 | 2.07148 |
| | 0.8 | 1.77434 | 1.95348 | 2.18295 | 2.35197 |
| 1.2 | 0.0 | 1.39413 | 1.51788 | 1.67736 | 1.79545 |
| | 0.4 | 1.63857 | 1.79589 | 1.99733 | 2.14565 |
| | 0.8 | 1.88135 | 2.06896 | 2.30798 | 2.48324 |
| | 1.2 | 2.05163 | 2.25800 | 2.51990 | 2.71125 |
| 1.6 | 0.0 | 1.45167 | 1.57341 | 1.72907 | 1.84355 |
| | 0.4 | 1.66480 | 1.81757 | 2.01200 | 2.15445 |
| | 0.8 | 1.90414 | 2.08862 | 2.32260 | 2.49349 |
| | 1.2 | 2.11747 | 2.32810 | 2.59449 | 2.78853 |
| | 1.6 | 2.24924 | 2.47419 | 2.75795 | 2.96418 |
| 2.0 | 0.0 | 1.49920 | 1.61826 | 1.76946 | 1.88005 |
| | 0.4 | 1.67627 | 1.82277 | 2.00828 | 2.14359 |
| | 0.8 | 1.88885 | 2.06510 | 2.28776 | 2.44982 |
| | 1.2 | 2.10761 | 2.31232 | 2.57042 | 2.75791 |
| | 1.6 | 2.28579 | 2.51220 | 2.79711 | 3.00372 |
| | 2.0 | 2.38492 | 2.62199 | 2.91977 | 3.13537 |
| 2.4 | 0.0 | 1.54341 | 1.66002 | 1.80723 | 1.91437 |
| | 0.4 | 1.68735 | 1.82791 | 2.00506 | 2.13378 |
| | 0.8 | 1.86666 | 2.03382 | 2.24421 | 2.39685 |
| | 1.2 | 2.06676 | 2.26134 | 2.50592 | 2.68315 |
| | 1.6 | 2.25792 | 2.47717 | 2.75242 | 2.95162 |
| | 2.0 | 2.40265 | 2.63950 | 2.93647 | 3.15113 |
| | 2.4 | 2.47650 | 2.72123 | 3.02767 | 3.24890 |
| 2.8 | 0.0 | 1.58666 | 1.70137 | 1.84540 | 1.94974 |
| | 0.4 | 1.70256 | 1.83815 | 2.00829 | 2.13145 |
| | 0.8 | 1.85033 | 2.00929 | 2.20866 | 2.35287 |
| | 1.2 | 2.02315 | 2.20710 | 2.43767 | 2.60432 |
| | 1.6 | 2.20421 | 2.41270 | 2.67385 | 2.86246 |
| | 2.0 | 2.36677 | 2.59620 | 2.88335 | 3.09058 |
| | 2.4 | 2.48288 | 2.72646 | 3.03105 | 3.25068 |
| | 2.8 | 2.53799 | 2.78741 | 3.09899 | 3.32344 |
| 3.2 | 0.0 | 1.62953 | 1.74297 | 1.88466 | 1.98685 |
| | 0.4 | 1.72242 | 1.85411 | 2.01866 | 2.13735 |
| | 0.8 | 1.84299 | 1.99514 | 2.18531 | 2.32246 |
| | 1.2 | 1.98822 | 2.16263 | 2.38062 | 2.53780 |
| | 1.6 | 2.14905 | 2.34637 | 2.59296 | 2.77070 |
| | 2.0 | 2.30870 | 2.52757 | 2.80100 | 2.99799 |
| | 2.4 | 2.44491 | 2.68137 | 2.97661 | 3.18922 |
| | 2.8 | 2.53781 | 2.78565 | 3.09490 | 3.31745 |
| | 3.2 | 2.57927 | 2.83146 | 3.14591 | 3.37203 |

FRACTILES $W_\alpha$ RHO = -0.80

| DELTAS | | $\alpha$: 0.9950 | 0.9975 | 0.9990 | 0.9995 |
|---|---|---|---|---|---|
| 3.6 | 0.0 | 1.67195 | 1.78472 | 1.92485 | 2.02551 |
| | 0.4 | 1.74614 | 1.87492 | 2.03518 | 2.15038 |
| | 0.8 | 1.84407 | 1.99080 | 2.17356 | 2.30499 |
| | 1.2 | 1.96459 | 2.13099 | 2.33838 | 2.48756 |
| | 1.6 | 2.10288 | 2.29006 | 2.52342 | 2.69129 |
| | 2.0 | 2.24896 | 2.45679 | 2.71592 | 2.90230 |
| | 2.4 | 2.38750 | 2.61403 | 2.89645 | 3.09953 |
| | 2.8 | 2.50090 | 2.74215 | 3.04282 | 3.25895 |
| | 3.2 | 2.57547 | 2.82591 | 3.13788 | 3.36203 |
| | 3.6 | 2.60698 | 2.86071 | 3.17658 | 3.40340 |
| 4.0 | 0.0 | 1.71361 | 1.82627 | 1.96557 | 2.06521 |
| | 0.4 | 1.77264 | 1.89940 | 2.05649 | 2.16902 |
| | 0.8 | 1.85200 | 1.99454 | 2.17148 | 2.29835 |
| | 1.2 | 1.95142 | 2.11133 | 2.31005 | 2.45265 |
| | 1.6 | 2.06839 | 2.24691 | 2.46893 | 2.62832 |
| | 2.0 | 2.19709 | 2.39471 | 2.64062 | 2.81719 |
| | 2.4 | 2.32762 | 2.54364 | 2.81250 | 3.00556 |
| | 2.8 | 2.44678 | 2.67893 | 2.96788 | 3.17534 |
| | 3.2 | 2.54110 | 2.78558 | 3.08980 | 3.30818 |
| | 3.6 | 2.60132 | 2.85327 | 3.16669 | 3.39158 |
| | 4.0 | 2.62556 | 2.88003 | 3.19640 | 3.42331 |
| 6.0 | 0.0 | 1.90072 | 2.01947 | 2.16286 | 2.26340 |
| | 0.4 | 1.91762 | 2.04379 | 2.19699 | 2.30487 |
| | 0.8 | 1.94418 | 2.07881 | 2.24302 | 2.35906 |
| | 1.2 | 1.98093 | 2.12504 | 2.30145 | 2.42644 |
| | 1.6 | 2.02807 | 2.18263 | 2.37238 | 2.50711 |
| | 2.0 | 2.08538 | 2.25126 | 2.45537 | 2.60056 |
| | 2.4 | 2.15196 | 2.32987 | 2.54920 | 2.70542 |
| | 2.8 | 2.22608 | 2.41649 | 2.65156 | 2.81918 |
| | 3.2 | 2.30499 | 2.50799 | 2.75889 | 2.93795 |
| | 3.6 | 2.38493 | 2.60015 | 2.86637 | 3.05647 |
| | 4.0 | 2.46133 | 2.68783 | 2.96817 | 3.16845 |
| | 6.0 | 2.65587 | 2.90913 | 3.22269 | 3.44671 |
| 12.0 | 0.0 | 2.20773 | 2.36417 | 2.54939 | 2.67646 |
| | 0.4 | 2.20642 | 2.36338 | 2.54958 | 2.67759 |
| | 0.8 | 2.20619 | 2.36397 | 2.55156 | 2.68082 |
| | 1.2 | 2.20722 | 2.36615 | 2.55556 | 2.68638 |
| | 1.6 | 2.20969 | 2.37011 | 2.56178 | 2.69448 |
| | 2.0 | 2.21377 | 2.37605 | 2.57042 | 2.70533 |
| | 2.4 | 2.21964 | 2.38413 | 2.58167 | 2.71910 |
| | 2.8 | 2.22744 | 2.39453 | 2.59568 | 2.73594 |
| | 3.2 | 2.23732 | 2.40736 | 2.61256 | 2.75596 |
| | 3.6 | 2.24938 | 2.42272 | 2.63240 | 2.77922 |
| | 4.0 | 2.26367 | 2.44065 | 2.65520 | 2.80572 |
| | 6.0 | 2.36702 | 2.56620 | 2.80974 | 2.98186 |
| | 12.0 | 2.63832 | 2.88390 | 3.18650 | 3.40172 |
| INFINITY | | 2.57583 | 2.80703 | 3.09024 | 3.29051 |

FRACTILES $W_\alpha$ RHO = -0.60

| DELTAS | | $\alpha$:0.0005 | 0.0010 | 0.0025 | 0.0050 | 0.0075 |
|---|---|---|---|---|---|---|
| 0.0 | 0.0 | -7.64498 | -6.76149 | -5.60449 | -4.73960 | -4.23881 |
| 0.4 | 0.0 | -7.57118 | -6.69964 | -5.55730 | -4.70263 | -4.20745 |
| | 0.4 | -7.33666 | -6.49222 | -5.38642 | -4.55990 | -4.08137 |
| 0.8 | 0.0 | -7.29399 | -6.47572 | -5.39797 | -4.58709 | -4.11526 |
| | 0.4 | -7.05367 | -6.24769 | -5.19143 | -4.40131 | -3.94362 |
| | 0.8 | -6.61567 | -5.86332 | -4.87832 | -4.14232 | -3.71634 |
| 1.2 | 0.0 | -6.86402 | -6.12558 | -5.14756 | -4.40665 | -3.97306 |
| | 0.4 | -6.74092 | -5.99162 | -5.00481 | -4.26251 | -3.83066 |
| | 0.8 | -6.30159 | -5.59206 | -4.66243 | -3.96732 | -3.56484 |
| | 1.2 | -5.81853 | -5.16981 | -4.32082 | -3.68680 | -3.32007 |
| 1.6 | 0.0 | -6.41542 | -5.75409 | -4.87445 | -4.20451 | -3.81065 |
| | 0.4 | -6.37558 | -5.69653 | -4.79720 | -4.11596 | -3.71733 |
| | 0.8 | -6.07426 | -5.40922 | -4.53357 | -3.87511 | -3.49219 |
| | 1.2 | -5.56567 | -4.95248 | -4.14944 | -3.54938 | -3.20217 |
| | 1.6 | -5.12476 | -4.56877 | -3.84155 | -3.29897 | -2.98542 |
| 2.0 | 0.0 | -6.01794 | -5.42100 | -4.62424 | -4.01482 | -3.65528 |
| | 0.4 | -6.01349 | -5.40002 | -4.58401 | -3.96249 | -3.59710 |
| | 0.8 | -5.83126 | -5.21934 | -4.40897 | -3.79521 | -3.43612 |
| | 1.2 | -5.44461 | -4.86084 | -4.09245 | -3.51496 | -3.17931 |
| | 1.6 | -4.94753 | -4.41777 | -3.72441 | -3.20684 | -2.90767 |
| | 2.0 | -4.57438 | -4.09486 | -3.46820 | -3.00129 | -2.73187 |
| 2.4 | 0.0 | -5.68590 | -5.14071 | -4.41080 | -3.85045 | -3.51885 |
| | 0.4 | -5.69274 | -5.13398 | -4.38810 | -3.81752 | -3.48084 |
| | 0.8 | -5.58109 | -5.01987 | -4.27330 | -3.70464 | -3.37032 |
| | 1.2 | -5.31729 | -4.77006 | -4.04543 | -3.49672 | -3.17579 |
| | 1.6 | -4.91039 | -4.39761 | -3.72298 | -3.21634 | -2.92215 |
| | 2.0 | -4.45931 | -3.99819 | -3.39524 | -2.94583 | -2.68651 |
| | 2.4 | -4.15273 | -3.73492 | -3.18958 | -2.78412 | -2.55076 |
| 2.8 | 0.0 | -5.41297 | -4.90928 | -4.23306 | -3.71220 | -3.40313 |
| | 0.4 | -5.42167 | -4.90724 | -4.21838 | -3.68942 | -3.37633 |
| | 0.8 | -5.34902 | -4.83122 | -4.13990 | -3.61095 | -3.29878 |
| | 1.2 | -5.16882 | -4.65840 | -3.97936 | -3.46211 | -3.15800 |
| | 1.6 | -4.87030 | -4.38087 | -3.73288 | -3.24236 | -2.95558 |
| | 2.0 | -4.47835 | -4.02521 | -3.42943 | -2.98255 | -2.72340 |
| | 2.4 | -4.08283 | -3.67758 | -3.14838 | -2.75488 | -2.52849 |
| | 2.8 | -3.83390 | -3.46601 | -2.98675 | -2.63174 | -2.42854 |
| 3.2 | 0.0 | -5.18839 | -4.71829 | -4.08555 | -3.59672 | -3.30593 |
| | 0.4 | -5.19579 | -4.71725 | -4.07465 | -3.57954 | -3.28567 |
| | 0.8 | -5.14520 | -4.66342 | -4.01814 | -3.52249 | -3.22904 |
| | 1.2 | -5.01781 | -4.54023 | -3.90248 | -3.41439 | -3.12629 |
| | 1.6 | -4.80107 | -4.33713 | -3.71986 | -3.24963 | -2.97317 |
| | 2.0 | -4.49741 | -4.05787 | -3.47608 | -3.03589 | -2.77870 |
| | 2.4 | -4.13623 | -3.73266 | -3.20258 | -2.80570 | -2.57610 |
| | 2.8 | -3.79534 | -3.43581 | -2.96731 | -2.62046 | -2.42232 |
| | 3.2 | -3.59446 | -3.26754 | -2.84311 | -2.53139 | -2.35466 |

FRACTILES $W_\alpha$ RHO = -0.60

| DELTAS | | $\alpha$:0.0005 | 0.0010 | 0.0025 | 0.0050 | 0.0075 |
|---|---|---|---|---|---|---|
| 3.6 | 0.0 | -5.00218 | -4.55961 | -3.96254 | -3.49997 | -3.22420 |
| | 0.4 | -5.00749 | -4.55829 | -3.95353 | -3.48612 | -3.20800 |
| | 0.8 | -4.97003 | -4.51802 | -3.91086 | -3.44287 | -3.16500 |
| | 1.2 | -4.87636 | -4.42693 | -3.82478 | -3.36209 | -3.08806 |
| | 1.6 | -4.71586 | -4.27583 | -3.68808 | -3.23812 | -2.97245 |
| | 2.0 | -4.48510 | -4.06235 | -3.49984 | -3.07130 | -2.81937 |
| | 2.4 | -4.19282 | -3.79581 | -3.27048 | -2.87334 | -2.64158 |
| | 2.8 | -3.86795 | -3.50570 | -3.03062 | -2.67610 | -2.47212 |
| | 3.2 | -3.57721 | -3.25567 | -2.83832 | -2.53221 | -2.35846 |
| | 3.6 | -3.41625 | -3.12378 | -2.74680 | -2.47196 | -2.31498 |
| 4.0 | 0.0 | -4.84625 | -4.42654 | -3.85909 | -3.41835 | -3.15506 |
| | 0.4 | -4.84953 | -4.42459 | -3.85112 | -3.40665 | -3.14158 |
| | 0.8 | -4.82032 | -4.39306 | -3.81762 | -3.37268 | -3.10783 |
| | 1.2 | -4.74893 | -4.32337 | -3.75152 | -3.31053 | -3.04859 |
| | 1.6 | -4.62696 | -4.20821 | -3.64697 | -3.21550 | -2.95986 |
| | 2.0 | -4.44982 | -4.04384 | -3.50140 | -3.08597 | -2.84061 |
| | 2.4 | -4.21940 | -3.83262 | -3.31791 | -2.92580 | -2.69534 |
| | 2.8 | -3.94681 | -3.58596 | -3.10879 | -2.74858 | -2.53894 |
| | 3.2 | -3.65889 | -3.33144 | -2.90321 | -2.58584 | -2.40432 |
| | 3.6 | -3.41328 | -3.12412 | -2.75156 | -2.47935 | -2.32323 |
| | 4.0 | -3.28609 | -3.02374 | -2.68729 | -2.43943 | -2.29504 |
| 6.0 | 0.0 | -4.34576 | -3.99843 | -3.52477 | -3.15317 | -2.92939 |
| | 0.4 | -4.34349 | -3.99444 | -3.51893 | -3.14632 | -2.92214 |
| | 0.8 | -4.33013 | -3.98040 | -3.50450 | -3.13205 | -2.90819 |
| | 1.2 | -4.30324 | -3.95416 | -3.47971 | -3.10889 | -2.88623 |
| | 1.6 | -4.26040 | -3.91359 | -3.44283 | -3.07543 | -2.85506 |
| | 2.0 | -4.19946 | -3.85683 | -3.39241 | -3.03052 | -2.81372 |
| | 2.4 | -4.11685 | -3.78259 | -3.32753 | -2.97354 | -2.76175 |
| | 2.8 | -4.01799 | -3.69051 | -3.24750 | -2.90463 | -2.69945 |
| | 3.2 | -3.89759 | -3.58142 | -3.15517 | -2.82504 | -2.62820 |
| | 3.6 | -3.75995 | -3.45768 | -3.05124 | -2.73750 | -2.55096 |
| | 4.0 | -3.60909 | -3.32339 | -2.94083 | -2.64717 | -2.47328 |
| | 6.0 | -3.06146 | -2.87668 | -2.62159 | -2.41621 | -2.28929 |
| 12.0 | 0.0 | -3.81277 | -3.54110 | -3.16538 | -2.86588 | -2.68331 |
| | 0.4 | -3.80985 | -3.53817 | -3.16255 | -2.86320 | -2.68076 |
| | 0.8 | -3.80525 | -3.53376 | -3.15847 | -2.85948 | -2.67728 |
| | 1.2 | -3.79880 | -3.52769 | -3.15303 | -2.85460 | -2.67277 |
| | 1.6 | -3.79031 | -3.51982 | -3.14609 | -2.84846 | -2.66715 |
| | 2.0 | -3.77961 | -3.50998 | -3.13752 | -2.84097 | -2.66033 |
| | 2.4 | -3.76651 | -3.49802 | -3.12721 | -2.83202 | -2.65224 |
| | 2.8 | -3.75086 | -3.48381 | -3.11506 | -2.82155 | -2.64279 |
| | 3.2 | -3.73252 | -3.46724 | -3.10098 | -2.80948 | -2.63196 |
| | 3.6 | -3.71138 | -3.44822 | -3.08491 | -2.79579 | -2.61971 |
| | 4.0 | -3.68738 | -3.42669 | -3.06684 | -2.78047 | -2.60606 |
| | 6.0 | -3.52614 | -3.28380 | -2.94905 | -2.68227 | -2.51955 |
| | 12.0 | -3.11880 | -2.94023 | -2.68566 | -2.47597 | -2.34500 |
| INFINITY | | -3.29051 | -3.09024 | -2.80703 | -2.57583 | -2.43238 |

FRACTILES $W_\alpha$ RHO = -0.60

| DELTAS | | α:0.0100 | 0.0175 | 0.0250 | 0.0375 | 0.0500 |
|---|---|---|---|---|---|---|
| 0.0 | 0.0 | -3.88620 | -3.20787 | -2.78173 | -2.30450 | -1.97143 |
| 0.4 | 0.0 | -3.85865 | -3.18734 | -2.76541 | -2.29267 | -1.96262 |
| | 0.4 | -3.74446 | -3.09644 | -2.68943 | -2.23374 | -1.91581 |
| 0.8 | 0.0 | -3.78193 | -3.13789 | -2.73125 | -2.27379 | -1.95314 |
| | 0.4 | -3.62128 | -3.00107 | -2.61141 | -2.17506 | -1.87061 |
| | 0.8 | -3.41652 | -2.84008 | -2.47830 | -2.07358 | -1.79153 |
| 1.2 | 0.0 | -3.66545 | -3.06758 | -2.68727 | -2.25630 | -1.95189 |
| | 0.4 | -3.52564 | -2.93647 | -2.56466 | -2.14662 | -1.85385 |
| | 0.8 | -3.28148 | -2.73659 | -2.39456 | -2.01199 | -1.74546 |
| | 1.2 | -3.06207 | -2.56649 | -2.25586 | -1.90896 | -1.66777 |
| 1.6 | 0.0 | -3.53028 | -2.98264 | -2.63203 | -2.23205 | -1.94745 |
| | 0.4 | -3.43456 | -2.88505 | -2.53562 | -2.13980 | -1.86040 |
| | 0.8 | -3.22180 | -2.69984 | -2.37073 | -2.00113 | -1.74268 |
| | 1.2 | -2.95787 | -2.48858 | -2.19448 | -1.86620 | -1.63817 |
| | 1.6 | -2.76502 | -2.34228 | -2.07794 | -1.78368 | -1.58006 |
| 2.0 | 0.0 | -3.39865 | -2.89548 | -2.57176 | -2.20058 | -1.93499 |
| | 0.4 | -3.33699 | -2.82894 | -2.50370 | -2.13271 | -1.86880 |
| | 0.8 | -3.18144 | -2.68668 | -2.37224 | -2.01633 | -1.76537 |
| | 1.2 | -2.94244 | -2.48558 | -2.19795 | -1.87560 | -1.65084 |
| | 1.6 | -2.69739 | -2.29412 | -2.04212 | -1.76198 | -1.56863 |
| | 2.0 | -2.54276 | -2.18097 | -1.95577 | -1.70696 | -1.53762 |
| 2.4 | 0.0 | -3.28164 | -2.81508 | -2.51372 | -2.16677 | -1.91742 |
| | 0.4 | -3.24052 | -2.76926 | -2.46603 | -2.11830 | -1.86947 |
| | 0.8 | -3.13233 | -2.66750 | -2.36997 | -2.03067 | -1.78942 |
| | 1.2 | -2.94822 | -2.50638 | -2.22582 | -1.90868 | -1.68551 |
| | 1.6 | -2.71469 | -2.31519 | -2.06432 | -1.78424 | -1.59022 |
| | 2.0 | -2.50452 | -2.15661 | -1.94048 | -1.70317 | -1.54268 |
| | 2.4 | -2.38737 | -2.07659 | -1.88589 | -1.67833 | -1.53562 |
| 2.8 | 0.0 | -3.18161 | -2.74470 | -2.46151 | -2.13436 | -1.89837 |
| | 0.4 | -3.15234 | -2.71167 | -2.42695 | -2.09907 | -1.86335 |
| | 0.8 | -3.07593 | -2.63887 | -2.35759 | -2.03500 | -1.80413 |
| | 1.2 | -2.94152 | -2.51875 | -2.24821 | -1.93984 | -1.72076 |
| | 1.6 | -2.75232 | -2.35800 | -2.10801 | -1.82613 | -1.62864 |
| | 2.0 | -2.54092 | -2.19044 | -1.97155 | -1.73000 | -1.56523 |
| | 2.4 | -2.37011 | -2.06985 | -1.88615 | -1.68465 | -1.54436 |
| | 2.8 | -2.28730 | -2.02157 | -1.85803 | -1.67464 | -1.54362 |
| 3.2 | 0.0 | -3.09716 | -2.68436 | -2.41598 | -2.10499 | -1.87994 |
| | 0.4 | -3.07500 | -2.65937 | -2.38987 | -2.07841 | -1.85364 |
| | 0.8 | -3.01905 | -2.60582 | -2.33872 | -2.03102 | -1.80973 |
| | 1.2 | -2.92058 | -2.51707 | -2.25733 | -1.95940 | -1.74619 |
| | 1.6 | -2.77639 | -2.39215 | -2.14639 | -1.86655 | -1.66812 |
| | 2.0 | -2.59655 | -2.24373 | -2.02076 | -1.77112 | -1.59814 |
| | 2.4 | -2.41484 | -2.10728 | -1.91752 | -1.70818 | -1.56221 |
| | 2.8 | -2.28479 | -2.02539 | -1.86445 | -1.68242 | -1.55148 |
| | 3.2 | -2.23195 | -1.99784 | -1.84936 | -1.67782 | -1.55220 |

FRACTILES $W_\alpha$ RHO = -0.60

| DELTAS | | α: 0.0100 | 0.0175 | 0.0250 | 0.0375 | 0.0500 |
|---|---|---|---|---|---|---|
| 3.6 | 0.0 | -3.02589 | -2.63290 | -2.37669 | -2.07901 | -1.86297 |
| | 0.4 | -3.00827 | -2.61321 | -2.35624 | -2.05833 | -1.84262 |
| | 0.8 | -2.96576 | -2.57251 | -2.31740 | -2.02240 | -1.80939 |
| | 1.2 | -2.89192 | -2.50579 | -2.25610 | -1.96832 | -1.76124 |
| | 1.6 | -2.78273 | -2.41049 | -2.17084 | -1.89594 | -1.69927 |
| | 2.0 | -2.64007 | -2.29013 | -2.06656 | -1.81272 | -1.63375 |
| | 2.4 | -2.47767 | -2.16135 | -1.96301 | -1.74186 | -1.58715 |
| | 2.8 | -2.32972 | -2.05950 | -1.89134 | -1.70146 | -1.56539 |
| | 3.2 | -2.23745 | -2.00527 | -1.85706 | -1.68509 | -1.55878 |
| | 3.6 | -2.20449 | -1.98875 | -1.84827 | -1.68277 | -1.55975 |
| 4.0 | 0.0 | -2.96544 | -2.58893 | -2.34285 | -2.05623 | -1.84769 |
| | 0.4 | -2.95091 | -2.57291 | -2.32634 | -2.03969 | -1.83154 |
| | 0.8 | -2.91756 | -2.54105 | -2.29600 | -2.01174 | -1.80576 |
| | 1.2 | -2.86070 | -2.48968 | -2.24883 | -1.97015 | -1.76877 |
| | 1.6 | -2.77682 | -2.41633 | -2.18308 | -1.91409 | -1.72044 |
| | 2.0 | -2.66537 | -2.32146 | -2.10003 | -1.84616 | -1.66485 |
| | 2.4 | -2.53138 | -2.21178 | -2.00827 | -1.77805 | -1.61548 |
| | 2.8 | -2.39116 | -2.10735 | -1.92923 | -1.72803 | -1.58445 |
| | 3.2 | -2.27752 | -2.03417 | -1.87934 | -1.70054 | -1.56989 |
| | 3.6 | -2.21299 | -1.99694 | -1.85579 | -1.68920 | -1.56526 |
| | 4.0 | -2.19184 | -1.98652 | -1.85038 | -1.68800 | -1.56621 |
| 6.0 | 0.0 | -2.76733 | -2.44300 | -2.22901 | -1.97749 | -1.79276 |
| | 0.4 | -2.75988 | -2.43543 | -2.22157 | -1.97041 | -1.78609 |
| | 0.8 | -2.74626 | -2.42275 | -2.20971 | -1.95975 | -1.77647 |
| | 1.2 | -2.72529 | -2.40403 | -2.19270 | -1.94497 | -1.76348 |
| | 1.6 | -2.69589 | -2.37848 | -2.16991 | -1.92565 | -1.74687 |
| | 2.0 | -2.65725 | -2.34554 | -2.14097 | -1.90163 | -1.72663 |
| | 2.4 | -2.60903 | -2.30515 | -2.10598 | -1.87322 | -1.70321 |
| | 2.8 | -2.55165 | -2.25796 | -2.06576 | -1.84147 | -1.67783 |
| | 3.2 | -2.48659 | -2.20574 | -2.02232 | -1.80859 | -1.65272 |
| | 3.6 | -2.41703 | -2.15206 | -1.97929 | -1.77784 | -1.63053 |
| | 4.0 | -2.34868 | -2.10222 | -1.94109 | -1.75219 | -1.61306 |
| | 6.0 | -2.19551 | -2.00237 | -1.87035 | -1.70978 | -1.58771 |
| 12.0 | 0.0 | -2.54995 | -2.28001 | -2.09947 | -1.88457 | -1.72471 |
| | 0.4 | -2.54751 | -2.27783 | -2.09749 | -1.88285 | -1.72320 |
| | 0.8 | -2.54422 | -2.27496 | -2.09493 | -1.88068 | -1.72133 |
| | 1.2 | -2.54000 | -2.27135 | -2.09175 | -1.87801 | -1.71906 |
| | 1.6 | -2.53477 | -2.26693 | -2.08788 | -1.87482 | -1.71637 |
| | 2.0 | -2.52845 | -2.26164 | -2.08329 | -1.87107 | -1.71324 |
| | 2.4 | -2.52098 | -2.25544 | -2.07795 | -1.86673 | -1.70965 |
| | 2.8 | -2.51230 | -2.24829 | -2.07182 | -1.86180 | -1.70559 |
| | 3.2 | -2.50236 | -2.24017 | -2.06490 | -1.85628 | -1.70108 |
| | 3.6 | -2.49117 | -2.23108 | -2.05718 | -1.85016 | -1.69612 |
| | 4.0 | -2.47871 | -2.22103 | -2.04870 | -1.84349 | -1.69075 |
| | 6.0 | -2.40058 | -2.15934 | -1.99755 | -1.80430 | -1.65997 |
| | 12.0 | -2.24777 | -2.04670 | -1.90892 | -1.74126 | -1.61387 |
| INFINITY | | -2.32635 | -2.10836 | -1.95996 | -1.78046 | -1.64485 |

FRACTILES $W_\alpha$ RHO = -0.60

| DELTAS | | $\alpha$:0.0750 | 0.1000 | 0.1500 | 0.2000 | 0.2500 |
|---|---|---|---|---|---|---|
| 0.0 | 0.0 | -1.51158 | -1.19362 | -0.76058 | -0.46716 | -0.24991 |
| 0.4 | 0.0 | -1.50678 | -1.19150 | -0.76202 | -0.47101 | -0.25559 |
| | 0.4 | -1.47710 | -1.17399 | -0.76171 | -0.48302 | -0.27732 |
| 0.8 | 0.0 | -1.50845 | -1.19954 | -0.77688 | -0.48927 | -0.27578 |
| | 0.4 | -1.45055 | -1.16046 | -0.76632 | -0.50052 | -0.30507 |
| | 0.8 | -1.40300 | -1.13528 | -0.77289 | -0.53013 | -0.35341 |
| 1.2 | 0.0 | -1.52591 | -1.22695 | -0.81295 | -0.52733 | -0.31295 |
| | 0.4 | -1.44830 | -1.16712 | -0.78373 | -0.52452 | -0.33394 |
| | 0.8 | -1.37859 | -1.12617 | -0.78564 | -0.55930 | -0.39702 |
| | 1.2 | -1.33674 | -1.11009 | -0.80728 | -0.61129 | -0.48224 |
| 1.6 | 0.0 | -1.54553 | -1.26023 | -0.85925 | -0.57726 | -0.36172 |
| | 0.4 | -1.46981 | -1.19613 | -0.81836 | -0.55941 | -0.36702 |
| | 0.8 | -1.38537 | -1.13902 | -0.80560 | -0.58440 | -0.42857 |
| | 1.2 | -1.32584 | -1.11287 | -0.83202 | -0.65852 | -0.52467 |
| | 1.6 | -1.30311 | -1.11734 | -0.88050 | -0.71424 | -0.57718 |
| 2.0 | 0.0 | -1.55723 | -1.28664 | -0.90160 | -0.62614 | -0.41180 |
| | 0.4 | -1.49628 | -1.23208 | -0.86149 | -0.60202 | -0.40525 |
| | 0.8 | -1.41521 | -1.17069 | -0.83561 | -0.61014 | -0.45190 |
| | 1.2 | -1.34188 | -1.13070 | -0.85478 | -0.67561 | -0.53460 |
| | 1.6 | -1.30769 | -1.13643 | -0.90509 | -0.73598 | -0.59498 |
| | 2.0 | -1.31291 | -1.15896 | -0.93777 | -0.76911 | -0.62586 |
| 2.4 | 0.0 | -1.56082 | -1.30364 | -0.93426 | -0.66658 | -0.45538 |
| | 0.4 | -1.51555 | -1.26207 | -0.90159 | -0.64419 | -0.44470 |
| | 0.8 | -1.44918 | -1.20830 | -0.87191 | -0.63936 | -0.46954 |
| | 1.2 | -1.37532 | -1.16043 | -0.87398 | -0.68305 | -0.53382 |
| | 1.6 | -1.32818 | -1.15343 | -0.91400 | -0.73884 | -0.59349 |
| | 2.0 | -1.32566 | -1.17329 | -0.95068 | -0.77942 | -0.63360 |
| | 2.4 | -1.33559 | -1.19001 | -0.97138 | -0.79995 | -0.65265 |
| 2.8 | 0.0 | -1.55937 | -1.31352 | -0.95785 | -0.69757 | -0.49013 |
| | 0.4 | -1.52613 | -1.28283 | -0.93327 | -0.67993 | -0.48026 |
| | 0.8 | -1.47579 | -1.24072 | -0.90697 | -0.67001 | -0.48986 |
| | 1.2 | -1.41237 | -1.19497 | -0.89619 | -0.69221 | -0.53373 |
| | 1.6 | -1.35751 | -1.17308 | -0.91973 | -0.73623 | -0.58565 |
| | 2.0 | -1.34072 | -1.18282 | -0.95295 | -0.77721 | -0.62843 |
| | 2.4 | -1.34527 | -1.19907 | -0.97833 | -0.80484 | -0.65574 |
| | 2.8 | -1.35315 | -1.21043 | -0.99172 | -0.81801 | -0.66792 |
| 3.2 | 0.0 | -1.55538 | -1.31891 | -0.97476 | -0.72092 | -0.51702 |
| | 0.4 | -1.53055 | -1.29608 | -0.95658 | -0.70785 | -0.50945 |
| | 0.8 | -1.49300 | -1.26438 | -0.93584 | -0.69802 | -0.51184 |
| | 1.2 | -1.44306 | -1.22632 | -0.92060 | -0.70578 | -0.53829 |
| | 1.6 | -1.39043 | -1.19696 | -0.92865 | -0.73559 | -0.57890 |
| | 2.0 | -1.36050 | -1.19365 | -0.95301 | -0.77127 | -0.61884 |
| | 2.4 | -1.35558 | -1.20456 | -0.97796 | -0.80104 | -0.64976 |
| | 2.8 | -1.36015 | -1.21632 | -0.99558 | -0.82022 | -0.66880 |
| | 3.2 | -1.36585 | -1.22407 | -1.00450 | -0.82894 | -0.67683 |

FRACTILES $W_\alpha$ RHO = -0.60

| DELTAS | | $\alpha$:0.0750 | 0.1000 | 0.1500 | 0.2000 | 0.2500 |
|---|---|---|---|---|---|---|
| 3.6 | 0.0 | -1.55034 | -1.32164 | -0.98708 | -0.73869 | -0.53790 |
| | 0.4 | -1.53132 | -1.30430 | -0.97349 | -0.72910 | -0.53245 |
| | 0.8 | -1.50298 | -1.28040 | -0.95775 | -0.72110 | -0.53246 |
| | 1.2 | -1.46472 | -1.25061 | -0.94330 | -0.72183 | -0.54709 |
| | 1.6 | -1.42014 | -1.22156 | -0.94142 | -0.73922 | -0.57615 |
| | 2.0 | -1.38424 | -1.20811 | -0.95545 | -0.76671 | -0.60996 |
| | 2.4 | -1.36900 | -1.21083 | -0.97591 | -0.79443 | -0.64043 |
| | 2.8 | -1.36761 | -1.21979 | -0.99429 | -0.81620 | -0.66309 |
| | 3.2 | -1.37106 | -1.22811 | -1.00672 | -0.82981 | -0.67666 |
| | 3.6 | -1.37516 | -1.23350 | -1.01283 | -0.83573 | -0.68209 |
| 4.0 | 0.0 | -1.54501 | -1.32279 | -0.99626 | -0.75248 | -0.55435 |
| | 0.4 | -1.53008 | -1.30931 | -0.98592 | -0.74536 | -0.55049 |
| | 0.8 | -1.50823 | -1.29100 | -0.97396 | -0.73924 | -0.55005 |
| | 1.2 | -1.47882 | -1.26797 | -0.96197 | -0.73751 | -0.55803 |
| | 1.6 | -1.44306 | -1.24293 | -0.95579 | -0.74638 | -0.57751 |
| | 2.0 | -1.40815 | -1.22485 | -0.96129 | -0.76548 | -0.60408 |
| | 2.4 | -1.38572 | -1.21978 | -0.97542 | -0.78861 | -0.63139 |
| | 2.8 | -1.37727 | -1.22374 | -0.99164 | -0.80995 | -0.65470 |
| | 3.2 | -1.37679 | -1.23053 | -1.00522 | -0.82607 | -0.67156 |
| | 3.6 | -1.37919 | -1.23642 | -1.01413 | -0.83592 | -0.68144 |
| | 4.0 | -1.38220 | -1.24027 | -1.01843 | -0.84005 | -0.68519 |
| 6.0 | 0.0 | -1.52168 | -1.32010 | -1.01928 | -0.79043 | -0.60120 |
| | 0.4 | -1.51586 | -1.31512 | -1.01583 | -0.78837 | -0.60044 |
| | 0.8 | -1.50800 | -1.30875 | -1.01199 | -0.78665 | -0.60057 |
| | 1.2 | -1.49789 | -1.30096 | -1.00791 | -0.78555 | -0.60198 |
| | 1.6 | -1.48548 | -1.29184 | -1.00390 | -0.78553 | -0.60521 |
| | 2.0 | -1.47099 | -1.28177 | -1.00056 | -0.78723 | -0.61081 |
| | 2.4 | -1.45510 | -1.27158 | -0.99875 | -0.79129 | -0.61904 |
| | 2.8 | -1.43918 | -1.26258 | -0.99925 | -0.79784 | -0.62952 |
| | 3.2 | -1.42509 | -1.25602 | -1.00223 | -0.80633 | -0.64129 |
| | 3.6 | -1.41424 | -1.25237 | -1.00707 | -0.81571 | -0.65320 |
| | 4.0 | -1.40698 | -1.25124 | -1.01278 | -0.82488 | -0.66417 |
| | 6.0 | -1.40141 | -1.25674 | -1.02974 | -0.84700 | -0.68849 |
| 12.0 | 0.0 | -1.48674 | -1.30693 | -1.03356 | -0.82108 | -0.64213 |
| | 0.4 | -1.48557 | -1.30603 | -1.03306 | -0.82091 | -0.64222 |
| | 0.8 | -1.48415 | -1.30496 | -1.03254 | -0.82079 | -0.64243 |
| | 1.2 | -1.48248 | -1.30374 | -1.03199 | -0.82073 | -0.64276 |
| | 1.6 | -1.48053 | -1.30235 | -1.03141 | -0.82075 | -0.64323 |
| | 2.0 | -1.47831 | -1.30079 | -1.03082 | -0.82085 | -0.64386 |
| | 2.4 | -1.47580 | -1.29907 | -1.03022 | -0.82105 | -0.64466 |
| | 2.8 | -1.47301 | -1.29719 | -1.02964 | -0.82137 | -0.64565 |
| | 3.2 | -1.46995 | -1.29517 | -1.02908 | -0.82181 | -0.64684 |
| | 3.6 | -1.46664 | -1.29304 | -1.02857 | -0.82241 | -0.64824 |
| | 4.0 | -1.46311 | -1.29080 | -1.02812 | -0.82317 | -0.64986 |
| | 6.0 | -1.44397 | -1.27966 | -1.02761 | -0.82949 | -0.66087 |
| | 12.0 | -1.41980 | -1.26949 | -1.03455 | -0.84633 | -0.68380 |
| INFINITY | | -1.43953 | -1.28155 | -1.03643 | -0.84162 | -0.67449 |

FRACTILES $W_\alpha$ RHO = −0.60

| DELTAS | | $\alpha$:0.3000 | 0.3500 | 0.4000 | 0.4500 | 0.5000 |
|---|---|---|---|---|---|---|
| 0.0 | 0.0 | −0.08086 | 0.05473 | 0.16554 | 0.25705 | 0.33287 |
| 0.4 | 0.0 | −0.08804 | 0.04624 | 0.15583 | 0.24616 | 0.32077 |
| | 0.4 | −0.11796 | 0.00906 | 0.11191 | 0.19563 | 0.26335 |
| 0.8 | 0.0 | −0.10951 | 0.02377 | 0.13238 | 0.22155 | 0.29461 |
| | 0.4 | −0.15451 | −0.03558 | 0.05925 | 0.13424 | 0.19074 |
| | 0.8 | −0.21954 | −0.11710 | −0.04205 | 0.01395 | 0.08158 |
| 1.2 | 0.0 | −0.14446 | −0.00846 | 0.10285 | 0.19427 | 0.26861 |
| | 0.4 | −0.18769 | −0.07332 | 0.01554 | 0.07824 | 0.13488 |
| | 0.8 | −0.27874 | −0.19667 | −0.11478 | −0.03259 | 0.05092 |
| | 1.2 | −0.37405 | −0.27410 | −0.17824 | −0.08378 | 0.01134 |
| 1.6 | 0.0 | −0.18934 | −0.04786 | 0.06973 | 0.16752 | 0.24729 |
| | 0.4 | −0.21842 | −0.10248 | −0.01762 | 0.05577 | 0.13063 |
| | 0.8 | −0.31886 | −0.22049 | −0.12792 | −0.03785 | 0.05203 |
| | 1.2 | −0.40822 | −0.30093 | −0.19844 | −0.09793 | 0.00277 |
| | 1.6 | −0.45528 | −0.34187 | −0.23306 | −0.12618 | −0.01910 |
| 2.0 | 0.0 | −0.23704 | −0.09055 | 0.03406 | 0.14033 | 0.22906 |
| | 0.4 | −0.25019 | −0.12728 | −0.03267 | 0.05417 | 0.13755 |
| | 0.8 | −0.32922 | −0.22209 | −0.12313 | −0.02825 | 0.06526 |
| | 1.2 | −0.41240 | −0.30055 | −0.19444 | −0.09104 | 0.01191 |
| | 1.6 | −0.46919 | −0.35217 | −0.24006 | −0.13017 | −0.02034 |
| | 2.0 | −0.49687 | −0.37629 | −0.26046 | −0.14679 | −0.03313 |
| 2.4 | 0.0 | −0.28049 | −0.13120 | −0.00136 | 0.11251 | 0.21128 |
| | 0.4 | −0.28346 | −0.15158 | −0.04367 | 0.05301 | 0.14380 |
| | 0.8 | −0.33554 | −0.22025 | −0.11535 | −0.01614 | 0.08049 |
| | 1.2 | −0.40594 | −0.29012 | −0.18123 | −0.07600 | 0.02803 |
| | 1.6 | −0.46452 | −0.34516 | −0.23134 | −0.12026 | −0.00969 |
| | 2.0 | −0.50228 | −0.37962 | −0.26194 | −0.14663 | −0.03153 |
| | 2.4 | −0.51933 | −0.39445 | −0.27445 | −0.15677 | −0.03927 |
| 2.8 | 0.0 | −0.31638 | −0.16611 | −0.03326 | 0.08584 | 0.19286 |
| | 0.4 | −0.31543 | −0.17631 | −0.05772 | 0.04786 | 0.14585 |
| | 0.8 | −0.34537 | −0.22186 | −0.11077 | −0.00694 | 0.09307 |
| | 1.2 | −0.39952 | −0.27935 | −0.16750 | −0.06034 | 0.04477 |
| | 1.6 | −0.45329 | −0.33174 | −0.21660 | −0.10487 | 0.00577 |
| | 2.0 | −0.49509 | −0.37105 | −0.25249 | −0.13669 | −0.02146 |
| | 2.4 | −0.52089 | −0.39470 | −0.27358 | −0.15495 | −0.03665 |
| | 2.8 | −0.53176 | −0.40412 | −0.28150 | −0.16132 | −0.04145 |
| 3.2 | 0.0 | −0.34481 | −0.19446 | −0.06003 | 0.06225 | 0.17470 |
| | 0.4 | −0.34327 | −0.19992 | −0.07423 | 0.03871 | 0.14331 |
| | 0.8 | −0.35923 | −0.22828 | −0.11108 | −0.00240 | 0.10137 |
| | 1.2 | −0.39741 | −0.27241 | −0.15713 | −0.04759 | 0.05905 |
| | 1.6 | −0.44255 | −0.31842 | −0.20170 | −0.08916 | 0.02166 |
| | 2.0 | −0.48324 | −0.35786 | −0.23862 | −0.12267 | −0.00774 |
| | 2.4 | −0.51349 | −0.38640 | −0.26478 | −0.14597 | −0.02777 |
| | 2.8 | −0.53156 | −0.40304 | −0.27969 | −0.15892 | −0.03860 |
| | 3.2 | −0.53870 | −0.40919 | −0.28482 | −0.16301 | −0.04163 |

FRACTILES $W_{\alpha}$ RHO = -0.60

| DELTAS | | $\alpha$:0.3000 | 0.3500 | 0.4000 | 0.4500 | 0.5000 |
|---|---|---|---|---|---|---|
| 3.6 | 0.0 | -0.36720 | -0.21710 | -0.08184 | 0.04239 | 0.15820 |
| | 0.4 | -0.36611 | -0.22073 | -0.09091 | 0.02730 | 0.13733 |
| | 0.8 | -0.37483 | -0.23817 | -0.11561 | -0.00231 | 0.10532 |
| | 1.2 | -0.40006 | -0.27021 | -0.15121 | -0.03890 | 0.06972 |
| | 1.6 | -0.43538 | -0.30821 | -0.18948 | -0.07571 | 0.03569 |
| | 2.0 | -0.47164 | -0.34461 | -0.22448 | -0.10824 | 0.00648 |
| | 2.4 | -0.50253 | -0.37453 | -0.25253 | -0.13376 | -0.01597 |
| | 2.8 | -0.52477 | -0.39560 | -0.27193 | -0.15111 | -0.03097 |
| | 3.2 | -0.53772 | -0.40757 | -0.28270 | -0.16051 | -0.03886 |
| | 3.6 | -0.54252 | -0.41168 | -0.28609 | -0.16316 | -0.04078 |
| 4.0 | 0.0 | -0.38502 | -0.23530 | -0.09957 | 0.02594 | 0.14393 |
| | 0.4 | -0.38447 | -0.23817 | -0.10608 | 0.01551 | 0.12955 |
| | 0.8 | -0.38984 | -0.24943 | -0.12278 | -0.00552 | 0.10570 |
| | 1.2 | -0.40614 | -0.27198 | -0.14939 | -0.03418 | 0.07671 |
| | 1.6 | -0.43236 | -0.30195 | -0.18088 | -0.06550 | 0.04690 |
| | 2.0 | -0.46267 | -0.33361 | -0.21225 | -0.09539 | 0.01943 |
| | 2.4 | -0.49150 | -0.36236 | -0.23983 | -0.12101 | -0.00357 |
| | 2.8 | -0.51514 | -0.38530 | -0.26139 | -0.14068 | -0.02094 |
| | 3.2 | -0.53176 | -0.40112 | -0.27602 | -0.15382 | -0.03236 |
| | 3.6 | -0.54123 | -0.40991 | -0.28395 | -0.16077 | -0.03822 |
| | 4.0 | -0.54452 | -0.41269 | -0.28622 | -0.16251 | -0.03943 |
| 6.0 | 0.0 | -0.43677 | -0.28899 | -0.15275 | -0.02457 | 0.09816 |
| | 0.4 | -0.43722 | -0.29056 | -0.15537 | -0.02817 | 0.09368 |
| | 0.8 | -0.43900 | -0.29383 | -0.15997 | -0.03394 | 0.08691 |
| | 1.2 | -0.44258 | -0.29928 | -0.16702 | -0.04228 | 0.07760 |
| | 1.6 | -0.44849 | -0.30736 | -0.17679 | -0.05326 | 0.06590 |
| | 2.0 | -0.45708 | -0.31817 | -0.18911 | -0.06647 | 0.05240 |
| | 2.4 | -0.46823 | -0.33124 | -0.20330 | -0.08109 | 0.03795 |
| | 2.8 | -0.48119 | -0.34564 | -0.21834 | -0.09613 | 0.02347 |
| | 3.2 | -0.49485 | -0.36023 | -0.23316 | -0.11060 | 0.00981 |
| | 3.6 | -0.50807 | -0.37395 | -0.24679 | -0.12370 | -0.00237 |
| | 4.0 | -0.51987 | -0.38595 | -0.25854 | -0.13483 | -0.01259 |
| | 6.0 | -0.54474 | -0.41030 | -0.28162 | -0.15608 | -0.03151 |
| 12.0 | 0.0 | -0.48401 | -0.33964 | -0.20454 | -0.07555 | 0.04977 |
| | 0.4 | -0.48433 | -0.34015 | -0.20520 | -0.07635 | 0.04886 |
| | 0.8 | -0.48480 | -0.34084 | -0.20608 | -0.07738 | 0.04771 |
| | 1.2 | -0.48545 | -0.34175 | -0.20719 | -0.07865 | 0.04632 |
| | 1.6 | -0.48628 | -0.34287 | -0.20855 | -0.08019 | 0.04466 |
| | 2.0 | -0.48732 | -0.34424 | -0.21017 | -0.08200 | 0.04272 |
| | 2.4 | -0.48859 | -0.34587 | -0.21208 | -0.08410 | 0.04049 |
| | 2.8 | -0.49009 | -0.34776 | -0.21426 | -0.08649 | 0.03798 |
| | 3.2 | -0.49184 | -0.34993 | -0.21674 | -0.08917 | 0.03519 |
| | 3.6 | -0.49383 | -0.35237 | -0.21949 | -0.09213 | 0.03213 |
| | 4.0 | -0.49608 | -0.35507 | -0.22251 | -0.09534 | 0.02883 |
| | 6.0 | -0.51035 | -0.37156 | -0.24038 | -0.11388 | 0.01027 |
| | 12.0 | -0.53700 | -0.40025 | -0.26986 | -0.14311 | -0.01781 |
| INFINITY | | -0.52440 | -0.38532 | -0.25335 | -0.12566 | 0.0 |

FRACTILES $w_\alpha$ RHO = -0.60

| DELTAS | | $\alpha$: 0.5500 | 0.6000 | 0.6500 | 0.7000 | 0.7500 |
|---|---|---|---|---|---|---|
| 0.0 | 0.0 | 0.39541 | 0.44619 | 0.48586 | 0.51302 | 0.53743 |
| 0.4 | 0.0 | 0.38198 | 0.43117 | 0.46861 | 0.49193 | 0.52469 |
| | 0.4 | 0.31669 | 0.35500 | 0.38460 | 0.43176 | 0.49494 |
| 0.8 | 0.0 | 0.35357 | 0.39909 | 0.42811 | 0.46414 | 0.51638 |
| | 0.4 | 0.22886 | 0.28063 | 0.34410 | 0.41896 | 0.50757 |
| | 0.8 | 0.15519 | 0.23484 | 0.32192 | 0.41879 | 0.52917 |
| 1.2 | 0.0 | 0.32686 | 0.36546 | 0.41105 | 0.46935 | 0.53980 |
| | 0.4 | 0.20069 | 0.27232 | 0.35062 | 0.43758 | 0.53649 |
| | 0.8 | 0.13727 | 0.22811 | 0.32549 | 0.43215 | 0.55206 |
| | 1.2 | 0.10896 | 0.21098 | 0.31965 | 0.43790 | 0.57002 |
| 1.6 | 0.0 | 0.30550 | 0.36074 | 0.42434 | 0.49504 | 0.57504 |
| | 0.4 | 0.20698 | 0.28641 | 0.37078 | 0.46251 | 0.56501 |
| | 0.8 | 0.14366 | 0.23893 | 0.34000 | 0.44962 | 0.57174 |
| | 1.2 | 0.10557 | 0.21245 | 0.32570 | 0.44831 | 0.58457 |
| | 1.6 | 0.09011 | 0.20347 | 0.32336 | 0.45286 | 0.59640 |
| 2.0 | 0.0 | 0.29876 | 0.36967 | 0.44319 | 0.52149 | 0.60756 |
| | 0.4 | 0.22028 | 0.30460 | 0.39269 | 0.48710 | 0.59124 |
| | 0.8 | 0.15955 | 0.25663 | 0.35870 | 0.46850 | 0.58987 |
| | 1.2 | 0.11640 | 0.22445 | 0.33834 | 0.46103 | 0.59669 |
| | 1.6 | 0.09139 | 0.20705 | 0.32903 | 0.46040 | 0.60558 |
| | 2.0 | 0.08247 | 0.20207 | 0.32808 | 0.46363 | 0.61322 |
| 2.4 | 0.0 | 0.29549 | 0.37692 | 0.45853 | 0.54335 | 0.63476 |
| | 0.4 | 0.23224 | 0.32098 | 0.41241 | 0.50919 | 0.61474 |
| | 0.8 | 0.17690 | 0.27525 | 0.37777 | 0.48719 | 0.60724 |
| | 1.2 | 0.13295 | 0.24079 | 0.35384 | 0.47497 | 0.60825 |
| | 1.6 | 0.10236 | 0.21795 | 0.33942 | 0.46980 | 0.61340 |
| | 2.0 | 0.08534 | 0.20603 | 0.33295 | 0.46922 | 0.61928 |
| | 2.4 | 0.08003 | 0.20319 | 0.33264 | 0.47153 | 0.62433 |
| 2.8 | 0.0 | 0.28865 | 0.37929 | 0.46853 | 0.55970 | 0.65641 |
| | 0.4 | 0.24003 | 0.33332 | 0.42832 | 0.52776 | 0.63511 |
| | 0.8 | 0.19189 | 0.29179 | 0.39508 | 0.50447 | 0.62363 |
| | 1.2 | 0.15006 | 0.25760 | 0.36970 | 0.48916 | 0.61991 |
| | 1.6 | 0.11739 | 0.23204 | 0.35205 | 0.48038 | 0.62118 |
| | 2.0 | 0.09521 | 0.21538 | 0.34143 | 0.47642 | 0.62470 |
| | 2.4 | 0.08329 | 0.20696 | 0.33675 | 0.47581 | 0.62857 |
| | 2.8 | 0.08008 | 0.20535 | 0.33678 | 0.47752 | 0.63203 |
| 3.2 | 0.0 | 0.27862 | 0.37710 | 0.47339 | 0.57075 | 0.67282 |
| | 0.4 | 0.24312 | 0.34111 | 0.43994 | 0.54244 | 0.65209 |
| | 0.8 | 0.20303 | 0.30497 | 0.40957 | 0.51956 | 0.63853 |
| | 1.2 | 0.16513 | 0.27282 | 0.38441 | 0.50269 | 0.63147 |
| | 1.6 | 0.13289 | 0.24662 | 0.36515 | 0.49138 | 0.62935 |
| | 2.0 | 0.10822 | 0.22726 | 0.35175 | 0.48468 | 0.63028 |
| | 2.4 | 0.09181 | 0.21484 | 0.34371 | 0.48150 | 0.63257 |
| | 2.8 | 0.08326 | 0.20874 | 0.34025 | 0.48091 | 0.63515 |
| | 3.2 | 0.08130 | 0.20786 | 0.34046 | 0.48224 | 0.63762 |

FRACTILES $w_{\alpha}$ RHO = -0.60

| DELTAS | | α:0.5500 | 0.6000 | 0.6500 | 0.7000 | 0.7500 |
|---|---|---|---|---|---|---|
| 3.6 | 0.0 | 0.26735 | 0.37186 | 0.47422 | 0.57730 | 0.68460 |
| | 0.4 | 0.24223 | 0.34475 | 0.44752 | 0.55334 | 0.66570 |
| | 0.8 | 0.21010 | 0.31448 | 0.42089 | 0.53206 | 0.65155 |
| | 1.2 | 0.17711 | 0.28548 | 0.39716 | 0.51491 | 0.64245 |
| | 1.6 | 0.14692 | 0.26012 | 0.37758 | 0.50215 | 0.63775 |
| | 2.0 | 0.12177 | 0.23970 | 0.36262 | 0.49344 | 0.63629 |
| | 2.4 | 0.10286 | 0.22480 | 0.35222 | 0.48814 | 0.63681 |
| | 2.8 | 0.09050 | 0.21536 | 0.34601 | 0.48553 | 0.63827 |
| | 3.2 | 0.08424 | 0.21085 | 0.34340 | 0.48499 | 0.64002 |
| | 3.6 | 0.08306 | 0.21042 | 0.34371 | 0.48605 | 0.64185 |
| 4.0 | 0.0 | 0.25642 | 0.36523 | 0.47240 | 0.58045 | 0.69260 |
| | 0.4 | 0.23862 | 0.34518 | 0.45170 | 0.56089 | 0.67618 |
| | 0.8 | 0.21359 | 0.32060 | 0.42914 | 0.54193 | 0.66249 |
| | 1.2 | 0.18578 | 0.29530 | 0.40760 | 0.52543 | 0.65244 |
| | 1.6 | 0.15860 | 0.27175 | 0.38866 | 0.51214 | 0.64601 |
| | 2.0 | 0.13436 | 0.25148 | 0.37313 | 0.50217 | 0.64262 |
| | 2.4 | 0.11452 | 0.23536 | 0.36129 | 0.49527 | 0.64145 |
| | 2.8 | 0.09983 | 0.22372 | 0.35309 | 0.49097 | 0.64164 |
| | 3.2 | 0.09037 | 0.21643 | 0.34822 | 0.48880 | 0.64254 |
| | 3.6 | 0.08570 | 0.21304 | 0.34623 | 0.48834 | 0.64376 |
| | 4.0 | 0.08501 | 0.21288 | 0.34658 | 0.48922 | 0.64516 |
| 6.0 | 0.0 | 0.21761 | 0.33570 | 0.45444 | 0.57607 | 0.70356 |
| | 0.4 | 0.21234 | 0.32980 | 0.44809 | 0.56958 | 0.69736 |
| | 0.8 | 0.20479 | 0.32175 | 0.43989 | 0.56170 | 0.69044 |
| | 1.2 | 0.19489 | 0.31167 | 0.43015 | 0.55290 | 0.68334 |
| | 1.6 | 0.18298 | 0.30009 | 0.41948 | 0.54379 | 0.67658 |
| | 2.0 | 0.16975 | 0.28770 | 0.40852 | 0.53492 | 0.67057 |
| | 2.4 | 0.15602 | 0.27524 | 0.39788 | 0.52672 | 0.66552 |
| | 2.8 | 0.14260 | 0.26336 | 0.38805 | 0.51947 | 0.66153 |
| | 3.2 | 0.13018 | 0.25260 | 0.37936 | 0.51335 | 0.65854 |
| | 3.6 | 0.11928 | 0.24330 | 0.37204 | 0.50839 | 0.65643 |
| | 4.0 | 0.11024 | 0.23571 | 0.36617 | 0.50458 | 0.65507 |
| | 6.0 | 0.09408 | 0.22275 | 0.35686 | 0.49946 | 0.65479 |
| 12.0 | 0.0 | 0.17348 | 0.29759 | 0.42420 | 0.55583 | 0.69585 |
| | 0.4 | 0.17249 | 0.29655 | 0.42314 | 0.55480 | 0.69492 |
| | 0.8 | 0.17127 | 0.29529 | 0.42188 | 0.55360 | 0.69386 |
| | 1.2 | 0.16980 | 0.29379 | 0.42041 | 0.55222 | 0.69266 |
| | 1.6 | 0.16806 | 0.29204 | 0.41871 | 0.55065 | 0.69133 |
| | 2.0 | 0.16606 | 0.29003 | 0.41679 | 0.54890 | 0.68988 |
| | 2.4 | 0.16378 | 0.28778 | 0.41464 | 0.54698 | 0.68832 |
| | 2.8 | 0.16123 | 0.28528 | 0.41229 | 0.54490 | 0.68666 |
| | 3.2 | 0.15842 | 0.28255 | 0.40975 | 0.54269 | 0.68494 |
| | 3.6 | 0.15537 | 0.27961 | 0.40705 | 0.54035 | 0.68316 |
| | 4.0 | 0.15210 | 0.27649 | 0.40420 | 0.53794 | 0.68136 |
| | 6.0 | 0.13413 | 0.25975 | 0.38938 | 0.52582 | 0.67294 |
| | 12.0 | 0.10805 | 0.23653 | 0.36993 | 0.51119 | 0.66441 |
| INFINITY | | 0.12566 | 0.25335 | 0.38532 | 0.52440 | 0.67449 |

FRACTILES $w_\alpha$ RHO = -0.60

| DELTAS | | $\alpha$: 0.8000 | 0.8500 | 0.9000 | 0.9250 | 0.9500 |
|---|---|---|---|---|---|---|
| 0.0 | 0.0 | 0.57942 | 0.64245 | 0.74129 | 0.81610 | 0.92613 |
| 0.4 | 0.0 | 0.57473 | 0.64788 | 0.76077 | 0.84538 | 0.96888 |
| | 0.4 | 0.57779 | 0.68915 | 0.85005 | 0.96537 | 1.12838 |
| 0.8 | 0.0 | 0.58647 | 0.68167 | 0.81996 | 0.91930 | 1.05985 |
| | 0.4 | 0.61494 | 0.75109 | 0.93869 | 1.06910 | 1.24963 |
| | 0.8 | 0.65925 | 0.82049 | 1.03819 | 1.18734 | 1.39162 |
| 1.2 | 0.0 | 0.62577 | 0.73499 | 0.88548 | 0.99002 | 1.13462 |
| | 0.4 | 0.65284 | 0.79681 | 0.99091 | 1.12373 | 1.30551 |
| | 0.8 | 0.69164 | 0.86261 | 1.09068 | 1.24546 | 1.45587 |
| | 1.2 | 0.72282 | 0.90873 | 1.15488 | 1.32089 | 1.54542 |
| 1.6 | 0.0 | 0.66859 | 0.78369 | 0.93804 | 1.04328 | 1.18693 |
| | 0.4 | 0.68373 | 0.82850 | 1.02086 | 1.15105 | 1.32770 |
| | 0.8 | 0.71265 | 0.88372 | 1.10978 | 1.26203 | 1.46776 |
| | 1.2 | 0.74134 | 0.93098 | 1.18050 | 1.34790 | 1.57332 |
| | 1.6 | 0.76104 | 0.95954 | 1.21966 | 1.39357 | 1.62703 |
| 2.0 | 0.0 | 0.70593 | 0.82459 | 0.98075 | 1.08577 | 1.22764 |
| | 0.4 | 0.71040 | 0.85403 | 1.04262 | 1.16907 | 1.33940 |
| | 0.8 | 0.72882 | 0.89622 | 1.11560 | 1.26238 | 1.45965 |
| | 1.2 | 0.75197 | 0.93882 | 1.18324 | 1.34644 | 1.56532 |
| | 1.6 | 0.77156 | 0.97099 | 1.23129 | 1.40471 | 1.63686 |
| | 2.0 | 0.78396 | 0.98867 | 1.25520 | 1.43237 | 1.66906 |
| 2.4 | 0.0 | 0.73742 | 0.85926 | 1.01712 | 1.12203 | 1.26250 |
| | 0.4 | 0.73424 | 0.87679 | 1.06197 | 1.18510 | 1.34988 |
| | 0.8 | 0.74369 | 0.90688 | 1.11911 | 1.26023 | 1.44896 |
| | 1.2 | 0.76002 | 0.94171 | 1.17803 | 1.33509 | 1.54496 |
| | 1.6 | 0.77701 | 0.97287 | 1.22746 | 1.39650 | 1.62214 |
| | 2.0 | 0.79019 | 0.99461 | 1.26001 | 1.43601 | 1.67063 |
| | 2.4 | 0.79817 | 1.00581 | 1.27493 | 1.45311 | 1.69030 |
| 2.8 | 0.0 | 0.76344 | 0.88871 | 1.04876 | 1.15399 | 1.29376 |
| | 0.4 | 0.75546 | 0.89764 | 1.08052 | 1.20118 | 1.36168 |
| | 0.8 | 0.75812 | 0.91784 | 1.12404 | 1.26034 | 1.44177 |
| | 1.2 | 0.76805 | 0.94446 | 1.17264 | 1.32361 | 1.52460 |
| | 1.6 | 0.78102 | 0.97161 | 1.21833 | 1.38158 | 1.59886 |
| | 2.0 | 0.79314 | 0.99406 | 1.25411 | 1.42611 | 1.65492 |
| | 2.4 | 0.80207 | 1.00894 | 1.27649 | 1.45330 | 1.68829 |
| | 2.8 | 0.80738 | 1.01625 | 1.28605 | 1.46414 | 1.70057 |
| 3.2 | 0.0 | 0.78444 | 0.91350 | 1.07636 | 1.18242 | 1.32222 |
| | 0.4 | 0.77392 | 0.91658 | 1.09838 | 1.21745 | 1.37493 |
| | 0.8 | 0.77192 | 0.92925 | 1.13093 | 1.26349 | 1.43914 |
| | 1.2 | 0.77662 | 0.94858 | 1.16977 | 1.31546 | 1.50872 |
| | 1.6 | 0.78535 | 0.97063 | 1.20943 | 1.36688 | 1.57585 |
| | 2.0 | 0.79520 | 0.99132 | 1.24433 | 1.41121 | 1.63271 |
| | 2.4 | 0.80381 | 1.00753 | 1.27038 | 1.44373 | 1.67372 |
| | 2.8 | 0.80997 | 1.01792 | 1.28607 | 1.46282 | 1.69718 |
| | 3.2 | 0.81364 | 1.02284 | 1.29237 | 1.46985 | 1.70500 |

FRACTILES $w_\alpha$ RHO = -0.60

| DELTAS | | $\alpha$:0.8000 | 0.8500 | 0.9000 | 0.9250 | 0.9500 |
|---|---|---|---|---|---|---|
| 3.6 | 0.0 | 0.80092 | 0.93410 | 1.10035 | 1.20767 | 1.34818 |
| | 0.4 | 0.78957 | 0.93347 | 1.11527 | 1.23354 | 1.38911 |
| | 0.8 | 0.78467 | 0.94068 | 1.13930 | 1.26913 | 1.44042 |
| | 1.2 | 0.78548 | 0.95407 | 1.16973 | 1.31115 | 1.49808 |
| | 1.6 | 0.79046 | 0.97109 | 1.20289 | 1.35518 | 1.55671 |
| | 2.0 | 0.79759 | 0.98880 | 1.23462 | 1.39630 | 1.61038 |
| | 2.4 | 0.80494 | 1.00448 | 1.26124 | 1.43019 | 1.65395 |
| | 2.8 | 0.81110 | 1.01632 | 1.28045 | 1.45424 | 1.68436 |
| | 3.2 | 0.81545 | 1.02373 | 1.29169 | 1.46793 | 1.70119 |
| | 3.6 | 0.81806 | 1.02714 | 1.29593 | 1.47259 | 1.70625 |
| 4.0 | 0.0 | 0.81351 | 0.95096 | 1.12102 | 1.22998 | 1.37176 |
| | 0.4 | 0.80249 | 0.94821 | 1.13092 | 1.24903 | 1.40360 |
| | 0.8 | 0.79605 | 0.95165 | 1.14847 | 1.27646 | 1.44461 |
| | 1.2 | 0.79421 | 0.96046 | 1.17201 | 1.31014 | 1.49208 |
| | 1.6 | 0.79618 | 0.97308 | 1.19909 | 1.34705 | 1.54230 |
| | 2.0 | 0.80070 | 0.98747 | 1.22673 | 1.38363 | 1.59090 |
| | 2.4 | 0.80633 | 1.00151 | 1.25194 | 1.41634 | 1.63364 |
| | 2.8 | 0.81179 | 1.01343 | 1.27237 | 1.44243 | 1.66727 |
| | 3.2 | 0.81626 | 1.02220 | 1.28673 | 1.46048 | 1.69016 |
| | 3.6 | 0.81940 | 1.02759 | 1.29495 | 1.47050 | 1.70249 |
| | 4.0 | 0.82131 | 1.03002 | 1.29787 | 1.47363 | 1.70577 |
| 6.0 | 0.0 | 0.84130 | 0.99691 | 1.18643 | 1.30562 | 1.45791 |
| | 0.4 | 0.83610 | 0.99382 | 1.18750 | 1.31022 | 1.46799 |
| | 0.8 | 0.83102 | 0.99192 | 1.19105 | 1.31805 | 1.48217 |
| | 1.2 | 0.82660 | 0.99160 | 1.19722 | 1.32908 | 1.50021 |
| | 1.6 | 0.82321 | 0.99302 | 1.20587 | 1.34299 | 1.52154 |
| | 2.0 | 0.82104 | 0.99612 | 1.21660 | 1.35916 | 1.54531 |
| | 2.4 | 0.82009 | 1.00060 | 1.22881 | 1.37679 | 1.57044 |
| | 2.8 | 0.82019 | 1.00606 | 1.24174 | 1.39492 | 1.59569 |
| | 3.2 | 0.82110 | 1.01198 | 1.25457 | 1.41252 | 1.61980 |
| | 3.6 | 0.82250 | 1.01784 | 1.26653 | 1.42865 | 1.64162 |
| | 4.0 | 0.82410 | 1.02319 | 1.27697 | 1.44256 | 1.66023 |
| | 6.0 | 0.82955 | 1.03565 | 1.29867 | 1.47038 | 1.69613 |
| 12.0 | 0.0 | 0.84937 | 1.02524 | 1.24201 | 1.37918 | 1.55466 |
| | 0.4 | 0.84863 | 1.02483 | 1.24222 | 1.37993 | 1.55626 |
| | 0.8 | 0.84781 | 1.02444 | 1.24260 | 1.38095 | 1.55830 |
| | 1.2 | 0.84693 | 1.02408 | 1.24316 | 1.38227 | 1.56082 |
| | 1.6 | 0.84599 | 1.02378 | 1.24394 | 1.38393 | 1.56383 |
| | 2.0 | 0.84501 | 1.02354 | 1.24496 | 1.38593 | 1.56735 |
| | 2.4 | 0.84401 | 1.02339 | 1.24622 | 1.38830 | 1.57139 |
| | 2.8 | 0.84299 | 1.02335 | 1.24774 | 1.39104 | 1.57594 |
| | 3.2 | 0.84199 | 1.02342 | 1.24953 | 1.39414 | 1.58099 |
| | 3.6 | 0.84101 | 1.02362 | 1.25158 | 1.39759 | 1.58650 |
| | 4.0 | 0.84009 | 1.02397 | 1.25388 | 1.40136 | 1.59240 |
| | 6.0 | 0.83670 | 1.02759 | 1.26800 | 1.42318 | 1.62523 |
| | 12.0 | 0.83599 | 1.03725 | 1.29243 | 1.45805 | 1.67466 |
| INFINITY | | 0.84162 | 1.03643 | 1.28155 | 1.43953 | 1.64485 |

FRACTILES $w_\alpha$ RHO = -0.60

| DELTAS | | $\alpha$: 0.9625 | 0.9750 | 0.9825 | 0.9900 | 0.9925 |
|---|---|---|---|---|---|---|
| 0.0 | 0.0 | 1.00670 | 1.12297 | 1.22736 | 1.39430 | 1.48134 |
| 0.4 | 0.0 | 1.05873 | 1.18766 | 1.30277 | 1.48573 | 1.58064 |
| | 0.4 | 1.24401 | 1.40667 | 1.54931 | 1.77217 | 1.88627 |
| 0.8 | 0.0 | 1.15960 | 1.29995 | 1.42305 | 1.61539 | 1.71386 |
| | 0.4 | 1.37570 | 1.55093 | 1.70298 | 1.93816 | 2.05763 |
| | 0.8 | 1.53304 | 1.72822 | 1.89649 | 2.15503 | 2.28569 |
| 1.2 | 0.0 | 1.23554 | 1.37572 | 1.49730 | 1.68526 | 1.78071 |
| | 0.4 | 1.43128 | 1.60479 | 1.75430 | 1.98395 | 2.09998 |
| | 0.8 | 1.60063 | 1.79936 | 1.96982 | 2.23040 | 2.36154 |
| | 1.2 | 1.69920 | 1.90952 | 2.08925 | 2.36296 | 2.50027 |
| 1.6 | 0.0 | 1.28613 | 1.42278 | 1.54037 | 1.72079 | 1.81186 |
| | 0.4 | 1.44905 | 1.61546 | 1.75807 | 1.97587 | 2.08542 |
| | 0.8 | 1.60856 | 1.80102 | 1.96540 | 2.21562 | 2.34111 |
| | 1.2 | 1.72713 | 1.93681 | 2.11542 | 2.38652 | 2.52215 |
| | 1.6 | 1.78590 | 2.00194 | 2.18554 | 2.46350 | 2.60227 |
| 2.0 | 0.0 | 1.32478 | 1.45769 | 1.57133 | 1.74458 | 1.83159 |
| | 0.4 | 1.45569 | 1.61438 | 1.74970 | 1.95539 | 2.05843 |
| | 0.8 | 1.59405 | 1.77705 | 1.93279 | 2.16893 | 2.28699 |
| | 1.2 | 1.71416 | 1.91646 | 2.08830 | 2.34832 | 2.47809 |
| | 1.6 | 1.79443 | 2.00822 | 2.18951 | 2.46332 | 2.59975 |
| | 2.0 | 1.82942 | 2.04664 | 2.23052 | 2.50775 | 2.64568 |
| 2.4 | 0.0 | 1.35798 | 1.48781 | 1.59821 | 1.76556 | 1.84924 |
| | 0.4 | 1.46177 | 1.61375 | 1.74279 | 1.93808 | 2.03556 |
| | 0.8 | 1.57701 | 1.75073 | 1.89807 | 2.12070 | 2.23169 |
| | 1.2 | 1.68721 | 1.88001 | 2.04333 | 2.28978 | 2.41249 |
| | 1.6 | 1.77491 | 1.98174 | 2.15675 | 2.42049 | 2.55166 |
| | 2.0 | 1.82929 | 2.04385 | 2.22518 | 2.49810 | 2.63368 |
| | 2.4 | 1.85047 | 2.06684 | 2.24947 | 2.52396 | 2.66017 |
| 2.8 | 0.0 | 1.38812 | 1.51575 | 1.62371 | 1.78654 | 1.86762 |
| | 0.4 | 1.47011 | 1.61676 | 1.74079 | 1.92771 | 2.02072 |
| | 0.8 | 1.56437 | 1.73016 | 1.87031 | 2.08140 | 2.18634 |
| | 1.2 | 1.66040 | 1.84396 | 1.99906 | 2.23248 | 2.34844 |
| | 1.6 | 1.74561 | 1.94387 | 2.11127 | 2.36300 | 2.48798 |
| | 2.0 | 1.80934 | 2.01782 | 2.19373 | 2.45802 | 2.58912 |
| | 2.4 | 1.84676 | 2.06054 | 2.24076 | 2.51127 | 2.64534 |
| | 2.8 | 1.85985 | 2.07453 | 2.25533 | 2.52643 | 2.66068 |
| 3.2 | 0.0 | 1.41604 | 1.54227 | 1.64855 | 1.80808 | 1.88721 |
| | 0.4 | 1.48081 | 1.62345 | 1.74362 | 1.92404 | 2.01353 |
| | 0.8 | 1.55738 | 1.71676 | 1.85108 | 2.05274 | 2.15275 |
| | 1.2 | 1.63889 | 1.81439 | 1.96231 | 2.18434 | 2.29442 |
| | 1.6 | 1.71663 | 1.90643 | 2.06636 | 2.30635 | 2.42527 |
| | 2.0 | 1.78190 | 1.98299 | 2.15237 | 2.40641 | 2.53224 |
| | 2.4 | 1.82858 | 2.03721 | 2.21285 | 2.47611 | 2.60644 |
| | 2.8 | 1.85488 | 2.06721 | 2.24585 | 2.51342 | 2.64579 |
| | 3.2 | 1.86310 | 2.07582 | 2.25465 | 2.52229 | 2.65459 |

FRACTILES $W_{\alpha}$ RHO = -0.60

| DELTAS | | $\alpha$:0.9625 | 0.9750 | 0.9825 | 0.9900 | 0.9925 |
|---|---|---|---|---|---|---|
| 3.6 | 0.0 | 1.44193 | 1.56749 | 1.67273 | 1.82999 | 1.90772 |
| | 0.4 | 1.49323 | 1.63296 | 1.75026 | 1.92572 | 2.01250 |
| | 0.8 | 1.55530 | 1.70967 | 1.83938 | 2.03354 | 2.12959 |
| | 1.2 | 1.62361 | 1.79243 | 1.93435 | 2.14685 | 2.25199 |
| | 1.6 | 1.69214 | 1.87435 | 2.02757 | 2.25700 | 2.37050 |
| | 2.0 | 1.75429 | 1.94793 | 2.11076 | 2.35454 | 2.47512 |
| | 2.4 | 1.80436 | 2.00672 | 2.17685 | 2.43148 | 2.55738 |
| | 2.8 | 1.83901 | 2.04701 | 2.22182 | 2.48334 | 2.61259 |
| | 3.2 | 1.85789 | 2.06853 | 2.24548 | 2.51004 | 2.64072 |
| | 3.6 | 1.86310 | 2.07384 | 2.25076 | 2.51510 | 2.64560 |
| 4.0 | 0.0 | 1.46586 | 1.59133 | 1.69607 | 1.85191 | 1.92868 |
| | 0.4 | 1.50661 | 1.64435 | 1.75959 | 1.93137 | 2.01608 |
| | 0.8 | 1.55698 | 1.70753 | 1.83367 | 2.02193 | 2.11483 |
| | 1.2 | 1.61390 | 1.77731 | 1.91436 | 2.11906 | 2.22013 |
| | 1.6 | 1.67318 | 1.84888 | 1.99634 | 2.21668 | 2.32550 |
| | 2.0 | 1.72994 | 1.91670 | 2.07348 | 2.30779 | 2.42352 |
| | 2.4 | 1.77947 | 1.97538 | 2.13986 | 2.38567 | 2.50706 |
| | 2.8 | 1.81816 | 2.02087 | 2.19104 | 2.44531 | 2.57084 |
| | 3.2 | 1.84428 | 2.05128 | 2.22501 | 2.48450 | 2.61257 |
| | 3.6 | 1.85810 | 2.06702 | 2.24229 | 2.50397 | 2.63307 |
| | 4.0 | 1.86141 | 2.07026 | 2.24538 | 2.50669 | 2.63554 |
| 6.0 | 0.0 | 1.55726 | 1.68769 | 1.79488 | 1.95178 | 2.02801 |
| | 0.4 | 1.57144 | 1.70783 | 1.82035 | 1.98567 | 2.06622 |
| | 0.8 | 1.59025 | 1.73322 | 1.85154 | 2.02588 | 2.11101 |
| | 1.2 | 1.61329 | 1.76329 | 1.88774 | 2.07152 | 2.16142 |
| | 1.6 | 1.63986 | 1.79714 | 1.92788 | 2.12131 | 2.21606 |
| | 2.0 | 1.66893 | 1.83355 | 1.97059 | 2.17363 | 2.27320 |
| | 2.4 | 1.69926 | 1.87103 | 2.01419 | 2.22654 | 2.33076 |
| | 2.8 | 1.72943 | 1.90794 | 2.05686 | 2.27793 | 2.38649 |
| | 3.2 | 1.75802 | 1.94265 | 2.09678 | 2.32571 | 2.43818 |
| | 3.6 | 1.78373 | 1.97367 | 2.13229 | 2.36801 | 2.48385 |
| | 4.0 | 1.80554 | 1.99983 | 2.16214 | 2.40339 | 2.52197 |
| | 6.0 | 1.84685 | 2.04831 | 2.21658 | 2.46656 | 2.58936 |
| 12.0 | 0.0 | 1.66890 | 1.81822 | 1.94011 | 2.11679 | 2.20178 |
| | 0.4 | 1.67119 | 1.82154 | 1.94441 | 2.12275 | 2.20864 |
| | 0.8 | 1.67402 | 1.82558 | 1.94957 | 2.12979 | 2.21669 |
| | 1.2 | 1.67745 | 1.83037 | 1.95563 | 2.13794 | 2.22596 |
| | 1.6 | 1.68148 | 1.83593 | 1.96260 | 2.14721 | 2.23644 |
| | 2.0 | 1.68615 | 1.84228 | 1.97048 | 2.15758 | 2.24811 |
| | 2.4 | 1.69144 | 1.84939 | 1.97925 | 2.16900 | 2.26092 |
| | 2.8 | 1.69734 | 1.85724 | 1.98885 | 2.18141 | 2.27479 |
| | 3.2 | 1.70381 | 1.86577 | 1.99923 | 2.19472 | 2.28962 |
| | 3.6 | 1.71081 | 1.87492 | 2.01029 | 2.20880 | 2.30525 |
| | 4.0 | 1.71827 | 1.88459 | 2.02192 | 2.22351 | 2.32155 |
| | 6.0 | 1.75896 | 1.93638 | 2.08344 | 2.30017 | 2.40593 |
| | 12.0 | 1.81856 | 2.01006 | 2.16926 | 2.40458 | 2.51967 |
| INFINITY | | 1.78046 | 1.95996 | 2.10836 | 2.32635 | 2.43238 |

FRACTILES $w_\alpha$ RHO = -0.60

| DELTAS | | $\alpha$: 0.9950 | 0.9975 | 0.9990 | 0.9995 |
|---|---|---|---|---|---|
| 0.0 | 0.0 | 1.60519 | 1.81955 | 2.10694 | 2.32672 |
| 0.4 | 0.0 | 1.71517 | 1.94674 | 2.25497 | 2.48921 |
| | 0.4 | 2.04654 | 2.31913 | 2.67696 | 2.94593 |
| 0.8 | 0.0 | 1.85218 | 2.08744 | 2.39625 | 2.62835 |
| | 0.4 | 2.22457 | 2.50648 | 2.87342 | 3.14733 |
| | 0.8 | 2.46758 | 2.77317 | 3.16841 | 3.46191 |
| 1.2 | 0.0 | 1.91404 | 2.13912 | 2.43197 | 2.65052 |
| | 0.4 | 2.26147 | 2.53271 | 2.88345 | 3.14383 |
| | 0.8 | 2.54357 | 2.84810 | 3.23995 | 3.52964 |
| | 1.2 | 2.69041 | 3.00750 | 3.41382 | 3.71316 |
| 1.6 | 0.0 | 1.93855 | 2.15117 | 2.42587 | 2.62968 |
| | 0.4 | 2.23740 | 2.49152 | 2.81830 | 3.05977 |
| | 0.8 | 2.51484 | 2.80447 | 3.17548 | 3.44873 |
| | 1.2 | 2.70959 | 3.02128 | 3.41926 | 3.71153 |
| | 1.6 | 2.79375 | 3.11143 | 3.51588 | 3.81216 |
| 2.0 | 0.0 | 1.95219 | 2.15360 | 2.41225 | 2.60318 |
| | 0.4 | 2.20099 | 2.43842 | 2.74224 | 2.96583 |
| | 0.8 | 2.45007 | 2.72107 | 3.06686 | 3.32067 |
| | 1.2 | 2.65711 | 2.95402 | 3.33190 | 3.60864 |
| | 1.6 | 2.78773 | 3.09896 | 3.49415 | 3.78297 |
| | 2.0 | 2.83550 | 3.14926 | 3.54680 | 3.83679 |
| 2.4 | 0.0 | 1.96484 | 2.15706 | 2.40258 | 2.58301 |
| | 0.4 | 2.17008 | 2.39333 | 2.67777 | 2.88631 |
| | 0.8 | 2.38469 | 2.63820 | 2.96049 | 3.19632 |
| | 1.2 | 2.58149 | 2.86112 | 3.21592 | 3.47509 |
| | 1.6 | 2.73214 | 3.03038 | 3.40811 | 3.68358 |
| | 2.0 | 2.82006 | 3.12765 | 3.51657 | 3.79976 |
| | 2.4 | 2.84725 | 3.15561 | 3.54486 | 3.82786 |
| 2.8 | 0.0 | 1.97931 | 2.16429 | 2.39938 | 2.57143 |
| | 0.4 | 2.14875 | 2.36055 | 2.62929 | 2.82565 |
| | 0.8 | 2.33073 | 2.56933 | 2.87164 | 3.09220 |
| | 1.2 | 2.50789 | 2.77114 | 3.10420 | 3.34689 |
| | 1.6 | 2.65970 | 2.94294 | 3.30083 | 3.56128 |
| | 2.0 | 2.76915 | 3.06580 | 3.44014 | 3.71224 |
| | 2.4 | 2.82932 | 3.13220 | 3.51391 | 3.79102 |
| | 2.8 | 2.84478 | 3.14755 | 3.52861 | 3.80493 |
| 3.2 | 0.0 | 1.99593 | 2.17530 | 2.40222 | 2.56766 |
| | 0.4 | 2.13645 | 2.33916 | 2.59539 | 2.78200 |
| | 0.8 | 2.29009 | 2.51646 | 2.80234 | 3.01034 |
| | 1.2 | 2.44554 | 2.69448 | 3.00860 | 3.23694 |
| | 1.6 | 2.58848 | 2.85719 | 3.19593 | 3.44195 |
| | 2.0 | 2.70485 | 2.98885 | 3.34653 | 3.60608 |
| | 2.4 | 2.78513 | 3.07893 | 3.44859 | 3.71657 |
| | 2.8 | 2.82720 | 3.12523 | 3.49982 | 3.77113 |
| | 3.2 | 2.83582 | 3.13330 | 3.50681 | 3.77707 |

FRACTILES $W_\alpha$ RHO = −0.60

| DELTAS | | $\alpha$:0.9950 | 0.9975 | 0.9990 | 0.9995 |
|---|---|---|---|---|---|
| 3.6 | 0.0 | 2.01423 | 2.18935 | 2.40994 | 2.57017 |
| | 0.4 | 2.13144 | 2.32701 | 2.57332 | 2.75215 |
| | 0.8 | 2.26125 | 2.47774 | 2.75029 | 2.94807 |
| | 1.2 | 2.39611 | 2.63302 | 2.93117 | 3.14742 |
| | 1.6 | 2.52606 | 2.78173 | 3.10329 | 3.33639 |
| | 2.0 | 2.64035 | 2.91179 | 3.25299 | 3.50017 |
| | 2.4 | 2.72986 | 3.01307 | 3.36883 | 3.62638 |
| | 2.8 | 2.78960 | 3.08009 | 3.44471 | 3.70847 |
| | 3.2 | 2.81962 | 3.11304 | 3.48104 | 3.74704 |
| | 3.6 | 2.82416 | 3.11684 | 3.48358 | 3.74845 |
| 4.0 | 0.0 | 2.03362 | 2.20559 | 2.42132 | 2.57748 |
| | 0.4 | 2.13196 | 2.32196 | 2.56041 | 2.73303 |
| | 0.8 | 2.24198 | 2.45052 | 2.71230 | 2.90178 |
| | 1.2 | 2.35847 | 2.58543 | 2.87031 | 3.07648 |
| | 1.6 | 2.47446 | 2.71884 | 3.02554 | 3.24743 |
| | 2.0 | 2.58193 | 2.84178 | 3.16779 | 3.40358 |
| | 2.4 | 2.67320 | 2.94568 | 3.28739 | 3.53441 |
| | 2.8 | 2.74264 | 3.02427 | 3.37729 | 3.63235 |
| | 3.2 | 2.78779 | 3.07492 | 3.43460 | 3.69433 |
| | 3.6 | 2.80963 | 3.09882 | 3.46085 | 3.72209 |
| | 4.0 | 2.81170 | 3.10007 | 3.46080 | 3.72093 |
| 6.0 | 0.0 | 2.13120 | 2.29794 | 2.50348 | 2.65005 |
| | 0.4 | 2.17545 | 2.35237 | 2.57102 | 2.72724 |
| | 0.8 | 2.22663 | 2.41424 | 2.64657 | 2.81280 |
| | 1.2 | 2.28365 | 2.48228 | 2.72865 | 2.90514 |
| | 1.6 | 2.34500 | 2.55478 | 2.81529 | 3.00208 |
| | 2.0 | 2.40878 | 2.62957 | 2.90401 | 3.10092 |
| | 2.4 | 2.47275 | 2.70412 | 2.99192 | 3.19851 |
| | 2.8 | 2.53446 | 2.77569 | 3.07592 | 3.29149 |
| | 3.2 | 2.59153 | 2.84161 | 3.15294 | 3.37654 |
| | 3.6 | 2.64181 | 2.89947 | 3.22031 | 3.45075 |
| | 4.0 | 2.68369 | 2.94749 | 3.27600 | 3.51195 |
| | 6.0 | 2.75677 | 3.02965 | 3.36907 | 3.61258 |
| 12.0 | 0.0 | 2.31584 | 2.49762 | 2.71720 | 2.87073 |
| | 0.4 | 2.32401 | 2.50814 | 2.73100 | 2.88710 |
| | 0.8 | 2.33353 | 2.52026 | 2.74669 | 2.90557 |
| | 1.2 | 2.34440 | 2.53396 | 2.76425 | 2.92611 |
| | 1.6 | 2.35663 | 2.54924 | 2.78364 | 2.94864 |
| | 2.0 | 2.37017 | 2.56602 | 2.80476 | 2.97306 |
| | 2.4 | 2.38496 | 2.58421 | 2.82749 | 2.99922 |
| | 2.8 | 2.40089 | 2.60370 | 2.85167 | 3.02693 |
| | 3.2 | 2.41786 | 2.62433 | 2.87710 | 3.05597 |
| | 3.6 | 2.43569 | 2.64589 | 2.90356 | 3.08608 |
| | 4.0 | 2.45421 | 2.66819 | 2.93078 | 3.11696 |
| | 6.0 | 2.54937 | 2.78150 | 3.06754 | 3.27105 |
| | 12.0 | 2.67602 | 2.92960 | 3.24287 | 3.46620 |
| INFINITY | | 2.57583 | 2.80703 | 3.09024 | 3.29051 |

FRACTILES $W_\alpha$ RHO = -0.40

| DELTAS | | $\alpha$: 0.0005 | 0.0010 | 0.0025 | 0.0050 | 0.0075 |
|---|---|---|---|---|---|---|
| 0.0 | 0.0 | -7.27408 | -6.43723 | -5.34138 | -4.52228 | -4.04805 |
| 0.4 | 0.0 | -7.18244 | -6.36021 | -5.28237 | -4.47584 | -4.00854 |
| | 0.4 | -6.78133 | -6.00643 | -4.99179 | -4.23352 | -3.79456 |
| 0.8 | 0.0 | -6.86325 | -6.10123 | -5.09678 | -4.34036 | -3.89989 |
| | 0.4 | -6.39662 | -5.67334 | -4.72537 | -4.01623 | -3.60545 |
| | 0.8 | -5.78811 | -5.13985 | -4.29129 | -3.65742 | -3.29066 |
| 1.2 | 0.0 | -6.40716 | -5.72818 | -4.82802 | -4.14526 | -3.74529 |
| | 0.4 | -6.06377 | -5.40069 | -4.52679 | -3.86886 | -3.48584 |
| | 0.8 | -5.43553 | -4.83570 | -4.04983 | -3.46230 | -3.12217 |
| | 1.2 | -4.87836 | -4.34961 | -3.65786 | -3.14157 | -2.84310 |
| 1.6 | 0.0 | -5.96046 | -5.35715 | -4.55385 | -3.94124 | -3.58071 |
| | 0.4 | -5.72536 | -5.12861 | -4.33750 | -3.73751 | -3.38606 |
| | 0.8 | -5.24419 | -4.68485 | -3.94788 | -3.39330 | -3.07062 |
| | 1.2 | -4.64390 | -4.14939 | -3.50191 | -3.01832 | -2.73865 |
| | 1.6 | -4.20268 | -3.76701 | -3.19756 | -2.77312 | -2.52811 |
| 2.0 | 0.0 | -5.58142 | -5.03911 | -4.31443 | -3.75934 | -3.43146 |
| | 0.4 | -5.41122 | -4.87287 | -4.15599 | -3.60923 | -3.28743 |
| | 0.8 | -5.06633 | -4.55099 | -3.86783 | -3.34979 | -3.04641 |
| | 1.2 | -4.57748 | -4.10557 | -3.48409 | -3.01677 | -2.74509 |
| | 1.6 | -4.06493 | -3.65157 | -3.11089 | -2.70770 | -2.47492 |
| | 2.0 | -3.72896 | -3.36320 | -2.88576 | -2.53071 | -2.32627 |
| 2.4 | 0.0 | -5.27410 | -4.77973 | -4.11699 | -3.60739 | -3.30543 |
| | 0.4 | -5.14254 | -4.65150 | -3.99520 | -3.49236 | -3.19527 |
| | 0.8 | -4.88929 | -4.41420 | -3.78147 | -3.29882 | -3.01471 |
| | 1.2 | -4.51988 | -4.07441 | -3.48394 | -3.03629 | -2.77422 |
| | 1.6 | -4.07670 | -3.67384 | -3.14366 | -2.74547 | -2.51430 |
| | 2.0 | -3.65540 | -3.30380 | -2.84460 | -2.50305 | -2.30644 |
| | 2.4 | -3.40181 | -3.08914 | -2.68190 | -2.38024 | -2.20741 |
| 2.8 | 0.0 | -5.02686 | -4.57035 | -3.95657 | -3.48298 | -3.20155 |
| | 0.4 | -4.92014 | -4.46684 | -3.85898 | -3.39139 | -3.11423 |
| | 0.8 | -4.72665 | -4.28543 | -3.69552 | -3.24339 | -2.97618 |
| | 1.2 | -4.44563 | -4.02616 | -3.46739 | -3.04107 | -2.79011 |
| | 1.6 | -4.09409 | -3.70566 | -3.19089 | -2.80080 | -2.57255 |
| | 2.0 | -3.71035 | -3.36153 | -2.90295 | -2.55922 | -2.36016 |
| | 2.4 | -3.36871 | -3.06488 | -2.66903 | -2.37596 | -2.20835 |
| | 2.8 | -3.17807 | -2.90683 | -2.55494 | -2.29659 | -2.15119 |
| 3.2 | 0.0 | -4.82675 | -4.40054 | -3.82596 | -3.38120 | -3.11623 |
| | 0.4 | -4.73728 | -4.31428 | -3.74533 | -3.30609 | -3.04496 |
| | 0.8 | -4.58414 | -4.17082 | -3.61632 | -3.18956 | -2.93647 |
| | 1.2 | -4.36506 | -3.96857 | -3.43829 | -3.03165 | -2.79123 |
| | 1.6 | -4.08778 | -3.71516 | -3.21870 | -2.83985 | -2.61680 |
| | 2.0 | -3.77064 | -3.42825 | -2.97463 | -2.63114 | -2.43038 |
| | 2.4 | -3.44505 | -3.13901 | -2.73740 | -2.43745 | -2.26467 |
| | 2.8 | -3.17014 | -2.90426 | -2.55946 | -2.30743 | -2.16570 |
| | 3.2 | -3.02837 | -2.79067 | -2.48547 | -2.26518 | -2.13942 |

FRACTILES $W_\alpha$ RHO = -0.40

| DELTAS | | α: 0.0005 | 0.0010 | 0.0025 | 0.0050 | 0.0075 |
|---|---|---|---|---|---|---|
| 3.6 | 0.0 | -4.66303 | -4.26143 | -3.71870 | -3.29736 | -3.04575 |
| | 0.4 | -4.58631 | -4.18791 | -3.65057 | -3.23438 | -2.98628 |
| | 0.8 | -4.46159 | -4.07128 | -3.54602 | -3.14025 | -2.89887 |
| | 1.2 | -4.28649 | -3.90971 | -3.40397 | -3.01446 | -2.78332 |
| | 1.6 | -4.06469 | -3.70694 | -3.22822 | -2.86090 | -2.64361 |
| | 2.0 | -3.80640 | -3.47271 | -3.02801 | -2.68859 | -2.48876 |
| | 2.4 | -3.52787 | -3.22281 | -2.81889 | -2.51350 | -2.33549 |
| | 2.8 | -3.25477 | -2.98331 | -2.62827 | -2.36571 | -2.21636 |
| | 3.2 | -3.03594 | -2.80143 | -2.50106 | -2.28305 | -2.15733 |
| | 3.6 | -2.93358 | -2.72514 | -2.45951 | -2.26228 | -2.14541 |
| 4.0 | 0.0 | -4.52748 | -4.14616 | -3.62967 | -3.22763 | -2.98702 |
| | 0.4 | -4.46060 | -4.08245 | -3.57113 | -3.17390 | -2.93654 |
| | 0.8 | -4.35670 | -3.98550 | -3.48455 | -3.09624 | -2.86461 |
| | 1.2 | -4.21366 | -3.85365 | -3.36887 | -2.99403 | -2.77089 |
| | 1.6 | -4.03322 | -3.68877 | -3.22611 | -2.86946 | -2.65765 |
| | 2.0 | -3.82143 | -3.49666 | -3.06182 | -2.72792 | -2.53028 |
| | 2.4 | -3.58827 | -3.28688 | -2.88513 | -2.57848 | -2.39804 |
| | 2.8 | -3.34743 | -3.07301 | -2.71013 | -2.43699 | -2.27894 |
| | 3.2 | -3.12079 | -2.87813 | -2.56356 | -2.33253 | -2.19907 |
| | 3.6 | -2.95070 | -2.74404 | -2.47927 | -2.28081 | -2.16258 |
| | 4.0 | -2.88220 | -2.69793 | -2.45707 | -2.27080 | -2.15755 |
| 6.0 | 0.0 | -4.10247 | -3.78441 | -3.34966 | -3.00763 | -2.80120 |
| | 0.4 | -4.06379 | -3.74840 | -3.31766 | -2.97910 | -2.77491 |
| | 0.8 | -4.01206 | -3.70075 | -3.27596 | -2.94238 | -2.74133 |
| | 1.2 | -3.94658 | -3.64088 | -3.22415 | -2.89722 | -2.70031 |
| | 1.6 | -3.86721 | -3.56878 | -3.16235 | -2.84382 | -2.65210 |
| | 2.0 | -3.77462 | -3.48514 | -3.09130 | -2.78297 | -2.59753 |
| | 2.4 | -3.67030 | -3.39143 | -3.01246 | -2.71612 | -2.53801 |
| | 2.8 | -3.55662 | -3.28996 | -2.92804 | -2.64540 | -2.47566 |
| | 3.2 | -3.43675 | -3.18379 | -2.84101 | -2.57373 | -2.41339 |
| | 3.6 | -3.31460 | -3.07677 | -2.75522 | -2.50507 | -2.35511 |
| | 4.0 | -3.19492 | -2.97386 | -2.67601 | -2.44446 | -2.30520 |
| | 6.0 | -2.89713 | -2.74237 | -2.52157 | -2.33867 | -2.22372 |
| 12.0 | 0.0 | -3.67190 | -3.41756 | -3.06480 | -2.78269 | -2.61028 |
| | 0.4 | -3.65855 | -3.40552 | -3.05460 | -2.77396 | -2.60243 |
| | 0.8 | -3.64327 | -3.39179 | -3.04302 | -2.76408 | -2.59359 |
| | 1.2 | -3.62603 | -3.37634 | -3.03004 | -2.75306 | -2.58375 |
| | 1.6 | -3.60681 | -3.35915 | -3.01566 | -2.74089 | -2.57291 |
| | 2.0 | -3.58560 | -3.34024 | -2.99990 | -2.72760 | -2.56110 |
| | 2.4 | -3.56245 | -3.31965 | -2.98280 | -2.71324 | -2.54837 |
| | 2.8 | -3.53745 | -3.29747 | -2.96446 | -2.69789 | -2.53480 |
| | 3.2 | -3.51071 | -3.27381 | -2.94498 | -2.68164 | -2.52048 |
| | 3.6 | -3.48239 | -3.24883 | -2.92450 | -2.66464 | -2.50553 |
| | 4.0 | -3.45272 | -3.22272 | -2.90321 | -2.64704 | -2.49010 |
| | 6.0 | -3.29435 | -3.08497 | -2.79280 | -2.55725 | -2.41228 |
| | 12.0 | -3.07001 | -2.89810 | -2.65208 | -2.44865 | -2.32125 |
| INFINITY | | -3.29051 | -3.09024 | -2.80703 | -2.57583 | -2.43238 |

FRACTILES $W_\alpha$ RHO = -0.40

| DELTAS | | $\alpha$:0.0100 | 0.0175 | 0.0250 | 0.0375 | 0.0500 |
|---|---|---|---|---|---|---|
| 0.0 | 0.0 | -3.71417 | -3.07197 | -2.66863 | -2.21702 | -1.90195 |
| 0.4 | 0.0 | -3.67937 | -3.04584 | -2.64767 | -2.20163 | -1.89028 |
| | 0.4 | -3.48556 | -2.89133 | -2.51822 | -2.10062 | -1.80941 |
| 0.8 | 0.0 | -3.58856 | -2.98662 | -2.60626 | -2.17807 | -1.87775 |
| | 0.4 | -3.31618 | -2.75964 | -2.41006 | -2.01872 | -1.74581 |
| | 0.8 | -3.03257 | -2.53658 | -2.22547 | -1.87768 | -1.63554 |
| 1.2 | 0.0 | -3.46132 | -2.90879 | -2.55685 | -2.15745 | -1.87494 |
| | 0.4 | -3.21519 | -2.69211 | -2.36180 | -1.99025 | -1.72995 |
| | 0.8 | -2.88276 | -2.42257 | -2.13388 | -1.81124 | -1.58672 |
| | 1.2 | -2.63322 | -2.23040 | -1.97822 | -1.69702 | -1.50191 |
| 1.6 | 0.0 | -3.32385 | -2.82155 | -2.49950 | -2.13153 | -1.86927 |
| | 0.4 | -3.13657 | -2.65124 | -2.34220 | -1.99170 | -1.74396 |
| | 0.8 | -2.84270 | -2.40257 | -2.12498 | -1.81326 | -1.59535 |
| | 1.2 | -2.54197 | -2.16450 | -1.92829 | -1.66514 | -1.48285 |
| | 1.6 | -2.35605 | -2.02657 | -1.82109 | -1.59316 | -1.43627 |
| 2.0 | 0.0 | -3.19724 | -2.73743 | -2.44114 | -2.10086 | -1.85697 |
| | 0.4 | -3.05815 | -2.60978 | -2.32230 | -1.99387 | -1.75984 |
| | 0.8 | -2.83108 | -2.41237 | -2.14594 | -1.84407 | -1.63103 |
| | 1.2 | -2.55334 | -2.18357 | -1.95089 | -1.69042 | -1.50918 |
| | 1.6 | -2.31148 | -1.99868 | -1.80386 | -1.58838 | -1.44088 |
| | 2.0 | -2.18304 | -1.91004 | -1.74124 | -1.55700 | -1.43592 |
| 2.4 | 0.0 | -3.08922 | -2.66338 | -2.38785 | -2.07009 | -1.84130 |
| | 0.4 | -2.98301 | -2.56622 | -2.29757 | -1.98896 | -1.76770 |
| | 0.8 | -2.81229 | -2.41643 | -2.16263 | -1.87274 | -1.66628 |
| | 1.2 | -2.58828 | -2.22696 | -1.99736 | -1.73771 | -1.55505 |
| | 1.6 | -2.35136 | -2.03793 | -1.84158 | -1.62334 | -1.47347 |
| | 2.0 | -2.16880 | -1.90697 | -1.74600 | -1.57385 | -1.45706 |
| | 2.4 | -2.08699 | -1.86112 | -1.72561 | -1.57500 | -1.46665 |
| 2.8 | 0.0 | -2.99963 | -2.60079 | -2.34179 | -2.04201 | -1.82534 |
| | 0.4 | -2.91572 | -2.52459 | -2.27139 | -1.97923 | -1.76875 |
| | 0.8 | -2.78522 | -2.41013 | -2.16827 | -1.89033 | -1.69100 |
| | 1.2 | -2.61129 | -2.26159 | -2.03744 | -1.78158 | -1.59956 |
| | 1.6 | -2.41071 | -2.09666 | -1.89760 | -1.67355 | -1.51740 |
| | 2.0 | -2.22020 | -1.95245 | -1.78697 | -1.60784 | -1.48429 |
| | 2.4 | -2.09198 | -1.87545 | -1.74280 | -1.59219 | -1.48243 |
| | 2.8 | -2.05117 | -1.86106 | -1.73950 | -1.59677 | -1.49026 |
| 3.2 | 0.0 | -2.92576 | -2.54855 | -2.30281 | -2.01747 | -1.81053 |
| | 0.4 | -2.85754 | -2.48714 | -2.24645 | -1.96767 | -1.76600 |
| | 0.8 | -2.75514 | -2.39767 | -2.16610 | -1.89871 | -1.70592 |
| | 1.2 | -2.61937 | -2.28164 | -2.06377 | -1.81331 | -1.63365 |
| | 1.6 | -2.45788 | -2.14716 | -1.94814 | -1.72134 | -1.56066 |
| | 2.0 | -2.28822 | -2.01320 | -1.84033 | -1.64892 | -1.51573 |
| | 2.4 | -2.14414 | -1.91727 | -1.77693 | -1.61758 | -1.50192 |
| | 2.8 | -2.06739 | -1.87799 | -1.75539 | -1.61054 | -1.50211 |
| | 3.2 | -2.05033 | -1.87374 | -1.75630 | -1.61511 | -1.50814 |

FRACTILES $W_{\alpha}$ RHO = −0.40

| DELTAS | | α:0.0100 | 0.0175 | 0.0250 | 0.0375 | 0.0500 |
|---|---|---|---|---|---|---|
| 3.6 | 0.0 | −2.86458 | −2.50495 | −2.26997 | −1.99635 | −1.79731 |
| | 0.4 | −2.80768 | −2.45435 | −2.22384 | −1.95596 | −1.76149 |
| | 0.8 | −2.72554 | −2.38276 | −2.15982 | −1.90135 | −1.71417 |
| | 1.2 | −2.61764 | −2.29080 | −2.07887 | −1.83394 | −1.65716 |
| | 1.6 | −2.48823 | −2.18274 | −1.98557 | −1.75886 | −1.59635 |
| | 2.0 | −2.34641 | −2.06833 | −1.89074 | −1.68990 | −1.54819 |
| | 2.4 | −2.20994 | −1.96942 | −1.81879 | −1.64780 | −1.52442 |
| | 2.8 | −2.11225 | −1.91150 | −1.78206 | −1.62999 | −1.51683 |
| | 3.2 | −2.06771 | −1.88906 | −1.76979 | −1.62619 | −1.51737 |
| | 3.6 | −2.06067 | −1.88869 | −1.77206 | −1.63029 | −1.52213 |
| 4.0 | 0.0 | −2.81351 | −2.46834 | −2.24222 | −1.97824 | −1.78569 |
| | 0.4 | −2.76555 | −2.42589 | −2.20377 | −1.94486 | −1.75632 |
| | 0.8 | −2.69795 | −2.36743 | −2.15171 | −1.90072 | −1.71829 |
| | 1.2 | −2.61056 | −2.29322 | −2.08657 | −1.84670 | −1.67274 |
| | 1.6 | −2.50575 | −2.20580 | −2.01109 | −1.78580 | −1.62302 |
| | 2.0 | −2.38889 | −2.11078 | −1.93127 | −1.72526 | −1.57783 |
| | 2.4 | −2.26964 | −2.01979 | −1.86078 | −1.67918 | −1.54826 |
| | 2.8 | −2.16780 | −1.95268 | −1.81437 | −1.65299 | −1.53383 |
| | 3.2 | −2.10414 | −1.91586 | −1.79099 | −1.64157 | −1.52896 |
| | 3.6 | −2.07665 | −1.90196 | −1.78342 | −1.63936 | −1.52954 |
| | 4.0 | −2.07433 | −1.90330 | −1.78616 | −1.64299 | −1.53341 |
| 6.0 | 0.0 | −2.65145 | −2.35110 | −2.15237 | −1.91814 | −1.74559 |
| | 0.4 | −2.62684 | −2.33002 | −2.13375 | −1.90251 | −1.73225 |
| | 0.8 | −2.59561 | −2.30364 | −2.11068 | −1.88345 | −1.71620 |
| | 1.2 | −2.55765 | −2.27196 | −2.08325 | −1.86111 | −1.69765 |
| | 1.6 | −2.51326 | −2.23537 | −2.05190 | −1.83597 | −1.67709 |
| | 2.0 | −2.46329 | −2.19472 | −2.01747 | −1.80890 | −1.65541 |
| | 2.4 | −2.40914 | −2.15141 | −1.98136 | −1.78123 | −1.63387 |
| | 2.8 | −2.35290 | −2.10748 | −1.94554 | −1.75480 | −1.61409 |
| | 3.2 | −2.29747 | −2.06571 | −1.91254 | −1.73157 | −1.59748 |
| | 3.6 | −2.24662 | −2.02913 | −1.88465 | −1.71287 | −1.58471 |
| | 4.0 | −2.20407 | −1.99994 | −1.86312 | −1.69907 | −1.57571 |
| | 6.0 | −2.13795 | −1.95931 | −1.83584 | −1.68436 | −1.56833 |
| 12.0 | 0.0 | −2.48411 | −2.22809 | −2.05637 | −1.85138 | −1.69846 |
| | 0.4 | −2.47692 | −2.22220 | −2.05134 | −1.84734 | −1.69515 |
| | 0.8 | −2.46883 | −2.21561 | −2.04573 | −1.84287 | −1.69149 |
| | 1.2 | −2.45984 | −2.20832 | −2.03954 | −1.83796 | −1.68751 |
| | 1.6 | −2.44996 | −2.20034 | −2.03280 | −1.83264 | −1.68320 |
| | 2.0 | −2.43921 | −2.19170 | −2.02552 | −1.82693 | −1.67861 |
| | 2.4 | −2.42766 | −2.18246 | −2.01776 | −1.82088 | −1.67376 |
| | 2.8 | −2.41536 | −2.17266 | −2.00958 | −1.81453 | −1.66871 |
| | 3.2 | −2.40241 | −2.16240 | −2.00104 | −1.80794 | −1.66349 |
| | 3.6 | −2.38893 | −2.15178 | −1.99224 | −1.80119 | −1.65819 |
| | 4.0 | −2.37505 | −2.14091 | −1.98327 | −1.79437 | −1.65286 |
| | 6.0 | −2.30563 | −2.08760 | −1.93998 | −1.76214 | −1.62821 |
| | 12.0 | −2.22651 | −2.03017 | −1.89532 | −1.73091 | −1.60575 |
| INFINITY | | −2.32635 | −2.10836 | −1.95996 | −1.78046 | −1.64485 |

FRACTILES $W_\alpha$ RHO = −0.40

| DELTAS | | α:0.0750 | 0.1000 | 0.1500 | 0.2000 | 0.2500 |
|---|---|---|---|---|---|---|
| 0.0 | 0.0 | -1.46716 | -1.16676 | -0.75812 | -0.48185 | -0.27787 |
| 0.4 | 0.0 | -1.46043 | -1.16332 | -0.75903 | -0.48566 | -0.28385 |
| | 0.4 | -1.40784 | -1.13067 | -0.75432 | -0.50067 | -0.31421 |
| 0.8 | 0.0 | -1.46100 | -1.17138 | -0.77506 | -0.50556 | -0.30587 |
| | 0.4 | -1.36954 | -1.11001 | -0.75812 | -0.52171 | -0.34878 |
| | 0.8 | -1.30242 | -1.07337 | -0.76445 | -0.55891 | -0.41079 |
| 1.2 | 0.0 | -1.47894 | -1.20047 | -0.81402 | -0.54685 | -0.34612 |
| | 0.4 | -1.36933 | -1.11938 | -0.77898 | -0.54957 | -0.38184 |
| | 0.8 | -1.27821 | -1.06655 | -0.78252 | -0.59576 | -0.46442 |
| | 1.2 | -1.23500 | -1.05322 | -0.81305 | -0.66228 | -0.56800 |
| 1.6 | 0.0 | -1.49813 | -1.23401 | -0.86158 | -0.59861 | -0.39691 |
| | 0.4 | -1.39715 | -1.15381 | -0.81756 | -0.58715 | -0.41654 |
| | 0.8 | -1.29458 | -1.08739 | -0.80872 | -0.62642 | -0.50272 |
| | 1.2 | -1.23433 | -1.06630 | -0.84977 | -0.72089 | -0.60549 |
| | 1.6 | -1.22506 | -1.08706 | -0.91164 | -0.77108 | -0.64693 |
| 2.0 | 0.0 | -1.50932 | -1.25964 | -0.90310 | -0.64685 | -0.44654 |
| | 0.4 | -1.42882 | -1.19347 | -0.86243 | -0.63003 | -0.45370 |
| | 0.8 | -1.33359 | -1.12591 | -0.84212 | -0.65356 | -0.52747 |
| | 1.2 | -1.26115 | -1.09326 | -0.88258 | -0.73596 | -0.61011 |
| | 1.6 | -1.24630 | -1.12164 | -0.93569 | -0.78759 | -0.65748 |
| | 2.0 | -1.27132 | -1.15027 | -0.96335 | -0.81168 | -0.67769 |
| 2.4 | 0.0 | -1.51337 | -1.27619 | -0.93429 | -0.68531 | -0.48791 |
| | 0.4 | -1.45225 | -1.22567 | -0.90227 | -0.67029 | -0.48988 |
| | 0.8 | -1.37456 | -1.16767 | -0.87853 | -0.67994 | -0.53828 |
| | 1.2 | -1.30165 | -1.12731 | -0.90072 | -0.73995 | -0.60531 |
| | 1.6 | -1.27668 | -1.14237 | -0.94436 | -0.78863 | -0.65328 |
| | 2.0 | -1.29095 | -1.16678 | -0.97416 | -0.81803 | -0.68055 |
| | 2.4 | -1.30626 | -1.18321 | -0.98945 | -0.83111 | -0.69131 |
| 2.8 | 0.0 | -1.51331 | -1.28634 | -0.95670 | -0.71421 | -0.51993 |
| | 0.4 | -1.46684 | -1.24831 | -0.93298 | -0.70318 | -0.52110 |
| | 0.8 | -1.40679 | -1.20269 | -0.91195 | -0.70545 | -0.54984 |
| | 1.2 | -1.34314 | -1.16272 | -0.91788 | -0.74385 | -0.60068 |
| | 1.6 | -1.30674 | -1.16106 | -0.94902 | -0.78510 | -0.64450 |
| | 2.0 | -1.30846 | -1.17779 | -0.97688 | -0.81562 | -0.67469 |
| | 2.4 | -1.31884 | -1.19302 | -0.99496 | -0.83350 | -0.69133 |
| | 2.8 | -1.32882 | -1.20320 | -1.00406 | -0.84106 | -0.69733 |
| 3.2 | 0.0 | -1.51130 | -1.29256 | -0.97288 | -0.73577 | -0.54424 |
| | 0.4 | -1.47529 | -1.26354 | -0.95546 | -0.72831 | -0.54595 |
| | 0.8 | -1.42913 | -1.22855 | -0.93890 | -0.72815 | -0.56311 |
| | 1.2 | -1.37756 | -1.19408 | -0.93605 | -0.75051 | -0.59931 |
| | 1.6 | -1.33751 | -1.18095 | -0.95469 | -0.78235 | -0.63635 |
| | 2.0 | -1.32700 | -1.18818 | -0.97746 | -0.81051 | -0.66594 |
| | 2.4 | -1.33065 | -1.19988 | -0.99563 | -0.83039 | -0.68571 |
| | 2.8 | -1.33759 | -1.20964 | -1.00712 | -0.84177 | -0.69630 |
| | 3.2 | -1.34444 | -1.21635 | -1.01285 | -0.84633 | -0.69975 |

FRACTILES $W_{\alpha}$ RHO = -0.40

| DELTAS | | α: 0.0750 | 0.1000 | 0.1500 | 0.2000 | 0.2500 |
|---|---|---|---|---|---|---|
| 3.6 | 0.0 | -1.50846 | -1.29642 | -0.98482 | -0.75210 | -0.56286 |
| | 0.4 | -1.47994 | -1.27380 | -0.97184 | -0.74718 | -0.56512 |
| | 0.8 | -1.44396 | -1.24679 | -0.95927 | -0.74674 | -0.57632 |
| | 1.2 | -1.40307 | -1.21880 | -0.95341 | -0.75927 | -0.60124 |
| | 1.6 | -1.36577 | -1.20105 | -0.96243 | -0.78212 | -0.63081 |
| | 2.0 | -1.34695 | -1.19981 | -0.97886 | -0.80593 | -0.65757 |
| | 2.4 | -1.34336 | -1.20650 | -0.99492 | -0.82543 | -0.67804 |
| | 2.8 | -1.34626 | -1.21442 | -1.00708 | -0.83880 | -0.69141 |
| | 3.2 | -1.35096 | -1.22091 | -1.01468 | -0.84631 | -0.69838 |
| | 3.6 | -1.35591 | -1.22557 | -1.01845 | -0.84915 | -0.70038 |
| 4.0 | 0.0 | -1.50537 | -1.29881 | -0.99384 | -0.76473 | -0.57738 |
| | 0.4 | -1.48233 | -1.28082 | -0.98397 | -0.76150 | -0.57992 |
| | 0.8 | -1.45372 | -1.25960 | -0.97439 | -0.76133 | -0.58808 |
| | 1.2 | -1.42115 | -1.23717 | -0.96832 | -0.76859 | -0.60531 |
| | 1.6 | -1.38902 | -1.21927 | -0.97138 | -0.78425 | -0.62808 |
| | 2.0 | -1.36679 | -1.21240 | -0.98192 | -0.80316 | -0.65099 |
| | 2.4 | -1.35724 | -1.21402 | -0.99471 | -0.82071 | -0.67048 |
| | 2.8 | -1.35567 | -1.21911 | -1.00607 | -0.83446 | -0.68497 |
| | 3.2 | -1.35768 | -1.22452 | -1.01442 | -0.84371 | -0.69425 |
| | 3.6 | -1.36098 | -1.22899 | -1.01961 | -0.84881 | -0.69897 |
| | 4.0 | -1.36473 | -1.23239 | -1.02219 | -0.85063 | -0.70011 |
| 6.0 | 0.0 | -1.49150 | -1.30176 | -1.01711 | -0.79910 | -0.61769 |
| | 0.4 | -1.48161 | -1.29451 | -1.01383 | -0.79882 | -0.61982 |
| | 0.8 | -1.47005 | -1.28631 | -1.01064 | -0.79933 | -0.62323 |
| | 1.2 | -1.45708 | -1.27750 | -1.00789 | -0.80095 | -0.62817 |
| | 1.6 | -1.44325 | -1.26859 | -1.00604 | -0.80402 | -0.63481 |
| | 2.0 | -1.42937 | -1.26034 | -1.00559 | -0.80870 | -0.64296 |
| | 2.4 | -1.41655 | -1.25358 | -1.00680 | -0.81476 | -0.65206 |
| | 2.8 | -1.40584 | -1.24887 | -1.00947 | -0.82162 | -0.66140 |
| | 3.2 | -1.39781 | -1.24621 | -1.01309 | -0.82860 | -0.67024 |
| | 3.6 | -1.39240 | -1.24521 | -1.01704 | -0.83507 | -0.67805 |
| | 4.0 | -1.38916 | -1.24532 | -1.02079 | -0.84060 | -0.68447 |
| | 6.0 | -1.38995 | -1.25044 | -1.03002 | -0.85132 | -0.69551 |
| 12.0 | 0.0 | -1.47007 | -1.29686 | -1.03236 | -0.82571 | -0.65087 |
| | 0.4 | -1.46779 | -1.29533 | -1.03187 | -0.82595 | -0.65165 |
| | 0.8 | -1.46531 | -1.29368 | -1.03138 | -0.82627 | -0.65256 |
| | 1.2 | -1.46262 | -1.29192 | -1.03091 | -0.82667 | -0.65361 |
| | 1.6 | -1.45976 | -1.29007 | -1.03047 | -0.82717 | -0.65479 |
| | 2.0 | -1.45674 | -1.28815 | -1.03005 | -0.82778 | -0.65612 |
| | 2.4 | -1.45358 | -1.28618 | -1.02969 | -0.82848 | -0.65759 |
| | 2.8 | -1.45033 | -1.28418 | -1.02939 | -0.82930 | -0.65919 |
| | 3.2 | -1.44703 | -1.28219 | -1.02916 | -0.83022 | -0.66091 |
| | 3.6 | -1.44371 | -1.28024 | -1.02902 | -0.83124 | -0.66273 |
| | 4.0 | -1.44042 | -1.27834 | -1.02896 | -0.83235 | -0.66464 |
| | 6.0 | -1.42595 | -1.27064 | -1.02994 | -0.83867 | -0.67446 |
| | 12.0 | -1.41470 | -1.26642 | -1.03416 | -0.84766 | -0.68630 |
| INFINITY | | -1.43953 | -1.28155 | -1.03643 | -0.84162 | -0.67449 |

FRACTILES $W_\alpha$ RHO = -0.40

| DELTAS | | $\alpha$:0.3000 | 0.3500 | 0.4000 | 0.4500 | 0.5000 |
|---|---|---|---|---|---|---|
| 0.0 | 0.0 | -0.11977 | 0.00636 | 0.10862 | 0.19209 | 0.25994 |
| 0.4 | 0.0 | -0.12749 | -0.00287 | 0.09805 | 0.18023 | 0.24677 |
| | 0.4 | -0.17055 | -0.05696 | 0.03390 | 0.10637 | 0.16273 |
| 0.8 | 0.0 | -0.15080 | -0.02710 | 0.07296 | 0.15409 | 0.21912 |
| | 0.4 | -0.21660 | -0.11343 | -0.03283 | 0.02821 | 0.06706 |
| | 0.8 | -0.30045 | -0.21885 | -0.16659 | -0.10916 | -0.03786 |
| 1.2 | 0.0 | -0.18842 | -0.06143 | 0.04196 | 0.12595 | 0.19263 |
| | 0.4 | -0.25433 | -0.15631 | -0.08302 | -0.03510 | 0.02504 |
| | 0.8 | -0.37324 | -0.30849 | -0.23041 | -0.14640 | -0.05714 |
| | 1.2 | -0.47315 | -0.37807 | -0.28232 | -0.18464 | -0.08361 |
| 1.6 | 0.0 | -0.23515 | -0.10221 | 0.00815 | 0.09931 | 0.17200 |
| | 0.4 | -0.28588 | -0.18609 | -0.11811 | -0.04764 | 0.03050 |
| | 0.8 | -0.41586 | -0.32451 | -0.23269 | -0.13944 | -0.04339 |
| | 1.2 | -0.49748 | -0.39322 | -0.29019 | -0.18647 | -0.08031 |
| | 1.6 | -0.53143 | -0.42045 | -0.31128 | -0.20186 | -0.09035 |
| 2.0 | 0.0 | -0.28249 | -0.14444 | -0.02666 | 0.07366 | 0.15600 |
| | 0.4 | -0.31538 | -0.20804 | -0.12611 | -0.04050 | 0.04672 |
| | 0.8 | -0.41948 | -0.31766 | -0.21886 | -0.12069 | -0.02119 |
| | 1.2 | -0.49493 | -0.38540 | -0.27841 | -0.17174 | -0.06348 |
| | 1.6 | -0.53717 | -0.42222 | -0.30973 | -0.19754 | -0.08374 |
| | 2.0 | -0.55365 | -0.43519 | -0.31943 | -0.20417 | -0.08748 |
| 2.4 | 0.0 | -0.32361 | -0.18263 | -0.05939 | 0.04908 | 0.14233 |
| | 0.4 | -0.34399 | -0.22652 | -0.12848 | -0.03330 | 0.06058 |
| | 0.8 | -0.41810 | -0.30836 | -0.20403 | -0.10196 | 0.00018 |
| | 1.2 | -0.48419 | -0.37052 | -0.26068 | -0.15214 | -0.04285 |
| | 1.6 | -0.52921 | -0.41154 | -0.29711 | -0.18361 | -0.06905 |
| | 2.0 | -0.55371 | -0.43298 | -0.31539 | -0.19865 | -0.08081 |
| | 2.4 | -0.56224 | -0.43942 | -0.31988 | -0.20133 | -0.08180 |
| 2.8 | 0.0 | -0.35632 | -0.21398 | -0.08735 | 0.02687 | 0.12948 |
| | 0.4 | -0.37016 | -0.24351 | -0.13347 | -0.03043 | 0.06938 |
| | 0.8 | -0.41949 | -0.30255 | -0.19301 | -0.08719 | 0.01755 |
| | 1.2 | -0.47380 | -0.35618 | -0.24366 | -0.13344 | -0.02328 |
| | 1.6 | -0.51691 | -0.39688 | -0.28096 | -0.16665 | -0.05187 |
| | 2.0 | -0.54543 | -0.42302 | -0.30429 | -0.18689 | -0.06878 |
| | 2.4 | -0.56041 | -0.43613 | -0.31545 | -0.19602 | -0.07586 |
| | 2.8 | -0.56495 | -0.43932 | -0.31739 | -0.19682 | -0.07562 |
| 3.2 | 0.0 | -0.38153 | -0.23857 | -0.10987 | 0.00811 | 0.11724 |
| | 0.4 | -0.39224 | -0.25926 | -0.14124 | -0.03168 | 0.07343 |
| | 0.8 | -0.42449 | -0.30112 | -0.18667 | -0.07716 | 0.03029 |
| | 1.2 | -0.46683 | -0.34524 | -0.22998 | -0.11793 | -0.00670 |
| | 1.6 | -0.50518 | -0.38277 | -0.26535 | -0.15025 | -0.03527 |
| | 2.0 | -0.53430 | -0.41036 | -0.29073 | -0.17292 | -0.05487 |
| | 2.4 | -0.55307 | -0.42763 | -0.30621 | -0.18639 | -0.06615 |
| | 2.8 | -0.56259 | -0.43595 | -0.31325 | -0.19211 | -0.07053 |
| | 3.2 | -0.56501 | -0.43744 | -0.31389 | -0.19199 | -0.06973 |

FRACTILES $W_\alpha$ RHO = -0.40

| DELTAS | | $\alpha$:0.3000 | 0.3500 | 0.4000 | 0.4500 | 0.5000 |
|---|---|---|---|---|---|---|
| 3.6 | 0.0 | -0.40099 | -0.25771 | -0.12763 | -0.00716 | 0.10613 |
| | 0.4 | -0.40995 | -0.27338 | -0.15018 | -0.03561 | 0.07396 |
| | 0.8 | -0.43172 | -0.30309 | -0.18430 | -0.07132 | 0.03882 |
| | 1.2 | -0.46361 | -0.33821 | -0.22016 | -0.10613 | 0.00637 |
| | 1.6 | -0.49601 | -0.37112 | -0.25208 | -0.13603 | -0.02067 |
| | 2.0 | -0.52345 | -0.39791 | -0.27734 | -0.15914 | -0.04114 |
| | 2.4 | -0.54365 | -0.41711 | -0.29507 | -0.17504 | -0.05493 |
| | 2.8 | -0.55640 | -0.42889 | -0.30565 | -0.18426 | -0.06267 |
| | 3.2 | -0.56264 | -0.43431 | -0.31020 | -0.18792 | -0.06541 |
| | 3.6 | -0.56388 | -0.43487 | -0.31016 | -0.18735 | -0.06438 |
| 4.0 | 0.0 | -0.41624 | -0.27277 | -0.14172 | -0.01951 | 0.09649 |
| | 0.4 | -0.42396 | -0.28535 | -0.15894 | -0.04079 | 0.07228 |
| | 0.8 | -0.43959 | -0.30702 | -0.18469 | -0.06866 | 0.04400 |
| | 1.2 | -0.46336 | -0.33452 | -0.21380 | -0.09778 | 0.01616 |
| | 1.6 | -0.48980 | -0.36243 | -0.24166 | -0.12449 | -0.00855 |
| | 2.0 | -0.51431 | -0.38708 | -0.26548 | -0.14675 | -0.02867 |
| | 2.4 | -0.53424 | -0.40655 | -0.28387 | -0.16361 | -0.04364 |
| | 2.8 | -0.54860 | -0.42024 | -0.29655 | -0.17502 | -0.05357 |
| | 3.2 | -0.55748 | -0.42847 | -0.30395 | -0.18149 | -0.05899 |
| | 3.6 | -0.56167 | -0.43208 | -0.30695 | -0.18385 | -0.06072 |
| | 4.0 | -0.56223 | -0.43213 | -0.30655 | -0.18306 | -0.05960 |
| 6.0 | 0.0 | -0.45907 | -0.31557 | -0.18239 | -0.05618 | 0.06562 |
| | 0.4 | -0.46318 | -0.32132 | -0.18948 | -0.06434 | 0.05669 |
| | 0.8 | -0.46890 | -0.32890 | -0.19850 | -0.07440 | 0.04599 |
| | 1.2 | -0.47641 | -0.33833 | -0.20930 | -0.08605 | 0.03397 |
| | 1.6 | -0.48561 | -0.34932 | -0.22141 | -0.09870 | 0.02130 |
| | 2.0 | -0.49607 | -0.36122 | -0.23408 | -0.11158 | 0.00871 |
| | 2.4 | -0.50706 | -0.37326 | -0.24655 | -0.12396 | -0.00317 |
| | 2.8 | -0.51780 | -0.38469 | -0.25812 | -0.13526 | -0.01383 |
| | 3.2 | -0.52763 | -0.39490 | -0.26829 | -0.14505 | -0.02295 |
| | 3.6 | -0.53606 | -0.40351 | -0.27674 | -0.15308 | -0.03035 |
| | 4.0 | -0.54285 | -0.41032 | -0.28335 | -0.15929 | -0.03601 |
| | 6.0 | -0.55359 | -0.42038 | -0.29246 | -0.16728 | -0.04272 |
| 12.0 | 0.0 | -0.49573 | -0.35350 | -0.21986 | -0.09176 | 0.03320 |
| | 0.4 | -0.49692 | -0.35500 | -0.22161 | -0.09367 | 0.03118 |
| | 0.8 | -0.49828 | -0.35670 | -0.22356 | -0.09580 | 0.02895 |
| | 1.2 | -0.49980 | -0.35859 | -0.22571 | -0.09813 | 0.02653 |
| | 1.6 | -0.50150 | -0.36067 | -0.22806 | -0.10065 | 0.02391 |
| | 2.0 | -0.50337 | -0.36293 | -0.23059 | -0.10336 | 0.02113 |
| | 2.4 | -0.50539 | -0.36534 | -0.23328 | -0.10621 | 0.01821 |
| | 2.8 | -0.50755 | -0.36791 | -0.23611 | -0.10920 | 0.01517 |
| | 3.2 | -0.50984 | -0.37059 | -0.23905 | -0.11228 | 0.01205 |
| | 3.6 | -0.51222 | -0.37336 | -0.24206 | -0.11542 | 0.00890 |
| | 4.0 | -0.51468 | -0.37618 | -0.24512 | -0.11858 | 0.00574 |
| | 6.0 | -0.52682 | -0.38981 | -0.25957 | -0.13330 | -0.00877 |
| | 12.0 | -0.54034 | -0.40418 | -0.27418 | -0.14766 | -0.02243 |
| INFINITY | | -0.52440 | -0.38532 | -0.25335 | -0.12566 | 0.0 |

FRACTILES $w_\alpha$ RHO = -0.40

| DELTAS | | $\alpha$:0.5500 | 0.6000 | 0.6500 | 0.7000 | 0.7500 |
|---|---|---|---|---|---|---|
| 0.0 | 0.0 | 0.31405 | 0.35477 | 0.38055 | 0.41905 | 0.47501 |
| 0.4 | 0.0 | 0.29939 | 0.33799 | 0.36445 | 0.40751 | 0.46834 |
| | 0.4 | 0.20233 | 0.23470 | 0.28627 | 0.35493 | 0.44377 |
| 0.8 | 0.0 | 0.26919 | 0.30117 | 0.33939 | 0.39538 | 0.46994 |
| | 0.4 | 0.11399 | 0.17719 | 0.25482 | 0.34835 | 0.46157 |
| | 0.8 | 0.04346 | 0.13477 | 0.23749 | 0.35444 | 0.49033 |
| 1.2 | 0.0 | 0.24115 | 0.27914 | 0.33645 | 0.40926 | 0.49935 |
| | 0.4 | 0.09712 | 0.17915 | 0.27190 | 0.37769 | 0.50067 |
| | 0.8 | 0.03815 | 0.14096 | 0.25348 | 0.37892 | 0.52219 |
| | 1.2 | 0.02236 | 0.13516 | 0.25726 | 0.39207 | 0.54469 |
| 1.6 | 0.0 | 0.22206 | 0.28281 | 0.35618 | 0.44135 | 0.54088 |
| | 0.4 | 0.11435 | 0.20468 | 0.30328 | 0.41288 | 0.53769 |
| | 0.8 | 0.05701 | 0.16358 | 0.27867 | 0.40549 | 0.54882 |
| | 1.2 | 0.03004 | 0.14657 | 0.27179 | 0.40910 | 0.56354 |
| | 1.6 | 0.02506 | 0.14643 | 0.27632 | 0.41817 | 0.57705 |
| 2.0 | 0.0 | 0.22128 | 0.29738 | 0.38095 | 0.47343 | 0.57811 |
| | 0.4 | 0.13680 | 0.23147 | 0.33285 | 0.44381 | 0.56848 |
| | 0.8 | 0.08146 | 0.18921 | 0.30442 | 0.43023 | 0.57124 |
| | 1.2 | 0.04822 | 0.16539 | 0.29051 | 0.42689 | 0.57941 |
| | 1.6 | 0.03354 | 0.15636 | 0.28729 | 0.42971 | 0.58861 |
| | 2.0 | 0.03252 | 0.15792 | 0.29129 | 0.43603 | 0.59711 |
| 2.4 | 0.0 | 0.22384 | 0.31019 | 0.40104 | 0.49893 | 0.60753 |
| | 0.4 | 0.15550 | 0.25360 | 0.35721 | 0.46924 | 0.59374 |
| | 0.8 | 0.10441 | 0.21278 | 0.32767 | 0.45214 | 0.59060 |
| | 1.2 | 0.06912 | 0.18584 | 0.30974 | 0.44407 | 0.59348 |
| | 1.6 | 0.04847 | 0.17104 | 0.30115 | 0.44215 | 0.59886 |
| | 2.0 | 0.04005 | 0.16601 | 0.29962 | 0.44423 | 0.60476 |
| | 2.4 | 0.04062 | 0.16804 | 0.30298 | 0.44882 | 0.61043 |
| 2.8 | 0.0 | 0.22340 | 0.31782 | 0.41490 | 0.51767 | 0.62999 |
| | 0.4 | 0.16878 | 0.27022 | 0.37615 | 0.48951 | 0.61431 |
| | 0.8 | 0.12343 | 0.23258 | 0.34740 | 0.47092 | 0.60740 |
| | 1.2 | 0.08885 | 0.20502 | 0.32766 | 0.45991 | 0.60627 |
| | 1.6 | 0.06532 | 0.18702 | 0.31569 | 0.45459 | 0.60837 |
| | 2.0 | 0.05196 | 0.17742 | 0.31012 | 0.45335 | 0.61189 |
| | 2.4 | 0.04697 | 0.17456 | 0.30944 | 0.45493 | 0.61583 |
| | 2.8 | 0.04817 | 0.17662 | 0.31226 | 0.45841 | 0.61984 |
| 3.2 | 0.0 | 0.21975 | 0.32083 | 0.42344 | 0.53078 | 0.64677 |
| | 0.4 | 0.17707 | 0.28182 | 0.39022 | 0.50525 | 0.63085 |
| | 0.8 | 0.13803 | 0.24829 | 0.36348 | 0.48661 | 0.62181 |
| | 1.2 | 0.10580 | 0.22170 | 0.34342 | 0.47403 | 0.61789 |
| | 1.6 | 0.08158 | 0.20240 | 0.32963 | 0.46643 | 0.61735 |
| | 2.0 | 0.06540 | 0.18997 | 0.32133 | 0.46270 | 0.61874 |
| | 2.4 | 0.05646 | 0.18354 | 0.31758 | 0.46186 | 0.62108 |
| | 2.8 | 0.05345 | 0.18192 | 0.31738 | 0.46312 | 0.62385 |
| | 3.2 | 0.05486 | 0.18386 | 0.31977 | 0.46585 | 0.62682 |

FRACTILES $W_\alpha$ RHO = -0.40

| DELTAS | | $\alpha$: 0.5500 | 0.6000 | 0.6500 | 0.7000 | 0.7500 |
|---|---|---|---|---|---|---|
| 3.6 | 0.0 | 0.21433 | 0.32067 | 0.42803 | 0.53953 | 0.65906 |
| | 0.4 | 0.18141 | 0.28932 | 0.40024 | 0.51715 | 0.64396 |
| | 0.8 | 0.14858 | 0.26021 | 0.37617 | 0.49940 | 0.63398 |
| | 1.2 | 0.11955 | 0.23556 | 0.35679 | 0.48629 | 0.62828 |
| | 1.6 | 0.09604 | 0.21622 | 0.34228 | 0.47733 | 0.62576 |
| | 2.0 | 0.07865 | 0.20232 | 0.33233 | 0.47184 | 0.62537 |
| | 2.4 | 0.06722 | 0.19351 | 0.32641 | 0.46912 | 0.62626 |
| | 2.8 | 0.06109 | 0.18909 | 0.32383 | 0.46855 | 0.62788 |
| | 3.2 | 0.05927 | 0.18821 | 0.32390 | 0.46958 | 0.62991 |
| | 3.6 | 0.06070 | 0.18998 | 0.32593 | 0.47179 | 0.63219 |
| 4.0 | 0.0 | 0.20842 | 0.31872 | 0.42996 | 0.54509 | 0.66789 |
| | 0.4 | 0.18296 | 0.29372 | 0.40706 | 0.52593 | 0.65417 |
| | 0.8 | 0.15577 | 0.26892 | 0.38589 | 0.50962 | 0.64409 |
| | 1.2 | 0.13025 | 0.24666 | 0.36782 | 0.49668 | 0.63741 |
| | 1.6 | 0.10828 | 0.22812 | 0.35336 | 0.48707 | 0.63352 |
| | 2.0 | 0.09079 | 0.21373 | 0.34258 | 0.48044 | 0.63173 |
| | 2.4 | 0.07804 | 0.20351 | 0.33523 | 0.47635 | 0.63137 |
| | 2.8 | 0.06978 | 0.19712 | 0.33090 | 0.47432 | 0.63195 |
| | 3.2 | 0.06548 | 0.19403 | 0.32911 | 0.47394 | 0.63311 |
| | 3.6 | 0.06440 | 0.19360 | 0.32934 | 0.47482 | 0.63466 |
| | 4.0 | 0.06579 | 0.19520 | 0.33109 | 0.47665 | 0.63647 |
| 6.0 | 0.0 | 0.18516 | 0.30449 | 0.42574 | 0.55149 | 0.68517 |
| | 0.4 | 0.17577 | 0.29500 | 0.41659 | 0.54321 | 0.67845 |
| | 0.8 | 0.16485 | 0.28433 | 0.40667 | 0.53465 | 0.67197 |
| | 1.2 | 0.15296 | 0.27306 | 0.39656 | 0.52629 | 0.66608 |
| | 1.6 | 0.14076 | 0.26182 | 0.38678 | 0.51853 | 0.66100 |
| | 2.0 | 0.12891 | 0.25114 | 0.37774 | 0.51162 | 0.65682 |
| | 2.4 | 0.11794 | 0.24145 | 0.36972 | 0.50571 | 0.65354 |
| | 2.8 | 0.10823 | 0.23302 | 0.36289 | 0.50084 | 0.65107 |
| | 3.2 | 0.10004 | 0.22600 | 0.35731 | 0.49700 | 0.64933 |
| | 3.6 | 0.09347 | 0.22045 | 0.35298 | 0.49413 | 0.64819 |
| | 4.0 | 0.08850 | 0.21632 | 0.34984 | 0.49214 | 0.64756 |
| | 6.0 | 0.08319 | 0.21254 | 0.34771 | 0.49180 | 0.64919 |
| 12.0 | 0.0 | 0.15709 | 0.28192 | 0.40987 | 0.54357 | 0.68660 |
| | 0.4 | 0.15503 | 0.27989 | 0.40795 | 0.54185 | 0.68521 |
| | 0.8 | 0.15278 | 0.27769 | 0.40587 | 0.54001 | 0.68374 |
| | 1.2 | 0.15033 | 0.27531 | 0.40365 | 0.53806 | 0.68221 |
| | 1.6 | 0.14772 | 0.27278 | 0.40131 | 0.53603 | 0.68064 |
| | 2.0 | 0.14495 | 0.27012 | 0.39887 | 0.53393 | 0.67904 |
| | 2.4 | 0.14205 | 0.26736 | 0.39635 | 0.53179 | 0.67744 |
| | 2.8 | 0.13907 | 0.26452 | 0.39379 | 0.52963 | 0.67585 |
| | 3.2 | 0.13602 | 0.26165 | 0.39121 | 0.52748 | 0.67430 |
| | 3.6 | 0.13295 | 0.25878 | 0.38864 | 0.52536 | 0.67279 |
| | 4.0 | 0.12989 | 0.25593 | 0.38612 | 0.52330 | 0.67135 |
| | 6.0 | 0.11607 | 0.24326 | 0.37511 | 0.51452 | 0.66554 |
| | 12.0 | 0.10351 | 0.23221 | 0.36599 | 0.50783 | 0.66187 |
| INFINITY | | 0.12566 | 0.25335 | 0.38532 | 0.52440 | 0.67449 |

FRACTILES $W_\alpha$ RHO = −0.40

| DELTAS | | $\alpha$: 0.8000 | 0.8500 | 0.9000 | 0.9250 | 0.9500 |
|---|---|---|---|---|---|---|
| 0.0 | 0.0 | 0.55253 | 0.66213 | 0.82894 | 0.95349 | 1.13537 |
| 0.4 | 0.0 | 0.55173 | 0.66886 | 0.84621 | 0.97808 | 1.17001 |
| | 0.4 | 0.55961 | 0.71615 | 0.94480 | 1.11034 | 1.34629 |
| 0.8 | 0.0 | 0.56854 | 0.70284 | 0.89996 | 1.04301 | 1.24715 |
| | 0.4 | 0.60168 | 0.78276 | 1.03678 | 1.21561 | 1.46549 |
| | 0.8 | 0.65322 | 0.85830 | 1.13942 | 1.33421 | 1.60328 |
| 1.2 | 0.0 | 0.61202 | 0.75843 | 0.96436 | 1.10946 | 1.31224 |
| | 0.4 | 0.64806 | 0.83355 | 1.08767 | 1.26366 | 1.50667 |
| | 0.8 | 0.69141 | 0.90158 | 1.18590 | 1.38094 | 1.64831 |
| | 1.2 | 0.72346 | 0.94366 | 1.23896 | 1.44012 | 1.71436 |
| 1.6 | 0.0 | 0.66029 | 0.81042 | 1.01574 | 1.15770 | 1.35347 |
| | 0.4 | 0.68473 | 0.86692 | 1.11281 | 1.28121 | 1.51179 |
| | 0.8 | 0.71645 | 0.92265 | 1.19883 | 1.38680 | 1.64289 |
| | 1.2 | 0.74325 | 0.96316 | 1.25597 | 1.45430 | 1.72340 |
| | 1.6 | 0.76116 | 0.98542 | 1.28253 | 1.48291 | 1.75383 |
| 2.0 | 0.0 | 0.70067 | 0.85168 | 1.05442 | 1.19275 | 1.38167 |
| | 0.4 | 0.71355 | 0.89121 | 1.12820 | 1.28903 | 1.50769 |
| | 0.8 | 0.73481 | 0.93440 | 1.19948 | 1.37868 | 1.62151 |
| | 1.2 | 0.75588 | 0.97056 | 1.25463 | 1.44605 | 1.70469 |
| | 1.6 | 0.77201 | 0.99447 | 1.28782 | 1.48491 | 1.75051 |
| | 2.0 | 0.78254 | 1.00681 | 1.30156 | 1.49901 | 1.76446 |
| 2.4 | 0.0 | 0.73256 | 0.88427 | 1.08501 | 1.22048 | 1.40397 |
| | 0.4 | 0.73713 | 0.91099 | 1.14055 | 1.29512 | 1.50395 |
| | 0.8 | 0.75008 | 0.94327 | 1.19790 | 1.36900 | 1.59971 |
| | 1.2 | 0.76545 | 0.97353 | 1.24726 | 1.43085 | 1.67795 |
| | 1.6 | 0.77903 | 0.99669 | 1.28245 | 1.47372 | 1.73071 |
| | 2.0 | 0.78903 | 1.01125 | 1.30233 | 1.49678 | 1.75757 |
| | 2.4 | 0.79561 | 1.01847 | 1.30970 | 1.50383 | 1.76373 |
| 2.8 | 0.0 | 0.75757 | 0.91044 | 1.11019 | 1.24372 | 1.42324 |
| | 0.4 | 0.75676 | 0.92796 | 1.15195 | 1.30168 | 1.50285 |
| | 0.8 | 0.76355 | 0.95147 | 1.19741 | 1.36176 | 1.58238 |
| | 1.2 | 0.77390 | 0.97569 | 1.23971 | 1.41599 | 1.65241 |
| | 1.6 | 0.78451 | 0.99646 | 1.27352 | 1.45833 | 1.70590 |
| | 2.0 | 0.79336 | 1.01155 | 1.29641 | 1.48616 | 1.74007 |
| | 2.4 | 0.79983 | 1.02077 | 1.30876 | 1.50033 | 1.75633 |
| | 2.8 | 0.80419 | 1.02522 | 1.31281 | 1.50381 | 1.75868 |
| 3.2 | 0.0 | 0.77710 | 0.93163 | 1.13134 | 1.26372 | 1.44054 |
| | 0.4 | 0.77310 | 0.94271 | 1.16279 | 1.30896 | 1.50433 |
| | 0.8 | 0.77557 | 0.95948 | 1.19863 | 1.35760 | 1.57012 |
| | 1.2 | 0.78186 | 0.97830 | 1.23399 | 1.40399 | 1.63120 |
| | 1.6 | 0.78955 | 0.99597 | 1.26469 | 1.44330 | 1.68193 |
| | 2.0 | 0.79683 | 1.01030 | 1.28807 | 1.47259 | 1.71893 |
| | 2.4 | 0.80275 | 1.02039 | 1.30335 | 1.49115 | 1.74164 |
| | 2.8 | 0.80711 | 1.02646 | 1.31129 | 1.50014 | 1.75177 |
| | 3.2 | 0.81015 | 1.02933 | 1.31353 | 1.50172 | 1.75220 |

FRACTILES $W_\alpha$ RHO = -0.40

| DELTAS | | α:0.8000 | 0.8500 | 0.9000 | 0.9250 | 0.9500 |
|---|---|---|---|---|---|---|
| 3.6 | 0.0 | 0.79225 | 0.94878 | 1.14923 | 1.28111 | 1.45623 |
| | 0.4 | 0.78659 | 0.95549 | 1.17304 | 1.31667 | 1.50774 |
| | 0.8 | 0.78619 | 0.96724 | 1.20126 | 1.35606 | 1.56220 |
| | 1.2 | 0.78941 | 0.98156 | 1.23043 | 1.39524 | 1.61481 |
| | 1.6 | 0.79452 | 0.99607 | 1.25738 | 1.43050 | 1.66115 |
| | 2.0 | 0.80009 | 1.00890 | 1.27970 | 1.45910 | 1.69806 |
| | 2.4 | 0.80515 | 1.01894 | 1.29615 | 1.47972 | 1.72413 |
| | 2.8 | 0.80923 | 1.02588 | 1.30663 | 1.49242 | 1.73962 |
| | 3.2 | 0.81229 | 1.03001 | 1.31189 | 1.49827 | 1.74605 |
| | 3.6 | 0.81450 | 1.03193 | 1.31310 | 1.49883 | 1.74551 |
| 4.0 | 0.0 | 0.80390 | 0.96266 | 1.16439 | 1.29627 | 1.47047 |
| | 0.4 | 0.79762 | 0.96648 | 1.18256 | 1.32445 | 1.51240 |
| | 0.8 | 0.79545 | 0.97457 | 1.20482 | 1.35646 | 1.55764 |
| | 1.2 | 0.79644 | 0.98529 | 1.22875 | 1.38937 | 1.60269 |
| | 1.6 | 0.79946 | 0.99693 | 1.25194 | 1.42036 | 1.64415 |
| | 2.0 | 0.80341 | 1.00797 | 1.27239 | 1.44710 | 1.67930 |
| | 2.4 | 0.80744 | 1.01735 | 1.28878 | 1.46813 | 1.70647 |
| | 2.8 | 0.81101 | 1.02452 | 1.30058 | 1.48293 | 1.72519 |
| | 3.2 | 0.81391 | 1.02942 | 1.30794 | 1.49183 | 1.73600 |
| | 3.6 | 0.81613 | 1.03231 | 1.31151 | 1.49571 | 1.74013 |
| | 4.0 | 0.81779 | 1.03364 | 1.31212 | 1.49570 | 1.73911 |
| 6.0 | 0.0 | 0.83201 | 1.00113 | 1.21194 | 1.34724 | 1.52303 |
| | 0.4 | 0.82777 | 1.00074 | 1.21772 | 1.35768 | 1.54024 |
| | 0.8 | 0.82432 | 1.00169 | 1.22536 | 1.37024 | 1.55978 |
| | 1.2 | 0.82182 | 1.00389 | 1.23448 | 1.38433 | 1.58086 |
| | 1.6 | 0.82028 | 1.00712 | 1.24458 | 1.39928 | 1.60257 |
| | 2.0 | 0.81960 | 1.01108 | 1.25508 | 1.41436 | 1.62396 |
| | 2.4 | 0.81963 | 1.01541 | 1.26541 | 1.42886 | 1.64417 |
| | 2.8 | 0.82015 | 1.01977 | 1.27507 | 1.44217 | 1.66246 |
| | 3.2 | 0.82098 | 1.02388 | 1.28365 | 1.45382 | 1.67827 |
| | 3.6 | 0.82196 | 1.02751 | 1.29089 | 1.46350 | 1.69126 |
| | 4.0 | 0.82296 | 1.03055 | 1.29666 | 1.47111 | 1.70132 |
| | 6.0 | 0.82676 | 1.03683 | 1.30589 | 1.48210 | 1.71441 |
| 12.0 | 0.0 | 0.84443 | 1.02663 | 1.25337 | 1.39819 | 1.58502 |
| | 0.4 | 0.84354 | 1.02650 | 1.25450 | 1.40030 | 1.58862 |
| | 0.8 | 0.84263 | 1.02643 | 1.25580 | 1.40267 | 1.59259 |
| | 1.2 | 0.84172 | 1.02644 | 1.25729 | 1.40530 | 1.59692 |
| | 1.6 | 0.84081 | 1.02653 | 1.25896 | 1.40816 | 1.60156 |
| | 2.0 | 0.83994 | 1.02671 | 1.26079 | 1.41125 | 1.60649 |
| | 2.4 | 0.83910 | 1.02698 | 1.26278 | 1.41453 | 1.61165 |
| | 2.8 | 0.83831 | 1.02734 | 1.26490 | 1.41796 | 1.61698 |
| | 3.2 | 0.83759 | 1.02779 | 1.26712 | 1.42150 | 1.62244 |
| | 3.6 | 0.83693 | 1.02831 | 1.26943 | 1.42512 | 1.62794 |
| | 4.0 | 0.83634 | 1.02890 | 1.27178 | 1.42877 | 1.63344 |
| | 6.0 | 0.83444 | 1.03237 | 1.28311 | 1.44577 | 1.65849 |
| | 12.0 | 0.83459 | 1.03751 | 1.29525 | 1.46281 | 1.68226 |
| INFINITY | | 0.84162 | 1.03643 | 1.28155 | 1.43953 | 1.64485 |

FRACTILES $w_\alpha$ RHO = −0.40

| DELTAS | | $\alpha$:0.9625 | 0.9750 | 0.9825 | 0.9900 | 0.9925 |
|---|---|---|---|---|---|---|
| 0.0 | 0.0 | 1.26794 | 1.45871 | 1.62961 | 1.90241 | 2.04447 |
| 0.4 | 0.0 | 1.30948 | 1.50962 | 1.68842 | 1.97290 | 2.12065 |
| | 0.4 | 1.51482 | 1.75322 | 1.96337 | 2.29333 | 2.46291 |
| 0.8 | 0.0 | 1.39306 | 1.59956 | 1.78165 | 2.06759 | 2.21457 |
| | 0.4 | 1.64126 | 1.88699 | 2.10135 | 2.43459 | 2.60456 |
| | 0.8 | 1.79087 | 2.05125 | 2.27691 | 2.62548 | 2.80238 |
| 1.2 | 0.0 | 1.45488 | 1.65427 | 1.82817 | 2.09848 | 2.23633 |
| | 0.4 | 1.67603 | 1.91105 | 2.11468 | 2.42916 | 2.58873 |
| | 0.8 | 1.83352 | 2.08925 | 2.30979 | 2.64876 | 2.82009 |
| | 1.2 | 1.90343 | 2.16345 | 2.38683 | 2.72882 | 2.90113 |
| 1.6 | 0.0 | 1.48977 | 1.67876 | 1.84239 | 2.09491 | 2.22298 |
| | 0.4 | 1.67137 | 1.89157 | 2.08134 | 2.37285 | 2.52013 |
| | 0.8 | 1.81936 | 2.06197 | 2.27031 | 2.58919 | 2.74982 |
| | 1.2 | 1.90818 | 2.16144 | 2.37830 | 2.70917 | 2.87542 |
| | 1.6 | 1.93930 | 2.19282 | 2.40934 | 2.73879 | 2.90395 |
| 2.0 | 0.0 | 1.51216 | 1.69193 | 1.84666 | 2.08405 | 2.20388 |
| | 0.4 | 1.65814 | 1.86473 | 2.04195 | 2.31293 | 2.44933 |
| | 0.8 | 1.78808 | 2.01620 | 2.21139 | 2.50901 | 2.65847 |
| | 1.2 | 1.88166 | 2.12347 | 2.32990 | 2.64392 | 2.80131 |
| | 1.6 | 1.93183 | 2.17907 | 2.38972 | 2.70946 | 2.86942 |
| | 2.0 | 1.94529 | 2.19138 | 2.40065 | 2.71766 | 2.87600 |
| 2.4 | 0.0 | 1.52985 | 1.70231 | 1.84998 | 2.07537 | 2.18867 |
| | 0.4 | 1.64689 | 1.84231 | 2.00928 | 2.26350 | 2.39104 |
| | 0.8 | 1.75734 | 1.97244 | 2.15589 | 2.43465 | 2.57426 |
| | 1.2 | 1.84648 | 2.07610 | 2.27160 | 2.56815 | 2.71643 |
| | 1.6 | 1.90569 | 2.14374 | 2.34611 | 2.65257 | 2.80560 |
| | 2.0 | 1.93484 | 2.17567 | 2.38009 | 2.68917 | 2.84329 |
| | 2.4 | 1.94012 | 2.17939 | 2.38220 | 2.68836 | 2.84084 |
| 2.8 | 0.0 | 1.54566 | 1.71256 | 1.85479 | 2.07087 | 2.17909 |
| | 0.4 | 1.63990 | 1.82652 | 1.98537 | 2.22633 | 2.34684 |
| | 0.8 | 1.73253 | 1.93679 | 2.11044 | 2.37351 | 2.50491 |
| | 1.2 | 1.81315 | 2.03158 | 2.21709 | 2.49774 | 2.63778 |
| | 1.6 | 1.87404 | 2.10230 | 2.29593 | 2.58852 | 2.73435 |
| | 2.0 | 1.91231 | 2.14588 | 2.34381 | 2.64251 | 2.79124 |
| | 2.4 | 1.92978 | 2.16473 | 2.36360 | 2.66337 | 2.81246 |
| | 2.8 | 1.93115 | 2.16451 | 2.36182 | 2.65885 | 2.80644 |
| 3.2 | 0.0 | 1.56046 | 1.72321 | 1.86132 | 2.07025 | 2.17453 |
| | 0.4 | 1.63685 | 1.81665 | 1.96916 | 2.19968 | 2.31465 |
| | 0.8 | 1.71424 | 1.90971 | 2.07542 | 2.32571 | 2.45043 |
| | 1.2 | 1.78523 | 1.99403 | 2.17093 | 2.43789 | 2.57082 |
| | 1.6 | 1.84359 | 2.06261 | 2.24802 | 2.52760 | 2.66671 |
| | 2.0 | 1.88570 | 2.11147 | 2.30246 | 2.59019 | 2.73323 |
| | 2.4 | 1.91109 | 2.14030 | 2.33403 | 2.62562 | 2.77047 |
| | 2.8 | 1.92182 | 2.15165 | 2.34575 | 2.63759 | 2.78245 |
| | 3.2 | 1.92131 | 2.14965 | 2.34230 | 2.63170 | 2.77522 |

FRACTILES $W_\alpha$ RHO = −0.40

| DELTAS | | $\alpha$: 0.9625 | 0.9750 | 0.9825 | 0.9900 | 0.9925 |
|---|---|---|---|---|---|---|
| 3.6 | 0.0 | 1.57442 | 1.73415 | 1.86918 | 2.07264 | 2.17387 |
| | 0.4 | 1.63681 | 1.81136 | 1.95894 | 2.18128 | 2.29187 |
| | 0.8 | 1.70154 | 1.88998 | 2.04930 | 2.28927 | 2.40858 |
| | 1.2 | 1.76323 | 1.96396 | 2.13363 | 2.38908 | 2.51603 |
| | 1.6 | 1.81705 | 2.02782 | 2.20592 | 2.47393 | 2.60705 |
| | 2.0 | 1.85953 | 2.07773 | 2.26202 | 2.53918 | 2.67677 |
| | 2.4 | 1.88918 | 2.11213 | 2.30032 | 2.58314 | 2.72347 |
| | 2.8 | 1.90646 | 2.13167 | 2.32165 | 2.60695 | 2.74842 |
| | 3.2 | 1.91316 | 2.13858 | 2.32860 | 2.61374 | 2.75503 |
| | 3.6 | 1.91174 | 2.13581 | 2.32455 | 2.60753 | 2.74765 |
| 4.0 | 0.0 | 1.58731 | 1.74512 | 1.87788 | 2.07719 | 2.17607 |
| | 0.4 | 1.63890 | 1.80943 | 1.95318 | 2.16909 | 2.27621 |
| | 0.8 | 1.69320 | 1.87606 | 2.03026 | 2.26192 | 2.37685 |
| | 1.2 | 1.74651 | 1.94057 | 2.10425 | 2.35012 | 2.47209 |
| | 1.6 | 1.79507 | 1.99873 | 2.17048 | 2.42844 | 2.55638 |
| | 2.0 | 1.83589 | 2.04716 | 2.22531 | 2.49278 | 2.62538 |
| | 2.4 | 1.86717 | 2.08393 | 2.26663 | 2.54083 | 2.67671 |
| | 2.8 | 1.88846 | 2.10861 | 2.29410 | 2.57232 | 2.71013 |
| | 3.2 | 1.90048 | 2.12215 | 2.30882 | 2.58865 | 2.72718 |
| | 3.6 | 1.90469 | 2.12633 | 2.31287 | 2.59232 | 2.73058 |
| | 4.0 | 1.90287 | 2.12330 | 2.30870 | 2.58624 | 2.72349 |
| 6.0 | 0.0 | 1.63937 | 1.79397 | 1.92250 | 2.11281 | 2.20614 |
| | 0.4 | 1.66144 | 1.82288 | 1.95739 | 2.15696 | 2.25497 |
| | 0.8 | 1.68594 | 1.85430 | 1.99480 | 2.20358 | 2.30623 |
| | 1.2 | 1.71191 | 1.88707 | 2.03343 | 2.25116 | 2.35831 |
| | 1.6 | 1.73832 | 1.91996 | 2.07189 | 2.29811 | 2.40949 |
| | 2.0 | 1.76409 | 1.95174 | 2.10881 | 2.34284 | 2.45812 |
| | 2.4 | 1.78824 | 1.98129 | 2.14296 | 2.38394 | 2.50269 |
| | 2.8 | 1.80995 | 2.00766 | 2.17329 | 2.42025 | 2.54197 |
| | 3.2 | 1.82860 | 2.03017 | 2.19908 | 2.45096 | 2.57511 |
| | 3.6 | 1.84384 | 2.04845 | 2.21992 | 2.47563 | 2.60167 |
| | 4.0 | 1.85556 | 2.06241 | 2.23574 | 2.49422 | 2.62161 |
| | 6.0 | 1.86988 | 2.07818 | 2.25254 | 2.51222 | 2.64005 |
| 12.0 | 0.0 | 1.70764 | 1.86913 | 2.00201 | 2.19633 | 2.29054 |
| | 0.4 | 1.71236 | 1.87549 | 2.00986 | 2.20659 | 2.30207 |
| | 0.8 | 1.71753 | 1.88239 | 2.01833 | 2.21759 | 2.31438 |
| | 1.2 | 1.72311 | 1.88978 | 2.02736 | 2.22923 | 2.32738 |
| | 1.6 | 1.72905 | 1.89761 | 2.03687 | 2.24143 | 2.34097 |
| | 2.0 | 1.73532 | 1.90581 | 2.04679 | 2.25407 | 2.35502 |
| | 2.4 | 1.74185 | 1.91429 | 2.05701 | 2.26703 | 2.36938 |
| | 2.8 | 1.74856 | 1.92297 | 2.06742 | 2.28017 | 2.38393 |
| | 3.2 | 1.75539 | 1.93174 | 2.07792 | 2.29336 | 2.39849 |
| | 3.6 | 1.76225 | 1.94052 | 2.08838 | 2.30645 | 2.41292 |
| | 4.0 | 1.76906 | 1.94920 | 2.09869 | 2.31930 | 2.42708 |
| | 6.0 | 1.79982 | 1.98794 | 2.14440 | 2.37579 | 2.48904 |
| | 12.0 | 1.82824 | 2.02275 | 2.18465 | 2.42428 | 2.54161 |
| INFINITY | | 1.78046 | 1.95996 | 2.10836 | 2.32635 | 2.43238 |

FRACTILES $W_\alpha$ RHO = -0.40

| DELTAS | | $\alpha$:0.9950 | 0.9975 | 0.9990 | 0.9995 |
|---|---|---|---|---|---|
| 0.0 | 0.0 | 2.24647 | 2.59579 | 3.06372 | 3.42136 |
| 0.4 | 0.0 | 2.33028 | 2.69168 | 3.17375 | 3.54080 |
| | 0.4 | 2.70175 | 3.10940 | 3.64672 | 4.05192 |
| 0.8 | 0.0 | 2.42158 | 2.77493 | 3.24070 | 3.59194 |
| | 0.4 | 2.84273 | 3.24645 | 3.77434 | 4.16988 |
| | 0.8 | 3.04938 | 3.46605 | 4.00768 | 4.41154 |
| 1.2 | 0.0 | 2.42947 | 2.75680 | 3.18471 | 3.50529 |
| | 0.4 | 2.81151 | 3.18727 | 3.67562 | 4.03972 |
| | 0.8 | 3.05864 | 3.45945 | 3.97790 | 4.36290 |
| | 1.2 | 3.14048 | 3.54135 | 4.05778 | 4.43996 |
| 1.6 | 0.0 | 2.40171 | 2.70304 | 3.09445 | 3.38615 |
| | 0.4 | 2.72513 | 3.06945 | 3.51465 | 3.84514 |
| | 0.8 | 2.97290 | 3.34644 | 3.82757 | 4.18356 |
| | 1.2 | 3.10587 | 3.49075 | 3.98483 | 4.34935 |
| | 1.6 | 3.13252 | 3.51336 | 4.00080 | 4.35950 |
| 2.0 | 0.0 | 2.37057 | 2.65031 | 3.01169 | 3.27978 |
| | 0.4 | 2.63867 | 2.95552 | 3.36332 | 3.66489 |
| | 0.8 | 2.86560 | 3.21136 | 3.65499 | 3.98218 |
| | 1.2 | 3.01909 | 3.38185 | 3.84605 | 4.18758 |
| | 1.6 | 3.09046 | 3.45797 | 3.92706 | 4.27145 |
| | 2.0 | 3.09452 | 3.45719 | 3.91905 | 4.25745 |
| 2.4 | 0.0 | 2.34581 | 2.60848 | 2.94612 | 3.19556 |
| | 0.4 | 2.56766 | 2.86222 | 3.23978 | 3.51802 |
| | 0.8 | 2.76734 | 3.08875 | 3.49971 | 3.80189 |
| | 1.2 | 2.92128 | 3.26170 | 3.69601 | 4.01473 |
| | 1.6 | 3.01676 | 3.36714 | 3.81324 | 4.14003 |
| | 2.0 | 3.05576 | 3.40779 | 3.85513 | 4.18229 |
| | 2.4 | 3.05082 | 3.39826 | 3.83898 | 4.16077 |
| 2.8 | 0.0 | 2.32878 | 2.57807 | 2.89708 | 3.13186 |
| | 0.4 | 2.51336 | 2.79022 | 3.14375 | 3.40343 |
| | 0.8 | 2.68631 | 2.98750 | 3.37135 | 3.65282 |
| | 1.2 | 2.83092 | 3.15119 | 3.55865 | 3.85696 |
| | 1.6 | 2.93533 | 3.26817 | 3.69094 | 3.99999 |
| | 2.0 | 2.99603 | 3.33481 | 3.76443 | 4.07807 |
| | 2.4 | 3.01761 | 3.35658 | 3.78580 | 4.09872 |
| | 2.8 | 3.00935 | 3.34424 | 3.76769 | 4.07599 |
| 3.2 | 0.0 | 2.31842 | 2.55723 | 2.86156 | 3.08476 |
| | 0.4 | 2.47318 | 2.73602 | 3.07044 | 3.31536 |
| | 0.8 | 2.62231 | 2.90699 | 3.26871 | 3.53325 |
| | 1.2 | 2.75389 | 3.05684 | 3.44124 | 3.72204 |
| | 1.6 | 2.85818 | 3.17471 | 3.57584 | 3.86851 |
| | 2.0 | 2.93000 | 3.25499 | 3.66633 | 3.96611 |
| | 2.4 | 2.96959 | 3.29817 | 3.71353 | 4.01590 |
| | 2.8 | 2.98145 | 3.30954 | 3.72378 | 4.02499 |
| | 3.2 | 2.97227 | 3.29682 | 3.70609 | 4.00337 |

FRACTILES $W_{\alpha}$ RHO = -0.40

| DELTAS | | $\alpha$:0.9950 | 0.9975 | 0.9990 | 0.9995 |
|---|---|---|---|---|---|
| 3.6 | 0.0 | 2.31324 | 2.54382 | 2.83651 | 3.05048 |
| | 0.4 | 2.44407 | 2.69574 | 3.01490 | 3.24798 |
| | 0.8 | 2.57274 | 2.84400 | 3.18767 | 3.43841 |
| | 1.2 | 2.69063 | 2.97897 | 3.34393 | 3.60995 |
| | 1.6 | 2.79006 | 3.09209 | 3.47402 | 3.75215 |
| | 2.0 | 2.86584 | 3.17766 | 3.57157 | 3.85817 |
| | 2.4 | 2.91619 | 3.23382 | 3.63467 | 3.92606 |
| | 2.8 | 2.94261 | 3.26242 | 3.66563 | 3.95846 |
| | 3.2 | 2.94888 | 3.26788 | 3.66966 | 3.96119 |
| | 3.6 | 2.93979 | 3.25573 | 3.65326 | 3.94143 |
| 4.0 | 0.0 | 2.31190 | 2.53599 | 2.81942 | 3.02598 |
| | 0.4 | 2.42338 | 2.66612 | 2.97298 | 3.19649 |
| | 0.8 | 2.53473 | 2.79506 | 3.12398 | 3.36340 |
| | 1.2 | 2.63961 | 2.91573 | 3.26440 | 3.51802 |
| | 1.6 | 2.73204 | 3.02147 | 3.38669 | 3.65218 |
| | 2.0 | 2.80740 | 3.10716 | 3.48515 | 3.75973 |
| | 2.4 | 2.86317 | 3.17008 | 3.55680 | 3.83751 |
| | 2.8 | 2.89917 | 3.21013 | 3.60166 | 3.88564 |
| | 3.2 | 2.91713 | 3.22942 | 3.62227 | 3.90701 |
| | 3.6 | 2.92009 | 3.23144 | 3.62278 | 3.90620 |
| | 4.0 | 2.91151 | 3.22021 | 3.60789 | 3.88843 |
| 6.0 | 0.0 | 2.33330 | 2.54062 | 2.79896 | 2.98485 |
| | 0.4 | 2.38863 | 2.60681 | 2.87900 | 3.07502 |
| | 0.8 | 2.44632 | 2.67518 | 2.96097 | 3.16689 |
| | 1.2 | 2.50461 | 2.74377 | 3.04261 | 3.25802 |
| | 1.6 | 2.56165 | 2.81049 | 3.12157 | 3.34586 |
| | 2.0 | 2.61564 | 2.87334 | 3.19557 | 3.42795 |
| | 2.4 | 2.66497 | 2.93051 | 3.26259 | 3.50207 |
| | 2.8 | 2.70832 | 2.98055 | 3.32100 | 3.56651 |
| | 3.2 | 2.74479 | 3.02247 | 3.36972 | 3.62009 |
| | 3.6 | 2.77393 | 3.05580 | 3.40823 | 3.66229 |
| | 4.0 | 2.79571 | 3.08055 | 3.43661 | 3.69322 |
| | 6.0 | 2.81459 | 3.09975 | 3.45553 | 3.71148 |
| 12.0 | 0.0 | 2.41773 | 2.62222 | 2.87217 | 3.04881 |
| | 0.4 | 2.43106 | 2.63868 | 2.89283 | 3.07266 |
| | 0.8 | 2.44525 | 2.65611 | 2.91456 | 3.09765 |
| | 1.2 | 2.46018 | 2.67436 | 2.93720 | 3.12360 |
| | 1.6 | 2.47573 | 2.69327 | 2.96055 | 3.15028 |
| | 2.0 | 2.49176 | 2.71269 | 2.98441 | 3.17747 |
| | 2.4 | 2.50811 | 2.73242 | 3.00855 | 3.20490 |
| | 2.8 | 2.52462 | 2.75226 | 3.03274 | 3.23233 |
| | 3.2 | 2.54112 | 2.77203 | 3.05675 | 3.25950 |
| | 3.6 | 2.55743 | 2.79152 | 3.08036 | 3.28615 |
| | 4.0 | 2.57340 | 2.81054 | 3.10333 | 3.31205 |
| | 6.0 | 2.64298 | 2.89289 | 3.20210 | 3.42289 |
| | 12.0 | 2.70117 | 2.96028 | 3.28096 | 3.50995 |
| INFINITY | | 2.57583 | 2.80703 | 3.09024 | 3.29051 |

FRACTILES $W_\alpha$ RHO = -0.20

| DELTAS | | $\alpha$:0.0005 | 0.0010 | 0.0025 | 0.0050 | 0.0075 |
|---|---|---|---|---|---|---|
| 0.0 | 0.0 | -6.63665 | -5.87937 | -4.88781 | -4.14679 | -3.71784 |
| 0.4 | 0.0 | -6.55669 | -5.81308 | -4.83811 | -4.10844 | -3.68563 |
| | 0.4 | -6.00394 | -5.32586 | -4.43810 | -3.77478 | -3.39086 |
| 0.8 | 0.0 | -6.25392 | -5.56989 | -4.66736 | -3.98692 | -3.59032 |
| | 0.4 | -5.58853 | -4.96639 | -4.15087 | -3.54075 | -3.18734 |
| | 0.8 | -4.89741 | -4.36078 | -3.65850 | -3.13407 | -2.83073 |
| 1.2 | 0.0 | -5.82544 | -5.22035 | -4.41724 | -3.80722 | -3.44943 |
| | 0.4 | -5.29891 | -4.73237 | -3.98500 | -3.42169 | -3.09348 |
| | 0.8 | -4.57767 | -4.08624 | -3.44237 | -2.96104 | -2.68243 |
| | 1.2 | -4.03089 | -3.61025 | -3.06017 | -2.64985 | -2.41278 |
| 1.6 | 0.0 | -5.41860 | -4.88284 | -4.16858 | -3.62305 | -3.30157 |
| | 0.4 | -5.02651 | -4.51718 | -3.84115 | -3.32769 | -3.02654 |
| | 0.8 | -4.45339 | -3.99449 | -3.38931 | -2.93344 | -2.66798 |
| | 1.2 | -3.85206 | -3.45968 | -2.94602 | -2.56253 | -2.34085 |
| | 1.6 | -3.45996 | -3.12173 | -2.67993 | -2.35100 | -2.16134 |
| 2.0 | 0.0 | -5.08239 | -4.60126 | -3.95743 | -3.46343 | -3.17121 |
| | 0.4 | -4.78158 | -4.32080 | -3.70639 | -3.23701 | -2.96036 |
| | 0.8 | -4.34823 | -3.92319 | -3.35905 | -2.93060 | -2.67936 |
| | 1.2 | -3.84336 | -3.46649 | -2.96976 | -2.59593 | -2.37847 |
| | 1.6 | -3.37416 | -3.05266 | -2.63236 | -2.31924 | -2.13869 |
| | 2.0 | -3.09688 | -2.81727 | -2.45277 | -2.18226 | -2.02686 |
| 2.4 | 0.0 | -4.81525 | -4.37647 | -3.78730 | -3.33340 | -3.06402 |
| | 0.4 | -4.57601 | -4.15389 | -3.58882 | -3.15507 | -2.89840 |
| | 0.8 | -4.24191 | -3.84708 | -3.32045 | -2.91802 | -2.68077 |
| | 1.2 | -3.84312 | -3.48431 | -3.00809 | -2.64649 | -2.43451 |
| | 1.6 | -3.42657 | -3.11036 | -2.69396 | -2.38107 | -2.19941 |
| | 2.0 | -3.06554 | -2.79566 | -2.44362 | -2.18235 | -2.03237 |
| | 2.4 | -2.86977 | -2.63339 | -2.32626 | -2.09975 | -1.97072 |
| 2.8 | 0.0 | -4.60381 | -4.19812 | -3.65171 | -3.22921 | -2.97770 |
| | 0.4 | -4.40798 | -4.01644 | -3.49047 | -3.08501 | -2.84424 |
| | 0.8 | -4.14303 | -3.77333 | -3.27818 | -2.89788 | -2.67271 |
| | 1.2 | -3.82606 | -3.48477 | -3.02941 | -2.68129 | -2.47601 |
| | 1.6 | -3.48298 | -3.17501 | -2.76633 | -2.45613 | -2.27440 |
| | 2.0 | -3.14640 | -2.87574 | -2.51983 | -2.25316 | -2.09890 |
| | 2.4 | -2.86973 | -2.63883 | -2.33879 | -2.11785 | -1.99255 |
| | 2.8 | -2.73320 | -2.53014 | -2.26811 | -2.07853 | -1.97440 |
| 3.2 | 0.0 | -4.43503 | -4.05562 | -3.54315 | -3.14555 | -2.90823 |
| | 0.4 | -4.27116 | -3.90405 | -3.40931 | -3.02644 | -2.79837 |
| | 0.8 | -4.05574 | -3.70667 | -3.23744 | -2.87541 | -2.66026 |
| | 1.2 | -3.79923 | -3.47332 | -3.03656 | -2.70082 | -2.50190 |
| | 1.6 | -3.51782 | -3.21908 | -2.82031 | -2.51531 | -2.33539 |
| | 2.0 | -3.23065 | -2.96203 | -2.60565 | -2.33539 | -2.17731 |
| | 2.4 | -2.96103 | -2.72552 | -2.41669 | -2.18671 | -2.05509 |
| | 2.8 | -2.75102 | -2.55086 | -2.29310 | -2.10876 | -2.00483 |
| | 3.2 | -2.65956 | -2.48385 | -2.26256 | -2.10164 | -2.00607 |

FRACTILES $W_\alpha$ RHO = -0.20

| DELTAS | | α: 0.0005 | 0.0010 | 0.0025 | 0.0050 | 0.0075 |
|---|---|---|---|---|---|---|
| 3.6 | 0.0 | -4.29864 | -3.94043 | -3.45534 | -3.07781 | -2.85190 |
| | 0.4 | -4.15915 | -3.81180 | -3.34231 | -2.97771 | -2.75990 |
| | 0.8 | -3.98035 | -3.64828 | -3.20039 | -2.85342 | -2.64653 |
| | 1.2 | -3.76900 | -3.45626 | -3.03549 | -2.71048 | -2.51714 |
| | 1.6 | -3.53582 | -3.24574 | -2.85668 | -2.55729 | -2.37974 |
| | 2.0 | -3.29375 | -3.02885 | -2.67508 | -2.40434 | -2.24461 |
| | 2.4 | -3.05664 | -2.81899 | -2.50394 | -2.26562 | -2.12694 |
| | 2.8 | -2.84263 | -2.63522 | -2.36506 | -2.16818 | -2.05581 |
| | 3.2 | -2.68739 | -2.51356 | -2.29420 | -2.13103 | -2.03303 |
| | 3.6 | -2.63328 | -2.48283 | -2.28718 | -2.13395 | -2.03925 |
| 4.0 | 0.0 | -4.18700 | -3.84615 | -3.38347 | -3.02235 | -2.80577 |
| | 0.4 | -4.06661 | -3.73548 | -3.28669 | -2.93704 | -2.72761 |
| | 0.8 | -3.91568 | -3.59771 | -3.16753 | -2.83305 | -2.63302 |
| | 1.2 | -3.73875 | -3.43721 | -3.03008 | -2.71425 | -2.52570 |
| | 1.6 | -3.54319 | -3.26087 | -2.88062 | -2.58648 | -2.41127 |
| | 2.0 | -3.33816 | -3.07729 | -2.72702 | -2.45711 | -2.29685 |
| | 2.4 | -3.13356 | -2.89584 | -2.57817 | -2.33500 | -2.19166 |
| | 2.8 | -2.93989 | -2.72717 | -2.44588 | -2.23528 | -2.11328 |
| | 3.2 | -2.77313 | -2.58971 | -2.35422 | -2.17822 | -2.07311 |
| | 3.6 | -2.66694 | -2.51595 | -2.31682 | -2.15952 | -2.06211 |
| | 4.0 | -2.64328 | -2.50654 | -2.31892 | -2.16595 | -2.06981 |
| 6.0 | 0.0 | -3.84655 | -3.55904 | -3.16496 | -2.85389 | -2.66565 |
| | 0.4 | -3.78047 | -3.49907 | -3.11358 | -2.80947 | -2.62550 |
| | 0.8 | -3.70358 | -3.42960 | -3.05449 | -2.75372 | -2.57985 |
| | 1.2 | -3.61696 | -3.35168 | -2.98869 | -2.70260 | -2.52964 |
| | 1.6 | -3.52227 | -3.26691 | -2.91766 | -2.64253 | -2.47621 |
| | 2.0 | -3.42175 | -3.17739 | -2.84337 | -2.58031 | -2.42129 |
| | 2.4 | -3.31805 | -3.08564 | -2.76811 | -2.51809 | -2.36692 |
| | 2.8 | -3.21416 | -2.99450 | -2.69452 | -2.45833 | -2.31545 |
| | 3.2 | -3.11326 | -2.90703 | -2.62552 | -2.40380 | -2.26950 |
| | 3.6 | -3.01880 | -2.82666 | -2.56448 | -2.35750 | -2.23155 |
| | 4.0 | -2.93487 | -2.75759 | -2.51494 | -2.32169 | -2.20304 |
| | 6.0 | -2.79513 | -2.65642 | -2.45494 | -2.28545 | -2.17791 |
| 12.0 | 0.0 | -3.52598 | -3.28963 | -2.96071 | -2.69664 | -2.53476 |
| | 0.4 | -3.50515 | -3.27111 | -2.94535 | -2.68372 | -2.52330 |
| | 0.8 | -3.48283 | -3.25131 | -2.92896 | -2.67000 | -2.51115 |
| | 1.2 | -3.45913 | -3.23033 | -2.91166 | -2.65554 | -2.49838 |
| | 1.6 | -3.43419 | -3.20830 | -2.89356 | -2.64046 | -2.48510 |
| | 2.0 | -3.40817 | -3.18537 | -2.87479 | -2.62489 | -2.47141 |
| | 2.4 | -3.38128 | -3.16173 | -2.85552 | -2.60896 | -2.45743 |
| | 2.8 | -3.35374 | -3.13759 | -2.83591 | -2.59281 | -2.44332 |
| | 3.2 | -3.32579 | -3.11316 | -2.81618 | -2.57663 | -2.42921 |
| | 3.6 | -3.29770 | -3.08869 | -2.79650 | -2.56058 | -2.41526 |
| | 4.0 | -3.26975 | -3.06443 | -2.77711 | -2.54483 | -2.40162 |
| | 6.0 | -3.14211 | -2.95502 | -2.69128 | -2.47629 | -2.34291 |
| | 12.0 | -3.03213 | -2.86521 | -2.62567 | -2.42703 | -2.30240 |
| INFINITY | | -3.29051 | -3.09024 | -2.80703 | -2.57583 | -2.43238 |

FRACTILES $w_\alpha$ RHO = -0.20

| DELTAS | | $\alpha$: 0.0100 | 0.0175 | 0.0250 | 0.0375 | 0.0500 |
|---|---|---|---|---|---|---|
| 0.0 | 0.0 | -3.41587 | -2.83518 | -2.47059 | -2.06253 | -1.77798 |
| 0.4 | 0.0 | -3.38779 | -2.81458 | -2.45435 | -2.05089 | -1.76935 |
| | 0.4 | -3.12065 | -2.60116 | -2.27511 | -1.91037 | -1.65617 |
| 0.8 | 0.0 | -3.30982 | -2.76699 | -2.42363 | -2.03673 | -1.76513 |
| | 0.4 | -2.93846 | -2.45970 | -2.15905 | -1.82262 | -1.58812 |
| | 0.8 | -2.61734 | -2.20742 | -1.95047 | -1.66347 | -1.46385 |
| 1.2 | 0.0 | -3.19517 | -2.69981 | -2.38375 | -2.02447 | -1.76987 |
| | 0.4 | -2.86141 | -2.41253 | -2.12882 | -1.80945 | -1.58555 |
| | 0.8 | -2.48637 | -2.10962 | -1.87343 | -1.60968 | -1.42636 |
| | 1.2 | -2.24617 | -1.92667 | -1.72690 | -1.50452 | -1.35055 |
| 1.6 | 0.0 | -3.07231 | -2.62338 | -2.33503 | -2.00497 | -1.76925 |
| | 0.4 | -2.81257 | -2.39577 | -2.12993 | -1.82790 | -1.61405 |
| | 0.8 | -2.48038 | -2.11789 | -1.88913 | -1.63215 | -1.45252 |
| | 1.2 | -2.18504 | -1.88627 | -1.69957 | -1.49200 | -1.34862 |
| | 1.6 | -2.02829 | -1.77394 | -1.61576 | -1.44099 | -1.32139 |
| 2.0 | 0.0 | -2.96224 | -2.55140 | -2.28617 | -1.98098 | -1.76178 |
| | 0.4 | -2.76306 | -2.37663 | -2.12840 | -1.84425 | -1.64134 |
| | 0.8 | -2.50087 | -2.15334 | -1.93184 | -1.68048 | -1.50281 |
| | 1.2 | -2.22495 | -1.92883 | -1.74253 | -1.53412 | -1.38937 |
| | 1.6 | -2.01205 | -1.77019 | -1.62012 | -1.45504 | -1.34313 |
| | 2.0 | -1.91823 | -1.71205 | -1.58558 | -1.44972 | -1.36467 |
| 2.4 | 0.0 | -2.87091 | -2.48996 | -2.24295 | -1.95751 | -1.75152 |
| | 0.4 | -2.71480 | -2.35368 | -2.12043 | -1.85191 | -1.65893 |
| | 0.8 | -2.51154 | -2.18005 | -1.96707 | -1.72329 | -1.54927 |
| | 1.2 | -2.28397 | -1.99109 | -1.80470 | -1.59370 | -1.44517 |
| | 1.6 | -2.07139 | -1.82534 | -1.67155 | -1.50141 | -1.38581 |
| | 2.0 | -1.92766 | -1.72969 | -1.60968 | -1.48557 | -1.39676 |
| | 2.4 | -1.88142 | -1.71800 | -1.61975 | -1.50405 | -1.41638 |
| 2.8 | 0.0 | -2.79702 | -2.43948 | -2.20678 | -1.93683 | -1.74124 |
| | 0.4 | -2.67157 | -2.33072 | -2.10955 | -1.85372 | -1.66893 |
| | 0.8 | -2.51159 | -2.19449 | -1.98952 | -1.75339 | -1.58357 |
| | 1.2 | -2.32955 | -2.04262 | -1.85828 | -1.64737 | -1.49699 |
| | 1.6 | -2.14545 | -1.89498 | -1.73617 | -1.55766 | -1.43403 |
| | 2.0 | -1.99062 | -1.78449 | -1.65923 | -1.52470 | -1.42744 |
| | 2.4 | -1.90667 | -1.74921 | -1.64841 | -1.52821 | -1.43680 |
| | 2.8 | -1.90164 | -1.75747 | -1.66038 | -1.54162 | -1.44999 |
| 3.2 | 0.0 | -2.73740 | -2.39842 | -2.17705 | -1.91937 | -1.73200 |
| | 0.4 | -2.63444 | -2.30979 | -2.09826 | -1.85260 | -1.67438 |
| | 0.8 | -2.50568 | -2.20087 | -2.00271 | -1.77324 | -1.60725 |
| | 1.2 | -2.35948 | -2.07898 | -1.89749 | -1.68822 | -1.53760 |
| | 1.6 | -2.20702 | -1.95552 | -1.79406 | -1.60975 | -1.47927 |
| | 2.0 | -2.06535 | -1.84903 | -1.71442 | -1.56537 | -1.45849 |
| | 2.4 | -1.96458 | -1.79402 | -1.68478 | -1.55550 | -1.45805 |
| | 2.8 | -1.93099 | -1.78281 | -1.68240 | -1.55948 | -1.46475 |
| | 3.2 | -1.93620 | -1.79227 | -1.69281 | -1.56979 | -1.47440 |

FRACTILES $W_\alpha$ RHO = -0.20

| DELTAS | | α:0.0100 | 0.0175 | 0.0250 | 0.0375 | 0.0500 |
|---|---|---|---|---|---|---|
| 3.6 | 0.0 | -2.68900 | -2.36495 | -2.15267 | -1.90482 | -1.72402 |
| | 0.4 | -2.60302 | -2.29143 | -2.08768 | -1.85020 | -1.67724 |
| | 0.8 | -2.49773 | -2.20270 | -2.01019 | -1.78627 | -1.62352 |
| | 1.2 | -2.37830 | -2.10367 | -1.92497 | -1.71768 | -1.56746 |
| | 1.6 | -2.25253 | -2.00174 | -1.83929 | -1.65185 | -1.51705 |
| | 2.0 | -2.13065 | -1.90769 | -1.76550 | -1.60452 | -1.48884 |
| | 2.4 | -2.02975 | -1.84354 | -1.72430 | -1.58439 | -1.48002 |
| | 2.8 | -1.97601 | -1.81677 | -1.70981 | -1.57995 | -1.48062 |
| | 3.2 | -1.96112 | -1.81274 | -1.71023 | -1.58363 | -1.48569 |
| | 3.6 | -1.96882 | -1.82171 | -1.71916 | -1.59190 | -1.49316 |
| 4.0 | 0.0 | -2.64934 | -2.33745 | -2.13256 | -1.89270 | -1.71721 |
| | 0.4 | -2.57649 | -2.27556 | -2.07813 | -1.84730 | -1.67862 |
| | 0.8 | -2.48882 | -2.20206 | -2.01424 | -1.79492 | -1.63487 |
| | 1.2 | -2.38995 | -2.12043 | -1.94422 | -1.73883 | -1.58918 |
| | 1.6 | -2.28532 | -2.03580 | -1.87310 | -1.68395 | -1.54659 |
| | 2.0 | -2.18193 | -1.95515 | -1.80829 | -1.63911 | -1.51658 |
| | 2.4 | -2.08972 | -1.89107 | -1.76293 | -1.61302 | -1.50191 |
| | 2.8 | -2.02652 | -1.85428 | -1.73966 | -1.60177 | -1.49720 |
| | 3.2 | -1.99640 | -1.83930 | -1.73166 | -1.59963 | -1.49807 |
| | 3.6 | -1.98965 | -1.83847 | -1.73326 | -1.60298 | -1.50211 |
| | 4.0 | -1.99780 | -1.84664 | -1.74096 | -1.60980 | -1.50811 |
| 6.0 | 0.0 | -2.52882 | -2.25364 | -2.07095 | -1.85489 | -1.69517 |
| | 0.4 | -2.49180 | -2.22297 | -2.04451 | -1.83347 | -1.67743 |
| | 0.8 | -2.44988 | -2.18856 | -2.01510 | -1.80992 | -1.65818 |
| | 1.2 | -2.40397 | -2.15129 | -1.98354 | -1.78503 | -1.63814 |
| | 1.6 | -2.35537 | -2.11236 | -1.95095 | -1.75982 | -1.61824 |
| | 2.0 | -2.30574 | -2.07327 | -1.91873 | -1.73552 | -1.59958 |
| | 2.4 | -2.25704 | -2.03581 | -1.88852 | -1.71352 | -1.58329 |
| | 2.8 | -2.21153 | -2.00194 | -1.86199 | -1.69503 | -1.57017 |
| | 3.2 | -2.17163 | -1.97349 | -1.84041 | -1.68066 | -1.56041 |
| | 3.6 | -2.13937 | -1.95149 | -1.82425 | -1.67037 | -1.55376 |
| | 4.0 | -2.11563 | -1.93598 | -1.81322 | -1.66371 | -1.54973 |
| | 6.0 | -2.09721 | -1.92803 | -1.81029 | -1.66505 | -1.55325 |
| 12.0 | 0.0 | -2.41605 | -2.17446 | -2.01186 | -1.81711 | -1.67136 |
| | 0.4 | -2.40564 | -2.16611 | -2.00484 | -1.81161 | -1.66693 |
| | 0.8 | -2.39462 | -2.15731 | -1.99746 | -1.80585 | -1.66232 |
| | 1.2 | -2.38306 | -2.14811 | -1.98977 | -1.79988 | -1.65757 |
| | 1.6 | -2.37105 | -2.13860 | -1.98185 | -1.79376 | -1.65271 |
| | 2.0 | -2.35870 | -2.12886 | -1.97376 | -1.78755 | -1.64782 |
| | 2.4 | -2.34612 | -2.11898 | -1.96561 | -1.78132 | -1.64292 |
| | 2.8 | -2.33344 | -2.10908 | -1.95746 | -1.77513 | -1.63810 |
| | 3.2 | -2.32080 | -2.09927 | -1.94941 | -1.76906 | -1.63339 |
| | 3.6 | -2.30833 | -2.08964 | -1.94156 | -1.76317 | -1.62885 |
| | 4.0 | -2.29617 | -2.08031 | -1.93398 | -1.75753 | -1.62453 |
| | 6.0 | -2.24427 | -2.04121 | -1.90269 | -1.73468 | -1.60738 |
| | 12.0 | -2.20959 | -2.01695 | -1.88440 | -1.72254 | -1.59915 |
| INFINITY | | -2.32635 | -2.10836 | -1.95996 | -1.78046 | -1.64485 |

FRACTILES $W_{\alpha}$ RHO = -0.20

| DELTAS | | $\alpha$:0.0750 | 0.1000 | 0.1500 | 0.2000 | 0.2500 |
|---|---|---|---|---|---|---|
| 0.0 | 0.0 | -1.38560 | -1.11479 | -0.74710 | -0.49930 | -0.31714 |
| 0.4 | 0.0 | -1.38085 | -1.11256 | -0.74807 | -0.50232 | -0.32164 |
| | 0.4 | -1.30595 | -1.06456 | -0.73754 | -0.51804 | -0.35758 |
| 0.8 | 0.0 | -1.38791 | -1.12556 | -0.76646 | -0.52244 | -0.34200 |
| | 0.4 | -1.26509 | -1.04260 | -0.74170 | -0.54050 | -0.39434 |
| | 0.8 | -1.18968 | -1.00162 | -0.74907 | -0.58241 | -0.46388 |
| 1.2 | 0.0 | -1.41221 | -1.16005 | -0.80907 | -0.56565 | -0.38239 |
| | 0.4 | -1.27523 | -1.06012 | -0.76744 | -0.57081 | -0.42794 |
| | 0.8 | -1.17493 | -1.00298 | -0.77356 | -0.62460 | -0.52241 |
| | 1.2 | -1.14067 | -0.99857 | -0.81319 | -0.70131 | -0.62987 |
| 1.6 | 0.0 | -1.43485 | -1.19612 | -0.85811 | -0.61815 | -0.43317 |
| | 0.4 | -1.31407 | -1.10310 | -0.81095 | -0.61051 | -0.46237 |
| | 0.8 | -1.20471 | -1.03432 | -0.80633 | -0.65941 | -0.56471 |
| | 1.2 | -1.15411 | -1.02383 | -0.86050 | -0.76283 | -0.66074 |
| | 1.6 | -1.16226 | -1.06171 | -0.92826 | -0.80544 | -0.69113 |
| 2.0 | 0.0 | -1.44852 | -1.22281 | -0.89912 | -0.66512 | -0.48111 |
| | 0.4 | -1.35356 | -1.14828 | -0.85833 | -0.65377 | -0.49808 |
| | 0.8 | -1.25438 | -1.08073 | -0.84375 | -0.68809 | -0.58763 |
| | 1.2 | -1.19210 | -1.05996 | -0.90059 | -0.77614 | -0.66133 |
| | 1.6 | -1.19997 | -1.10719 | -0.95083 | -0.81759 | -0.69604 |
| | 2.0 | -1.23836 | -1.13803 | -0.97400 | -0.83473 | -0.70824 |
| 2.4 | 0.0 | -1.45545 | -1.24059 | -0.92949 | -0.70156 | -0.51973 |
| | 0.4 | -1.38299 | -1.18405 | -0.89869 | -0.69264 | -0.53140 |
| | 0.8 | -1.30270 | -1.12726 | -0.88139 | -0.71311 | -0.59407 |
| | 1.2 | -1.23933 | -1.09862 | -0.91879 | -0.77851 | -0.65436 |
| | 1.6 | -1.23851 | -1.13053 | -0.95967 | -0.81773 | -0.69026 |
| | 2.0 | -1.26211 | -1.15588 | -0.98355 | -0.83860 | -0.70792 |
| | 2.4 | -1.28050 | -1.17209 | -0.99550 | -0.84697 | -0.71330 |
| 2.8 | 0.0 | -1.45873 | -1.25247 | -0.95148 | -0.72864 | -0.54894 |
| | 0.4 | -1.40299 | -1.20970 | -0.92927 | -0.72336 | -0.55894 |
| | 0.8 | -1.34057 | -1.16531 | -0.91431 | -0.73535 | -0.59993 |
| | 1.2 | -1.28471 | -1.13543 | -0.93354 | -0.78009 | -0.64766 |
| | 1.6 | -1.26948 | -1.14957 | -0.96464 | -0.81441 | -0.68147 |
| | 2.0 | -1.28179 | -1.16842 | -0.98701 | -0.83638 | -0.70181 |
| | 2.4 | -1.29528 | -1.18276 | -1.00044 | -0.84801 | -0.71151 |
| | 2.8 | -1.30694 | -1.19269 | -1.00731 | -0.85240 | -0.71386 |
| 3.2 | 0.0 | -1.46019 | -1.26071 | -0.96765 | -0.74880 | -0.57080 |
| | 0.4 | -1.41651 | -1.22784 | -0.95166 | -0.74633 | -0.58007 |
| | 0.8 | -1.36796 | -1.19367 | -0.94021 | -0.75422 | -0.60718 |
| | 1.2 | -1.32205 | -1.16701 | -0.94811 | -0.78340 | -0.64351 |
| | 1.6 | -1.29878 | -1.16794 | -0.96959 | -0.81142 | -0.67329 |
| | 2.0 | -1.30060 | -1.17952 | -0.98863 | -0.83229 | -0.69396 |
| | 2.4 | -1.30862 | -1.19094 | -1.00204 | -0.84550 | -0.70619 |
| | 2.8 | -1.31726 | -1.19987 | -1.01019 | -0.85238 | -0.71175 |
| | 3.2 | -1.32536 | -1.20652 | -1.01444 | -0.85477 | -0.71265 |

FRACTILES $W_\alpha$ RHO = -0.20

| DELTAS | | $\alpha$:0.0750 | 0.1000 | 0.1500 | 0.2000 | 0.2500 |
|---|---|---|---|---|---|---|
| 3.6 | 0.0 | -1.46073 | -1.26663 | -0.97983 | -0.76408 | -0.58735 |
| | 0.4 | -1.42584 | -1.24087 | -0.96812 | -0.76337 | -0.59589 |
| | 0.8 | -1.38750 | -1.21427 | -0.95961 | -0.76932 | -0.61488 |
| | 1.2 | -1.35041 | -1.19200 | -0.96190 | -0.78823 | -0.64194 |
| | 1.6 | -1.32538 | -1.18562 | -0.97546 | -0.80999 | -0.66699 |
| | 2.0 | -1.31922 | -1.19060 | -0.99038 | -0.82845 | -0.68646 |
| | 2.4 | -1.32175 | -1.19844 | -1.00255 | -0.84178 | -0.69968 |
| | 2.8 | -1.32710 | -1.20577 | -1.01105 | -0.85009 | -0.70734 |
| | 3.2 | -1.33306 | -1.21173 | -1.01626 | -0.85434 | -0.71062 |
| | 3.6 | -1.33905 | -1.21647 | -1.01905 | -0.85567 | -0.71080 |
| 4.0 | 0.0 | -1.46081 | -1.27102 | -0.98919 | -0.77586 | -0.60009 |
| | 0.4 | -1.43244 | -1.25043 | -0.98043 | -0.77617 | -0.60783 |
| | 0.8 | -1.40159 | -1.22934 | -0.97405 | -0.78102 | -0.62201 |
| | 1.2 | -1.37144 | -1.21098 | -0.97395 | -0.79368 | -0.64223 |
| | 1.6 | -1.34793 | -1.20166 | -0.98193 | -0.81011 | -0.66273 |
| | 2.0 | -1.33703 | -1.20168 | -0.99286 | -0.82567 | -0.68018 |
| | 2.4 | -1.33489 | -1.20598 | -1.00312 | -0.83815 | -0.69330 |
| | 2.8 | -1.33694 | -1.21130 | -1.01118 | -0.84696 | -0.70205 |
| | 3.2 | -1.34068 | -1.21624 | -1.01680 | -0.85239 | -0.70700 |
| | 3.6 | -1.34505 | -1.22044 | -1.02029 | -0.85510 | -0.70896 |
| | 4.0 | -1.34969 | -1.22400 | -1.02221 | -0.85583 | -0.70876 |
| 6.0 | 0.0 | -1.45896 | -1.28170 | -1.01405 | -0.80743 | -0.63420 |
| | 0.4 | -1.44660 | -1.27328 | -1.01130 | -0.80873 | -0.63856 |
| | 0.8 | -1.43357 | -1.26474 | -1.00917 | -0.81105 | -0.64415 |
| | 1.2 | -1.42048 | -1.25664 | -1.00800 | -0.81454 | -0.65092 |
| | 1.6 | -1.40812 | -1.24957 | -1.00808 | -0.81917 | -0.65855 |
| | 2.0 | -1.39730 | -1.24409 | -1.00945 | -0.82462 | -0.66652 |
| | 2.4 | -1.38867 | -1.24044 | -1.01184 | -0.83039 | -0.67425 |
| | 2.8 | -1.38243 | -1.23847 | -1.01481 | -0.83597 | -0.68127 |
| | 3.2 | -1.37836 | -1.23779 | -1.01791 | -0.84099 | -0.68728 |
| | 3.6 | -1.37606 | -1.23800 | -1.02083 | -0.84520 | -0.69211 |
| | 4.0 | -1.37509 | -1.23875 | -1.02335 | -0.84851 | -0.69576 |
| | 6.0 | -1.38054 | -1.24482 | -1.02931 | -0.85373 | -0.70005 |
| 12.0 | 0.0 | -1.45285 | -1.28643 | -1.03102 | -0.83033 | -0.65967 |
| | 0.4 | -1.44993 | -1.28458 | -1.03061 | -0.83089 | -0.66092 |
| | 0.8 | -1.44693 | -1.28269 | -1.03025 | -0.83153 | -0.66228 |
| | 1.2 | -1.44386 | -1.28079 | -1.02993 | -0.83225 | -0.66373 |
| | 1.6 | -1.44076 | -1.27890 | -1.02966 | -0.83304 | -0.66527 |
| | 2.0 | -1.43767 | -1.27705 | -1.02946 | -0.83392 | -0.66688 |
| | 2.4 | -1.43462 | -1.27526 | -1.02933 | -0.83485 | -0.66855 |
| | 2.8 | -1.43165 | -1.27355 | -1.02928 | -0.83584 | -0.67025 |
| | 3.2 | -1.42880 | -1.27194 | -1.02929 | -0.83688 | -0.67197 |
| | 3.6 | -1.42608 | -1.27044 | -1.02937 | -0.83794 | -0.67368 |
| | 4.0 | -1.42354 | -1.26908 | -1.02952 | -0.83901 | -0.67536 |
| | 6.0 | -1.41388 | -1.26430 | -1.03085 | -0.84400 | -0.68268 |
| | 12.0 | -1.41050 | -1.26385 | -1.03375 | -0.84863 | -0.68824 |
| INFINITY | | -1.43953 | -1.28155 | -1.03643 | -0.84162 | -0.67449 |

FRACTILES $w_\alpha$ RHO = -0.20

| DELTAS | | α:0.3000 | 0.3500 | 0.4000 | 0.4500 | 0.5000 |
|---|---|---|---|---|---|---|
| 0.0 | 0.0 | -0.17679 | -0.06580 | 0.02303 | 0.09397 | 0.14935 |
| 0.4 | 0.0 | -0.18245 | -0.07243 | 0.01552 | 0.08562 | 0.14007 |
| | 0.4 | -0.23495 | -0.13918 | -0.06415 | -0.00668 | 0.03271 |
| 0.8 | 0.0 | -0.20242 | -0.09181 | -0.00332 | 0.06700 | 0.12101 |
| | 0.4 | -0.28380 | -0.19904 | -0.13516 | -0.09251 | -0.05272 |
| | 0.8 | -0.37768 | -0.31795 | -0.27568 | -0.21395 | -0.13880 |
| 1.2 | 0.0 | -0.23835 | -0.12259 | -0.02892 | 0.04604 | 0.10317 |
| | 0.4 | -0.32056 | -0.23995 | -0.18411 | -0.13890 | -0.07395 |
| | 0.8 | -0.45838 | -0.39806 | -0.32235 | -0.23776 | -0.14521 |
| | 1.2 | -0.54461 | -0.45414 | -0.35974 | -0.26099 | -0.15690 |
| 1.6 | 0.0 | -0.28414 | -0.16124 | -0.05909 | 0.02480 | 0.08964 |
| | 0.4 | -0.34982 | -0.26607 | -0.20975 | -0.13939 | -0.05841 |
| | 0.8 | -0.49154 | -0.40567 | -0.31518 | -0.22042 | -0.12062 |
| | 1.2 | -0.55964 | -0.45868 | -0.35649 | -0.25168 | -0.14280 |
| | 1.6 | -0.58130 | -0.47331 | -0.36519 | -0.25525 | -0.14187 |
| 2.0 | 0.0 | -0.32946 | -0.20101 | -0.09075 | 0.00350 | 0.07967 |
| | 0.4 | -0.37609 | -0.28344 | -0.20958 | -0.12441 | -0.03397 |
| | 0.8 | -0.49009 | -0.39244 | -0.29428 | -0.19429 | -0.09098 |
| | 1.2 | -0.55200 | -0.44518 | -0.33870 | -0.23082 | -0.11985 |
| | 1.6 | -0.58075 | -0.46848 | -0.35698 | -0.24439 | -0.12897 |
| | 2.0 | -0.58880 | -0.47299 | -0.35843 | -0.24319 | -0.12547 |
| 2.4 | 0.0 | -0.36737 | -0.23568 | -0.11961 | -0.01660 | 0.07184 |
| | 0.4 | -0.40043 | -0.29641 | -0.20474 | -0.10994 | -0.01312 |
| | 0.8 | -0.48353 | -0.37782 | -0.27417 | -0.17039 | -0.06462 |
| | 1.2 | -0.53874 | -0.42750 | -0.31796 | -0.20806 | -0.09597 |
| | 1.6 | -0.57072 | -0.45538 | -0.34168 | -0.22761 | -0.11133 |
| | 2.0 | -0.58527 | -0.46698 | -0.35050 | -0.23380 | -0.11504 |
| | 2.4 | -0.58815 | -0.46773 | -0.34943 | -0.23118 | -0.11112 |
| 2.8 | 0.0 | -0.39655 | -0.26295 | -0.14305 | -0.03379 | 0.06505 |
| | 0.4 | -0.42160 | -0.30648 | -0.20239 | -0.10030 | 0.00170 |
| | 0.8 | -0.47945 | -0.36701 | -0.25859 | -0.15146 | -0.04346 |
| | 1.2 | -0.52641 | -0.41126 | -0.29904 | -0.18743 | -0.07445 |
| | 1.6 | -0.55818 | -0.44024 | -0.32481 | -0.20971 | -0.09302 |
| | 2.0 | -0.57641 | -0.45615 | -0.33831 | -0.22076 | -0.10159 |
| | 2.4 | -0.58421 | -0.46215 | -0.34262 | -0.22347 | -0.10281 |
| | 2.8 | -0.58488 | -0.46141 | -0.34070 | -0.22056 | -0.09910 |
| 3.2 | 0.0 | -0.41847 | -0.28355 | -0.16096 | -0.04738 | 0.05884 |
| | 0.4 | -0.43861 | -0.31541 | -0.20308 | -0.09517 | 0.01126 |
| | 0.8 | -0.47871 | -0.36046 | -0.24776 | -0.13752 | -0.02735 |
| | 1.2 | -0.51709 | -0.39829 | -0.28355 | -0.17028 | -0.05637 |
| | 1.6 | -0.54645 | -0.42609 | -0.30908 | -0.19306 | -0.07604 |
| | 2.0 | -0.56598 | -0.44398 | -0.32502 | -0.20685 | -0.08753 |
| | 2.4 | -0.57692 | -0.45348 | -0.33301 | -0.21331 | -0.09243 |
| | 2.8 | -0.58122 | -0.45659 | -0.33501 | -0.21428 | -0.09244 |
| | 3.2 | -0.58090 | -0.45527 | -0.33288 | -0.21148 | -0.08912 |

FRACTILES $W_\alpha$ RHO = -0.20

| DELTAS | | α:0.3000 | 0.3500 | 0.4000 | 0.4500 | 0.5000 |
|---|---|---|---|---|---|---|
| 3.6 | 0.0 | -0.43502 | -0.29905 | -0.17446 | -0.05780 | 0.05326 |
| | 0.4 | -0.45171 | -0.32350 | -0.20561 | -0.09315 | 0.01699 |
| | 0.8 | -0.48029 | -0.35724 | -0.24082 | -0.12774 | -0.01549 |
| | 1.2 | -0.51090 | -0.38875 | -0.27160 | -0.15668 | -0.04177 |
| | 1.6 | -0.53680 | -0.41412 | -0.29555 | -0.17861 | -0.06120 |
| | 2.0 | -0.55600 | -0.43232 | -0.31231 | -0.19359 | -0.07415 |
| | 2.4 | -0.56851 | -0.44380 | -0.32253 | -0.20241 | -0.08147 |
| | 2.8 | -0.57531 | -0.44964 | -0.32738 | -0.20625 | -0.08428 |
| | 3.2 | -0.57769 | -0.45118 | -0.32815 | -0.20632 | -0.08372 |
| | 3.6 | -0.57694 | -0.44969 | -0.32606 | -0.20376 | -0.08079 |
| 4.0 | 0.0 | -0.44771 | -0.31090 | -0.18475 | -0.06585 | 0.04841 |
| | 0.4 | -0.46174 | -0.33042 | -0.20883 | -0.09297 | 0.02015 |
| | 0.8 | -0.48303 | -0.35621 | -0.23668 | -0.12112 | -0.00696 |
| | 1.2 | -0.50724 | -0.38211 | -0.26273 | -0.14622 | -0.03025 |
| | 1.6 | -0.52943 | -0.40455 | -0.28446 | -0.16656 | -0.04867 |
| | 2.0 | -0.54732 | -0.42203 | -0.30097 | -0.18168 | -0.06207 |
| | 2.4 | -0.56027 | -0.43432 | -0.31230 | -0.19181 | -0.07083 |
| | 2.8 | -0.56855 | -0.44190 | -0.31903 | -0.19760 | -0.07560 |
| | 3.2 | -0.57289 | -0.44557 | -0.32201 | -0.19988 | -0.07718 |
| | 3.6 | -0.57417 | -0.44623 | -0.32211 | -0.19949 | -0.07635 |
| | 4.0 | -0.57325 | -0.44474 | -0.32016 | -0.19718 | -0.07378 |
| 6.0 | 0.0 | -0.48160 | -0.34251 | -0.21241 | -0.08812 | 0.03286 |
| | 0.4 | -0.48832 | -0.35104 | -0.22228 | -0.09889 | 0.02162 |
| | 0.8 | -0.49634 | -0.36082 | -0.23325 | -0.11053 | 0.00978 |
| | 1.2 | -0.50540 | -0.37140 | -0.24474 | -0.12240 | -0.00199 |
| | 1.6 | -0.51498 | -0.38217 | -0.25609 | -0.13385 | -0.01311 |
| | 2.0 | -0.52446 | -0.39248 | -0.26670 | -0.14434 | -0.02314 |
| | 2.4 | -0.53330 | -0.40182 | -0.27613 | -0.15352 | -0.03178 |
| | 2.8 | -0.54106 | -0.40987 | -0.28412 | -0.16118 | -0.03890 |
| | 3.2 | -0.54752 | -0.41644 | -0.29054 | -0.16727 | -0.04448 |
| | 3.6 | -0.55260 | -0.42150 | -0.29541 | -0.17181 | -0.04859 |
| | 4.0 | -0.55631 | -0.42512 | -0.29883 | -0.17494 | -0.05136 |
| | 6.0 | -0.55964 | -0.42748 | -0.30025 | -0.17546 | -0.05103 |
| 12.0 | 0.0 | -0.50753 | -0.36745 | -0.23527 | -0.10803 | 0.01661 |
| | 0.4 | -0.50929 | -0.36958 | -0.23767 | -0.11060 | 0.01396 |
| | 0.8 | -0.51117 | -0.37184 | -0.24018 | -0.11328 | 0.01122 |
| | 1.2 | -0.51316 | -0.37420 | -0.24280 | -0.11604 | 0.00839 |
| | 1.6 | -0.51522 | -0.37664 | -0.24549 | -0.11887 | 0.00552 |
| | 2.0 | -0.51736 | -0.37914 | -0.24822 | -0.12173 | 0.00264 |
| | 2.4 | -0.51953 | -0.38166 | -0.25096 | -0.12459 | -0.00023 |
| | 2.8 | -0.52172 | -0.38418 | -0.25369 | -0.12741 | -0.00306 |
| | 3.2 | -0.52391 | -0.38668 | -0.25637 | -0.13017 | -0.00580 |
| | 3.6 | -0.52606 | -0.38911 | -0.25897 | -0.13283 | -0.00844 |
| | 4.0 | -0.52815 | -0.39146 | -0.26146 | -0.13537 | -0.01095 |
| | 6.0 | -0.53696 | -0.40118 | -0.27163 | -0.14562 | -0.02094 |
| | 12.0 | -0.54296 | -0.40729 | -0.27762 | -0.15129 | -0.02613 |
| INFINITY | | -0.52440 | -0.38532 | -0.25335 | -0.12566 | 0.0 |

FRACTILES $W_\alpha$ RHO = -0.20

| DELTAS | | $\alpha$: 0.5500 | 0.6000 | 0.6500 | 0.7000 | 0.7500 |
|---|---|---|---|---|---|---|
| 0.0 | 0.0 | 0.18904 | 0.21848 | 0.26641 | 0.33187 | 0.41840 |
| 0.4 | 0.0 | 0.17832 | 0.20922 | 0.25926 | 0.32698 | 0.41613 |
| | 0.4 | 0.06690 | 0.12054 | 0.19107 | 0.28088 | 0.39514 |
| 0.8 | 0.0 | 0.15589 | 0.19468 | 0.25206 | 0.32767 | 0.42549 |
| | 0.4 | 0.00641 | 0.08103 | 0.17203 | 0.28227 | 0.41697 |
| | 0.8 | -0.05160 | 0.04820 | 0.16244 | 0.29447 | 0.44991 |
| 1.2 | 0.0 | 0.14027 | 0.19248 | 0.26243 | 0.35034 | 0.45978 |
| | 0.4 | 0.00414 | 0.09478 | 0.19922 | 0.32034 | 0.46318 |
| | 0.8 | -0.04418 | 0.06676 | 0.18995 | 0.32899 | 0.48953 |
| | 1.2 | -0.04603 | 0.07351 | 0.20437 | 0.35029 | 0.51701 |
| 1.6 | 0.0 | 0.13786 | 0.20602 | 0.28931 | 0.38810 | 0.50577 |
| | 0.4 | 0.03129 | 0.13025 | 0.24028 | 0.36445 | 0.50770 |
| | 0.8 | -0.01446 | 0.09985 | 0.22483 | 0.36404 | 0.52293 |
| | 1.2 | -0.02819 | 0.09417 | 0.22694 | 0.37383 | 0.54043 |
| | 1.6 | -0.02329 | 0.10257 | 0.23839 | 0.38789 | 0.55657 |
| 2.0 | 0.0 | 0.14479 | 0.22673 | 0.31985 | 0.42546 | 0.54732 |
| | 0.4 | 0.06206 | 0.16509 | 0.27728 | 0.40179 | 0.54343 |
| | 0.8 | 0.01727 | 0.13241 | 0.25696 | 0.39437 | 0.54985 |
| | 1.2 | -0.00405 | 0.11865 | 0.25087 | 0.39621 | 0.56002 |
| | 1.6 | -0.00893 | 0.11784 | 0.25400 | 0.40318 | 0.57076 |
| | 2.0 | -0.00345 | 0.12500 | 0.26252 | 0.41272 | 0.58091 |
| 2.4 | 0.0 | 0.15341 | 0.24528 | 0.34511 | 0.45522 | 0.57967 |
| | 0.4 | 0.08725 | 0.19305 | 0.30663 | 0.43117 | 0.57131 |
| | 0.8 | 0.04496 | 0.16039 | 0.28416 | 0.41964 | 0.57180 |
| | 1.2 | 0.02015 | 0.14236 | 0.27326 | 0.41631 | 0.57667 |
| | 1.6 | 0.00899 | 0.13545 | 0.27068 | 0.41822 | 0.58327 |
| | 2.0 | 0.00764 | 0.13635 | 0.27372 | 0.42328 | 0.59026 |
| | 2.4 | 0.01264 | 0.14221 | 0.28020 | 0.43012 | 0.59712 |
| 2.8 | 0.0 | 0.15914 | 0.25833 | 0.36336 | 0.47713 | 0.60386 |
| | 0.4 | 0.10576 | 0.21406 | 0.32904 | 0.45389 | 0.59313 |
| | 0.8 | 0.06738 | 0.18318 | 0.30642 | 0.44039 | 0.58987 |
| | 1.2 | 0.04182 | 0.16346 | 0.29302 | 0.43388 | 0.59099 |
| | 1.6 | 0.02714 | 0.15288 | 0.28677 | 0.43227 | 0.59441 |
| | 2.0 | 0.02107 | 0.14935 | 0.28582 | 0.43397 | 0.59886 |
| | 2.4 | 0.02126 | 0.15084 | 0.28854 | 0.43780 | 0.60369 |
| | 2.8 | 0.02560 | 0.15564 | 0.29360 | 0.44292 | 0.60861 |
| 3.2 | 0.0 | 0.16149 | 0.26639 | 0.37578 | 0.49283 | 0.62180 |
| | 0.4 | 0.11867 | 0.22936 | 0.34587 | 0.47136 | 0.61025 |
| | 0.8 | 0.08435 | 0.20123 | 0.32430 | 0.45728 | 0.60479 |
| | 1.2 | 0.06015 | 0.18141 | 0.30991 | 0.44896 | 0.60335 |
| | 1.6 | 0.04391 | 0.16891 | 0.30149 | 0.44501 | 0.60437 |
| | 2.0 | 0.03488 | 0.16246 | 0.29779 | 0.44425 | 0.60679 |
| | 2.4 | 0.03154 | 0.16071 | 0.29764 | 0.44573 | 0.60995 |
| | 2.8 | 0.03240 | 0.16237 | 0.30002 | 0.44874 | 0.61347 |
| | 3.2 | 0.03611 | 0.16633 | 0.30407 | 0.45272 | 0.61715 |

FRACTILES $w_\alpha$ RHO = −0.20

| DELTAS | | α:0.5500 | 0.6000 | 0.6500 | 0.7000 | 0.7500 |
|---|---|---|---|---|---|---|
| 3.6 | 0.0 | 0.16157 | 0.27091 | 0.38397 | 0.50395 | 0.63509 |
| | 0.4 | 0.12734 | 0.24028 | 0.35836 | 0.48471 | 0.62369 |
| | 0.8 | 0.09811 | 0.21527 | 0.33848 | 0.47092 | 0.61709 |
| | 1.2 | 0.07518 | 0.19630 | 0.32408 | 0.46176 | 0.61398 |
| | 1.6 | 0.05865 | 0.18305 | 0.31451 | 0.45634 | 0.61327 |
| | 2.0 | 0.04796 | 0.17485 | 0.30903 | 0.45384 | 0.61410 |
| | 2.4 | 0.04224 | 0.17082 | 0.30680 | 0.45353 | 0.61587 |
| | 2.8 | 0.04045 | 0.17006 | 0.30707 | 0.45483 | 0.61822 |
| | 3.2 | 0.04159 | 0.17170 | 0.30914 | 0.45726 | 0.62089 |
| | 3.6 | 0.04476 | 0.17501 | 0.31246 | 0.46044 | 0.62375 |
| 4.0 | 0.0 | 0.16048 | 0.27322 | 0.38928 | 0.51180 | 0.64496 |
| | 0.4 | 0.13300 | 0.24796 | 0.36756 | 0.49490 | 0.63425 |
| | 0.8 | 0.10803 | 0.22607 | 0.34965 | 0.48189 | 0.62721 |
| | 1.2 | 0.08725 | 0.20845 | 0.33581 | 0.47251 | 0.62309 |
| | 1.6 | 0.07121 | 0.19521 | 0.32582 | 0.46626 | 0.62117 |
| | 2.0 | 0.05981 | 0.18609 | 0.31926 | 0.46259 | 0.62078 |
| | 2.4 | 0.05260 | 0.18057 | 0.31559 | 0.46096 | 0.62144 |
| | 2.8 | 0.04890 | 0.17802 | 0.31426 | 0.46093 | 0.62280 |
| | 3.2 | 0.04802 | 0.17782 | 0.31474 | 0.46208 | 0.62461 |
| | 3.6 | 0.04924 | 0.17937 | 0.31656 | 0.46408 | 0.62671 |
| | 4.0 | 0.05197 | 0.18217 | 0.31931 | 0.46668 | 0.62900 |
| 6.0 | 0.0 | 0.15270 | 0.27349 | 0.39755 | 0.52768 | 0.66773 |
| | 0.4 | 0.14142 | 0.26265 | 0.38765 | 0.51931 | 0.66159 |
| | 0.8 | 0.12985 | 0.25182 | 0.37806 | 0.51150 | 0.65622 |
| | 1.2 | 0.11861 | 0.24155 | 0.36920 | 0.50454 | 0.65175 |
| | 1.6 | 0.10820 | 0.23222 | 0.36134 | 0.49857 | 0.64818 |
| | 2.0 | 0.09897 | 0.22409 | 0.35463 | 0.49365 | 0.64546 |
| | 2.4 | 0.09113 | 0.21729 | 0.34913 | 0.48974 | 0.64349 |
| | 2.8 | 0.08475 | 0.21185 | 0.34482 | 0.48678 | 0.64216 |
| | 3.2 | 0.07982 | 0.20771 | 0.34162 | 0.48469 | 0.64138 |
| | 3.6 | 0.07625 | 0.20477 | 0.33943 | 0.48336 | 0.64105 |
| | 4.0 | 0.07390 | 0.20292 | 0.33812 | 0.48268 | 0.64107 |
| | 6.0 | 0.07502 | 0.20476 | 0.34061 | 0.48572 | 0.64454 |
| 12.0 | 0.0 | 0.14071 | 0.26631 | 0.39563 | 0.53144 | 0.67751 |
| | 0.4 | 0.13807 | 0.26377 | 0.39330 | 0.52942 | 0.67596 |
| | 0.8 | 0.13535 | 0.26116 | 0.39091 | 0.52738 | 0.67441 |
| | 1.2 | 0.13256 | 0.25851 | 0.38850 | 0.52533 | 0.67289 |
| | 1.6 | 0.12975 | 0.25584 | 0.38609 | 0.52330 | 0.67140 |
| | 2.0 | 0.12693 | 0.25319 | 0.38370 | 0.52131 | 0.66996 |
| | 2.4 | 0.12414 | 0.25057 | 0.38137 | 0.51939 | 0.66859 |
| | 2.8 | 0.12141 | 0.24803 | 0.37912 | 0.51754 | 0.66729 |
| | 3.2 | 0.11876 | 0.24558 | 0.37696 | 0.51578 | 0.66608 |
| | 3.6 | 0.11623 | 0.24325 | 0.37491 | 0.51413 | 0.66496 |
| | 4.0 | 0.11384 | 0.24105 | 0.37300 | 0.51260 | 0.66393 |
| | 6.0 | 0.10441 | 0.23250 | 0.36566 | 0.50686 | 0.66027 |
| | 12.0 | 0.09985 | 0.22871 | 0.36279 | 0.50507 | 0.65976 |
| INFINITY | | 0.12566 | 0.25335 | 0.38532 | 0.52440 | 0.67449 |

FRACTILES $W_\alpha$ RHO = -0.20

| DELTAS | | $\alpha$:0.8000 | 0.8500 | 0.9000 | 0.9250 | 0.9500 |
|---|---|---|---|---|---|---|
| 0.0 | 0.0 | 0.53373 | 0.69339 | 0.93322 | 1.11107 | 1.36979 |
| 0.4 | 0.0 | 0.53465 | 0.69836 | 0.94371 | 1.12528 | 1.38896 |
| | 0.4 | 0.54324 | 0.74332 | 1.03656 | 1.24978 | 1.55497 |
| 0.8 | 0.0 | 0.55353 | 0.72756 | 0.98369 | 1.17034 | 1.43780 |
| | 0.4 | 0.58545 | 0.80561 | 1.11794 | 1.33966 | 1.65139 |
| | 0.8 | 0.63843 | 0.87834 | 1.21063 | 1.44263 | 1.76494 |
| 1.2 | 0.0 | 0.59814 | 0.78007 | 1.03910 | 1.22328 | 1.48241 |
| | 0.4 | 0.63657 | 0.85727 | 1.16297 | 1.37636 | 1.67279 |
| | 0.8 | 0.68104 | 0.92116 | 1.24907 | 1.47565 | 1.78797 |
| | 1.2 | 0.71399 | 0.95869 | 1.28973 | 1.51679 | 1.82800 |
| 1.6 | 0.0 | 0.64926 | 0.83223 | 1.08574 | 1.26265 | 1.50828 |
| | 0.4 | 0.67840 | 0.89217 | 1.18373 | 1.38499 | 1.66221 |
| | 0.8 | 0.71047 | 0.94326 | 1.25792 | 1.47362 | 1.76911 |
| | 1.2 | 0.73585 | 0.97689 | 1.30053 | 1.52118 | 1.82214 |
| | 1.6 | 0.75346 | 0.99503 | 1.31755 | 1.53643 | 1.83383 |
| 2.0 | 0.0 | 0.69228 | 0.87336 | 1.11967 | 1.28933 | 1.52266 |
| | 0.4 | 0.71010 | 0.91640 | 1.19451 | 1.38480 | 1.64513 |
| | 0.8 | 0.73185 | 0.95590 | 1.25619 | 1.46066 | 1.73928 |
| | 1.2 | 0.75102 | 0.98516 | 1.29749 | 1.50931 | 1.79699 |
| | 1.6 | 0.76548 | 1.00329 | 1.31918 | 1.53267 | 1.82176 |
| | 2.0 | 0.77570 | 1.01278 | 1.32649 | 1.53781 | 1.82320 |
| 2.4 | 0.0 | 0.72525 | 0.90446 | 1.14483 | 1.30870 | 1.53234 |
| | 0.4 | 0.73459 | 0.93475 | 1.20195 | 1.38339 | 1.63013 |
| | 0.8 | 0.74864 | 0.96479 | 1.25233 | 1.44695 | 1.71090 |
| | 1.2 | 0.76264 | 0.98935 | 1.29000 | 1.49293 | 1.76747 |
| | 1.6 | 0.77425 | 1.00651 | 1.31359 | 1.52033 | 1.79941 |
| | 2.0 | 0.78303 | 1.01689 | 1.32519 | 1.53224 | 1.81114 |
| | 2.4 | 0.78948 | 1.02225 | 1.32826 | 1.53327 | 1.80887 |
| 2.8 | 0.0 | 0.75023 | 0.92831 | 1.16441 | 1.32396 | 1.54022 |
| | 0.4 | 0.75398 | 0.94952 | 1.20832 | 1.38287 | 1.61899 |
| | 0.8 | 0.76249 | 0.97215 | 1.24917 | 1.43567 | 1.68751 |
| | 1.2 | 0.77229 | 0.99221 | 1.28228 | 1.47721 | 1.74001 |
| | 1.6 | 0.78130 | 1.00766 | 1.30565 | 1.50555 | 1.77461 |
| | 2.0 | 0.78865 | 1.01816 | 1.31967 | 1.52157 | 1.79287 |
| | 2.4 | 0.79433 | 1.02445 | 1.32613 | 1.52776 | 1.79828 |
| | 2.8 | 0.79863 | 1.02768 | 1.32724 | 1.52708 | 1.79477 |
| 3.2 | 0.0 | 0.76927 | 0.94693 | 1.18017 | 1.33658 | 1.54733 |
| | 0.4 | 0.76953 | 0.96176 | 1.21426 | 1.38352 | 1.61139 |
| | 0.8 | 0.77417 | 0.97872 | 1.24734 | 1.42729 | 1.66934 |
| | 1.2 | 0.78069 | 0.99482 | 1.27584 | 1.46392 | 1.71665 |
| | 1.6 | 0.78739 | 1.00823 | 1.29775 | 1.49132 | 1.75113 |
| | 2.0 | 0.79334 | 1.01822 | 1.31267 | 1.50928 | 1.77287 |
| | 2.4 | 0.79823 | 1.02493 | 1.32131 | 1.51895 | 1.78359 |
| | 2.8 | 0.80211 | 1.02894 | 1.32500 | 1.52214 | 1.78578 |
| | 3.2 | 0.80520 | 1.03098 | 1.32517 | 1.52077 | 1.78201 |

FRACTILES $W_\alpha$ RHO = −0.20

| DELTAS | | $\alpha$: 0.8000 | 0.8500 | 0.9000 | 0.9250 | 0.9500 |
|---|---|---|---|---|---|---|
| 3.6 | 0.0 | 0.78386 | 0.96166 | 1.19311 | 1.34729 | 1.55393 |
| | 0.4 | 0.78208 | 0.97203 | 1.21985 | 1.38508 | 1.60656 |
| | 0.8 | 0.78408 | 0.98470 | 1.24671 | 1.42145 | 1.65563 |
| | 1.2 | 0.78812 | 0.99748 | 1.27098 | 1.45334 | 1.69763 |
| | 1.6 | 0.79288 | 1.00884 | 1.29087 | 1.47883 | 1.73045 |
| | 2.0 | 0.79750 | 1.01795 | 1.30565 | 1.49724 | 1.75353 |
| | 2.4 | 0.80157 | 1.02462 | 1.31544 | 1.50893 | 1.76754 |
| | 2.8 | 0.80496 | 1.02907 | 1.32094 | 1.51492 | 1.77393 |
| | 3.2 | 0.80774 | 1.03173 | 1.32307 | 1.51647 | 1.77444 |
| | 3.6 | 0.81004 | 1.03308 | 1.32278 | 1.51485 | 1.77079 |
| 4.0 | 0.0 | 0.79510 | 0.97339 | 1.20384 | 1.35647 | 1.56008 |
| | 0.4 | 0.79223 | 0.98067 | 1.22506 | 1.38723 | 1.60376 |
| | 0.8 | 0.79249 | 0.99015 | 1.24699 | 1.41759 | 1.64546 |
| | 1.2 | 0.79471 | 1.00022 | 1.26755 | 1.44518 | 1.68245 |
| | 1.6 | 0.79789 | 1.00968 | 1.28526 | 1.46837 | 1.71290 |
| | 2.0 | 0.80133 | 1.01774 | 1.29930 | 1.48632 | 1.73597 |
| | 2.4 | 0.80459 | 1.02407 | 1.30949 | 1.49896 | 1.75174 |
| | 2.8 | 0.80747 | 1.02866 | 1.31609 | 1.50677 | 1.76097 |
| | 3.2 | 0.80992 | 1.03171 | 1.31967 | 1.51053 | 1.76478 |
| | 3.6 | 0.81199 | 1.03354 | 1.32088 | 1.51115 | 1.76440 |
| | 4.0 | 0.81375 | 1.03446 | 1.32037 | 1.50950 | 1.76100 |
| 6.0 | 0.0 | 0.82362 | 1.00580 | 1.23663 | 1.38681 | 1.58409 |
| | 0.4 | 0.82062 | 1.00725 | 1.24474 | 1.39975 | 1.60383 |
| | 0.8 | 0.81854 | 1.00967 | 1.25369 | 1.41334 | 1.62392 |
| | 1.2 | 0.81731 | 1.01277 | 1.26295 | 1.42694 | 1.64352 |
| | 1.6 | 0.81679 | 1.01626 | 1.27206 | 1.43997 | 1.66192 |
| | 2.0 | 0.81683 | 1.01986 | 1.28061 | 1.45193 | 1.67854 |
| | 2.4 | 0.81725 | 1.02334 | 1.28828 | 1.46246 | 1.69297 |
| | 2.8 | 0.81792 | 1.02652 | 1.29487 | 1.47137 | 1.70499 |
| | 3.2 | 0.81872 | 1.02929 | 1.30028 | 1.47854 | 1.71453 |
| | 3.6 | 0.81957 | 1.03160 | 1.30450 | 1.48403 | 1.72168 |
| | 4.0 | 0.82042 | 1.03344 | 1.30759 | 1.48793 | 1.72661 |
| | 6.0 | 0.82413 | 1.03710 | 1.31062 | 1.49020 | 1.72746 |
| 12.0 | 0.0 | 0.83965 | 1.02810 | 1.26463 | 1.41686 | 1.61462 |
| | 0.4 | 0.83878 | 1.02823 | 1.26632 | 1.41973 | 1.61925 |
| | 0.8 | 0.83794 | 1.02842 | 1.26812 | 1.42275 | 1.62404 |
| | 1.2 | 0.83715 | 1.02869 | 1.27002 | 1.42587 | 1.62894 |
| | 1.6 | 0.83642 | 1.02903 | 1.27200 | 1.42907 | 1.63391 |
| | 2.0 | 0.83574 | 1.02943 | 1.27403 | 1.43231 | 1.63889 |
| | 2.4 | 0.83513 | 1.02988 | 1.27608 | 1.43554 | 1.64382 |
| | 2.8 | 0.83458 | 1.03038 | 1.27814 | 1.43875 | 1.64866 |
| | 3.2 | 0.83410 | 1.03091 | 1.28018 | 1.44188 | 1.65335 |
| | 3.6 | 0.83369 | 1.03147 | 1.28216 | 1.44490 | 1.65785 |
| | 4.0 | 0.83333 | 1.03203 | 1.28408 | 1.44779 | 1.66212 |
| | 6.0 | 0.83237 | 1.03471 | 1.29202 | 1.45950 | 1.67912 |
| | 12.0 | 0.83339 | 1.03763 | 1.29744 | 1.46655 | 1.68829 |
| INFINITY | | 0.84162 | 1.03643 | 1.28155 | 1.43953 | 1.64485 |

FRACTILES $w_\alpha$ RHO = -0.20

| DELTAS | | $\alpha$: 0.9625 | 0.9750 | 0.9825 | 0.9900 | 0.9925 |
|---|---|---|---|---|---|---|
| 0.0 | 0.0 | 1.55790 | 1.82814 | 2.06994 | 2.45549 | 2.65614 |
| 0.4 | 0.0 | 1.58035 | 1.85486 | 2.10007 | 2.49031 | 2.69305 |
| | 0.4 | 1.77377 | 2.08429 | 2.35887 | 2.79130 | 3.01409 |
| 0.8 | 0.0 | 1.62971 | 1.90217 | 2.14318 | 2.52284 | 2.71847 |
| | 0.4 | 1.87174 | 2.18099 | 2.45171 | 2.87400 | 3.08997 |
| | 0.8 | 1.99069 | 2.30521 | 2.57874 | 3.00272 | 3.21848 |
| 1.2 | 0.0 | 1.66565 | 1.92285 | 2.14801 | 2.49925 | 2.67887 |
| | 0.4 | 1.88037 | 2.16954 | 2.42099 | 2.81070 | 3.00900 |
| | 0.8 | 2.00533 | 2.30659 | 2.56731 | 2.96949 | 3.17336 |
| | 1.2 | 2.04353 | 2.34105 | 2.59756 | 2.99168 | 3.19084 |
| 1.6 | 0.0 | 1.68020 | 1.91960 | 2.12769 | 2.45006 | 2.61405 |
| | 0.4 | 1.85503 | 2.12216 | 2.35325 | 2.70958 | 2.89016 |
| | 0.8 | 1.97370 | 2.25604 | 2.49939 | 2.87323 | 3.06211 |
| | 1.2 | 2.02972 | 2.31528 | 2.56065 | 2.93639 | 3.12574 |
| | 1.6 | 2.03829 | 2.31877 | 2.55912 | 2.92615 | 3.11068 |
| 2.0 | 0.0 | 1.68475 | 1.90909 | 2.10300 | 2.40176 | 2.55306 |
| | 0.4 | 1.82517 | 2.07344 | 2.28728 | 2.61554 | 2.78131 |
| | 0.8 | 1.93131 | 2.19534 | 2.42210 | 2.76917 | 2.94401 |
| | 1.2 | 1.99470 | 2.26584 | 2.49814 | 2.85277 | 3.03104 |
| | 1.6 | 2.01992 | 2.29108 | 2.52288 | 2.87594 | 3.05308 |
| | 2.0 | 2.01836 | 2.28485 | 2.51221 | 2.85776 | 3.03083 |
| 2.4 | 0.0 | 1.68670 | 1.89925 | 2.08208 | 2.36240 | 2.50383 |
| | 0.4 | 1.79993 | 2.03311 | 2.23317 | 2.53907 | 2.69305 |
| | 0.8 | 1.89209 | 2.14037 | 2.35291 | 2.67716 | 2.84006 |
| | 1.2 | 1.95554 | 2.21272 | 2.43248 | 2.76703 | 2.93482 |
| | 1.6 | 1.99019 | 2.25064 | 2.47278 | 2.81034 | 2.97938 |
| | 2.0 | 2.00144 | 2.26080 | 2.48165 | 2.81666 | 2.98416 |
| | 2.4 | 1.99657 | 2.25198 | 2.46914 | 2.79799 | 2.96219 |
| 2.8 | 0.0 | 1.68867 | 1.89213 | 2.06639 | 2.33243 | 2.46619 |
| | 0.4 | 1.78076 | 2.00208 | 2.19129 | 2.47957 | 2.62427 |
| | 0.8 | 1.85976 | 2.09506 | 2.29590 | 2.60136 | 2.75445 |
| | 1.2 | 1.91947 | 2.16427 | 2.37291 | 2.68974 | 2.84831 |
| | 1.6 | 1.95807 | 2.20799 | 2.42070 | 2.74322 | 2.90443 |
| | 2.0 | 1.97759 | 2.22889 | 2.44250 | 2.76592 | 2.92738 |
| | 2.4 | 1.98220 | 2.23208 | 2.44421 | 2.76495 | 2.92489 |
| | 2.8 | 1.97650 | 2.22311 | 2.43219 | 2.74791 | 2.90516 |
| 3.2 | 0.0 | 1.69127 | 1.88774 | 2.05536 | 2.31026 | 2.43801 |
| | 0.4 | 1.76688 | 1.97889 | 2.15957 | 2.43393 | 2.57127 |
| | 0.8 | 1.83434 | 2.05909 | 2.25041 | 2.54058 | 2.68567 |
| | 1.2 | 1.88876 | 2.12296 | 2.32210 | 2.62378 | 2.77447 |
| | 1.6 | 1.92787 | 2.16813 | 2.37222 | 2.68103 | 2.83512 |
| | 2.0 | 1.95198 | 2.19522 | 2.40162 | 2.71358 | 2.86909 |
| | 2.4 | 1.96321 | 2.20689 | 2.41346 | 2.72532 | 2.88063 |
| | 2.8 | 1.96453 | 2.20677 | 2.41190 | 2.72124 | 2.87515 |
| | 3.2 | 1.95892 | 2.19843 | 2.40104 | 2.70624 | 2.85795 |

FRACTILES $W_\alpha$ RHO = −0.20

| DELTAS | | α: 0.9625 | 0.9750 | 0.9825 | 0.9900 | 0.9925 |
|---|---|---|---|---|---|---|
| 3.6 | 0.0 | 1.69444 | 1.88552 | 2.04797 | 2.29412 | 2.41713 |
| | 0.4 | 1.75714 | 1.96181 | 2.13572 | 2.39901 | 2.53048 |
| | 0.8 | 1.81478 | 2.03098 | 2.21456 | 2.49226 | 2.63082 |
| | 1.2 | 1.86354 | 2.08881 | 2.27995 | 2.56885 | 2.71288 |
| | 1.6 | 1.90123 | 2.13295 | 2.32941 | 2.62610 | 2.77391 |
| | 2.0 | 1.92735 | 2.16301 | 2.36265 | 2.66388 | 2.81383 |
| | 2.4 | 1.94278 | 2.18018 | 2.38114 | 2.68408 | 2.83476 |
| | 2.8 | 1.94928 | 2.18665 | 2.38741 | 2.68978 | 2.84006 |
| | 3.2 | 1.94894 | 2.18493 | 2.38437 | 2.68446 | 2.83349 |
| | 3.6 | 1.94373 | 2.17743 | 2.37476 | 2.67141 | 2.81862 |
| 4.0 | 0.0 | 1.69798 | 1.88489 | 2.04329 | 2.28252 | 2.40176 |
| | 0.4 | 1.75047 | 1.94933 | 2.11783 | 2.37223 | 2.49898 |
| | 0.8 | 1.79987 | 2.00913 | 2.18640 | 2.45391 | 2.58712 |
| | 1.2 | 1.84320 | 2.06100 | 2.24542 | 2.52358 | 2.66202 |
| | 1.6 | 1.87851 | 2.10281 | 2.29263 | 2.57877 | 2.72111 |
| | 2.0 | 1.90497 | 2.13373 | 2.32723 | 2.61871 | 2.76362 |
| | 2.4 | 1.92276 | 2.15412 | 2.34970 | 2.64412 | 2.79039 |
| | 2.8 | 1.93284 | 2.16520 | 2.36150 | 2.65679 | 2.80339 |
| | 3.2 | 1.93655 | 2.16863 | 2.36456 | 2.65905 | 2.80518 |
| | 3.6 | 1.93536 | 2.16619 | 2.36092 | 2.65339 | 2.79841 |
| | 4.0 | 1.93064 | 2.15952 | 2.35247 | 2.64204 | 2.78553 |
| 6.0 | 0.0 | 1.71589 | 1.89239 | 2.04023 | 2.26077 | 2.36958 |
| | 0.4 | 1.74042 | 1.92358 | 2.07716 | 2.30651 | 2.41973 |
| | 0.8 | 1.76503 | 1.95446 | 2.11341 | 2.35095 | 2.46828 |
| | 1.2 | 1.78880 | 1.98395 | 2.14780 | 2.39277 | 2.51381 |
| | 1.6 | 1.81091 | 2.01114 | 2.17932 | 2.43084 | 2.55513 |
| | 2.0 | 1.83073 | 2.03532 | 2.20721 | 2.46431 | 2.59137 |
| | 2.4 | 1.84782 | 2.05602 | 2.23096 | 2.49263 | 2.62195 |
| | 2.8 | 1.86195 | 2.07301 | 2.25035 | 2.51559 | 2.64667 |
| | 3.2 | 1.87308 | 2.08627 | 2.26538 | 2.53325 | 2.66559 |
| | 3.6 | 1.88133 | 2.09597 | 2.27628 | 2.54589 | 2.67906 |
| | 4.0 | 1.88693 | 2.10243 | 2.28342 | 2.55397 | 2.68758 |
| | 6.0 | 1.88655 | 2.10007 | 2.27910 | 2.54625 | 2.67797 |
| 12.0 | 0.0 | 1.74526 | 1.91832 | 2.06157 | 2.27245 | 2.37526 |
| | 0.4 | 1.75117 | 1.92607 | 2.07097 | 2.28446 | 2.38862 |
| | 0.8 | 1.75725 | 1.93401 | 2.08056 | 2.29666 | 2.40217 |
| | 1.2 | 1.76345 | 1.94205 | 2.09024 | 2.30892 | 2.41575 |
| | 1.6 | 1.76970 | 1.95012 | 2.09991 | 2.32112 | 2.42925 |
| | 2.0 | 1.77593 | 1.95812 | 2.10949 | 2.33315 | 2.44253 |
| | 2.4 | 1.78208 | 1.96599 | 2.11887 | 2.34489 | 2.45548 |
| | 2.8 | 1.78809 | 1.97365 | 2.12797 | 2.35624 | 2.46798 |
| | 3.2 | 1.79389 | 1.98102 | 2.13671 | 2.36711 | 2.47992 |
| | 3.6 | 1.79944 | 1.98804 | 2.14501 | 2.37740 | 2.49123 |
| | 4.0 | 1.80469 | 1.99466 | 2.15283 | 2.38706 | 2.50183 |
| | 6.0 | 1.82540 | 2.02055 | 2.18320 | 2.42435 | 2.54259 |
| | 12.0 | 1.83596 | 2.03290 | 2.19699 | 2.44013 | 2.55929 |
| INFINITY | | 1.78046 | 1.95996 | 2.10836 | 2.32635 | 2.43238 |

FRACTILES $W_\alpha$ RHO = −0.20

| DELTAS | | $\alpha$:0.9950 | 0.9975 | 0.9990 | 0.9995 |
|---|---|---|---|---|---|
| 0.0 | 0.0 | 2.94130 | 3.43420 | 4.09412 | 4.59832 |
| 0.4 | 0.0 | 2.98082 | 3.47721 | 4.13998 | 4.64508 |
| | 0.4 | 3.32840 | 3.86609 | 4.57674 | 5.11382 |
| 0.8 | 0.0 | 2.99449 | 3.46672 | 4.09091 | 4.56266 |
| | 0.4 | 3.39313 | 3.90830 | 4.58391 | 5.09135 |
| | 0.8 | 3.52031 | 4.03084 | 4.69661 | 5.19437 |
| 1.2 | 0.0 | 2.93101 | 3.35943 | 3.92122 | 4.34314 |
| | 0.4 | 3.28639 | 3.75552 | 4.36725 | 4.82456 |
| | 0.8 | 3.45779 | 3.93703 | 4.55910 | 5.02239 |
| | 1.2 | 3.46805 | 3.93368 | 4.53570 | 4.98256 |
| 1.6 | 0.0 | 2.84338 | 3.23113 | 3.73657 | 4.11433 |
| | 0.4 | 3.14204 | 3.56634 | 4.11692 | 4.52688 |
| | 0.8 | 3.32499 | 3.76646 | 4.33717 | 4.76073 |
| | 1.2 | 3.38877 | 3.82934 | 4.39698 | 4.81707 |
| | 1.6 | 3.36659 | 3.79422 | 4.34355 | 4.74905 |
| 2.0 | 0.0 | 2.76403 | 3.11921 | 3.57983 | 3.92263 |
| | 0.4 | 3.01194 | 3.39909 | 3.89932 | 4.27043 |
| | 0.8 | 3.18682 | 3.59341 | 4.11707 | 4.50451 |
| | 1.2 | 3.27823 | 3.69123 | 4.22168 | 4.61318 |
| | 1.6 | 3.29837 | 3.70736 | 4.23132 | 4.61719 |
| | 2.0 | 3.27017 | 3.66852 | 4.17765 | 4.55185 |
| 2.4 | 0.0 | 2.70049 | 3.03034 | 3.45615 | 3.77184 |
| | 0.4 | 2.90680 | 3.26448 | 3.72483 | 4.06522 |
| | 0.8 | 3.06587 | 3.44295 | 3.92699 | 4.28409 |
| | 1.2 | 3.16711 | 3.55430 | 4.05015 | 4.41522 |
| | 1.6 | 3.21311 | 3.60206 | 4.09910 | 4.46435 |
| | 2.0 | 3.21554 | 3.59997 | 4.09024 | 4.44989 |
| | 2.4 | 3.18876 | 3.56466 | 4.04315 | 4.39358 |
| 2.8 | 0.0 | 2.65171 | 2.96184 | 3.36052 | 3.65507 |
| | 0.4 | 2.82471 | 3.15915 | 3.58803 | 3.90421 |
| | 0.8 | 2.96628 | 3.31914 | 3.77068 | 4.10293 |
| | 1.2 | 3.06749 | 3.43206 | 3.89767 | 4.23969 |
| | 1.6 | 3.12705 | 3.49683 | 3.96824 | 4.31396 |
| | 2.0 | 3.15014 | 3.51968 | 3.98998 | 4.33437 |
| | 2.4 | 3.14537 | 3.51065 | 3.97478 | 4.31416 |
| | 2.8 | 3.12176 | 3.48017 | 3.93487 | 4.26690 |
| 3.2 | 0.0 | 2.61481 | 2.90944 | 3.28673 | 3.56458 |
| | 0.4 | 2.76116 | 3.07716 | 3.48104 | 3.77796 |
| | 0.8 | 2.88610 | 3.21920 | 3.64423 | 3.95620 |
| | 1.2 | 2.98246 | 3.32773 | 3.76757 | 4.08996 |
| | 1.6 | 3.04765 | 3.40005 | 3.84830 | 4.17641 |
| | 2.0 | 3.08342 | 3.43841 | 3.88931 | 4.21894 |
| | 2.4 | 3.09452 | 3.44843 | 3.89733 | 4.22508 |
| | 2.8 | 3.08697 | 3.43708 | 3.88058 | 4.20399 |
| | 3.2 | 3.06660 | 3.41112 | 3.84696 | 4.16443 |

FRACTILES $W_{\alpha}$ RHO = -0.20

| DELTAS | | $\alpha$: 0.9950 | 0.9975 | 0.9990 | 0.9995 |
|---|---|---|---|---|---|
| 3.6 | 0.0 | 2.58702 | 2.86932 | 3.22955 | 3.49403 |
| | 0.4 | 2.71193 | 3.01314 | 3.39694 | 3.67635 |
| | 0.8 | 2.82195 | 3.13889 | 3.54221 | 3.83757 |
| | 1.2 | 2.91143 | 3.24039 | 3.65847 | 3.96428 |
| | 1.6 | 2.97753 | 3.31459 | 3.74244 | 4.05504 |
| | 2.0 | 3.02029 | 3.36174 | 3.79464 | 4.11059 |
| | 2.4 | 3.04210 | 3.38471 | 3.81858 | 4.13490 |
| | 2.8 | 3.04672 | 3.38791 | 3.81950 | 4.13382 |
| | 3.2 | 3.03832 | 3.37619 | 3.80309 | 4.11369 |
| | 3.6 | 3.02081 | 3.35407 | 3.77466 | 4.08037 |
| 4.0 | 0.0 | 2.56612 | 2.83852 | 3.18496 | 3.43861 |
| | 0.4 | 2.67362 | 2.96285 | 3.33033 | 3.59912 |
| | 0.8 | 2.77059 | 3.07425 | 3.45967 | 3.74130 |
| | 1.2 | 2.85261 | 3.16783 | 3.56753 | 3.85933 |
| | 1.6 | 2.91697 | 3.24068 | 3.65075 | 3.94985 |
| | 2.0 | 2.96293 | 3.29210 | 3.70870 | 4.01227 |
| | 2.4 | 2.99149 | 3.32337 | 3.74298 | 4.04848 |
| | 2.8 | 3.00485 | 3.33709 | 3.75676 | 4.06203 |
| | 3.2 | 3.00587 | 3.33661 | 3.75397 | 4.05730 |
| | 3.6 | 2.99749 | 3.32532 | 3.73863 | 4.03875 |
| | 4.0 | 2.98240 | 3.30636 | 3.71440 | 4.01044 |
| 6.0 | 0.0 | 2.51846 | 2.76262 | 3.06905 | 3.29087 |
| | 0.4 | 2.57471 | 2.82899 | 3.14825 | 3.37938 |
| | 0.8 | 2.62892 | 2.89254 | 3.22361 | 3.46330 |
| | 1.2 | 2.67955 | 2.95160 | 3.29325 | 3.54060 |
| | 1.6 | 2.72535 | 3.00475 | 3.35563 | 3.60962 |
| | 2.0 | 2.76537 | 3.05099 | 3.40962 | 3.66917 |
| | 2.4 | 2.79904 | 3.08969 | 3.45456 | 3.71856 |
| | 2.8 | 2.82614 | 3.12066 | 3.49028 | 3.75764 |
| | 3.2 | 2.84679 | 3.14407 | 3.51703 | 3.78671 |
| | 3.6 | 2.86137 | 3.16039 | 3.53540 | 3.80646 |
| | 4.0 | 2.87045 | 3.17030 | 3.54620 | 3.81780 |
| | 6.0 | 2.85803 | 3.15275 | 3.52131 | 3.78700 |
| 12.0 | 0.0 | 2.51465 | 2.74014 | 3.01799 | 3.21576 |
| | 0.4 | 2.52992 | 2.75869 | 3.04083 | 3.24183 |
| | 0.8 | 2.54536 | 2.77735 | 3.06374 | 3.26791 |
| | 1.2 | 2.56081 | 2.79598 | 3.08650 | 3.29375 |
| | 1.6 | 2.57613 | 2.81438 | 3.10892 | 3.31916 |
| | 2.0 | 2.59117 | 2.83239 | 3.13080 | 3.34391 |
| | 2.4 | 2.60581 | 2.84987 | 3.15197 | 3.36780 |
| | 2.8 | 2.61991 | 2.86668 | 3.17226 | 3.39067 |
| | 3.2 | 2.63336 | 2.88267 | 3.19153 | 3.41236 |
| | 3.6 | 2.64608 | 2.89776 | 3.20966 | 3.43273 |
| | 4.0 | 2.65798 | 2.91184 | 3.22654 | 3.45167 |
| | 6.0 | 2.70358 | 2.96548 | 3.29045 | 3.52308 |
| | 12.0 | 2.72144 | 2.98507 | 3.31179 | 3.54541 |
| INFINITY | | 2.57583 | 2.80703 | 3.09024 | 3.29051 |

FRACTILES $W_\alpha$ RHO = 0.0

| DELTAS | | $\alpha$: 0.0005 | 0.0010 | 0.0025 | 0.0050 | 0.0075 |
|---|---|---|---|---|---|---|
| 0.0 | 0.0 | -5.71869 | -5.07546 | -4.23338 | -3.60421 | -3.24009 |
| 0.4 | 0.0 | -5.68910 | -5.05386 | -4.22073 | -3.59704 | -3.23559 |
| | 0.4 | -5.04974 | -4.49017 | -3.75772 | -3.21059 | -2.89401 |
| 0.8 | 0.0 | -5.47270 | -4.88725 | -4.11385 | -3.52988 | -3.18911 |
| | 0.4 | -4.68714 | -4.17759 | -3.50949 | -3.00955 | -2.71994 |
| | 0.8 | -4.00730 | -3.58221 | -3.02605 | -2.61090 | -2.37087 |
| 1.2 | 0.0 | -5.12870 | -4.61044 | -3.92160 | -3.39747 | -3.08959 |
| | 0.4 | -4.48946 | -4.02453 | -3.41045 | -2.94692 | -2.67652 |
| | 0.8 | -3.76689 | -3.37800 | -2.86842 | -2.48745 | -2.26696 |
| | 1.2 | -3.28294 | -2.95799 | -2.53324 | -2.21662 | -2.03381 |
| 1.6 | 0.0 | -4.79812 | -4.33822 | -3.72413 | -3.25419 | -2.97682 |
| | 0.4 | -4.30660 | -3.88647 | -3.32803 | -2.90310 | -2.65349 |
| | 0.8 | -3.71980 | -3.35407 | -2.87117 | -2.50688 | -2.29450 |
| | 1.2 | -3.17299 | -2.86860 | -2.47015 | -2.17276 | -2.00093 |
| | 1.6 | -2.84967 | -2.59206 | -2.25585 | -2.00583 | -1.86184 |
| 2.0 | 0.0 | -4.52658 | -4.11235 | -3.55704 | -3.13005 | -2.87702 |
| | 0.4 | -4.14154 | -3.75877 | -3.24752 | -2.85614 | -2.62506 |
| | 0.8 | -3.68376 | -3.34208 | -2.88785 | -2.54220 | -2.33918 |
| | 1.2 | -3.21605 | -2.92082 | -2.53123 | -2.23762 | -2.06666 |
| | 1.6 | -2.81561 | -2.56910 | -2.24696 | -2.00719 | -1.86909 |
| | 2.0 | -2.60042 | -2.38938 | -2.11465 | -1.91123 | -1.79467 |
| 2.4 | 0.0 | -4.31299 | -3.93398 | -3.42407 | -3.03030 | -2.79613 |
| | 0.4 | -4.00353 | -3.65032 | -3.17657 | -2.81205 | -2.59591 |
| | 0.8 | -3.64066 | -3.32010 | -2.89175 | -2.56366 | -2.36986 |
| | 1.2 | -3.26079 | -2.97687 | -2.59939 | -2.31215 | -2.14346 |
| | 1.6 | -2.90005 | -2.65484 | -2.33157 | -2.08840 | -1.94712 |
| | 2.0 | -2.60712 | -2.40179 | -2.13425 | -1.93616 | -1.82279 |
| | 2.4 | -2.46537 | -2.28821 | -2.05868 | -1.89027 | -1.79502 |
| 2.8 | 0.0 | -4.14588 | -3.79429 | -3.31973 | -2.95181 | -2.73233 |
| | 0.4 | -3.89157 | -3.56168 | -3.11754 | -2.77424 | -2.56992 |
| | 0.8 | -3.59764 | -3.29455 | -2.88773 | -2.57441 | -2.38848 |
| | 1.2 | -3.28734 | -3.01430 | -2.64923 | -2.36939 | -2.20400 |
| | 1.6 | -2.98333 | -2.74194 | -2.42098 | -2.17680 | -2.03347 |
| | 2.0 | -2.70696 | -2.49838 | -2.22390 | -2.01818 | -1.89927 |
| | 2.4 | -2.49364 | -2.31910 | -2.09296 | -1.92754 | -1.83476 |
| | 2.8 | -2.40434 | -2.25334 | -2.05985 | -1.92303 | -1.84908 |
| 3.2 | 0.0 | -4.01417 | -3.68426 | -3.23761 | -2.89009 | -2.68218 |
| | 0.4 | -3.80130 | -3.49001 | -3.06946 | -2.74303 | -2.54810 |
| | 0.8 | -3.55859 | -3.26981 | -2.88062 | -2.57941 | -2.39994 |
| | 1.2 | -3.30160 | -3.03803 | -2.68390 | -2.41080 | -2.24855 |
| | 1.6 | -3.04611 | -2.80925 | -2.49231 | -2.24912 | -2.10528 |
| | 2.0 | -2.80596 | -2.59656 | -2.31818 | -2.10655 | -1.98254 |
| | 2.4 | -2.59543 | -2.41476 | -2.17790 | -2.00204 | -1.90228 |
| | 2.8 | -2.44368 | -2.29359 | -2.10219 | -1.96893 | -1.89066 |
| | 3.2 | -2.39446 | -2.26578 | -2.10733 | -1.98592 | -1.90989 |

FRACTILES $W_\alpha$ RHO = 0.0

| DELTAS | | α:0.0005 | 0.0010 | 0.0025 | 0.0050 | 0.0075 |
|---|---|---|---|---|---|---|
| 3.6 | 0.0 | -3.90918 | -3.59668 | -3.17242 | -2.84123 | -2.64255 |
| | 0.4 | -3.72830 | -3.43201 | -3.03047 | -2.71760 | -2.53019 |
| | 0.8 | -3.52456 | -3.24754 | -2.87283 | -2.58152 | -2.40732 |
| | 1.2 | -3.30883 | -3.05330 | -2.70849 | -2.44115 | -2.28161 |
| | 1.6 | -3.09244 | -2.85983 | -2.54688 | -2.30511 | -2.16125 |
| | 2.0 | -2.88575 | -2.67680 | -2.39690 | -2.18187 | -2.05458 |
| | 2.4 | -2.69781 | -2.51311 | -2.26775 | -2.08184 | -1.97388 |
| | 2.8 | -2.54005 | -2.38168 | -2.17660 | -2.02915 | -1.94221 |
| | 3.2 | -2.43920 | -2.31190 | -2.15057 | -2.02317 | -1.94311 |
| | 3.6 | -2.42551 | -2.31631 | -2.16616 | -2.04109 | -1.96090 |
| 4.0 | 0.0 | -3.82447 | -3.52615 | -3.12010 | -2.80216 | -2.61096 |
| | 0.4 | -3.66883 | -3.38481 | -2.99876 | -2.69692 | -2.51560 |
| | 0.8 | -3.49543 | -3.22813 | -2.86536 | -2.58220 | -2.41230 |
| | 1.2 | -3.31209 | -3.06338 | -2.72643 | -2.46394 | -2.30667 |
| | 1.6 | -3.12718 | -2.89832 | -2.58897 | -2.34856 | -2.20478 |
| | 2.0 | -2.94865 | -2.74042 | -2.45979 | -2.24246 | -2.11285 |
| | 2.4 | -2.78356 | -2.59644 | -2.34548 | -2.15249 | -2.03839 |
| | 2.8 | -2.63862 | -2.47356 | -2.25525 | -2.09232 | -1.99602 |
| | 3.2 | -2.52539 | -2.38740 | -2.20916 | -2.06959 | -1.98289 |
| | 3.6 | -2.47242 | -2.35958 | -2.20259 | -2.07164 | -1.98788 |
| | 4.0 | -2.48034 | -2.37374 | -2.21970 | -2.08840 | -2.00375 |
| 6.0 | 0.0 | -3.57674 | -3.32122 | -2.96979 | -2.69124 | -2.52212 |
| | 0.4 | -3.49301 | -3.24603 | -2.90640 | -2.63724 | -2.47380 |
| | 0.8 | -3.40329 | -3.16578 | -2.83922 | -2.58040 | -2.42321 |
| | 1.2 | -3.30975 | -3.08250 | -2.77006 | -2.52238 | -2.37190 |
| | 1.6 | -3.21478 | -2.99843 | -2.70095 | -2.46502 | -2.32158 |
| | 2.0 | -3.12090 | -2.91592 | -2.63399 | -2.41021 | -2.27402 |
| | 2.4 | -3.03062 | -2.83732 | -2.57130 | -2.35988 | -2.23100 |
| | 2.8 | -2.94638 | -2.76493 | -2.51498 | -2.31590 | -2.19422 |
| | 3.2 | -2.87056 | -2.70104 | -2.46712 | -2.28003 | -2.16507 |
| | 3.6 | -2.80563 | -2.64807 | -2.42965 | -2.25335 | -2.14409 |
| | 4.0 | -2.75447 | -2.60836 | -2.40354 | -2.23582 | -2.13082 |
| | 6.0 | -2.71772 | -2.59010 | -2.40250 | -2.24295 | -2.14102 |
| 12.0 | 0.0 | -3.37428 | -3.15672 | -2.85268 | -2.60742 | -2.45652 |
| | 0.4 | -3.34918 | -3.13459 | -2.83455 | -2.59236 | -2.44327 |
| | 0.8 | -3.32346 | -3.11197 | -2.81609 | -2.57707 | -2.42984 |
| | 1.2 | -3.29733 | -3.08904 | -2.79744 | -2.56169 | -2.41637 |
| | 1.6 | -3.27102 | -3.06602 | -2.77880 | -2.54637 | -2.40299 |
| | 2.0 | -3.24475 | -3.04310 | -2.76032 | -2.53125 | -2.38982 |
| | 2.4 | -3.21878 | -3.02051 | -2.74220 | -2.51649 | -2.37701 |
| | 2.8 | -3.19333 | -2.99845 | -2.72460 | -2.50222 | -2.36466 |
| | 3.2 | -3.16864 | -2.97713 | -2.70768 | -2.48857 | -2.35289 |
| | 3.6 | -3.14493 | -2.95674 | -2.69160 | -2.47567 | -2.34180 |
| | 4.0 | -3.12240 | -2.93744 | -2.67649 | -2.46361 | -2.33148 |
| | 6.0 | -3.03306 | -2.86202 | -2.61862 | -2.41826 | -2.29310 |
| | 12.0 | -2.99978 | -2.83707 | -2.60301 | -2.40845 | -2.28617 |
| INFINITY | | -3.29051 | -3.09024 | -2.80703 | -2.57583 | -2.43238 |

FRACTILES $W_\alpha$ RHO = 0.0

| DELTAS | | $\alpha$: 0.0100 | 0.0175 | 0.0250 | 0.0375 | 0.0500 |
|---|---|---|---|---|---|---|
| 0.0 | 0.0 | -2.98381 | -2.49114 | -2.18195 | -1.83610 | -1.59510 |
| 0.4 | 0.0 | -2.98097 | -2.49091 | -2.18297 | -1.83815 | -1.59762 |
| | 0.4 | -2.67124 | -2.24311 | -1.97456 | -1.67432 | -1.46526 |
| 0.8 | 0.0 | -2.94788 | -2.48050 | -2.18443 | -1.85037 | -1.61556 |
| | 0.4 | -2.51600 | -2.12371 | -1.87742 | -1.60193 | -1.41003 |
| | 0.8 | -2.20206 | -1.87798 | -1.67500 | -1.44852 | -1.29120 |
| 1.2 | 0.0 | -2.87056 | -2.44315 | -2.16987 | -1.85857 | -1.63746 |
| | 0.4 | -2.48516 | -2.11461 | -1.88008 | -1.61575 | -1.43024 |
| | 0.8 | -2.11181 | -1.81379 | -1.62708 | -1.41876 | -1.27415 |
| | 1.2 | -1.90541 | -1.65941 | -1.50583 | -1.33519 | -1.21734 |
| 1.6 | 0.0 | -2.77879 | -2.39034 | -2.14030 | -1.85345 | -1.64809 |
| | 0.4 | -2.47592 | -2.12946 | -1.90800 | -1.65583 | -1.47686 |
| | 0.8 | -2.14430 | -1.85376 | -1.67020 | -1.46382 | -1.31947 |
| | 1.2 | -1.88020 | -1.64891 | -1.50460 | -1.34450 | -1.23425 |
| | 1.6 | -1.76095 | -1.56844 | -1.44910 | -1.31782 | -1.22859 |
| 2.0 | 0.0 | -2.69584 | -2.33898 | -2.10805 | -1.84171 | -1.64991 |
| | 0.4 | -2.46005 | -2.13626 | -1.92777 | -1.68851 | -1.51718 |
| | 0.8 | -2.19476 | -1.91306 | -1.73311 | -1.52842 | -1.38338 |
| | 1.2 | -1.94589 | -1.71277 | -1.56602 | -1.40190 | -1.28803 |
| | 1.6 | -1.77234 | -1.58798 | -1.47403 | -1.34949 | -1.26601 |
| | 2.0 | -1.71340 | -1.55988 | -1.46662 | -1.36845 | -1.30999 |
| 2.4 | 0.0 | -2.62804 | -2.29574 | -2.07973 | -1.82945 | -1.64834 |
| | 0.4 | -2.44109 | -2.13592 | -1.93829 | -1.71013 | -1.54568 |
| | 0.8 | -2.23142 | -1.95966 | -1.78458 | -1.58360 | -1.43967 |
| | 1.2 | -2.02350 | -1.78966 | -1.64049 | -1.47127 | -1.35192 |
| | 1.6 | -1.84754 | -1.65623 | -1.53684 | -1.40538 | -1.31713 |
| | 2.0 | -1.74389 | -1.59577 | -1.50761 | -1.41926 | -1.34947 |
| | 2.4 | -1.72965 | -1.61426 | -1.54153 | -1.44922 | -1.37574 |
| 2.8 | 0.0 | -2.57440 | -2.26120 | -2.05679 | -1.81899 | -1.64616 |
| | 0.4 | -2.42316 | -2.13276 | -1.94375 | -1.72446 | -1.56552 |
| | 0.8 | -2.25520 | -1.99227 | -1.82176 | -1.62468 | -1.48241 |
| | 1.2 | -2.08580 | -1.85364 | -1.70400 | -1.53221 | -1.40926 |
| | 1.6 | -1.93162 | -1.73345 | -1.60759 | -1.46613 | -1.36874 |
| | 2.0 | -1.81593 | -1.65811 | -1.56452 | -1.46206 | -1.38276 |
| | 2.4 | -1.77245 | -1.65749 | -1.57792 | -1.47826 | -1.39959 |
| | 2.8 | -1.79464 | -1.68070 | -1.60001 | -1.49779 | -1.41674 |
| 3.2 | 0.0 | -2.53226 | -2.23407 | -2.03874 | -1.81069 | -1.64431 |
| | 0.4 | -2.40773 | -2.12901 | -1.94680 | -1.73447 | -1.57986 |
| | 0.8 | -2.27091 | -2.01524 | -1.84854 | -1.65475 | -1.51397 |
| | 1.2 | -2.13215 | -1.90219 | -1.75280 | -1.57981 | -1.45471 |
| | 1.6 | -2.00243 | -1.80033 | -1.67008 | -1.52090 | -1.41516 |
| | 2.0 | -1.89463 | -1.72490 | -1.62078 | -1.50263 | -1.41403 |
| | 2.4 | -1.83533 | -1.70455 | -1.61622 | -1.50728 | -1.42246 |
| | 2.8 | -1.83292 | -1.71190 | -1.62645 | -1.51877 | -1.43385 |
| | 3.2 | -1.85256 | -1.73045 | -1.64339 | -1.53327 | -1.44634 |

FRACTILES $W_{\alpha}$ RHO = 0.0

| DELTAS | | $\alpha$:0.0100 | 0.0175 | 0.0250 | 0.0375 | 0.0500 |
|---|---|---|---|---|---|---|
| 3.6 | 0.0 | -2.49902 | -2.21278 | -2.02465 | -1.80429 | -1.64297 |
| | 0.4 | -2.39494 | -2.12553 | -1.94872 | -1.74187 | -1.59062 |
| | 0.8 | -2.28174 | -2.03197 | -1.86833 | -1.67716 | -1.53756 |
| | 1.2 | -2.16677 | -1.93879 | -1.78975 | -1.61600 | -1.48938 |
| | 1.6 | -2.05790 | -1.85336 | -1.72018 | -1.56565 | -1.45391 |
| | 2.0 | -1.96356 | -1.78491 | -1.67118 | -1.54034 | -1.44329 |
| | 2.4 | -1.89898 | -1.75230 | -1.65460 | -1.53587 | -1.44461 |
| | 2.8 | -1.87858 | -1.74678 | -1.65493 | -1.54038 | -1.45086 |
| | 3.2 | -1.88279 | -1.75471 | -1.66381 | -1.54934 | -1.45935 |
| | 3.6 | -1.90003 | -1.77000 | -1.67742 | -1.56071 | -1.46899 |
| 4.0 | 0.0 | -2.47260 | -2.19598 | -2.01363 | -1.79941 | -1.64211 |
| | 0.4 | -2.38449 | -2.12257 | -1.95007 | -1.74757 | -1.59895 |
| | 0.8 | -2.28954 | -2.04454 | -1.88335 | -1.69425 | -1.55554 |
| | 1.2 | -2.19313 | -1.96680 | -1.81805 | -1.64369 | -1.51584 |
| | 1.6 | -2.10111 | -1.89475 | -1.75936 | -1.60089 | -1.48491 |
| | 2.0 | -2.01959 | -1.83462 | -1.71420 | -1.57376 | -1.46972 |
| | 2.4 | -1.95732 | -1.79723 | -1.69102 | -1.56309 | -1.46572 |
| | 2.8 | -1.92598 | -1.78249 | -1.68376 | -1.56193 | -1.46758 |
| | 3.2 | -1.91808 | -1.78175 | -1.68592 | -1.56612 | -1.47255 |
| | 3.6 | -1.92445 | -1.78945 | -1.69370 | -1.57345 | -1.47924 |
| | 4.0 | -1.93944 | -1.80229 | -1.70493 | -1.58267 | -1.48696 |
| 6.0 | 0.0 | -2.39890 | -2.15023 | -1.98445 | -1.78757 | -1.64139 |
| | 0.4 | -2.35471 | -2.11430 | -1.95395 | -1.76340 | -1.62181 |
| | 0.8 | -2.30864 | -2.07726 | -1.92282 | -1.73911 | -1.60243 |
| | 1.2 | -2.26218 | -2.04042 | -1.89223 | -1.71571 | -1.58416 |
| | 1.6 | -2.21693 | -2.00518 | -1.86344 | -1.69427 | -1.56788 |
| | 2.0 | -2.17457 | -1.97300 | -1.83773 | -1.67580 | -1.55438 |
| | 2.4 | -2.13675 | -1.94522 | -1.81618 | -1.66100 | -1.54404 |
| | 2.8 | -2.10500 | -1.92290 | -1.79946 | -1.65008 | -1.53679 |
| | 3.2 | -2.08040 | -1.90643 | -1.78757 | -1.64274 | -1.53224 |
| | 3.6 | -2.06312 | -1.89547 | -1.78001 | -1.63845 | -1.52989 |
| | 4.0 | -2.05253 | -1.88928 | -1.77609 | -1.63662 | -1.52925 |
| | 6.0 | -2.06421 | -1.90237 | -1.78915 | -1.64888 | -1.54049 |
| 12.0 | 0.0 | -2.34557 | -2.11900 | -1.96586 | -1.78175 | -1.64341 |
| | 0.4 | -2.33361 | -2.10954 | -1.95800 | -1.77568 | -1.63860 |
| | 0.8 | -2.32151 | -2.10002 | -1.95011 | -1.76963 | -1.63383 |
| | 1.2 | -2.30940 | -2.09052 | -1.94227 | -1.76365 | -1.62914 |
| | 1.6 | -2.29738 | -2.08116 | -1.93457 | -1.75781 | -1.62459 |
| | 2.0 | -2.28560 | -2.07202 | -1.92709 | -1.75216 | -1.62021 |
| | 2.4 | -2.27415 | -2.06319 | -1.91989 | -1.74676 | -1.61606 |
| | 2.8 | -2.26314 | -2.05475 | -1.91304 | -1.74167 | -1.61216 |
| | 3.2 | -2.25268 | -2.04678 | -1.90660 | -1.73691 | -1.60853 |
| | 3.6 | -2.24286 | -2.03934 | -1.90062 | -1.73252 | -1.60522 |
| | 4.0 | -2.23374 | -2.03248 | -1.89514 | -1.72852 | -1.60222 |
| | 6.0 | -2.20013 | -2.00768 | -1.87559 | -1.71457 | -1.59198 |
| | 12.0 | -2.19501 | -2.00553 | -1.87496 | -1.71529 | -1.59341 |
| INFINITY | | -2.32635 | -2.10836 | -1.95996 | -1.78046 | -1.64485 |

FRACTILES $w_\alpha$ RHO = 0.0

| DELTAS | | $\alpha$: 0.0750 | 0.1000 | 0.1500 | 0.2000 | 0.2500 |
|---|---|---|---|---|---|---|
| 0.0 | 0.0 | -1.26313 | -1.03438 | -0.72464 | -0.51688 | -0.36517 |
| 0.4 | 0.0 | -1.26596 | -1.03718 | -0.72708 | -0.51886 | -0.36670 |
| | 0.4 | -1.17759 | -0.97969 | -0.71245 | -0.53413 | -0.40486 |
| 0.8 | 0.0 | -1.28900 | -1.06161 | -0.75013 | -0.53855 | -0.38249 |
| | 0.4 | -1.14598 | -0.96442 | -0.71970 | -0.55712 | -0.44017 |
| | 0.8 | -1.07554 | -0.92808 | -0.73115 | -0.60266 | -0.51305 |
| 1.2 | 0.0 | -1.32595 | -1.10558 | -0.79754 | -0.58285 | -0.42063 |
| | 0.4 | -1.17286 | -0.99434 | -0.75155 | -0.58895 | -0.47169 |
| | 0.8 | -1.07625 | -0.94139 | -0.76277 | -0.64872 | -0.57352 |
| | 1.2 | -1.05736 | -0.94981 | -0.81180 | -0.73410 | -0.67363 |
| 1.6 | 0.0 | -1.35585 | -1.14641 | -0.84831 | -0.63518 | -0.46971 |
| | 0.4 | -1.22506 | -1.04738 | -0.80035 | -0.63027 | -0.50457 |
| | 0.8 | -1.12034 | -0.98359 | -0.80154 | -0.68630 | -0.61831 |
| | 1.2 | -1.08554 | -0.98702 | -0.86824 | -0.79036 | -0.69869 |
| | 1.6 | -1.11157 | -1.04116 | -0.93575 | -0.82659 | -0.72045 |
| 2.0 | 0.0 | -1.37494 | -1.17603 | -0.88924 | -0.68039 | -0.51488 |
| | 0.4 | -1.27336 | -1.09867 | -0.85045 | -0.67395 | -0.53869 |
| | 0.8 | -1.18002 | -1.03749 | -0.84274 | -0.71606 | -0.63482 |
| | 1.2 | -1.13345 | -1.03104 | -0.91109 | -0.80309 | -0.69722 |
| | 1.6 | -1.16394 | -1.09228 | -0.95749 | -0.83587 | -0.72149 |
| | 2.0 | -1.20993 | -1.12433 | -0.97744 | -0.84808 | -0.72794 |
| 2.4 | 0.0 | -1.38713 | -1.19676 | -0.91957 | -0.71494 | -0.55041 |
| | 0.4 | -1.30962 | -1.13863 | -0.89172 | -0.71177 | -0.56956 |
| | 0.8 | -1.23490 | -1.08848 | -0.88201 | -0.74042 | -0.63887 |
| | 1.2 | -1.18639 | -1.07377 | -0.93012 | -0.80519 | -0.68944 |
| | 1.6 | -1.20729 | -1.11784 | -0.96687 | -0.83576 | -0.71492 |
| | 2.0 | -1.23662 | -1.14334 | -0.98642 | -0.85046 | -0.72557 |
| | 2.4 | -1.25730 | -1.15972 | -0.99641 | -0.85584 | -0.72741 |
| 2.8 | 0.0 | -1.39559 | -1.21183 | -0.94202 | -0.74066 | -0.57694 |
| | 0.4 | -1.33583 | -1.16801 | -0.92276 | -0.74086 | -0.59400 |
| | 0.8 | -1.27787 | -1.12944 | -0.91503 | -0.76063 | -0.64133 |
| | 1.2 | -1.23497 | -1.11208 | -0.94406 | -0.80597 | -0.68198 |
| | 1.6 | -1.23936 | -1.13773 | -0.97264 | -0.83306 | -0.70654 |
| | 2.0 | -1.25815 | -1.15731 | -0.99070 | -0.84862 | -0.71949 |
| | 2.4 | -1.27393 | -1.17135 | -1.00126 | -0.85618 | -0.72451 |
| | 2.8 | -1.28719 | -1.18157 | -1.00700 | -0.85870 | -0.72457 |
| 3.2 | 0.0 | -1.40198 | -1.22326 | -0.95900 | -0.75995 | -0.59662 |
| | 0.4 | -1.35511 | -1.18971 | -0.94568 | -0.76226 | -0.61199 |
| | 0.8 | -1.30993 | -1.16026 | -0.94042 | -0.77684 | -0.64459 |
| | 1.2 | -1.27455 | -1.14403 | -0.95676 | -0.80770 | -0.67655 |
| | 1.6 | -1.26798 | -1.15572 | -0.97778 | -0.83048 | -0.69878 |
| | 2.0 | -1.27769 | -1.16938 | -0.99338 | -0.84550 | -0.71244 |
| | 2.4 | -1.28867 | -1.18072 | -1.00369 | -0.85419 | -0.71946 |
| | 2.8 | -1.29890 | -1.18956 | -1.00996 | -0.85833 | -0.72183 |
| | 3.2 | -1.30827 | -1.19657 | -1.01355 | -0.85950 | -0.72118 |

FRACTILES $W_\alpha$ RHO = 0.0

| DELTAS | | $\alpha$: 0.0750 | 0.1000 | 0.1500 | 0.2000 | 0.2500 |
|---|---|---|---|---|---|---|
| 3.6 | 0.0 | -1.40705 | -1.23221 | -0.97208 | -0.77461 | -0.61136 |
| | 0.4 | -1.36965 | -1.20604 | -0.96274 | -0.77796 | -0.62495 |
| | 0.8 | -1.33387 | -1.18321 | -0.95928 | -0.78928 | -0.64835 |
| | 1.2 | -1.30504 | -1.16919 | -0.96836 | -0.81041 | -0.67316 |
| | 1.6 | -1.29342 | -1.17228 | -0.98309 | -0.82885 | -0.69248 |
| | 2.0 | -1.29600 | -1.18065 | -0.99580 | -0.84248 | -0.70576 |
| | 2.4 | -1.30248 | -1.18911 | -1.00519 | -0.85141 | -0.71378 |
| | 2.8 | -1.30980 | -1.19640 | -1.01153 | -0.85655 | -0.71773 |
| | 3.2 | -1.31706 | -1.20245 | -1.01553 | -0.85891 | -0.71876 |
| | 3.6 | -1.32410 | -1.20758 | -1.01793 | -0.85940 | -0.71784 |
| 4.0 | 0.0 | -1.41122 | -1.23937 | -0.98233 | -0.78591 | -0.62254 |
| | 0.4 | -1.38089 | -1.21859 | -0.97565 | -0.78962 | -0.63438 |
| | 0.8 | -1.35202 | -1.20049 | -0.97329 | -0.79865 | -0.65199 |
| | 1.2 | -1.32828 | -1.18854 | -0.97847 | -0.81358 | -0.67130 |
| | 1.6 | -1.31518 | -1.18708 | -0.98848 | -0.82821 | -0.68770 |
| | 2.0 | -1.31293 | -1.19128 | -0.99840 | -0.84009 | -0.70000 |
| | 2.4 | -1.31559 | -1.19702 | -1.00651 | -0.84866 | -0.70829 |
| | 2.8 | -1.32020 | -1.20266 | -1.01251 | -0.85423 | -0.71318 |
| | 3.2 | -1.32548 | -1.20769 | -1.01663 | -0.85738 | -0.71542 |
| | 3.6 | -1.33095 | -1.21209 | -1.01932 | -0.85875 | -0.71574 |
| | 4.0 | -1.33646 | -1.21602 | -1.02101 | -0.85890 | -0.71476 |
| 6.0 | 0.0 | -1.42407 | -1.25996 | -1.01022 | -0.81554 | -0.65083 |
| | 0.4 | -1.41105 | -1.25163 | -1.00845 | -0.81823 | -0.65672 |
| | 0.8 | -1.39865 | -1.24415 | -1.00770 | -0.82190 | -0.66343 |
| | 1.2 | -1.38755 | -1.23801 | -1.00812 | -0.82642 | -0.67059 |
| | 1.6 | -1.37835 | -1.23356 | -1.00965 | -0.83142 | -0.67769 |
| | 2.0 | -1.37142 | -1.23083 | -1.01195 | -0.83647 | -0.68429 |
| | 2.4 | -1.36673 | -1.22959 | -1.01463 | -0.84117 | -0.69006 |
| | 2.8 | -1.36395 | -1.22944 | -1.01734 | -0.84527 | -0.69484 |
| | 3.2 | -1.36268 | -1.23003 | -1.01985 | -0.84864 | -0.69858 |
| | 3.6 | -1.36252 | -1.23105 | -1.02203 | -0.85127 | -0.70133 |
| | 4.0 | -1.36317 | -1.23233 | -1.02384 | -0.85319 | -0.70319 |
| | 6.0 | -1.37239 | -1.23977 | -1.02835 | -0.85539 | -0.70353 |
| 12.0 | 0.0 | -1.43510 | -1.27567 | -1.02959 | -0.83496 | -0.66854 |
| | 0.4 | -1.43204 | -1.27381 | -1.02933 | -0.83574 | -0.67005 |
| | 0.8 | -1.42903 | -1.27201 | -1.02913 | -0.83658 | -0.67160 |
| | 1.2 | -1.42611 | -1.27029 | -1.02900 | -0.83747 | -0.67318 |
| | 1.6 | -1.42331 | -1.26868 | -1.02894 | -0.83839 | -0.67476 |
| | 2.0 | -1.42065 | -1.26718 | -1.02895 | -0.83933 | -0.67634 |
| | 2.4 | -1.41817 | -1.26580 | -1.02901 | -0.84029 | -0.67789 |
| | 2.8 | -1.41586 | -1.26456 | -1.02913 | -0.84124 | -0.67939 |
| | 3.2 | -1.41376 | -1.26346 | -1.02929 | -0.84218 | -0.68083 |
| | 3.6 | -1.41186 | -1.26249 | -1.02950 | -0.84309 | -0.68219 |
| | 4.0 | -1.41018 | -1.26166 | -1.02973 | -0.84396 | -0.68347 |
| | 6.0 | -1.40473 | -1.25927 | -1.03107 | -0.84746 | -0.68829 |
| | 12.0 | -1.40683 | -1.26159 | -1.03335 | -0.84943 | -0.68988 |
| INFINITY | | -1.43953 | -1.28155 | -1.03643 | -0.84162 | -0.67449 |

FRACTILES $W_\alpha$ RHO = 0.0

| DELTAS | | $\alpha$: 0.3000 | 0.3500 | 0.4000 | 0.4500 | 0.5000 |
|---|---|---|---|---|---|---|
| 0.0 | 0.0 | -0.24939 | -0.15917 | -0.08873 | -0.03519 | 0.0 |
| 0.4 | 0.0 | -0.25051 | -0.15992 | -0.08917 | -0.03538 | 0.0 |
| | 0.4 | -0.30733 | -0.23280 | -0.17688 | -0.13990 | -0.10503 |
| 0.8 | 0.0 | -0.26239 | -0.16812 | -0.09406 | -0.03744 | 0.0 |
| | 0.4 | -0.35315 | -0.28854 | -0.24418 | -0.20997 | -0.15653 |
| | 0.8 | -0.45073 | -0.41328 | -0.36354 | -0.29832 | -0.22000 |
| 1.2 | 0.0 | -0.29289 | -0.19036 | -0.10802 | -0.04365 | 0.0 |
| | 0.4 | -0.38496 | -0.32246 | -0.28329 | -0.22930 | -0.15970 |
| | 0.8 | -0.52823 | -0.46746 | -0.39334 | -0.30898 | -0.21494 |
| | 1.2 | -0.59542 | -0.50940 | -0.41721 | -0.31887 | -0.21362 |
| 1.6 | 0.0 | -0.33545 | -0.22400 | -0.13097 | -0.05487 | 0.0 |
| | 0.4 | -0.40982 | -0.34211 | -0.29053 | -0.21918 | -0.13585 |
| | 0.8 | -0.54966 | -0.46827 | -0.37970 | -0.28478 | -0.18304 |
| | 1.2 | -0.60378 | -0.50634 | -0.40579 | -0.30114 | -0.19115 |
| | 1.6 | -0.61577 | -0.51092 | -0.40447 | -0.29502 | -0.18107 |
| 2.0 | 0.0 | -0.37733 | -0.25970 | -0.15767 | -0.06960 | 0.0 |
| | 0.4 | -0.43236 | -0.35380 | -0.28321 | -0.19823 | -0.10541 |
| | 0.8 | -0.54529 | -0.45149 | -0.35456 | -0.25387 | -0.14829 |
| | 1.2 | -0.59312 | -0.48920 | -0.38397 | -0.27600 | -0.16378 |
| | 1.6 | -0.61077 | -0.50135 | -0.39138 | -0.27925 | -0.16337 |
| | 2.0 | -0.61269 | -0.49960 | -0.38664 | -0.27207 | -0.15421 |
| 2.4 | 0.0 | -0.41136 | -0.28996 | -0.18170 | -0.08423 | 0.0 |
| | 0.4 | -0.45290 | -0.36079 | -0.27286 | -0.17823 | -0.07902 |
| | 0.8 | -0.53589 | -0.43368 | -0.33103 | -0.22643 | -0.11835 |
| | 1.2 | -0.57865 | -0.47000 | -0.36143 | -0.25121 | -0.13770 |
| | 1.6 | -0.59956 | -0.48674 | -0.37432 | -0.26051 | -0.14362 |
| | 2.0 | -0.60677 | -0.49096 | -0.37595 | -0.25987 | -0.14098 |
| | 2.4 | -0.60580 | -0.48775 | -0.37092 | -0.25339 | -0.13339 |
| 2.8 | 0.0 | -0.43685 | -0.31279 | -0.20010 | -0.09588 | 0.0 |
| | 0.4 | -0.46980 | -0.36467 | -0.26478 | -0.16311 | -0.05912 |
| | 0.8 | -0.52841 | -0.41962 | -0.31235 | -0.20454 | -0.09437 |
| | 1.2 | -0.56552 | -0.45287 | -0.34151 | -0.22947 | -0.11494 |
| | 1.6 | -0.58718 | -0.47152 | -0.35714 | -0.24209 | -0.12458 |
| | 2.0 | -0.59763 | -0.47961 | -0.36303 | -0.24593 | -0.12649 |
| | 2.4 | -0.60049 | -0.48062 | -0.36243 | -0.24394 | -0.12333 |
| | 2.8 | -0.59863 | -0.47723 | -0.35784 | -0.23841 | -0.11711 |
| 3.2 | 0.0 | -0.45554 | -0.32927 | -0.21315 | -0.10398 | 0.0 |
| | 0.4 | -0.48241 | -0.36791 | -0.25977 | -0.15273 | -0.04486 |
| | 0.8 | -0.52389 | -0.40964 | -0.29842 | -0.18779 | -0.07573 |
| | 1.2 | -0.55510 | -0.43890 | -0.32506 | -0.21137 | -0.09591 |
| | 1.6 | -0.57582 | -0.45763 | -0.34153 | -0.22542 | -0.10742 |
| | 2.0 | -0.58782 | -0.46787 | -0.34997 | -0.23207 | -0.11230 |
| | 2.4 | -0.59325 | -0.47181 | -0.35253 | -0.23335 | -0.11239 |
| | 2.8 | -0.59412 | -0.47142 | -0.35107 | -0.23098 | -0.10928 |
| | 3.2 | -0.59207 | -0.46826 | -0.34705 | -0.22632 | -0.10415 |

FRACTILES $w_\alpha$ RHO = 0.0

| DELTAS | | α:0.3000 | 0.3500 | 0.4000 | 0.4500 | 0.5000 |
|---|---|---|---|---|---|---|
| 3.6 | 0.0 | -0.46928 | -0.34111 | -0.22222 | -0.10934 | 0.0 |
| | 0.4 | -0.49141 | -0.37078 | -0.25699 | -0.14581 | -0.03478 |
| | 0.8 | -0.52162 | -0.40288 | -0.28829 | -0.17519 | -0.06138 |
| | 1.2 | -0.54736 | -0.42801 | -0.31191 | -0.19669 | -0.08032 |
| | 1.6 | -0.56624 | -0.44574 | -0.32806 | -0.21097 | -0.09251 |
| | 2.0 | -0.57859 | -0.45687 | -0.33780 | -0.21920 | -0.09916 |
| | 2.4 | -0.58554 | -0.46268 | -0.34245 | -0.22271 | -0.10154 |
| | 2.8 | -0.58838 | -0.46449 | -0.34333 | -0.22274 | -0.10080 |
| | 3.2 | -0.58830 | -0.46350 | -0.34158 | -0.22036 | -0.09792 |
| | 3.6 | -0.58630 | -0.46067 | -0.33811 | -0.21642 | -0.09366 |
| 4.0 | 0.0 | -0.47950 | -0.34970 | -0.22859 | -0.11289 | 0.0 |
| | 0.4 | -0.49782 | -0.37319 | -0.25552 | -0.14118 | -0.02761 |
| | 0.8 | -0.52071 | -0.39836 | -0.28097 | -0.16573 | -0.05037 |
| | 1.2 | -0.54178 | -0.41968 | -0.30156 | -0.18493 | -0.06768 |
| | 1.6 | -0.55851 | -0.43590 | -0.31676 | -0.19873 | -0.07980 |
| | 2.0 | -0.57048 | -0.44711 | -0.32694 | -0.20769 | -0.08739 |
| | 2.4 | -0.57814 | -0.45396 | -0.33286 | -0.21262 | -0.09129 |
| | 2.8 | -0.58226 | -0.45729 | -0.33540 | -0.21439 | -0.09231 |
| | 3.2 | -0.58368 | -0.45796 | -0.33541 | -0.21381 | -0.09120 |
| | 3.6 | -0.58315 | -0.45673 | -0.33362 | -0.21156 | -0.08860 |
| | 4.0 | -0.58130 | -0.45423 | -0.33062 | -0.20820 | -0.08501 |
| 6.0 | 0.0 | -0.50444 | -0.36982 | -0.24279 | -0.12032 | 0.0 |
| | 0.4 | -0.51266 | -0.37969 | -0.25372 | -0.13180 | -0.01157 |
| | 0.8 | -0.52144 | -0.38981 | -0.26459 | -0.14293 | -0.02252 |
| | 1.2 | -0.53027 | -0.39960 | -0.27482 | -0.15317 | -0.03239 |
| | 1.6 | -0.53860 | -0.40856 | -0.28396 | -0.16214 | -0.04089 |
| | 2.0 | -0.54603 | -0.41634 | -0.29174 | -0.16965 | -0.04791 |
| | 2.4 | -0.55231 | -0.42278 | -0.29807 | -0.17565 | -0.05343 |
| | 2.8 | -0.55736 | -0.42783 | -0.30294 | -0.18021 | -0.05754 |
| | 3.2 | -0.56119 | -0.43157 | -0.30646 | -0.18342 | -0.06038 |
| | 3.6 | -0.56389 | -0.43411 | -0.30878 | -0.18546 | -0.06210 |
| | 4.0 | -0.56560 | -0.43562 | -0.31006 | -0.18650 | -0.06288 |
| | 6.0 | -0.56441 | -0.43317 | -0.30656 | -0.18216 | -0.05788 |
| 12.0 | 0.0 | -0.51943 | -0.38149 | -0.25075 | -0.12435 | 0.0 |
| | 0.4 | -0.52146 | -0.38388 | -0.25338 | -0.12712 | -0.00280 |
| | 0.8 | -0.52351 | -0.38628 | -0.25601 | -0.12986 | -0.00558 |
| | 1.2 | -0.52557 | -0.38867 | -0.25861 | -0.13257 | -0.00829 |
| | 1.6 | -0.52762 | -0.39102 | -0.26115 | -0.13520 | -0.01092 |
| | 2.0 | -0.52962 | -0.39331 | -0.26360 | -0.13773 | -0.01344 |
| | 2.4 | -0.53157 | -0.39552 | -0.26596 | -0.14014 | -0.01583 |
| | 2.8 | -0.53343 | -0.39761 | -0.26818 | -0.14241 | -0.01807 |
| | 3.2 | -0.53520 | -0.39959 | -0.27027 | -0.14453 | -0.02015 |
| | 3.6 | -0.53686 | -0.40142 | -0.27220 | -0.14648 | -0.02206 |
| | 4.0 | -0.53839 | -0.40311 | -0.27396 | -0.14825 | -0.02379 |
| | 6.0 | -0.54402 | -0.40919 | -0.28022 | -0.15447 | -0.02976 |
| | 12.0 | -0.54520 | -0.40995 | -0.28057 | -0.15441 | -0.02933 |
| INFINITY | | -0.52440 | -0.38532 | -0.25335 | -0.12566 | 0.0 |

FRACTILES $W_\alpha$ RHO = 0.0

| DELTAS | | $\alpha$: 0.5500 | 0.6000 | 0.6500 | 0.7000 | 0.7500 |
|---|---|---|---|---|---|---|
| 0.0 | 0.0 | 0.03519 | 0.08873 | 0.15917 | 0.24939 | 0.36517 |
| 0.4 | 0.0 | 0.03538 | 0.08917 | 0.15992 | 0.25051 | 0.36670 |
| | 0.4 | -0.05171 | 0.01754 | 0.10450 | 0.21304 | 0.34974 |
| 0.8 | 0.0 | 0.03744 | 0.09406 | 0.16812 | 0.26239 | 0.38249 |
| | 0.4 | -0.08777 | -0.00328 | 0.09907 | 0.22311 | 0.37526 |
| | 0.8 | -0.12862 | -0.02301 | 0.09916 | 0.24177 | 0.41125 |
| 1.2 | 0.0 | 0.04365 | 0.10802 | 0.19036 | 0.29289 | 0.42063 |
| | 0.4 | -0.07638 | 0.02110 | 0.13463 | 0.26767 | 0.42613 |
| | 0.8 | -0.11066 | 0.00538 | 0.13567 | 0.28414 | 0.45702 |
| | 1.2 | -0.10016 | 0.02343 | 0.15989 | 0.31325 | 0.48970 |
| 1.6 | 0.0 | 0.05487 | 0.13097 | 0.22400 | 0.33545 | 0.46971 |
| | 0.4 | -0.04177 | 0.06374 | 0.18263 | 0.31833 | 0.47643 |
| | 0.8 | -0.07333 | 0.04613 | 0.17798 | 0.32606 | 0.49636 |
| | 1.2 | -0.07423 | 0.05166 | 0.18930 | 0.34262 | 0.51762 |
| | 1.6 | -0.06092 | 0.06753 | 0.20705 | 0.36155 | 0.53688 |
| 2.0 | 0.0 | 0.06960 | 0.15767 | 0.25970 | 0.37733 | 0.51488 |
| | 0.4 | -0.00482 | 0.10479 | 0.22565 | 0.36119 | 0.51681 |
| | 0.8 | -0.03633 | 0.08395 | 0.21521 | 0.36117 | 0.52750 |
| | 1.2 | -0.04564 | 0.08052 | 0.21742 | 0.36887 | 0.54062 |
| | 1.6 | -0.04195 | 0.08710 | 0.22654 | 0.38016 | 0.55366 |
| | 2.0 | -0.03127 | 0.09888 | 0.23897 | 0.39274 | 0.56576 |
| 2.4 | 0.0 | 0.08423 | 0.18170 | 0.28996 | 0.41136 | 0.55041 |
| | 0.4 | 0.02576 | 0.13782 | 0.25958 | 0.39447 | 0.54767 |
| | 0.8 | -0.00508 | 0.11541 | 0.24572 | 0.38948 | 0.55210 |
| | 1.2 | -0.01910 | 0.10667 | 0.24228 | 0.39143 | 0.55962 |
| | 1.6 | -0.02184 | 0.10695 | 0.24545 | 0.39735 | 0.56816 |
| | 2.0 | -0.01747 | 0.11280 | 0.25251 | 0.40534 | 0.57673 |
| | 2.4 | -0.00907 | 0.12170 | 0.26159 | 0.41421 | 0.58493 |
| 2.8 | 0.0 | 0.09588 | 0.20010 | 0.31279 | 0.43685 | 0.57694 |
| | 0.4 | 0.04888 | 0.16290 | 0.28544 | 0.41990 | 0.57129 |
| | 0.8 | 0.02000 | 0.14067 | 0.27023 | 0.41219 | 0.57175 |
| | 1.2 | 0.00391 | 0.12919 | 0.26353 | 0.41053 | 0.57546 |
| | 1.6 | -0.00277 | 0.12546 | 0.26276 | 0.41274 | 0.58072 |
| | 2.0 | -0.00288 | 0.12703 | 0.26590 | 0.41732 | 0.58660 |
| | 2.4 | 0.00127 | 0.13199 | 0.27145 | 0.42324 | 0.59260 |
| | 2.8 | 0.00796 | 0.13890 | 0.27834 | 0.42981 | 0.59849 |
| 3.2 | 0.0 | 0.10398 | 0.21315 | 0.32927 | 0.45554 | 0.59662 |
| | 0.4 | 0.06588 | 0.18166 | 0.30504 | 0.43939 | 0.58957 |
| | 0.8 | 0.03973 | 0.16072 | 0.28980 | 0.43043 | 0.58761 |
| | 1.2 | 0.02321 | 0.14812 | 0.28141 | 0.42657 | 0.58873 |
| | 1.6 | 0.01434 | 0.14198 | 0.27812 | 0.42628 | 0.59161 |
| | 2.0 | 0.01122 | 0.14061 | 0.27849 | 0.42837 | 0.59544 |
| | 2.4 | 0.01221 | 0.14259 | 0.28135 | 0.43201 | 0.59970 |
| | 2.8 | 0.01593 | 0.14676 | 0.28581 | 0.43657 | 0.60413 |
| | 3.2 | 0.02133 | 0.15226 | 0.29120 | 0.44161 | 0.60855 |

FRACTILES $W_\alpha$ RHO = 0.0

| DELTAS | | $\alpha$: 0.5500 | 0.6000 | 0.6500 | 0.7000 | 0.7500 |
|---|---|---|---|---|---|---|
| 3.6 | 0.0 | 0.10934 | 0.22222 | 0.34111 | 0.46928 | 0.61136 |
| | 0.4 | 0.07830 | 0.19569 | 0.31996 | 0.45443 | 0.60385 |
| | 0.8 | 0.05515 | 0.17657 | 0.30543 | 0.44512 | 0.60050 |
| | 1.2 | 0.03914 | 0.16384 | 0.29632 | 0.44002 | 0.59991 |
| | 1.6 | 0.02924 | 0.15638 | 0.29151 | 0.43807 | 0.60108 |
| | 2.0 | 0.02423 | 0.15308 | 0.28997 | 0.43838 | 0.60334 |
| | 2.4 | 0.02295 | 0.15288 | 0.29084 | 0.44028 | 0.60623 |
| | 2.8 | 0.02438 | 0.15491 | 0.29338 | 0.44323 | 0.60946 |
| | 3.2 | 0.02764 | 0.15844 | 0.29703 | 0.44685 | 0.61286 |
| | 3.6 | 0.03208 | 0.16290 | 0.30136 | 0.45084 | 0.61629 |
| 4.0 | 0.0 | 0.11289 | 0.22859 | 0.34970 | 0.47950 | 0.62254 |
| | 0.4 | 0.08745 | 0.20626 | 0.33140 | 0.46614 | 0.61513 |
| | 0.8 | 0.06718 | 0.18910 | 0.31794 | 0.45700 | 0.61104 |
| | 1.2 | 0.05219 | 0.17681 | 0.30872 | 0.45129 | 0.60934 |
| | 1.6 | 0.04200 | 0.16877 | 0.30307 | 0.44828 | 0.60931 |
| | 2.0 | 0.03588 | 0.16425 | 0.30027 | 0.44734 | 0.61039 |
| | 2.4 | 0.03306 | 0.16253 | 0.29969 | 0.44793 | 0.61220 |
| | 2.8 | 0.03276 | 0.16293 | 0.30075 | 0.44962 | 0.61446 |
| | 3.2 | 0.03433 | 0.16487 | 0.30299 | 0.45207 | 0.61700 |
| | 3.6 | 0.03718 | 0.16787 | 0.30603 | 0.45501 | 0.61968 |
| | 4.0 | 0.04088 | 0.17156 | 0.30957 | 0.45825 | 0.62243 |
| 6.0 | 0.0 | 0.12032 | 0.24279 | 0.36982 | 0.50444 | 0.65083 |
| | 0.4 | 0.10912 | 0.23240 | 0.36071 | 0.49712 | 0.64591 |
| | 0.8 | 0.09874 | 0.22300 | 0.35268 | 0.49089 | 0.64200 |
| | 1.2 | 0.08957 | 0.21485 | 0.34587 | 0.48579 | 0.63903 |
| | 1.6 | 0.08180 | 0.20806 | 0.34033 | 0.48178 | 0.63690 |
| | 2.0 | 0.07549 | 0.20264 | 0.33601 | 0.47878 | 0.63549 |
| | 2.4 | 0.07060 | 0.19853 | 0.33282 | 0.47668 | 0.63467 |
| | 2.8 | 0.06703 | 0.19560 | 0.33064 | 0.47535 | 0.63434 |
| | 3.2 | 0.06464 | 0.19372 | 0.32933 | 0.47469 | 0.63440 |
| | 3.6 | 0.06327 | 0.19273 | 0.32876 | 0.47457 | 0.63478 |
| | 4.0 | 0.06276 | 0.19250 | 0.32881 | 0.47491 | 0.63540 |
| | 6.0 | 0.06823 | 0.19825 | 0.33462 | 0.48052 | 0.64049 |
| 12.0 | 0.0 | 0.12435 | 0.25075 | 0.38149 | 0.51943 | 0.66854 |
| | 0.4 | 0.12160 | 0.24815 | 0.37914 | 0.51745 | 0.66709 |
| | 0.8 | 0.11889 | 0.24560 | 0.37685 | 0.51555 | 0.66570 |
| | 1.2 | 0.11626 | 0.24313 | 0.37465 | 0.51372 | 0.66440 |
| | 1.6 | 0.11371 | 0.24076 | 0.37255 | 0.51200 | 0.66319 |
| | 2.0 | 0.11129 | 0.23851 | 0.37057 | 0.51039 | 0.66208 |
| | 2.4 | 0.10900 | 0.23640 | 0.36872 | 0.50889 | 0.66106 |
| | 2.8 | 0.10686 | 0.23444 | 0.36700 | 0.50752 | 0.66014 |
| | 3.2 | 0.10488 | 0.23263 | 0.36544 | 0.50628 | 0.65932 |
| | 3.6 | 0.10308 | 0.23099 | 0.36403 | 0.50517 | 0.65860 |
| | 4.0 | 0.10145 | 0.22952 | 0.36276 | 0.50418 | 0.65797 |
| | 6.0 | 0.09590 | 0.22457 | 0.35861 | 0.50104 | 0.65611 |
| | 12.0 | 0.09669 | 0.22568 | 0.36001 | 0.50268 | 0.65791 |
| INFINITY | | 0.12566 | 0.25335 | 0.38532 | 0.52440 | 0.67449 |

FRACTILES $W_\alpha$ RHO = 0.0

| DELTAS | | $\alpha$:0.8000 | 0.8500 | 0.9000 | 0.9250 | 0.9500 |
|---|---|---|---|---|---|---|
| 0.0 | 0.0 | 0.51688 | 0.72464 | 1.03438 | 1.26313 | 1.59510 |
| 0.4 | 0.0 | 0.51886 | 0.72708 | 1.03718 | 1.26596 | 1.59762 |
| | 0.4 | 0.52608 | 0.76398 | 1.11299 | 1.36730 | 1.73211 |
| 0.8 | 0.0 | 0.53855 | 0.75013 | 1.06161 | 1.28900 | 1.61556 |
| | 0.4 | 0.56667 | 0.81847 | 1.17835 | 1.43532 | 1.79821 |
| | 0.8 | 0.61858 | 0.88458 | 1.25591 | 1.51667 | 1.88051 |
| 1.2 | 0.0 | 0.58285 | 0.79754 | 1.10558 | 1.32595 | 1.63746 |
| | 0.4 | 0.62023 | 0.86944 | 1.21746 | 1.46187 | 1.80290 |
| | 0.8 | 0.66484 | 0.92728 | 1.28826 | 1.53902 | 1.88613 |
| | 1.2 | 0.69955 | 0.96192 | 1.31922 | 1.56555 | 1.90451 |
| 1.6 | 0.0 | 0.63518 | 0.84831 | 1.14641 | 1.35585 | 1.64809 |
| | 0.4 | 0.66646 | 0.90636 | 1.23613 | 1.46510 | 1.78189 |
| | 0.8 | 0.69878 | 0.95172 | 1.29599 | 1.53325 | 1.85964 |
| | 1.2 | 0.72415 | 0.98045 | 1.32675 | 1.56403 | 1.88895 |
| | 1.6 | 0.74267 | 0.99658 | 1.33759 | 1.57012 | 1.88726 |
| 2.0 | 0.0 | 0.68039 | 0.88924 | 1.17603 | 1.37494 | 1.64991 |
| | 0.4 | 0.70146 | 0.93131 | 1.24481 | 1.46025 | 1.75636 |
| | 0.8 | 0.72354 | 0.96650 | 1.29436 | 1.51882 | 1.82597 |
| | 1.2 | 0.74205 | 0.99043 | 1.32380 | 1.55101 | 1.86082 |
| | 1.6 | 0.75630 | 1.00511 | 1.33750 | 1.56316 | 1.86987 |
| | 2.0 | 0.76711 | 1.01338 | 1.34099 | 1.56262 | 1.86299 |
| 2.4 | 0.0 | 0.71494 | 0.91957 | 1.19676 | 1.38713 | 1.64834 |
| | 0.4 | 0.72768 | 0.95014 | 1.24955 | 1.45413 | 1.73370 |
| | 0.8 | 0.74240 | 0.97659 | 1.29029 | 1.50380 | 1.79462 |
| | 1.2 | 0.75580 | 0.99637 | 1.31734 | 1.53507 | 1.83079 |
| | 1.6 | 0.76681 | 1.00962 | 1.33245 | 1.55077 | 1.84654 |
| | 2.0 | 0.77551 | 1.01776 | 1.33877 | 1.55524 | 1.84784 |
| | 2.4 | 0.78240 | 1.02239 | 1.33939 | 1.55261 | 1.84015 |
| 2.8 | 0.0 | 0.74066 | 0.94202 | 1.21183 | 1.39559 | 1.64616 |
| | 0.4 | 0.74773 | 0.96406 | 1.25283 | 1.44888 | 1.71544 |
| | 0.8 | 0.75732 | 0.98427 | 1.28630 | 1.49080 | 1.76818 |
| | 1.2 | 0.76689 | 1.00047 | 1.31046 | 1.51983 | 1.80322 |
| | 1.6 | 0.77530 | 1.01219 | 1.32576 | 1.53704 | 1.82246 |
| | 2.0 | 0.78229 | 1.02000 | 1.33385 | 1.54486 | 1.82936 |
| | 2.4 | 0.78800 | 1.02484 | 1.33674 | 1.54599 | 1.82762 |
| | 2.8 | 0.79271 | 1.02761 | 1.33621 | 1.54283 | 1.82040 |
| 3.2 | 0.0 | 0.75995 | 0.95900 | 1.22326 | 1.40198 | 1.64431 |
| | 0.4 | 0.76339 | 0.97503 | 1.25555 | 1.44490 | 1.70121 |
| | 0.8 | 0.76942 | 0.99054 | 1.28312 | 1.48028 | 1.74671 |
| | 1.2 | 0.77611 | 1.00372 | 1.30433 | 1.50656 | 1.77941 |
| | 1.6 | 0.78245 | 1.01390 | 1.31902 | 1.52394 | 1.79999 |
| | 2.0 | 0.78800 | 1.02118 | 1.32799 | 1.53369 | 1.81038 |
| | 2.4 | 0.79271 | 1.02604 | 1.33245 | 1.53753 | 1.81299 |
| | 2.8 | 0.79668 | 1.02907 | 1.33365 | 1.53716 | 1.81011 |
| | 3.2 | 0.80007 | 1.03083 | 1.33267 | 1.53402 | 1.80369 |

FRACTILES $W_{\alpha}$ RHO = 0.0

| DELTAS | | $\alpha$:0.8000 | 0.8500 | 0.9000 | 0.9250 | 0.9500 |
|---|---|---|---|---|---|---|
| 3.6 | 0.0 | 0.77461 | 0.97208 | 1.23221 | 1.40705 | 1.64297 |
| | 0.4 | 0.77578 | 0.98389 | 1.25798 | 1.44208 | 1.69024 |
| | 0.8 | 0.77937 | 0.99586 | 1.28081 | 1.47201 | 1.72951 |
| | 1.2 | 0.78393 | 1.00654 | 1.29926 | 1.49547 | 1.75942 |
| | 1.6 | 0.78860 | 1.01526 | 1.31294 | 1.51223 | 1.78004 |
| | 2.0 | 0.79294 | 1.02187 | 1.32213 | 1.52290 | 1.79237 |
| | 2.4 | 0.79678 | 1.02659 | 1.32755 | 1.52852 | 1.79795 |
| | 2.8 | 0.80012 | 1.02975 | 1.33002 | 1.53026 | 1.79839 |
| | 3.2 | 0.80302 | 1.03174 | 1.33036 | 1.52922 | 1.79519 |
| | 3.6 | 0.80556 | 1.03290 | 1.32923 | 1.52631 | 1.78956 |
| 4.0 | 0.0 | 0.78591 | 0.98233 | 1.23937 | 1.41122 | 1.64211 |
| | 0.4 | 0.78571 | 0.99114 | 1.26022 | 1.44013 | 1.68179 |
| | 0.8 | 0.78763 | 1.00044 | 1.27923 | 1.46558 | 1.71577 |
| | 1.2 | 0.79062 | 1.00910 | 1.29522 | 1.48638 | 1.74283 |
| | 1.6 | 0.79398 | 1.01649 | 1.30771 | 1.50212 | 1.76274 |
| | 2.0 | 0.79731 | 1.02239 | 1.31671 | 1.51303 | 1.77597 |
| | 2.4 | 0.80040 | 1.02684 | 1.32262 | 1.51971 | 1.78346 |
| | 2.8 | 0.80317 | 1.03000 | 1.32596 | 1.52295 | 1.78632 |
| | 3.2 | 0.80564 | 1.03214 | 1.32731 | 1.52355 | 1.78566 |
| | 3.6 | 0.80783 | 1.03350 | 1.32719 | 1.52223 | 1.78247 |
| | 4.0 | 0.80981 | 1.03429 | 1.32605 | 1.51959 | 1.77757 |
| 6.0 | 0.0 | 0.81554 | 1.01022 | 1.25996 | 1.42407 | 1.64139 |
| | 0.4 | 0.81380 | 1.01276 | 1.26862 | 1.43706 | 1.66036 |
| | 0.8 | 0.81287 | 1.01576 | 1.27716 | 1.44945 | 1.67804 |
| | 1.2 | 0.81258 | 1.01896 | 1.28518 | 1.46079 | 1.69391 |
| | 1.6 | 0.81278 | 1.02212 | 1.29240 | 1.47078 | 1.70765 |
| | 2.0 | 0.81329 | 1.02507 | 1.29864 | 1.47925 | 1.71909 |
| | 2.4 | 0.81402 | 1.02772 | 1.30383 | 1.48613 | 1.72822 |
| | 2.8 | 0.81487 | 1.02999 | 1.30797 | 1.49148 | 1.73515 |
| | 3.2 | 0.81577 | 1.03189 | 1.31110 | 1.49540 | 1.74005 |
| | 3.6 | 0.81668 | 1.03341 | 1.31333 | 1.49804 | 1.74316 |
| | 4.0 | 0.81759 | 1.03460 | 1.31478 | 1.49959 | 1.74475 |
| | 6.0 | 0.82171 | 1.03704 | 1.31425 | 1.49662 | 1.73798 |
| 12.0 | 0.0 | 0.83496 | 1.02959 | 1.27567 | 1.43510 | 1.64341 |
| | 0.4 | 0.83424 | 1.02991 | 1.27757 | 1.43818 | 1.64820 |
| | 0.8 | 0.83358 | 1.03028 | 1.27949 | 1.44125 | 1.65291 |
| | 1.2 | 0.83300 | 1.03069 | 1.28141 | 1.44427 | 1.65752 |
| | 1.6 | 0.83248 | 1.03114 | 1.28330 | 1.44721 | 1.66197 |
| | 2.0 | 0.83203 | 1.03162 | 1.28514 | 1.45005 | 1.66622 |
| | 2.4 | 0.83165 | 1.03210 | 1.28692 | 1.45275 | 1.67024 |
| | 2.8 | 0.83133 | 1.03260 | 1.28861 | 1.45530 | 1.67401 |
| | 3.2 | 0.83107 | 1.03309 | 1.29020 | 1.45768 | 1.67750 |
| | 3.6 | 0.83086 | 1.03357 | 1.29168 | 1.45988 | 1.68071 |
| | 4.0 | 0.83070 | 1.03404 | 1.29305 | 1.46189 | 1.68361 |
| | 6.0 | 0.83047 | 1.03597 | 1.29802 | 1.46901 | 1.69369 |
| | 12.0 | 0.83233 | 1.03770 | 1.29928 | 1.46973 | 1.69345 |
| INFINITY | | 0.84162 | 1.03643 | 1.28155 | 1.43953 | 1.64485 |

FRACTILES $W_\alpha$ RHO = 0.0

| DELTAS | | $\alpha$:0.9625 | 0.9750 | 0.9825 | 0.9900 | 0.9925 |
|---|---|---|---|---|---|---|
| 0.0 | 0.0 | 1.83610 | 2.18195 | 2.49114 | 2.98381 | 3.24009 |
| 0.4 | 0.0 | 1.83815 | 2.18297 | 2.49091 | 2.98097 | 3.23559 |
| | 0.4 | 1.99425 | 2.36702 | 2.69730 | 3.21854 | 3.48752 |
| 0.8 | 0.0 | 1.85037 | 2.18443 | 2.48050 | 2.94788 | 3.18911 |
| | 0.4 | 2.05568 | 2.41805 | 2.73610 | 3.23349 | 3.48837 |
| | 0.8 | 2.13622 | 2.49350 | 2.80500 | 3.28907 | 3.53591 |
| 1.2 | 0.0 | 1.85857 | 2.16987 | 2.44315 | 2.87056 | 3.08959 |
| | 0.4 | 2.04257 | 2.37741 | 2.66933 | 3.12294 | 3.35423 |
| | 0.8 | 2.12852 | 2.46541 | 2.75774 | 3.20985 | 3.43951 |
| | 1.2 | 2.14006 | 2.46613 | 2.74798 | 3.18221 | 3.40210 |
| 1.6 | 0.0 | 1.85345 | 2.14030 | 2.39034 | 2.77879 | 2.97682 |
| | 0.4 | 2.00304 | 2.31031 | 2.57687 | 2.98903 | 3.19835 |
| | 0.8 | 2.08639 | 2.40023 | 2.67147 | 3.08928 | 3.30084 |
| | 1.2 | 2.11381 | 2.42401 | 2.69126 | 3.10160 | 3.30884 |
| | 1.6 | 2.10600 | 2.40689 | 2.66540 | 3.06122 | 3.26065 |
| 2.0 | 0.0 | 1.84171 | 2.10805 | 2.33898 | 2.69584 | 2.87702 |
| | 0.4 | 1.96193 | 2.24628 | 2.49192 | 2.87011 | 3.06154 |
| | 0.8 | 2.03842 | 2.33138 | 2.58370 | 2.97098 | 3.16652 |
| | 1.2 | 2.07443 | 2.36822 | 2.62058 | 3.00692 | 3.20155 |
| | 1.6 | 2.08078 | 2.37014 | 2.61814 | 2.99688 | 3.18731 |
| | 2.0 | 2.06902 | 2.35109 | 2.59233 | 2.95994 | 3.14444 |
| 2.4 | 0.0 | 1.82945 | 2.07973 | 2.29574 | 2.62804 | 2.79613 |
| | 0.4 | 1.92687 | 2.19302 | 2.42207 | 2.77341 | 2.95071 |
| | 0.8 | 1.99499 | 2.27039 | 2.50683 | 2.86859 | 3.05077 |
| | 1.2 | 2.03404 | 2.31278 | 2.55160 | 2.91618 | 3.09945 |
| | 1.6 | 2.04937 | 2.32700 | 2.56440 | 2.92611 | 3.10762 |
| | 2.0 | 2.04807 | 2.32165 | 2.55518 | 2.91033 | 3.08828 |
| | 2.4 | 2.03655 | 2.30444 | 2.53274 | 2.87932 | 3.05273 |
| 2.8 | 0.0 | 1.81899 | 2.05679 | 2.26120 | 2.57440 | 2.73233 |
| | 0.4 | 1.89884 | 2.15063 | 2.36661 | 2.69676 | 2.86290 |
| | 0.8 | 1.95862 | 2.21958 | 2.44300 | 2.78384 | 2.95508 |
| | 1.2 | 1.99740 | 2.26305 | 2.49009 | 2.83582 | 3.00926 |
| | 1.6 | 2.01768 | 2.28432 | 2.51186 | 2.85777 | 3.03105 |
| | 2.0 | 2.02363 | 2.28857 | 2.51432 | 2.85699 | 3.02842 |
| | 2.4 | 2.01961 | 2.28109 | 2.50357 | 2.84077 | 3.00926 |
| | 2.8 | 2.00935 | 2.26632 | 2.48469 | 2.81517 | 2.98011 |
| 3.2 | 0.0 | 1.81069 | 2.03874 | 2.23407 | 2.53226 | 2.68218 |
| | 0.4 | 1.87688 | 2.11730 | 2.32289 | 2.63619 | 2.79346 |
| | 0.8 | 1.92904 | 2.17821 | 2.39098 | 2.71471 | 2.87700 |
| | 1.2 | 1.96587 | 2.22034 | 2.43734 | 2.76703 | 2.93209 |
| | 1.6 | 1.98837 | 2.24515 | 2.46383 | 2.79562 | 2.96155 |
| | 2.0 | 1.99894 | 2.25565 | 2.47401 | 2.80488 | 2.97015 |
| | 2.4 | 2.00045 | 2.25537 | 2.47196 | 2.79972 | 2.96327 |
| | 2.8 | 1.99563 | 2.24762 | 2.46147 | 2.78469 | 2.94581 |
| | 3.2 | 1.98674 | 2.23511 | 2.44565 | 2.76350 | 2.92179 |

FRACTILES $W_\alpha$ RHO = 0.0

| DELTAS | | α: 0.9625 | 0.9750 | 0.9825 | 0.9900 | 0.9925 |
|---|---|---|---|---|---|---|
| 3.6 | 0.0 | 1.80429 | 2.02465 | 2.21278 | 2.49902 | 2.64255 |
| | 0.4 | 1.85976 | 2.09107 | 2.28833 | 2.58808 | 2.73820 |
| | 0.8 | 1.90521 | 2.14473 | 2.34876 | 2.65843 | 2.81336 |
| | 1.2 | 1.93934 | 2.18436 | 2.39286 | 2.70896 | 2.86694 |
| | 1.6 | 1.96239 | 2.21047 | 2.42137 | 2.74075 | 2.90021 |
| | 2.0 | 1.97566 | 2.22479 | 2.43636 | 2.75639 | 2.91604 |
| | 2.4 | 1.98102 | 2.22959 | 2.44049 | 2.75917 | 2.91800 |
| | 2.8 | 1.98038 | 2.22725 | 2.43651 | 2.75237 | 2.90966 |
| | 3.2 | 1.97551 | 2.21990 | 2.42685 | 2.73893 | 2.89419 |
| | 3.6 | 1.96786 | 2.20928 | 2.41353 | 2.72122 | 2.87418 |
| 4.0 | 0.0 | 1.79941 | 2.01363 | 2.19598 | 2.47260 | 2.61096 |
| | 0.4 | 1.84635 | 2.07031 | 2.26081 | 2.54955 | 2.69385 |
| | 0.8 | 1.88603 | 2.11759 | 2.31441 | 2.61246 | 2.76129 |
| | 1.2 | 1.91723 | 2.15426 | 2.35558 | 2.66016 | 2.81213 |
| | 1.6 | 1.93983 | 2.18035 | 2.38446 | 2.69300 | 2.84684 |
| | 2.0 | 1.95450 | 2.19677 | 2.40221 | 2.71249 | 2.86707 |
| | 2.4 | 1.96238 | 2.20500 | 2.41057 | 2.72076 | 2.87518 |
| | 2.8 | 1.96483 | 2.20670 | 2.41146 | 2.72017 | 2.87373 |
| | 3.2 | 1.96316 | 2.20347 | 2.40675 | 2.71296 | 2.86516 |
| | 3.6 | 1.95855 | 2.19674 | 2.39807 | 2.70108 | 2.85158 |
| | 4.0 | 1.95195 | 2.18767 | 2.38676 | 2.68612 | 2.83471 |
| 6.0 | 0.0 | 1.78757 | 1.98445 | 2.15023 | 2.39890 | 2.52212 |
| | 0.4 | 1.81069 | 2.01326 | 2.18391 | 2.43995 | 2.56684 |
| | 0.8 | 1.83201 | 2.03956 | 2.21445 | 2.47687 | 2.60693 |
| | 1.2 | 1.85098 | 2.06276 | 2.24121 | 2.50899 | 2.64169 |
| | 1.6 | 1.86727 | 2.08249 | 2.26384 | 2.53594 | 2.67077 |
| | 2.0 | 1.88072 | 2.09863 | 2.28223 | 2.55766 | 2.69411 |
| | 2.4 | 1.89134 | 2.11124 | 2.29649 | 2.57432 | 2.71192 |
| | 2.8 | 1.89930 | 2.12055 | 2.30688 | 2.58627 | 2.72460 |
| | 3.2 | 1.90483 | 2.12685 | 2.31379 | 2.59399 | 2.73269 |
| | 3.6 | 1.90821 | 2.13053 | 2.31766 | 2.59803 | 2.73677 |
| | 4.0 | 1.90976 | 2.13197 | 2.31893 | 2.59896 | 2.73747 |
| | 6.0 | 1.90009 | 2.11797 | 2.30092 | 2.57433 | 2.70932 |
| 12.0 | 0.0 | 1.78175 | 1.96586 | 2.11900 | 2.34557 | 2.45652 |
| | 0.4 | 1.78776 | 1.97363 | 2.12830 | 2.35730 | 2.46948 |
| | 0.8 | 1.79367 | 1.98121 | 2.13737 | 2.36867 | 2.48204 |
| | 1.2 | 1.79941 | 1.98855 | 2.14611 | 2.37961 | 2.49409 |
| | 1.6 | 1.80493 | 1.99558 | 2.15447 | 2.39002 | 2.50555 |
| | 2.0 | 1.81019 | 2.00226 | 2.16238 | 2.39985 | 2.51635 |
| | 2.4 | 1.81515 | 2.00853 | 2.16979 | 2.40903 | 2.52643 |
| | 2.8 | 1.81978 | 2.01436 | 2.17667 | 2.41753 | 2.53574 |
| | 3.2 | 1.82405 | 2.01973 | 2.18299 | 2.42531 | 2.54426 |
| | 3.6 | 1.82796 | 2.02462 | 2.18873 | 2.43236 | 2.55197 |
| | 4.0 | 1.83150 | 2.02903 | 2.19389 | 2.43867 | 2.55886 |
| | 6.0 | 1.84363 | 2.04399 | 2.21126 | 2.45970 | 2.58171 |
| | 12.0 | 1.84258 | 2.04162 | 2.20760 | 2.45377 | 2.57452 |
| INFINITY | | 1.78046 | 1.95996 | 2.10836 | 2.32635 | 2.43238 |

FRACTILES $W_\alpha$ RHO = 0.0

| DELTAS | | $\alpha$:0.9950 | 0.9975 | 0.9990 | 0.9995 |
|---|---|---|---|---|---|
| 0.0 | 0.0 | 3.60421 | 4.23338 | 5.07546 | 5.71869 |
| 0.4 | 0.0 | 3.59704 | 4.22073 | 5.05386 | 5.68910 |
| | 0.4 | 3.86746 | 4.51849 | 5.38062 | 6.03320 |
| 0.8 | 0.0 | 3.52988 | 4.11385 | 4.88725 | 5.47270 |
| | 0.4 | 3.84664 | 4.45656 | 5.25816 | 5.86129 |
| | 0.8 | 3.88170 | 4.46769 | 5.23365 | 5.80742 |
| 1.2 | 0.0 | 3.39747 | 3.92160 | 4.61044 | 5.12870 |
| | 0.4 | 3.67823 | 4.22725 | 4.94486 | 5.48237 |
| | 0.8 | 3.76040 | 4.30219 | 5.00723 | 5.53340 |
| | 1.2 | 3.70866 | 4.22466 | 4.89357 | 5.39118 |
| 1.6 | 0.0 | 3.25419 | 3.72413 | 4.33822 | 4.79812 |
| | 0.4 | 3.49078 | 3.98444 | 4.62672 | 5.10597 |
| | 0.8 | 3.59575 | 4.09209 | 4.73545 | 5.21401 |
| | 1.2 | 3.59718 | 4.08119 | 4.70651 | 5.17035 |
| | 1.6 | 3.53767 | 4.00158 | 4.59916 | 5.04131 |
| 2.0 | 0.0 | 3.13005 | 3.55704 | 4.11235 | 4.52658 |
| | 0.4 | 3.32833 | 3.77720 | 4.35881 | 4.79132 |
| | 0.8 | 3.43854 | 3.89505 | 4.48469 | 4.92198 |
| | 1.2 | 3.47187 | 3.92452 | 4.50752 | 4.93884 |
| | 1.6 | 3.45141 | 3.89276 | 4.45974 | 4.87828 |
| | 2.0 | 3.39999 | 3.82624 | 4.37253 | 4.77498 |
| 2.4 | 0.0 | 3.03030 | 3.42407 | 3.93398 | 4.31299 |
| | 0.4 | 3.19726 | 3.61085 | 4.14477 | 4.54057 |
| | 0.8 | 3.30374 | 3.72717 | 4.27232 | 4.67551 |
| | 1.2 | 3.35358 | 3.77815 | 4.32343 | 4.72587 |
| | 1.6 | 3.35901 | 3.77828 | 4.31555 | 4.71132 |
| | 2.0 | 3.33445 | 3.74435 | 4.26853 | 4.65397 |
| | 2.4 | 3.29236 | 3.69078 | 4.19929 | 4.57257 |
| 2.8 | 0.0 | 2.95181 | 3.31973 | 3.79429 | 4.14588 |
| | 0.4 | 3.09350 | 3.47925 | 3.97555 | 4.34241 |
| | 0.8 | 3.19244 | 3.58882 | 4.09761 | 4.47296 |
| | 1.2 | 3.24940 | 3.64976 | 4.16259 | 4.54024 |
| | 1.6 | 3.27073 | 3.66972 | 4.17982 | 4.55482 |
| | 2.0 | 3.26530 | 3.65909 | 4.16164 | 4.53051 |
| | 2.4 | 3.24186 | 3.62802 | 4.12001 | 4.48059 |
| | 2.8 | 3.20761 | 3.58484 | 4.06466 | 4.41583 |
| 3.2 | 0.0 | 2.89009 | 3.23761 | 3.68426 | 4.01417 |
| | 0.4 | 3.01134 | 3.37491 | 3.84120 | 4.18498 |
| | 0.8 | 3.10160 | 3.47584 | 3.95489 | 4.30747 |
| | 1.2 | 3.16033 | 3.54011 | 4.02540 | 4.38201 |
| | 1.6 | 3.19077 | 3.57171 | 4.05767 | 4.41426 |
| | 2.0 | 3.19829 | 3.57698 | 4.05931 | 4.41275 |
| | 2.4 | 3.18885 | 3.56286 | 4.03853 | 4.38663 |
| | 2.8 | 3.16787 | 3.53564 | 4.00272 | 4.34410 |
| | 3.2 | 3.13978 | 3.50043 | 3.95784 | 4.29176 |

FRACTILES $W_\alpha$ RHC = 0.0

| DELTAS | | $\alpha$: 0.9950 | 0.9975 | 0.9990 | 0.9995 |
|---|---|---|---|---|---|
| 3.6 | 0.0 | 2.84123 | 3.17242 | 3.59668 | 3.90918 |
| | 0.4 | 2.94584 | 3.29150 | 3.73355 | 4.05864 |
| | 0.8 | 3.02747 | 3.38349 | 3.83804 | 4.17186 |
| | 1.2 | 3.08510 | 3.44745 | 3.90940 | 4.24820 |
| | 1.6 | 3.12026 | 3.48536 | 3.95016 | 4.29062 |
| | 2.0 | 3.13619 | 3.50108 | 3.96499 | 4.30439 |
| | 2.4 | 3.13686 | 3.49927 | 3.95944 | 4.29571 |
| | 2.8 | 3.12626 | 3.48457 | 3.93897 | 4.27066 |
| | 3.2 | 3.10787 | 3.46103 | 3.90834 | 4.23451 |
| | 3.6 | 3.08455 | 3.43192 | 3.87139 | 4.19152 |
| 4.0 | 0.0 | 2.80216 | 3.12010 | 3.52615 | 3.82447 |
| | 0.4 | 2.89312 | 3.22415 | 3.64634 | 3.95612 |
| | 0.8 | 2.96670 | 3.30759 | 3.74180 | 4.06001 |
| | 1.2 | 3.02175 | 3.36931 | 3.81146 | 4.13513 |
| | 1.6 | 3.05888 | 3.41017 | 3.85652 | 4.18292 |
| | 2.0 | 3.08002 | 3.43249 | 3.87984 | 4.20664 |
| | 2.4 | 3.08779 | 3.43940 | 3.88516 | 4.21046 |
| | 2.8 | 3.08504 | 3.43422 | 3.87641 | 4.19880 |
| | 3.2 | 3.07448 | 3.42010 | 3.85732 | 4.17577 |
| | 3.6 | 3.05845 | 3.39974 | 3.83104 | 4.14489 |
| | 4.0 | 3.03884 | 3.37535 | 3.80018 | 4.10904 |
| 6.0 | 0.0 | 2.69124 | 2.96979 | 3.32122 | 3.57674 |
| | 0.4 | 2.74100 | 3.02784 | 3.38967 | 3.65267 |
| | 0.8 | 2.78544 | 3.07938 | 3.45009 | 3.71945 |
| | 1.2 | 2.82382 | 3.12366 | 3.50167 | 3.77624 |
| | 1.6 | 2.85578 | 3.16031 | 3.54408 | 3.82272 |
| | 2.0 | 2.88132 | 3.18937 | 3.57742 | 3.85904 |
| | 2.4 | 2.90067 | 3.21117 | 3.60212 | 3.88572 |
| | 2.8 | 2.91432 | 3.22629 | 3.61890 | 3.90357 |
| | 3.2 | 2.92285 | 3.23544 | 3.62863 | 3.91359 |
| | 3.6 | 2.92694 | 3.23943 | 3.63228 | 3.91685 |
| | 4.0 | 2.92729 | 3.23907 | 3.63082 | 3.91444 |
| | 6.0 | 2.89404 | 3.19682 | 3.57618 | 3.85014 |
| 12.0 | 0.0 | 2.60742 | 2.85268 | 3.15672 | 3.37428 |
| | 0.4 | 2.62212 | 2.87032 | 3.17817 | 3.39857 |
| | 0.8 | 2.63633 | 2.88732 | 3.19879 | 3.42186 |
| | 1.2 | 2.64994 | 2.90356 | 3.21843 | 3.44402 |
| | 1.6 | 2.66287 | 2.91895 | 3.23698 | 3.46490 |
| | 2.0 | 2.67502 | 2.93338 | 3.25434 | 3.48441 |
| | 2.4 | 2.68635 | 2.94679 | 3.27043 | 3.50246 |
| | 2.8 | 2.69680 | 2.95914 | 3.28520 | 3.51901 |
| | 3.2 | 2.70634 | 2.97038 | 3.29862 | 3.53402 |
| | 3.6 | 2.71496 | 2.98052 | 3.31069 | 3.54749 |
| | 4.0 | 2.72265 | 2.98955 | 3.32140 | 3.55942 |
| | 6.0 | 2.74800 | 3.01899 | 3.35595 | 3.59763 |
| | 12.0 | 2.73893 | 3.00646 | 3.33844 | 3.57608 |
| INFINITY | | 2.57583 | 2.80703 | 3.09024 | 3.29051 |

FRACTILES $W_\alpha$ RHO = 0.20

| DELTAS | | $\alpha$: 0.0005 | 0.0010 | 0.0025 | 0.0050 | 0.0075 |
|---|---|---|---|---|---|---|
| 0.0 | 0.0 | -4.59832 | -4.09412 | -3.43420 | -2.94130 | -2.65614 |
| 0.4 | 0.0 | -4.64508 | -4.13998 | -3.47721 | -2.98082 | -2.69305 |
| | 0.4 | -4.01257 | -3.58194 | -3.01842 | -2.59763 | -2.35425 |
| 0.8 | 0.0 | -4.56266 | -4.09091 | -3.46672 | -2.99449 | -2.71847 |
| | 0.4 | -3.76419 | -3.37011 | -2.85318 | -2.46620 | -2.24198 |
| | 0.8 | -3.16825 | -2.84868 | -2.43073 | -2.11890 | -1.93871 |
| 1.2 | 0.0 | -4.34314 | -3.92122 | -3.35943 | -2.93101 | -2.67887 |
| | 0.4 | -3.67856 | -3.31515 | -2.83438 | -2.47075 | -2.25827 |
| | 0.8 | -3.02643 | -2.73163 | -2.34520 | -2.05623 | -1.88897 |
| | 1.2 | -2.63165 | -2.39057 | -2.07562 | -1.84103 | -1.70569 |
| 1.6 | 0.0 | -4.11433 | -3.73657 | -3.23113 | -2.84338 | -2.61405 |
| | 0.4 | -3.59042 | -3.25837 | -2.81616 | -2.47886 | -2.28033 |
| | 0.8 | -3.05124 | -2.77064 | -2.39951 | -2.11894 | -1.95510 |
| | 1.2 | -2.59028 | -2.36201 | -2.06313 | -1.84007 | -1.71123 |
| | 1.6 | -2.34090 | -2.15120 | -1.90382 | -1.72012 | -1.61448 |
| 2.0 | 0.0 | -3.92263 | -3.57983 | -3.11921 | -2.76403 | -2.55306 |
| | 0.4 | -3.50532 | -3.19936 | -2.78980 | -2.47542 | -2.28938 |
| | 0.8 | -3.07409 | -2.80878 | -2.45537 | -2.18572 | -2.02699 |
| | 1.2 | -2.67510 | -2.45080 | -2.15428 | -1.93034 | -1.79972 |
| | 1.6 | -2.35358 | -2.16984 | -1.92979 | -1.75124 | -1.64851 |
| | 2.0 | -2.19782 | -2.04320 | -1.84224 | -1.69384 | -1.60907 |
| 2.4 | 0.0 | -3.77184 | -3.45615 | -3.03034 | -2.70049 | -2.50383 |
| | 0.4 | -3.43325 | -3.14800 | -2.76443 | -2.46837 | -2.29237 |
| | 0.8 | -3.08402 | -2.83204 | -2.49450 | -2.23516 | -2.08157 |
| | 1.2 | -2.75330 | -2.53497 | -2.24402 | -2.02197 | -1.89122 |
| | 1.6 | -2.46183 | -2.27649 | -2.03171 | -1.84718 | -1.73982 |
| | 2.0 | -2.23733 | -2.08496 | -1.88663 | -1.74012 | -1.65654 |
| | 2.4 | -2.14420 | -2.01506 | -1.84827 | -1.72666 | -1.65848 |
| 2.8 | 0.0 | -3.65507 | -3.36052 | -2.96184 | -2.65171 | -2.46619 |
| | 0.4 | -3.37544 | -3.10657 | -2.74354 | -2.46194 | -2.29386 |
| | 0.8 | -3.08796 | -2.84694 | -2.52250 | -2.27172 | -2.12245 |
| | 1.2 | -2.81242 | -2.59981 | -2.31472 | -2.09540 | -1.96536 |
| | 1.6 | -2.56312 | -2.37840 | -2.13214 | -1.94413 | -1.83344 |
| | 2.0 | -2.35063 | -2.19341 | -1.98616 | -1.83059 | -1.74064 |
| | 2.4 | -2.19639 | -2.06744 | -1.90092 | -1.78004 | -1.71321 |
| | 2.8 | -2.14806 | -2.03878 | -1.90000 | -1.80521 | -1.75317 |
| 3.2 | 0.0 | -3.56458 | -3.28673 | -2.90944 | -2.61481 | -2.43801 |
| | 0.4 | -3.33005 | -3.07423 | -2.72752 | -2.45736 | -2.29551 |
| | 0.8 | -3.08984 | -2.85776 | -2.54396 | -2.30008 | -2.15426 |
| | 1.2 | -2.85789 | -2.65016 | -2.37011 | -2.15319 | -2.02384 |
| | 1.6 | -2.64496 | -2.46139 | -2.21490 | -2.02492 | -1.91210 |
| | 2.0 | -2.45850 | -2.29857 | -2.08527 | -1.92245 | -1.82672 |
| | 2.4 | -2.30550 | -2.16953 | -1.99114 | -1.85897 | -1.78478 |
| | 2.8 | -2.20545 | -2.09543 | -1.95696 | -1.86275 | -1.80253 |
| | 3.2 | -2.19200 | -2.10073 | -1.99008 | -1.89621 | -1.83403 |

FRACTILES $W_\alpha$ RHO = 0.20

| DELTAS | | $\alpha$:0.0005 | 0.0010 | 0.0025 | 0.0050 | 0.0075 |
|---|---|---|---|---|---|---|
| 3.6 | 0.0 | -3.49403 | -3.22955 | -2.86932 | -2.58702 | -2.41713 |
| | 0.4 | -3.29471 | -3.04939 | -2.71575 | -2.45470 | -2.29778 |
| | 0.8 | -3.09136 | -2.86657 | -2.56138 | -2.32303 | -2.17994 |
| | 1.2 | -2.89410 | -2.69044 | -2.41451 | -2.19951 | -2.07064 |
| | 1.6 | -2.71118 | -2.52864 | -2.28203 | -2.09046 | -1.97591 |
| | 2.0 | -2.54870 | -2.38697 | -2.16938 | -2.00126 | -1.90121 |
| | 2.4 | -2.41124 | -2.27013 | -2.08198 | -1.93884 | -1.85579 |
| | 2.8 | -2.30519 | -2.18632 | -2.03351 | -1.92368 | -1.85481 |
| | 3.2 | -2.25101 | -2.16107 | -2.04148 | -1.93912 | -1.87187 |
| | 3.6 | -2.27092 | -2.19093 | -2.07228 | -1.96729 | -1.89775 |
| 4.0 | 0.0 | -3.43861 | -3.18496 | -2.83852 | -2.56612 | -2.40176 |
| | 0.4 | -3.26730 | -3.03050 | -2.70742 | -2.45367 | -2.30066 |
| | 0.8 | -3.09319 | -2.87436 | -2.57615 | -2.34221 | -2.20124 |
| | 1.2 | -2.92385 | -2.72352 | -2.45092 | -2.23735 | -2.10876 |
| | 1.6 | -2.76566 | -2.58394 | -2.33709 | -2.14403 | -2.02792 |
| | 2.0 | -2.62367 | -2.46038 | -2.23913 | -2.06654 | -1.96293 |
| | 2.4 | -2.50176 | -2.35671 | -2.16105 | -2.00943 | -1.91925 |
| | 2.8 | -2.40359 | -2.27726 | -2.10991 | -1.98392 | -1.90633 |
| | 3.2 | -2.33761 | -2.23604 | -2.09953 | -1.98560 | -1.91202 |
| | 3.6 | -2.32770 | -2.24144 | -2.11369 | -2.00164 | -1.92796 |
| | 4.0 | -2.35709 | -2.27118 | -2.14090 | -2.02557 | -1.94962 |
| 6.0 | 0.0 | -3.29087 | -3.06905 | -2.76262 | -2.51846 | -2.36958 |
| | 0.4 | -3.20003 | -2.98812 | -2.69528 | -2.46183 | -2.31938 |
| | 0.8 | -3.10920 | -2.90761 | -2.62891 | -2.40652 | -2.27071 |
| | 1.2 | -3.02067 | -2.82965 | -2.56536 | -2.35423 | -2.22511 |
| | 1.6 | -2.93658 | -2.75621 | -2.50640 | -2.30648 | -2.18400 |
| | 2.0 | -2.85886 | -2.68909 | -2.45359 | -2.26465 | -2.14861 |
| | 2.4 | -2.78923 | -2.62987 | -2.40832 | -2.22991 | -2.11993 |
| | 2.8 | -2.72918 | -2.57994 | -2.37174 | -2.20314 | -2.09858 |
| | 3.2 | -2.68008 | -2.54055 | -2.34475 | -2.18465 | -2.08451 |
| | 3.6 | -2.64331 | -2.51276 | -2.32751 | -2.17392 | -2.07693 |
| | 4.0 | -2.61991 | -2.49668 | -2.31912 | -2.16981 | -2.07472 |
| | 6.0 | -2.65312 | -2.53430 | -2.35791 | -2.20654 | -2.10928 |
| 12.0 | 0.0 | -3.21576 | -3.01799 | -2.74014 | -2.51465 | -2.37526 |
| | 0.4 | -3.18993 | -2.99541 | -2.72189 | -2.49967 | -2.36219 |
| | 0.8 | -3.16454 | -2.97329 | -2.70410 | -2.48514 | -2.34955 |
| | 1.2 | -3.13984 | -2.95184 | -2.68692 | -2.47118 | -2.33744 |
| | 1.6 | -3.11601 | -2.93122 | -2.67051 | -2.45790 | -2.32597 |
| | 2.0 | -3.09326 | -2.91161 | -2.65499 | -2.44541 | -2.31522 |
| | 2.4 | -3.07175 | -2.89314 | -2.64048 | -2.43380 | -2.30526 |
| | 2.8 | -3.05163 | -2.87595 | -2.62706 | -2.42313 | -2.29615 |
| | 3.2 | -3.03302 | -2.86013 | -2.61480 | -2.41344 | -2.28791 |
| | 3.6 | -3.01602 | -2.84575 | -2.60375 | -2.40478 | -2.28057 |
| | 4.0 | -3.00071 | -2.83287 | -2.59393 | -2.39713 | -2.27413 |
| | 6.0 | -2.94971 | -2.79083 | -2.56281 | -2.37353 | -2.25459 |
| | 12.0 | -2.97098 | -2.81199 | -2.58279 | -2.39186 | -2.27166 |
| INFINITY | | -3.29051 | -3.09024 | -2.80703 | -2.57583 | -2.43238 |

FRACTILES $W_\alpha$ RHO = 0.20

| DELTAS | | $\alpha$:0.0100 | 0.0175 | 0.0250 | 0.0375 | 0.0500 |
|---|---|---|---|---|---|---|
| 0.0 | 0.0 | -2.45549 | -2.06994 | -1.82814 | -1.55790 | -1.36979 |
| 0.4 | 0.0 | -2.49031 | -2.10007 | -1.85486 | -1.58035 | -1.38896 |
| | 0.4 | -2.18304 | -1.85418 | -1.64805 | -1.41782 | -1.25770 |
| 0.8 | 0.0 | -2.52284 | -2.14318 | -1.90217 | -1.62971 | -1.43780 |
| | 0.4 | -2.08406 | -1.78033 | -1.58967 | -1.37649 | -1.22811 |
| | 0.8 | -1.81204 | -1.56902 | -1.41698 | -1.24755 | -1.13008 |
| 1.2 | 0.0 | -2.49925 | -2.14801 | -1.92285 | -1.66565 | -1.48241 |
| | 0.4 | -2.10772 | -1.81568 | -1.63047 | -1.42132 | -1.27424 |
| | 0.8 | -1.77129 | -1.54531 | -1.40380 | -1.24608 | -1.13676 |
| | 1.2 | -1.61070 | -1.42893 | -1.31566 | -1.19009 | -1.10365 |
| 1.6 | 0.0 | -2.45006 | -2.12769 | -1.91960 | -1.68020 | -1.50828 |
| | 0.4 | -2.13889 | -1.86230 | -1.68499 | -1.48250 | -1.33829 |
| | 0.8 | -1.83908 | -1.61428 | -1.47198 | -1.31169 | -1.19941 |
| | 1.2 | -1.62074 | -1.44753 | -1.33962 | -1.22020 | -1.13826 |
| | 1.6 | -1.54054 | -1.39981 | -1.31290 | -1.21781 | -1.15374 |
| 2.0 | 0.0 | -2.40176 | -2.10300 | -1.90909 | -1.68475 | -1.52266 |
| | 0.4 | -2.15630 | -1.89454 | -1.72545 | -1.53079 | -1.39089 |
| | 0.8 | -1.91389 | -1.69272 | -1.55097 | -1.38919 | -1.27413 |
| | 1.2 | -1.70733 | -1.52874 | -1.41613 | -1.29008 | -1.20264 |
| | 1.6 | -1.57663 | -1.43997 | -1.35589 | -1.26468 | -1.20442 |
| | 2.0 | -1.55012 | -1.43945 | -1.37302 | -1.30507 | -1.26543 |
| 2.4 | 0.0 | -2.36240 | -2.08208 | -1.89925 | -1.68670 | -1.53234 |
| | 0.4 | -2.16606 | -1.91640 | -1.75414 | -1.56614 | -1.43009 |
| | 0.8 | -1.97165 | -1.75524 | -1.61530 | -1.45401 | -1.33799 |
| | 1.2 | -1.79806 | -1.61595 | -1.49934 | -1.36654 | -1.27251 |
| | 1.6 | -1.66408 | -1.51846 | -1.42764 | -1.32802 | -1.26202 |
| | 2.0 | -1.59858 | -1.49072 | -1.42803 | -1.36605 | -1.31005 |
| | 2.4 | -1.61222 | -1.53410 | -1.47943 | -1.40406 | -1.34118 |
| 2.8 | 0.0 | -2.33243 | -2.06639 | -1.89213 | -1.68867 | -1.54022 |
| | 0.4 | -2.17286 | -1.93270 | -1.77577 | -1.59296 | -1.45988 |
| | 0.8 | -2.01521 | -1.80293 | -1.66468 | -1.50415 | -1.38767 |
| | 1.2 | -1.87221 | -1.68860 | -1.56969 | -1.43248 | -1.33369 |
| | 1.6 | -1.75460 | -1.60071 | -1.50258 | -1.39199 | -1.31612 |
| | 2.0 | -1.67765 | -1.55896 | -1.49100 | -1.41166 | -1.34550 |
| | 2.4 | -1.66969 | -1.58566 | -1.52131 | -1.43676 | -1.36775 |
| | 2.8 | -1.71157 | -1.61921 | -1.55059 | -1.46089 | -1.38806 |
| 3.2 | 0.0 | -2.31026 | -2.05536 | -1.88774 | -1.69127 | -1.54733 |
| | 0.4 | -2.17869 | -1.94592 | -1.79309 | -1.61420 | -1.48330 |
| | 0.8 | -2.04916 | -1.84010 | -1.70311 | -1.54304 | -1.42609 |
| | 1.2 | -1.93077 | -1.74613 | -1.62551 | -1.48497 | -1.38261 |
| | 1.6 | -1.83119 | -1.67146 | -1.56787 | -1.44841 | -1.36328 |
| | 2.0 | -1.75870 | -1.62716 | -1.54763 | -1.45238 | -1.37724 |
| | 2.4 | -1.73653 | -1.63451 | -1.56124 | -1.46733 | -1.39206 |
| | 2.8 | -1.75616 | -1.65474 | -1.58044 | -1.48439 | -1.40713 |
| | 3.2 | -1.78577 | -1.67984 | -1.60225 | -1.50220 | -1.42199 |

FRACTILES $w_\alpha$ RHO = 0.20

| DELTAS | | $\alpha$:0.0100 | 0.0175 | 0.0250 | 0.0375 | 0.0500 |
|---|---|---|---|---|---|---|
| 3.6 | 0.0 | -2.29412 | -2.04797 | -1.88552 | -1.69444 | -1.55393 |
| | 0.4 | -2.18423 | -1.95722 | -1.80753 | -1.63156 | -1.50219 |
| | 0.8 | -2.07648 | -1.86985 | -1.73372 | -1.57377 | -1.45622 |
| | 1.2 | -1.97756 | -1.79190 | -1.66973 | -1.52630 | -1.42092 |
| | 1.6 | -1.89332 | -1.72891 | -1.62101 | -1.49470 | -1.40241 |
| | 2.0 | -1.82937 | -1.68755 | -1.59706 | -1.48916 | -1.40601 |
| | 2.4 | -1.79870 | -1.68122 | -1.59918 | -1.49604 | -1.41466 |
| | 2.8 | -1.80266 | -1.69068 | -1.61004 | -1.50713 | -1.42522 |
| | 3.2 | -1.82001 | -1.70710 | -1.62511 | -1.52012 | -1.43648 |
| | 3.6 | -1.84402 | -1.72707 | -1.64228 | -1.53399 | -1.44794 |
| 4.0 | 0.0 | -2.28252 | -2.04329 | -1.88489 | -1.69798 | -1.56008 |
| | 0.4 | -2.18970 | -1.96717 | -1.81986 | -1.64603 | -1.51772 |
| | 0.8 | -2.09905 | -1.89417 | -1.75856 | -1.59848 | -1.48023 |
| | 1.2 | -2.01558 | -1.82882 | -1.70519 | -1.55915 | -1.45110 |
| | 1.6 | -1.94382 | -1.77532 | -1.66376 | -1.53183 | -1.43408 |
| | 2.0 | -1.88799 | -1.73810 | -1.63952 | -1.52161 | -1.43166 |
| | 2.4 | -1.85503 | -1.72427 | -1.63426 | -1.52260 | -1.43552 |
| | 2.8 | -1.84830 | -1.72558 | -1.63857 | -1.52881 | -1.44230 |
| | 3.2 | -1.85583 | -1.73489 | -1.64803 | -1.53774 | -1.45047 |
| | 3.6 | -1.87131 | -1.74871 | -1.66039 | -1.54814 | -1.45934 |
| | 4.0 | -1.89123 | -1.76505 | -1.67433 | -1.55930 | -1.46850 |
| 6.0 | 0.0 | -2.26077 | -2.04023 | -1.89239 | -1.71589 | -1.58409 |
| | 0.4 | -2.21521 | -2.00387 | -1.86203 | -1.69241 | -1.56553 |
| | 0.8 | -2.17131 | -1.96940 | -1.83363 | -1.67095 | -1.54897 |
| | 1.2 | -2.13052 | -1.93802 | -1.80826 | -1.65235 | -1.53508 |
| | 1.6 | -2.09413 | -1.91080 | -1.78681 | -1.63726 | -1.52430 |
| | 2.0 | -2.06328 | -1.88860 | -1.76990 | -1.62601 | -1.51673 |
| | 2.4 | -2.03879 | -1.87189 | -1.75773 | -1.61846 | -1.51205 |
| | 2.8 | -2.02109 | -1.86058 | -1.74994 | -1.61410 | -1.50972 |
| | 3.2 | -2.00984 | -1.85406 | -1.74589 | -1.61232 | -1.50923 |
| | 3.6 | -2.00420 | -1.85149 | -1.74481 | -1.61254 | -1.51012 |
| | 4.0 | -2.00311 | -1.85203 | -1.74605 | -1.61428 | -1.51203 |
| | 6.0 | -2.03573 | -1.88008 | -1.77070 | -1.63467 | -1.52920 |
| 12.0 | 0.0 | -2.27245 | -2.06157 | -1.91832 | -1.74526 | -1.61462 |
| | 0.4 | -2.26073 | -2.05245 | -1.91083 | -1.73959 | -1.61020 |
| | 0.8 | -2.24942 | -2.04370 | -1.90367 | -1.73420 | -1.60603 |
| | 1.2 | -2.23862 | -2.03538 | -1.89691 | -1.72914 | -1.60214 |
| | 1.6 | -2.22841 | -2.02758 | -1.89058 | -1.72445 | -1.59856 |
| | 2.0 | -2.21887 | -2.02033 | -1.88474 | -1.72014 | -1.59530 |
| | 2.4 | -2.21006 | -2.01368 | -1.87941 | -1.71625 | -1.59236 |
| | 2.8 | -2.20202 | -2.00765 | -1.87461 | -1.71276 | -1.58976 |
| | 3.2 | -2.19478 | -2.00227 | -1.87034 | -1.70970 | -1.58749 |
| | 3.6 | -2.18835 | -1.99753 | -1.86661 | -1.70704 | -1.58555 |
| | 4.0 | -2.18273 | -1.99342 | -1.86339 | -1.70477 | -1.58390 |
| | 6.0 | -2.16592 | -1.98153 | -1.85435 | -1.69867 | -1.57970 |
| | 12.0 | -2.18197 | -1.99531 | -1.86649 | -1.70878 | -1.58826 |
| INFINITY | | -2.32635 | -2.10836 | -1.95996 | -1.78046 | -1.64485 |

FRACTILES $W_\alpha$ RHO = 0.20

| DELTAS | | $\alpha$:0.0750 | 0.1000 | 0.1500 | 0.2000 | 0.2500 |
|---|---|---|---|---|---|---|
| 0.0 | 0.0 | -1.11107 | -0.93322 | -0.69339 | -0.53373 | -0.41840 |
| 0.4 | 0.0 | -1.12528 | -0.94371 | -0.69836 | -0.53465 | -0.41613 |
| | 0.4 | -1.03776 | -0.88637 | -0.68411 | -0.55003 | -0.45418 |
| 0.8 | 0.0 | -1.17034 | -0.98369 | -0.72756 | -0.55353 | -0.42549 |
| | 0.4 | -1.02421 | -0.88434 | -0.69668 | -0.57318 | -0.48574 |
| | 0.8 | -0.96947 | -0.86012 | -0.71526 | -0.62242 | -0.55999 |
| 1.2 | 0.0 | -1.22328 | -1.03910 | -0.78007 | -0.59814 | -0.45978 |
| | 0.4 | -1.06981 | -0.92776 | -0.73448 | -0.60542 | -0.51325 |
| | 0.8 | -0.98754 | -0.88633 | -0.75360 | -0.67105 | -0.62153 |
| | 1.2 | -0.98694 | -0.90926 | -0.81206 | -0.76258 | -0.70361 |
| 1.6 | 0.0 | -1.26265 | -1.08574 | -0.83223 | -0.64926 | -0.50577 |
| | 0.4 | -1.13458 | -0.99008 | -0.78789 | -0.64766 | -0.54365 |
| | 0.8 | -1.04436 | -0.93789 | -0.79686 | -0.70976 | -0.65901 |
| | 1.2 | -1.02850 | -0.95687 | -0.87705 | -0.80768 | -0.72428 |
| | 1.6 | -1.07138 | -1.02659 | -0.93717 | -0.83912 | -0.74010 |
| 2.0 | 0.0 | -1.28933 | -1.11967 | -0.87336 | -0.69228 | -0.54732 |
| | 0.4 | -1.19086 | -1.04670 | -0.84013 | -0.69149 | -0.57616 |
| | 0.8 | -1.11206 | -0.99785 | -0.84100 | -0.73965 | -0.67157 |
| | 1.2 | -1.08428 | -1.00683 | -0.91626 | -0.82081 | -0.72259 |
| | 1.6 | -1.13514 | -1.07697 | -0.95901 | -0.84701 | -0.73885 |
| | 2.0 | -1.18443 | -1.11016 | -0.97699 | -0.85590 | -0.74128 |
| 2.4 | 0.0 | -1.30870 | -1.14483 | -0.90446 | -0.72525 | -0.57967 |
| | 0.4 | -1.23383 | -1.09074 | -0.88227 | -0.72836 | -0.60491 |
| | 0.8 | -1.17196 | -1.05231 | -0.88175 | -0.76345 | -0.67473 |
| | 1.2 | -1.14159 | -1.05258 | -0.93651 | -0.82378 | -0.71513 |
| | 1.6 | -1.18026 | -1.10470 | -0.96937 | -0.84721 | -0.73212 |
| | 2.0 | -1.21356 | -1.13038 | -0.98589 | -0.85750 | -0.73767 |
| | 2.4 | -1.23619 | -1.14722 | -0.99479 | -0.86097 | -0.73712 |
| 2.8 | 0.0 | -1.32396 | -1.16441 | -0.92831 | -0.75023 | -0.60386 |
| | 0.4 | -1.26648 | -1.12417 | -0.91417 | -0.75623 | -0.62673 |
| | 0.8 | -1.21909 | -1.09568 | -0.91505 | -0.78238 | -0.67551 |
| | 1.2 | -1.19257 | -1.09218 | -0.95069 | -0.82471 | -0.70772 |
| | 1.6 | -1.21359 | -1.12571 | -0.97614 | -0.84534 | -0.72435 |
| | 2.0 | -1.23674 | -1.14567 | -0.99103 | -0.85615 | -0.73177 |
| | 2.4 | -1.25446 | -1.15984 | -0.99983 | -0.86100 | -0.73357 |
| | 2.8 | -1.26923 | -1.17062 | -1.00504 | -0.86239 | -0.73214 |
| 3.2 | 0.0 | -1.33658 | -1.18017 | -0.94693 | -0.76927 | -0.62180 |
| | 0.4 | -1.29184 | -1.14985 | -0.93812 | -0.77659 | -0.64208 |
| | 0.8 | -1.25529 | -1.12876 | -0.94021 | -0.79670 | -0.67627 |
| | 1.2 | -1.23377 | -1.12449 | -0.96258 | -0.82586 | -0.70187 |
| | 1.6 | -1.24206 | -1.14388 | -0.98184 | -0.84341 | -0.71722 |
| | 2.0 | -1.25712 | -1.15872 | -0.99469 | -0.85389 | -0.72543 |
| | 2.4 | -1.27045 | -1.17027 | -1.00299 | -0.85949 | -0.72878 |
| | 2.8 | -1.28222 | -1.17943 | -1.00825 | -0.86191 | -0.72901 |
| | 3.2 | -1.29282 | -1.18700 | -1.01160 | -0.86237 | -0.72737 |

FRACTILES $W_\alpha$ RHO = 0.20

| DELTAS | | $\alpha$:0.0750 | 0.1000 | 0.1500 | 0.2000 | 0.2500 |
|---|---|---|---|---|---|---|
| 3.6 | 0.0 | -1.34729 | -1.19311 | -0.96166 | -0.78386 | -0.63509 |
| | 0.4 | -1.31194 | -1.16989 | -0.95624 | -0.79140 | -0.65261 |
| | 0.8 | -1.28327 | -1.15396 | -0.95875 | -0.80707 | -0.67729 |
| | 1.2 | -1.26581 | -1.14973 | -0.97294 | -0.82749 | -0.69758 |
| | 1.6 | -1.26682 | -1.16001 | -0.98717 | -0.84201 | -0.71122 |
| | 2.0 | -1.27555 | -1.17043 | -0.99783 | -0.85164 | -0.71949 |
| | 2.4 | -1.28498 | -1.17948 | -1.00532 | -0.85745 | -0.72373 |
| | 2.8 | -1.29405 | -1.18708 | -1.01040 | -0.86052 | -0.72515 |
| | 3.2 | -1.30259 | -1.19353 | -1.01380 | -0.86173 | -0.72471 |
| | 3.6 | -1.31063 | -1.19916 | -1.01609 | -0.86176 | -0.72310 |
| 4.0 | 0.0 | -1.35647 | -1.20384 | -0.97339 | -0.79510 | -0.64496 |
| | 0.4 | -1.32812 | -1.18575 | -0.97008 | -0.80225 | -0.65984 |
| | 0.8 | -1.30519 | -1.17336 | -0.97250 | -0.81454 | -0.67835 |
| | 1.2 | -1.29070 | -1.16925 | -0.98179 | -0.82934 | -0.69452 |
| | 1.6 | -1.28800 | -1.17417 | -0.99217 | -0.84119 | -0.70638 |
| | 2.0 | -1.29216 | -1.18108 | -1.00078 | -0.84975 | -0.71426 |
| | 2.4 | -1.29832 | -1.18787 | -1.00731 | -0.85540 | -0.71891 |
| | 2.8 | -1.30502 | -1.19398 | -1.01203 | -0.85879 | -0.72111 |
| | 3.2 | -1.31171 | -1.19939 | -1.01535 | -0.86051 | -0.72156 |
| | 3.6 | -1.31825 | -1.20420 | -1.01768 | -0.86109 | -0.72083 |
| | 4.0 | -1.32459 | -1.20855 | -1.01932 | -0.86092 | -0.71934 |
| 6.0 | 0.0 | -1.38681 | -1.23663 | -1.00580 | -0.82362 | -0.66773 |
| | 0.4 | -1.37517 | -1.22982 | -1.00548 | -0.82748 | -0.67440 |
| | 0.8 | -1.36539 | -1.22466 | -1.00633 | -0.83198 | -0.68118 |
| | 1.2 | -1.35786 | -1.22133 | -1.00817 | -0.83673 | -0.68760 |
| | 1.6 | -1.35270 | -1.21970 | -1.01062 | -0.84130 | -0.69330 |
| | 2.0 | -1.34970 | -1.21940 | -1.01329 | -0.84539 | -0.69807 |
| | 2.4 | -1.34841 | -1.22003 | -1.01589 | -0.84885 | -0.70187 |
| | 2.8 | -1.34838 | -1.22123 | -1.01825 | -0.85162 | -0.70472 |
| | 3.2 | -1.34926 | -1.22277 | -1.02029 | -0.85373 | -0.70671 |
| | 3.6 | -1.35076 | -1.22448 | -1.02200 | -0.85525 | -0.70797 |
| | 4.0 | -1.35269 | -1.22627 | -1.02341 | -0.85626 | -0.70861 |
| | 6.0 | -1.36509 | -1.23517 | -1.02731 | -0.85666 | -0.70641 |
| 12.0 | 0.0 | -1.41686 | -1.26463 | -1.02810 | -0.83965 | -0.67751 |
| | 0.4 | -1.41415 | -1.26307 | -1.02805 | -0.84054 | -0.67904 |
| | 0.8 | -1.41163 | -1.26165 | -1.02806 | -0.84145 | -0.68054 |
| | 1.2 | -1.40931 | -1.26038 | -1.02813 | -0.84235 | -0.68199 |
| | 1.6 | -1.40721 | -1.25926 | -1.02826 | -0.84324 | -0.68338 |
| | 2.0 | -1.40533 | -1.25829 | -1.02843 | -0.84410 | -0.68469 |
| | 2.4 | -1.40367 | -1.25746 | -1.02864 | -0.84493 | -0.68591 |
| | 2.8 | -1.40223 | -1.25676 | -1.02888 | -0.84572 | -0.68704 |
| | 3.2 | -1.40099 | -1.25620 | -1.02913 | -0.84645 | -0.68806 |
| | 3.6 | -1.39996 | -1.25575 | -1.02940 | -0.84712 | -0.68899 |
| | 4.0 | -1.39912 | -1.25541 | -1.02967 | -0.84774 | -0.68981 |
| | 6.0 | -1.39728 | -1.25502 | -1.03093 | -0.84988 | -0.69243 |
| | 12.0 | -1.40352 | -1.25954 | -1.03296 | -0.85013 | -0.69133 |
| INFINITY | | -1.43953 | -1.28155 | -1.03643 | -0.84162 | -0.67449 |

FRACTILES $w_\alpha$ RHO = 0.20

| DELTAS | | $\alpha$:0.3000 | 0.3500 | 0.4000 | 0.4500 | 0.5000 |
|---|---|---|---|---|---|---|
| 0.0 | 0.0 | -0.33187 | -0.26641 | -0.21848 | -0.18904 | -0.14935 |
| 0.4 | 0.0 | -0.32698 | -0.25926 | -0.20922 | -0.17832 | -0.14007 |
| | 0.4 | -0.38357 | -0.33217 | -0.30023 | -0.26526 | -0.21424 |
| 0.8 | 0.0 | -0.32767 | -0.25206 | -0.19468 | -0.15589 | -0.12101 |
| | 0.4 | -0.42262 | -0.37943 | -0.34942 | -0.30218 | -0.24051 |
| | 0.8 | -0.52264 | -0.48475 | -0.43015 | -0.36243 | -0.28191 |
| 1.2 | 0.0 | -0.35034 | -0.26243 | -0.19248 | -0.14027 | -0.10317 |
| | 0.4 | -0.44684 | -0.40423 | -0.36346 | -0.30414 | -0.23068 |
| | 0.8 | -0.57987 | -0.51925 | -0.44638 | -0.36275 | -0.26844 |
| | 1.2 | -0.63067 | -0.54875 | -0.45924 | -0.36224 | -0.25712 |
| 1.6 | 0.0 | -0.38810 | -0.28931 | -0.20602 | -0.13786 | -0.08964 |
| | 0.4 | -0.46585 | -0.41516 | -0.35954 | -0.28691 | -0.20171 |
| | 0.8 | -0.59353 | -0.51579 | -0.42943 | -0.33524 | -0.23288 |
| | 1.2 | -0.63500 | -0.54119 | -0.44279 | -0.33910 | -0.22902 |
| | 1.6 | -0.64016 | -0.53848 | -0.43401 | -0.32556 | -0.21177 |
| 2.0 | 0.0 | -0.42546 | -0.31985 | -0.22673 | -0.14479 | -0.07967 |
| | 0.4 | -0.48453 | -0.41859 | -0.34736 | -0.26242 | -0.16792 |
| | 0.8 | -0.58822 | -0.49805 | -0.40278 | -0.30224 | -0.19552 |
| | 1.2 | -0.62331 | -0.52238 | -0.41879 | -0.31139 | -0.19882 |
| | 1.6 | -0.63233 | -0.52571 | -0.41751 | -0.30628 | -0.19053 |
| | 2.0 | -0.62985 | -0.51939 | -0.40815 | -0.29455 | -0.17700 |
| 2.4 | 0.0 | -0.45522 | -0.34511 | -0.24528 | -0.15341 | -0.07184 |
| | 0.4 | -0.50181 | -0.41899 | -0.33332 | -0.23883 | -0.13781 |
| | 0.8 | -0.57796 | -0.47900 | -0.37764 | -0.27289 | -0.16343 |
| | 1.2 | -0.60871 | -0.50268 | -0.39545 | -0.28554 | -0.17142 |
| | 1.6 | -0.62059 | -0.51028 | -0.39937 | -0.28626 | -0.16934 |
| | 2.0 | -0.62235 | -0.50896 | -0.39551 | -0.28031 | -0.16169 |
| | 2.4 | -0.61872 | -0.50292 | -0.38759 | -0.27095 | -0.15130 |
| 2.8 | 0.0 | -0.47713 | -0.36336 | -0.25833 | -0.15914 | -0.06505 |
| | 0.4 | -0.51508 | -0.41808 | -0.32105 | -0.21962 | -0.11397 |
| | 0.8 | -0.56872 | -0.46312 | -0.35711 | -0.24913 | -0.13757 |
| | 1.2 | -0.59552 | -0.48535 | -0.37517 | -0.26327 | -0.14798 |
| | 1.6 | -0.60860 | -0.49523 | -0.38215 | -0.26760 | -0.14987 |
| | 2.0 | -0.61315 | -0.49732 | -0.38213 | -0.26574 | -0.14643 |
| | 2.4 | -0.61251 | -0.49471 | -0.37790 | -0.26022 | -0.13991 |
| | 2.8 | -0.60893 | -0.48946 | -0.37138 | -0.25275 | -0.13178 |
| 3.2 | 0.0 | -0.49283 | -0.37578 | -0.26639 | -0.16149 | -0.05884 |
| | 0.4 | -0.52385 | -0.41665 | -0.31154 | -0.20502 | -0.09585 |
| | 0.8 | -0.56182 | -0.45092 | -0.34111 | -0.23044 | -0.11710 |
| | 1.2 | -0.58473 | -0.47101 | -0.35830 | -0.24469 | -0.12840 |
| | 1.6 | -0.59777 | -0.48174 | -0.36680 | -0.25103 | -0.13267 |
| | 2.0 | -0.60388 | -0.48595 | -0.36928 | -0.25194 | -0.13214 |
| | 2.4 | -0.60535 | -0.48581 | -0.36778 | -0.24928 | -0.12852 |
| | 2.8 | -0.60386 | -0.48295 | -0.36381 | -0.24445 | -0.12304 |
| | 3.2 | -0.60062 | -0.47850 | -0.35844 | -0.23841 | -0.11656 |

FRACTILES $W_\alpha$ RHO = 0.20

| DELTAS | | $\alpha$: 0.3000 | 0.3500 | 0.4000 | 0.4500 | 0.5000 |
|---|---|---|---|---|---|---|
| 3.6 | 0.0 | -0.50395 | -0.38397 | -0.27091 | -0.16157 | -0.05326 |
| | 0.4 | -0.52917 | -0.41509 | -0.30442 | -0.19407 | -0.08212 |
| | 0.8 | -0.55692 | -0.44173 | -0.32876 | -0.21580 | -0.10091 |
| | 1.2 | -0.57622 | -0.45942 | -0.34451 | -0.22939 | -0.11219 |
| | 1.6 | -0.58850 | -0.47013 | -0.35354 | -0.23670 | -0.11777 |
| | 2.0 | -0.59532 | -0.47551 | -0.35754 | -0.23938 | -0.11919 |
| | 2.4 | -0.59814 | -0.47707 | -0.35796 | -0.23878 | -0.11769 |
| | 2.8 | -0.59821 | -0.47600 | -0.35595 | -0.23599 | -0.11428 |
| | 3.2 | -0.59647 | -0.47324 | -0.35239 | -0.23182 | -0.10967 |
| | 3.6 | -0.59363 | -0.46947 | -0.34793 | -0.22687 | -0.10440 |
| 4.0 | 0.0 | -0.51180 | -0.38928 | -0.27322 | -0.16048 | -0.04841 |
| | 0.4 | -0.53230 | -0.41355 | -0.29900 | -0.18573 | -0.07157 |
| | 0.8 | -0.55340 | -0.43474 | -0.31916 | -0.20426 | -0.08802 |
| | 1.2 | -0.56957 | -0.45013 | -0.33329 | -0.21683 | -0.09881 |
| | 1.6 | -0.58077 | -0.46031 | -0.34226 | -0.22446 | -0.10501 |
| | 2.0 | -0.58772 | -0.46623 | -0.34710 | -0.22820 | -0.10766 |
| | 2.4 | -0.59135 | -0.46888 | -0.34881 | -0.22904 | -0.10768 |
| | 2.8 | -0.59251 | -0.46912 | -0.34826 | -0.22780 | -0.10584 |
| | 3.2 | -0.59193 | -0.46770 | -0.34614 | -0.22512 | -0.10274 |
| | 3.6 | -0.59019 | -0.46517 | -0.34301 | -0.22154 | -0.09884 |
| | 4.0 | -0.58772 | -0.46197 | -0.33926 | -0.21741 | -0.09447 |
| 6.0 | 0.0 | -0.52768 | -0.39755 | -0.27349 | -0.15270 | -0.03286 |
| | 0.4 | -0.53621 | -0.40723 | -0.28376 | -0.16308 | -0.04296 |
| | 0.8 | -0.54437 | -0.41613 | -0.29292 | -0.17213 | -0.05157 |
| | 1.2 | -0.55171 | -0.42387 | -0.30069 | -0.17964 | -0.05859 |
| | 1.6 | -0.55795 | -0.43027 | -0.30697 | -0.18558 | -0.06403 |
| | 2.0 | -0.56298 | -0.43528 | -0.31177 | -0.19004 | -0.06802 |
| | 2.4 | -0.56682 | -0.43899 | -0.31523 | -0.19316 | -0.07072 |
| | 2.8 | -0.56957 | -0.44153 | -0.31750 | -0.19510 | -0.07232 |
| | 3.2 | -0.57135 | -0.44307 | -0.31876 | -0.19607 | -0.07298 |
| | 3.6 | -0.57233 | -0.44377 | -0.31918 | -0.19623 | -0.07290 |
| | 4.0 | -0.57265 | -0.44378 | -0.31893 | -0.19575 | -0.07221 |
| | 6.0 | -0.56844 | -0.43803 | -0.31200 | -0.18795 | -0.06383 |
| 12.0 | 0.0 | -0.53144 | -0.39563 | -0.26631 | -0.14071 | -0.01661 |
| | 0.4 | -0.53341 | -0.39789 | -0.26875 | -0.14323 | -0.01912 |
| | 0.8 | -0.53531 | -0.40006 | -0.27107 | -0.14562 | -0.02149 |
| | 1.2 | -0.53713 | -0.40211 | -0.27325 | -0.14785 | -0.02370 |
| | 1.6 | -0.53884 | -0.40403 | -0.27529 | -0.14992 | -0.02574 |
| | 2.0 | -0.54044 | -0.40581 | -0.27716 | -0.15181 | -0.02759 |
| | 2.4 | -0.54192 | -0.40744 | -0.27886 | -0.15353 | -0.02926 |
| | 2.8 | -0.54326 | -0.40891 | -0.28039 | -0.15506 | -0.03074 |
| | 3.2 | -0.54447 | -0.41022 | -0.28175 | -0.15641 | -0.03204 |
| | 3.6 | -0.54555 | -0.41138 | -0.28293 | -0.15758 | -0.03317 |
| | 4.0 | -0.54649 | -0.41238 | -0.28395 | -0.15858 | -0.03412 |
| | 6.0 | -0.54935 | -0.41532 | -0.28685 | -0.16134 | -0.03666 |
| | 12.0 | -0.54718 | -0.41232 | -0.28320 | -0.15721 | -0.03218 |
| INFINITY | | -0.52440 | -0.38532 | -0.25335 | -0.12566 | 0.0 |

FRACTILES $W_\alpha$ RHO = 0.20

| DELTAS | | $\alpha$:0.5500 | 0.6000 | 0.6500 | 0.7000 | 0.7500 |
|---|---|---|---|---|---|---|
| 0.0 | 0.0 | -0.09397 | -0.02303 | 0.06580 | 0.17679 | 0.31714 |
| 0.4 | 0.0 | -0.08562 | -0.01552 | 0.07243 | 0.18245 | 0.32164 |
| | 0.4 | -0.14877 | -0.06751 | 0.03241 | 0.15568 | 0.30992 |
| 0.8 | 0.0 | -0.06700 | 0.00332 | 0.09181 | 0.20242 | 0.34200 |
| | 0.4 | -0.16438 | -0.07222 | 0.03881 | 0.17324 | 0.33841 |
| | 0.8 | -0.18787 | -0.07864 | 0.04853 | 0.19801 | 0.37684 |
| 1.2 | 0.0 | -0.04604 | 0.02892 | 0.12259 | 0.23835 | 0.38239 |
| | 0.4 | -0.14324 | -0.04062 | 0.07958 | 0.22140 | 0.39148 |
| | 0.8 | -0.16267 | -0.04380 | 0.09085 | 0.24548 | 0.42678 |
| | 1.2 | -0.14265 | -0.01689 | 0.12297 | 0.28117 | 0.46424 |
| 1.6 | 0.0 | -0.02480 | 0.05909 | 0.16124 | 0.28414 | 0.43317 |
| | 0.4 | -0.10447 | 0.00580 | 0.13131 | 0.27580 | 0.44544 |
| | 0.8 | -0.12127 | 0.00139 | 0.13783 | 0.29213 | 0.47066 |
| | 1.2 | -0.11103 | 0.01690 | 0.15763 | 0.31529 | 0.49616 |
| | 1.6 | -0.09098 | 0.03892 | 0.18078 | 0.33864 | 0.51863 |
| 2.0 | 0.0 | -0.00350 | 0.09075 | 0.20101 | 0.32946 | 0.48111 |
| | 0.4 | -0.06397 | 0.05066 | 0.17830 | 0.32265 | 0.48959 |
| | 0.8 | -0.08124 | 0.04256 | 0.17860 | 0.33085 | 0.50535 |
| | 1.2 | -0.07943 | 0.04888 | 0.18890 | 0.34463 | 0.52209 |
| | 1.6 | -0.06853 | 0.06184 | 0.20339 | 0.36007 | 0.53778 |
| | 2.0 | -0.05373 | 0.07738 | 0.21913 | 0.37537 | 0.55187 |
| 2.4 | 0.0 | 0.01660 | 0.11961 | 0.23568 | 0.36737 | 0.51973 |
| | 0.4 | -0.02959 | 0.08748 | 0.21589 | 0.35929 | 0.52334 |
| | 0.8 | -0.04766 | 0.07645 | 0.21162 | 0.36166 | 0.53237 |
| | 1.2 | -0.05137 | 0.07672 | 0.21561 | 0.36914 | 0.54312 |
| | 1.6 | -0.04684 | 0.08337 | 0.22404 | 0.37901 | 0.55400 |
| | 2.0 | -0.03786 | 0.09332 | 0.23458 | 0.38970 | 0.56432 |
| | 2.4 | -0.02680 | 0.10467 | 0.24582 | 0.40037 | 0.57385 |
| 2.8 | 0.0 | 0.03379 | 0.14305 | 0.26295 | 0.39655 | 0.54894 |
| | 0.4 | -0.00274 | 0.11600 | 0.24482 | 0.38734 | 0.54900 |
| | 0.8 | -0.02069 | 0.10359 | 0.23795 | 0.38611 | 0.55362 |
| | 1.2 | -0.02751 | 0.10024 | 0.23800 | 0.38950 | 0.56031 |
| | 1.6 | -0.02716 | 0.10265 | 0.24229 | 0.39547 | 0.56775 |
| | 2.0 | -0.02239 | 0.10852 | 0.24900 | 0.40275 | 0.57526 |
| | 2.4 | -0.01512 | 0.11627 | 0.25694 | 0.41055 | 0.58251 |
| | 2.8 | -0.00662 | 0.12486 | 0.26532 | 0.41838 | 0.58933 |
| 3.2 | 0.0 | 0.04738 | 0.16096 | 0.28355 | 0.41847 | 0.57080 |
| | 0.4 | 0.01774 | 0.13783 | 0.26703 | 0.40891 | 0.56873 |
| | 0.8 | 0.00075 | 0.12523 | 0.25898 | 0.40564 | 0.57056 |
| | 1.2 | -0.00758 | 0.11988 | 0.25666 | 0.40640 | 0.57449 |
| | 1.6 | -0.00985 | 0.11953 | 0.25816 | 0.40968 | 0.57946 |
| | 2.0 | -0.00804 | 0.12248 | 0.26210 | 0.41445 | 0.58486 |
| | 2.4 | -0.00364 | 0.12748 | 0.26750 | 0.42002 | 0.59033 |
| | 2.8 | 0.00228 | 0.13364 | 0.27368 | 0.42594 | 0.59567 |
| | 3.2 | 0.00899 | 0.14035 | 0.28015 | 0.43191 | 0.60079 |

FRACTILES $W_\alpha$ RHO = 0.20

| DELTAS | | $\alpha$:0.5500 | 0.6000 | 0.6500 | 0.7000 | 0.7500 |
|---|---|---|---|---|---|---|
| 3.6 | 0.0 | 0.05780 | 0.17446 | 0.29905 | 0.43502 | 0.58735 |
| | 0.4 | 0.03340 | 0.15465 | 0.28425 | 0.42571 | 0.58416 |
| | 0.8 | 0.01783 | 0.14255 | 0.27589 | 0.42138 | 0.58424 |
| | 1.2 | 0.00897 | 0.13622 | 0.27220 | 0.42048 | 0.58629 |
| | 1.6 | 0.00513 | 0.13413 | 0.27186 | 0.42190 | 0.58948 |
| | 2.0 | 0.00491 | 0.13504 | 0.27382 | 0.42483 | 0.59329 |
| | 2.4 | 0.00719 | 0.13797 | 0.27730 | 0.42870 | 0.59736 |
| | 2.8 | 0.01109 | 0.14221 | 0.28171 | 0.43310 | 0.60152 |
| | 3.2 | 0.01597 | 0.14721 | 0.28664 | 0.43774 | 0.60561 |
| | 3.6 | 0.02138 | 0.15258 | 0.29178 | 0.44244 | 0.60958 |
| 4.0 | 0.0 | 0.06585 | 0.18475 | 0.31090 | 0.44771 | 0.60009 |
| | 0.4 | 0.04554 | 0.16779 | 0.29779 | 0.43899 | 0.59639 |
| | 0.8 | 0.03153 | 0.15653 | 0.28959 | 0.43419 | 0.59540 |
| | 1.2 | 0.02270 | 0.14983 | 0.28519 | 0.43227 | 0.59618 |
| | 1.6 | 0.01800 | 0.14668 | 0.28365 | 0.43242 | 0.59809 |
| | 2.0 | 0.01643 | 0.14618 | 0.28421 | 0.43401 | 0.60069 |
| | 2.4 | 0.01716 | 0.14760 | 0.28624 | 0.43657 | 0.60368 |
| | 2.8 | 0.01950 | 0.15034 | 0.28927 | 0.43976 | 0.60687 |
| | 3.2 | 0.02292 | 0.15394 | 0.29293 | 0.44332 | 0.61013 |
| | 3.6 | 0.02700 | 0.15807 | 0.29695 | 0.44706 | 0.61336 |
| | 4.0 | 0.03145 | 0.16247 | 0.30113 | 0.45084 | 0.61652 |
| 6.0 | 0.0 | 0.08812 | 0.21241 | 0.34251 | 0.48160 | 0.63420 |
| | 0.4 | 0.07867 | 0.20396 | 0.33540 | 0.47622 | 0.63099 |
| | 0.8 | 0.07078 | 0.19705 | 0.32975 | 0.47211 | 0.62877 |
| | 1.2 | 0.06447 | 0.19165 | 0.32545 | 0.46914 | 0.62738 |
| | 1.6 | 0.05967 | 0.18764 | 0.32238 | 0.46716 | 0.62668 |
| | 2.0 | 0.05625 | 0.18488 | 0.32037 | 0.46601 | 0.62651 |
| | 2.4 | 0.05402 | 0.18318 | 0.31926 | 0.46555 | 0.62677 |
| | 2.8 | 0.05281 | 0.18238 | 0.31890 | 0.46565 | 0.62735 |
| | 3.2 | 0.05245 | 0.18233 | 0.31914 | 0.46619 | 0.62818 |
| | 3.6 | 0.05277 | 0.18286 | 0.31987 | 0.46707 | 0.62919 |
| | 4.0 | 0.05363 | 0.18385 | 0.32096 | 0.46822 | 0.63033 |
| | 6.0 | 0.06230 | 0.19253 | 0.32933 | 0.47590 | 0.63685 |
| 12.0 | 0.0 | 0.10803 | 0.23527 | 0.36745 | 0.50753 | 0.65967 |
| | 0.4 | 0.10560 | 0.23301 | 0.36545 | 0.50590 | 0.65852 |
| | 0.8 | 0.10332 | 0.23090 | 0.36360 | 0.50440 | 0.65749 |
| | 1.2 | 0.10121 | 0.22896 | 0.36190 | 0.50303 | 0.65657 |
| | 1.6 | 0.09928 | 0.22719 | 0.36036 | 0.50181 | 0.65576 |
| | 2.0 | 0.09752 | 0.22559 | 0.35899 | 0.50073 | 0.65506 |
| | 2.4 | 0.09595 | 0.22417 | 0.35777 | 0.49978 | 0.65445 |
| | 2.8 | 0.09456 | 0.22292 | 0.35671 | 0.49896 | 0.65395 |
| | 3.2 | 0.09335 | 0.22184 | 0.35580 | 0.49827 | 0.65353 |
| | 3.6 | 0.09231 | 0.22092 | 0.35503 | 0.49769 | 0.65320 |
| | 4.0 | 0.09144 | 0.22015 | 0.35440 | 0.49723 | 0.65295 |
| | 6.0 | 0.08919 | 0.21826 | 0.35295 | 0.49630 | 0.65263 |
| | 12.0 | 0.09385 | 0.22296 | 0.35751 | 0.50052 | 0.65624 |
| INFINITY | | 0.12566 | 0.25335 | 0.38532 | 0.52440 | 0.67449 |

FRACTILES $w_{\alpha}$ RHO = 0.20

| DELTAS | | $\alpha$:0.8000 | 0.8500 | 0.9000 | 0.9250 | 0.9500 |
|---|---|---|---|---|---|---|
| 0.0 | 0.0 | 0.49930 | 0.74710 | 1.11479 | 1.38560 | 1.77798 |
| 0.4 | 0.0 | 0.50232 | 0.74807 | 1.11256 | 1.38085 | 1.76935 |
| | 0.4 | 0.50819 | 0.77524 | 1.16710 | 1.45293 | 1.86352 |
| 0.8 | 0.0 | 0.52244 | 0.76646 | 1.12556 | 1.38791 | 1.76513 |
| | 0.4 | 0.54684 | 0.82222 | 1.21782 | 1.50150 | 1.90348 |
| | 0.8 | 0.59707 | 0.88142 | 1.28089 | 1.56272 | 1.95734 |
| 1.2 | 0.0 | 0.56565 | 0.80907 | 1.16005 | 1.41221 | 1.76987 |
| | 0.4 | 0.60124 | 0.87232 | 1.25336 | 1.52226 | 1.89879 |
| | 0.8 | 0.64609 | 0.92469 | 1.31011 | 1.57902 | 1.95248 |
| | 1.2 | 0.68314 | 0.95826 | 1.33489 | 1.59562 | 1.95554 |
| 1.6 | 0.0 | 0.61815 | 0.85811 | 1.19612 | 1.43485 | 1.76925 |
| | 0.4 | 0.65078 | 0.91167 | 1.27255 | 1.52428 | 1.87381 |
| | 0.8 | 0.68408 | 0.95222 | 1.31918 | 1.57315 | 1.92368 |
| | 1.2 | 0.71069 | 0.97821 | 1.34151 | 1.59143 | 1.93475 |
| | 1.6 | 0.73084 | 0.99388 | 1.34884 | 1.59180 | 1.92420 |
| 2.0 | 0.0 | 0.66512 | 0.89912 | 1.22281 | 1.44852 | 1.76178 |
| | 0.4 | 0.68900 | 0.93931 | 1.28155 | 1.51823 | 1.84469 |
| | 0.8 | 0.71215 | 0.96980 | 1.31941 | 1.55978 | 1.88981 |
| | 1.2 | 0.73122 | 0.99032 | 1.33982 | 1.57897 | 1.90607 |
| | 1.6 | 0.74625 | 1.00331 | 1.34831 | 1.58341 | 1.90390 |
| | 2.0 | 0.75809 | 1.01133 | 1.34968 | 1.57940 | 1.89161 |
| 2.4 | 0.0 | 0.70156 | 0.92949 | 1.24059 | 1.45545 | 1.75152 |
| | 0.4 | 0.71739 | 0.95872 | 1.28561 | 1.51007 | 1.81799 |
| | 0.8 | 0.73324 | 0.98176 | 1.31654 | 1.54539 | 1.85819 |
| | 1.2 | 0.74700 | 0.99819 | 1.33501 | 1.56440 | 1.87695 |
| | 1.6 | 0.75835 | 1.00920 | 1.34421 | 1.57159 | 1.88057 |
| | 2.0 | 0.76760 | 1.01630 | 1.34723 | 1.57116 | 1.87467 |
| | 2.4 | 0.77521 | 1.02081 | 1.34650 | 1.56627 | 1.86346 |
| 2.8 | 0.0 | 0.72864 | 0.95148 | 1.25247 | 1.45873 | 1.74124 |
| | 0.4 | 0.73870 | 0.97283 | 1.28749 | 1.50222 | 1.79537 |
| | 0.8 | 0.74953 | 0.99046 | 1.31295 | 1.53228 | 1.83087 |
| | 1.2 | 0.75951 | 1.00372 | 1.32947 | 1.55036 | 1.85032 |
| | 1.6 | 0.76814 | 1.01311 | 1.33883 | 1.55911 | 1.85755 |
| | 2.0 | 0.77542 | 1.01947 | 1.34300 | 1.56125 | 1.85631 |
| | 2.4 | 0.78156 | 1.02366 | 1.34370 | 1.55910 | 1.84973 |
| | 2.8 | 0.78679 | 1.02637 | 1.34226 | 1.55439 | 1.84007 |
| 3.2 | 0.0 | 0.74880 | 0.96765 | 1.26071 | 1.46019 | 1.73200 |
| | 0.4 | 0.75505 | 0.98350 | 1.28842 | 1.49539 | 1.77672 |
| | 0.8 | 0.76243 | 0.99714 | 1.30954 | 1.52105 | 1.80794 |
| | 1.2 | 0.76967 | 1.00791 | 1.32417 | 1.53781 | 1.82702 |
| | 1.6 | 0.77624 | 1.01591 | 1.33329 | 1.54722 | 1.83629 |
| | 2.0 | 0.78199 | 1.02159 | 1.33813 | 1.55106 | 1.83827 |
| | 2.4 | 0.78697 | 1.02550 | 1.33990 | 1.55098 | 1.83522 |
| | 2.8 | 0.79130 | 1.02812 | 1.33960 | 1.54831 | 1.82892 |
| | 3.2 | 0.79508 | 1.02985 | 1.33797 | 1.54407 | 1.82073 |

FRACTILES $W_\alpha$ RHO = 0.20

| DELTAS | | $\alpha$:0.8000 | 0.8500 | 0.9000 | 0.9250 | 0.9500 |
|---|---|---|---|---|---|---|
| 3.6 | 0.0 | 0.76408 | 0.97983 | 1.26663 | 1.46073 | 1.72402 |
| | 0.4 | 0.76784 | 0.99178 | 1.28892 | 1.48964 | 1.76144 |
| | 0.8 | 0.77284 | 1.00246 | 1.30657 | 1.51164 | 1.78887 |
| | 1.2 | 0.77808 | 1.01125 | 1.31947 | 1.52695 | 1.80701 |
| | 1.6 | 0.78307 | 1.01806 | 1.32810 | 1.53646 | 1.81729 |
| | 2.0 | 0.78761 | 1.02311 | 1.33326 | 1.54133 | 1.82139 |
| | 2.4 | 0.79165 | 1.02674 | 1.33575 | 1.54273 | 1.82093 |
| | 2.8 | 0.79523 | 1.02926 | 1.33632 | 1.54167 | 1.81729 |
| | 3.2 | 0.79842 | 1.03098 | 1.33557 | 1.53895 | 1.81158 |
| | 3.6 | 0.80127 | 1.03213 | 1.33397 | 1.53521 | 1.80462 |
| 4.0 | 0.0 | 0.77586 | 0.98919 | 1.27102 | 1.46081 | 1.71721 |
| | 0.4 | 0.77800 | 0.99836 | 1.28921 | 1.48486 | 1.74888 |
| | 0.8 | 0.78135 | 1.00682 | 1.30409 | 1.50381 | 1.77302 |
| | 1.2 | 0.78512 | 1.01402 | 1.31542 | 1.51767 | 1.78996 |
| | 1.6 | 0.78890 | 1.01982 | 1.32346 | 1.52694 | 1.80058 |
| | 2.0 | 0.79247 | 1.02428 | 1.32866 | 1.53238 | 1.80600 |
| | 2.4 | 0.79575 | 1.02761 | 1.33160 | 1.53479 | 1.80739 |
| | 2.8 | 0.79872 | 1.03002 | 1.33282 | 1.53494 | 1.80579 |
| | 3.2 | 0.80140 | 1.03172 | 1.33277 | 1.53346 | 1.80210 |
| | 3.6 | 0.80384 | 1.03289 | 1.33185 | 1.53088 | 1.79701 |
| | 4.0 | 0.80606 | 1.03368 | 1.33034 | 1.52760 | 1.79107 |
| 6.0 | 0.0 | 0.80743 | 1.01405 | 1.28170 | 1.45896 | 1.69517 |
| | 0.4 | 0.80694 | 1.01709 | 1.28960 | 1.47019 | 1.71090 |
| | 0.8 | 0.80706 | 1.02016 | 1.29668 | 1.47999 | 1.72434 |
| | 1.2 | 0.80759 | 1.02309 | 1.30280 | 1.48823 | 1.73539 |
| | 1.6 | 0.80839 | 1.02575 | 1.30787 | 1.49489 | 1.74411 |
| | 2.0 | 0.80937 | 1.02808 | 1.31193 | 1.50004 | 1.75064 |
| | 2.4 | 0.81043 | 1.03007 | 1.31503 | 1.50381 | 1.75521 |
| | 2.8 | 0.81152 | 1.03172 | 1.31727 | 1.50636 | 1.75807 |
| | 3.2 | 0.81263 | 1.03306 | 1.31877 | 1.50787 | 1.75948 |
| | 3.6 | 0.81371 | 1.03412 | 1.31965 | 1.50852 | 1.75969 |
| | 4.0 | 0.81476 | 1.03496 | 1.32001 | 1.50846 | 1.75895 |
| | 6.0 | 0.81946 | 1.03683 | 1.31723 | 1.50203 | 1.74698 |
| 12.0 | 0.0 | 0.83033 | 1.03102 | 1.28643 | 1.45285 | 1.67136 |
| | 0.4 | 0.82985 | 1.03146 | 1.28823 | 1.45564 | 1.67556 |
| | 0.8 | 0.82944 | 1.03192 | 1.28995 | 1.45828 | 1.67950 |
| | 1.2 | 0.82911 | 1.03238 | 1.29159 | 1.46076 | 1.68317 |
| | 1.6 | 0.82884 | 1.03285 | 1.29313 | 1.46306 | 1.68655 |
| | 2.0 | 0.82863 | 1.03331 | 1.29455 | 1.46518 | 1.68963 |
| | 2.4 | 0.82848 | 1.03376 | 1.29586 | 1.46710 | 1.69241 |
| | 2.8 | 0.82837 | 1.03419 | 1.29705 | 1.46882 | 1.69487 |
| | 3.2 | 0.82831 | 1.03459 | 1.29812 | 1.47035 | 1.69704 |
| | 3.6 | 0.82829 | 1.03498 | 1.29907 | 1.47169 | 1.69892 |
| | 4.0 | 0.82830 | 1.03533 | 1.29990 | 1.47285 | 1.70052 |
| | 6.0 | 0.82873 | 1.03667 | 1.30244 | 1.47619 | 1.70488 |
| | 12.0 | 0.83135 | 1.03774 | 1.30089 | 1.47254 | 1.69804 |
| INFINITY | | 0.84162 | 1.03643 | 1.28155 | 1.43953 | 1.64485 |

FRACTILES $W_\alpha$ RHO = 0.20

| DELTAS | | $\alpha$:0.9625 | 0.9750 | 0.9825 | 0.9900 | 0.9925 |
|---|---|---|---|---|---|---|
| 0.0 | 0.0 | 2.06253 | 2.47059 | 2.83518 | 3.41587 | 3.71784 |
| 0.4 | 0.0 | 2.05089 | 2.45435 | 2.81458 | 3.38779 | 3.68563 |
| | 0.4 | 2.15897 | 2.57967 | 2.95292 | 3.54285 | 3.84765 |
| 0.8 | 0.0 | 2.03673 | 2.42363 | 2.76699 | 3.30982 | 3.59032 |
| | 0.4 | 2.18946 | 2.59288 | 2.94770 | 3.50372 | 3.78909 |
| | 0.8 | 2.23546 | 2.62492 | 2.96518 | 3.49499 | 3.76557 |
| 1.2 | 0.0 | 2.02447 | 2.38375 | 2.69981 | 3.19517 | 3.44943 |
| | 0.4 | 2.16418 | 2.53578 | 2.86045 | 3.36596 | 3.62412 |
| | 0.8 | 2.21397 | 2.57822 | 2.89493 | 3.38576 | 3.63551 |
| | 1.2 | 2.20631 | 2.55421 | 2.85556 | 3.32080 | 3.55680 |
| 1.6 | 0.0 | 2.00497 | 2.33503 | 2.62338 | 3.07231 | 3.30157 |
| | 0.4 | 2.11850 | 2.45928 | 2.75554 | 3.21460 | 3.44814 |
| | 0.8 | 2.16787 | 2.50660 | 2.79997 | 3.25285 | 3.48256 |
| | 1.2 | 2.17297 | 2.50231 | 2.78666 | 3.22420 | 3.44555 |
| | 1.6 | 2.15405 | 2.47091 | 2.74372 | 3.16230 | 3.37357 |
| 2.0 | 0.0 | 1.98098 | 2.28617 | 2.55140 | 2.96224 | 3.17121 |
| | 0.4 | 2.07201 | 2.38723 | 2.66015 | 3.08131 | 3.29489 |
| | 0.8 | 2.11873 | 2.43514 | 2.70825 | 3.12840 | 3.34091 |
| | 1.2 | 2.13222 | 2.44393 | 2.71227 | 3.12395 | 3.33172 |
| | 1.6 | 2.12484 | 2.42863 | 2.68953 | 3.08882 | 3.28993 |
| | 2.0 | 2.10629 | 2.40081 | 2.65321 | 3.03864 | 3.23241 |
| 2.4 | 0.0 | 1.95751 | 2.24295 | 2.48996 | 2.87091 | 3.06402 |
| | 0.4 | 2.03141 | 2.32622 | 2.58057 | 2.97166 | 3.16941 |
| | 0.8 | 2.07433 | 2.37212 | 2.62838 | 3.02138 | 3.21967 |
| | 1.2 | 2.09234 | 2.38842 | 2.64263 | 3.03159 | 3.22747 |
| | 1.6 | 2.09298 | 2.38436 | 2.63404 | 3.01526 | 3.20690 |
| | 2.0 | 2.08287 | 2.36793 | 2.61175 | 2.98329 | 3.16978 |
| | 2.4 | 2.06690 | 2.34496 | 2.58239 | 2.94355 | 3.12455 |
| 2.8 | 0.0 | 1.93683 | 2.20678 | 2.43948 | 2.79702 | 2.97770 |
| | 0.4 | 1.99773 | 2.27632 | 2.51591 | 2.88310 | 3.06828 |
| | 0.8 | 2.03648 | 2.31894 | 2.56135 | 2.93206 | 3.11867 |
| | 1.2 | 2.05643 | 2.33903 | 2.58111 | 2.95058 | 3.13627 |
| | 1.6 | 2.06220 | 2.34233 | 2.58187 | 2.94683 | 3.12997 |
| | 2.0 | 2.05828 | 2.33429 | 2.56993 | 2.92834 | 3.10795 |
| | 2.4 | 2.04831 | 2.31929 | 2.55029 | 2.90110 | 3.07667 |
| | 2.8 | 2.03496 | 2.30052 | 2.52658 | 2.86940 | 3.04075 |
| 3.2 | 0.0 | 1.91937 | 2.17705 | 2.39842 | 2.73740 | 2.90823 |
| | 0.4 | 1.97023 | 2.23582 | 2.46357 | 2.81159 | 2.98668 |
| | 0.8 | 2.00489 | 2.27474 | 2.50575 | 2.85812 | 3.03513 |
| | 1.2 | 2.02521 | 2.29634 | 2.52807 | 2.88097 | 3.05800 |
| | 1.6 | 2.03406 | 2.30422 | 2.53479 | 2.88538 | 3.06103 |
| | 2.0 | 2.03446 | 2.30210 | 2.53022 | 2.87657 | 3.04988 |
| | 2.4 | 2.02909 | 2.29323 | 2.51806 | 2.85898 | 3.02938 |
| | 2.8 | 2.02005 | 2.28013 | 2.50125 | 2.83609 | 3.00327 |
| | 3.2 | 2.00892 | 2.26470 | 2.48190 | 2.81042 | 2.97427 |

FRACTILES $W_\alpha$ RHO = 0.20

| DELTAS | | α:0.9625 | 0.9750 | 0.9825 | 0.9900 | 0.9925 |
|---|---|---|---|---|---|---|
| 3.6 | 0.0 | 1.90482 | 2.15267 | 2.36495 | 2.68900 | 2.85190 |
| | 0.4 | 1.94778 | 2.20285 | 2.42099 | 2.75345 | 2.92034 |
| | 0.8 | 1.97866 | 2.23809 | 2.45966 | 2.79684 | 2.96589 |
| | 1.2 | 1.99847 | 2.25985 | 2.48279 | 2.82160 | 2.99127 |
| | 1.6 | 2.00901 | 2.27043 | 2.49315 | 2.83117 | 3.00025 |
| | 2.0 | 2.01233 | 2.27239 | 2.49369 | 2.82914 | 2.99677 |
| | 2.4 | 2.01036 | 2.26809 | 2.48716 | 2.81883 | 2.98441 |
| | 2.8 | 2.00475 | 2.25951 | 2.47583 | 2.80298 | 2.96614 |
| | 3.2 | 1.99678 | 2.24822 | 2.46151 | 2.78371 | 2.94426 |
| | 3.6 | 1.98744 | 2.23539 | 2.44551 | 2.76260 | 2.92046 |
| 4.0 | 0.0 | 1.89270 | 2.13256 | 2.33745 | 2.64934 | 2.80577 |
| | 0.4 | 1.92937 | 2.17580 | 2.38606 | 2.70571 | 2.86586 |
| | 0.8 | 1.95685 | 2.20759 | 2.42128 | 2.74577 | 2.90817 |
| | 1.2 | 1.97569 | 2.22876 | 2.44422 | 2.77101 | 2.93440 |
| | 1.6 | 1.98702 | 2.24081 | 2.45666 | 2.78370 | 2.94706 |
| | 2.0 | 1.99223 | 2.24549 | 2.46068 | 2.78638 | 2.94892 |
| | 2.4 | 1.99273 | 2.24453 | 2.45829 | 2.78147 | 2.94263 |
| | 2.8 | 1.98976 | 2.23947 | 2.45125 | 2.77113 | 2.93050 |
| | 3.2 | 1.98438 | 2.23157 | 2.44103 | 2.75710 | 2.91445 |
| | 3.6 | 1.97741 | 2.22184 | 2.42878 | 2.74076 | 2.89595 |
| | 4.0 | 1.96949 | 2.21104 | 2.41537 | 2.72314 | 2.87612 |
| 6.0 | 0.0 | 1.85489 | 2.07095 | 2.25364 | 2.52682 | 2.66565 |
| | 0.4 | 1.87368 | 2.09388 | 2.28005 | 2.56041 | 2.69977 |
| | 0.8 | 1.88957 | 2.11306 | 2.30196 | 2.58637 | 2.72770 |
| | 1.2 | 1.90250 | 2.12847 | 2.31943 | 2.60682 | 2.74959 |
| | 1.6 | 1.91256 | 2.14029 | 2.33268 | 2.62210 | 2.76582 |
| | 2.0 | 1.91998 | 2.14882 | 2.34208 | 2.63269 | 2.77694 |
| | 2.4 | 1.92502 | 2.15443 | 2.34808 | 2.63916 | 2.78358 |
| | 2.8 | 1.92802 | 2.15752 | 2.35117 | 2.64210 | 2.78640 |
| | 3.2 | 1.92928 | 2.15849 | 2.35182 | 2.64212 | 2.78604 |
| | 3.6 | 1.92913 | 2.15774 | 2.35047 | 2.63975 | 2.78310 |
| | 4.0 | 1.92783 | 2.15560 | 2.34754 | 2.63548 | 2.77811 |
| | 6.0 | 1.91172 | 2.13340 | 2.31979 | 2.59870 | 2.73656 |
| 12.0 | 0.0 | 1.81711 | 2.01186 | 2.17446 | 2.41605 | 2.53476 |
| | 0.4 | 1.82231 | 2.01846 | 2.18229 | 2.42578 | 2.54545 |
| | 0.8 | 1.82718 | 2.02462 | 2.18958 | 2.43479 | 2.55534 |
| | 1.2 | 1.83169 | 2.03031 | 2.19628 | 2.44306 | 2.56440 |
| | 1.6 | 1.83583 | 2.03550 | 2.20238 | 2.45056 | 2.57260 |
| | 2.0 | 1.83958 | 2.04019 | 2.20788 | 2.45730 | 2.57996 |
| | 2.4 | 1.84295 | 2.04438 | 2.21277 | 2.46327 | 2.58647 |
| | 2.8 | 1.84594 | 2.04807 | 2.21708 | 2.46850 | 2.59216 |
| | 3.2 | 1.84855 | 2.05129 | 2.22081 | 2.47301 | 2.59706 |
| | 3.6 | 1.85080 | 2.05404 | 2.22399 | 2.47683 | 2.60120 |
| | 4.0 | 1.85270 | 2.05636 | 2.22665 | 2.48000 | 2.60462 |
| | 6.0 | 1.85773 | 2.06224 | 2.23322 | 2.48752 | 2.61256 |
| | 12.0 | 1.84846 | 2.04939 | 2.21706 | 2.46595 | 2.58811 |
| INFINITY | | 1.78046 | 1.95996 | 2.10836 | 2.32635 | 2.43238 |

FRACTILES $w_\alpha$ RHO = 0.20

| DELTAS | | $\alpha$:0.9950 | 0.9975 | 0.9990 | 0.9995 |
|---|---|---|---|---|---|
| 0.0 | 0.0 | 4.14679 | 4.88781 | 5.87937 | 6.63665 |
| 0.4 | 0.0 | 4.10844 | 4.83811 | 5.81308 | 6.55669 |
| | 0.4 | 4.27856 | 5.01781 | 5.99823 | 6.74126 |
| 0.8 | 0.0 | 3.98692 | 4.66736 | 5.56989 | 6.25392 |
| | 0.4 | 4.19064 | 4.87525 | 5.77654 | 6.45561 |
| | 0.8 | 4.14504 | 4.78907 | 5.63242 | 6.26510 |
| 1.2 | 0.0 | 3.80722 | 4.41724 | 5.22035 | 5.82544 |
| | 0.4 | 3.98617 | 4.60062 | 5.40519 | 6.00875 |
| | 0.8 | 3.98484 | 4.57562 | 5.34589 | 5.92169 |
| | 1.2 | 3.88620 | 4.44157 | 5.16300 | 5.70061 |
| 1.6 | 0.0 | 3.62305 | 4.16858 | 4.88284 | 5.41860 |
| | 0.4 | 3.77480 | 4.32715 | 5.04724 | 5.58545 |
| | 0.8 | 3.80316 | 4.34367 | 5.04574 | 5.56888 |
| | 1.2 | 3.75390 | 4.27241 | 4.94373 | 5.44259 |
| | 1.6 | 3.66739 | 4.16031 | 4.79664 | 5.26832 |
| 2.0 | 0.0 | 3.46343 | 3.95743 | 4.60126 | 5.08239 |
| | 0.4 | 3.59292 | 4.09526 | 4.74759 | 5.23358 |
| | 0.8 | 3.63693 | 4.13460 | 4.77882 | 5.25748 |
| | 1.2 | 3.62066 | 4.10534 | 4.73097 | 5.19469 |
| | 1.6 | 3.56920 | 4.03671 | 4.63862 | 5.08379 |
| | 2.0 | 3.50114 | 3.95015 | 4.52688 | 4.95256 |
| 2.4 | 0.0 | 3.33340 | 3.78730 | 4.37647 | 4.81525 |
| | 0.4 | 3.44478 | 3.90761 | 4.50650 | 4.95136 |
| | 0.8 | 3.49536 | 3.95772 | 4.55436 | 4.99650 |
| | 1.2 | 3.49943 | 3.95463 | 4.54058 | 4.97388 |
| | 1.6 | 3.47265 | 3.91666 | 4.48691 | 4.90778 |
| | 2.0 | 3.42808 | 3.85892 | 4.41110 | 4.81789 |
| | 2.4 | 3.37499 | 3.79207 | 4.32558 | 4.71795 |
| 2.8 | 0.0 | 3.22921 | 3.65171 | 4.19812 | 4.60381 |
| | 0.4 | 3.32567 | 3.75714 | 4.31365 | 4.72590 |
| | 0.8 | 3.37771 | 3.81113 | 4.36884 | 4.78112 |
| | 1.2 | 3.39371 | 3.82374 | 4.37588 | 4.78328 |
| | 1.6 | 3.38361 | 3.80661 | 4.34865 | 4.74791 |
| | 2.0 | 3.35645 | 3.77027 | 4.29955 | 4.68879 |
| | 2.4 | 3.31935 | 3.72293 | 4.23822 | 4.61659 |
| | 2.8 | 3.27739 | 3.67043 | 4.17143 | 4.53878 |
| 3.2 | 0.0 | 3.14555 | 3.54315 | 4.05562 | 4.43503 |
| | 0.4 | 3.22963 | 3.63593 | 4.15841 | 4.54448 |
| | 0.8 | 3.28047 | 3.69009 | 4.21579 | 4.60353 |
| | 1.2 | 3.30311 | 3.71177 | 4.23522 | 4.62068 |
| | 1.6 | 3.30400 | 3.70852 | 4.22578 | 4.60609 |
| | 2.0 | 3.28941 | 3.68769 | 4.19613 | 4.56942 |
| | 2.4 | 3.26468 | 3.65546 | 4.15355 | 4.51874 |
| | 2.8 | 3.23394 | 3.61660 | 4.10361 | 4.46023 |
| | 3.2 | 3.20018 | 3.57453 | 4.05031 | 4.39826 |

FRACTILES $W_{\alpha}$ RHO = 0.20

| DELTAS | | $\alpha$:0.9950 | 0.9975 | 0.9990 | 0.9995 |
|---|---|---|---|---|---|
| 3.6 | 0.0 | 3.07781 | 3.45534 | 3.94043 | 4.29364 |
| | 0.4 | 3.15155 | 3.53736 | 4.03215 | 4.39690 |
| | 0.8 | 3.19987 | 3.58977 | 4.08892 | 4.45632 |
| | 1.2 | 3.22591 | 3.61640 | 4.11549 | 4.48231 |
| | 1.6 | 3.23389 | 3.62226 | 4.11788 | 4.48165 |
| | 2.0 | 3.22821 | 3.61251 | 4.10221 | 4.46117 |
| | 2.4 | 3.21285 | 3.59176 | 4.07393 | 4.42694 |
| | 2.8 | 3.19109 | 3.56382 | 4.03750 | 4.38390 |
| | 3.2 | 3.16545 | 3.53160 | 3.99634 | 4.33582 |
| | 3.6 | 3.13781 | 3.49727 | 3.95295 | 4.28546 |
| 4.0 | 0.0 | 3.02235 | 3.38347 | 3.84615 | 4.18700 |
| | 0.4 | 3.08740 | 3.45634 | 3.92827 | 4.27542 |
| | 0.8 | 3.13266 | 3.50606 | 3.98299 | 4.33335 |
| | 1.2 | 3.16010 | 3.53510 | 4.01339 | 4.36431 |
| | 1.6 | 3.17255 | 3.54683 | 4.02356 | 4.37291 |
| | 2.0 | 3.17312 | 3.54491 | 4.01786 | 4.36405 |
| | 2.4 | 3.16477 | 3.53279 | 4.00038 | 4.34226 |
| | 2.8 | 3.15006 | 3.51345 | 3.97462 | 4.31146 |
| | 3.2 | 3.13108 | 3.48933 | 3.94345 | 4.27481 |
| | 3.6 | 3.10949 | 3.46232 | 3.90908 | 4.23475 |
| | 4.0 | 3.08649 | 3.43382 | 3.87314 | 4.19310 |
| 6.0 | 0.0 | 2.85389 | 3.16496 | 3.55904 | 3.84655 |
| | 0.4 | 2.89146 | 3.20810 | 3.60901 | 3.90132 |
| | 0.8 | 2.92204 | 3.24294 | 3.64898 | 3.94486 |
| | 1.2 | 2.94585 | 3.26977 | 3.67940 | 3.97770 |
| | 1.6 | 2.96334 | 3.28919 | 3.70098 | 4.00069 |
| | 2.0 | 2.97514 | 3.30194 | 3.71468 | 4.01490 |
| | 2.4 | 2.98195 | 3.30889 | 3.72153 | 4.02150 |
| | 2.8 | 2.98453 | 3.31092 | 3.72259 | 4.02168 |
| | 3.2 | 2.98359 | 3.30886 | 3.71887 | 4.01657 |
| | 3.6 | 2.97980 | 3.30352 | 3.71131 | 4.00722 |
| | 4.0 | 2.97375 | 3.29558 | 3.70073 | 3.99455 |
| | 6.0 | 2.92537 | 3.23523 | 3.62411 | 3.90536 |
| 12.0 | 0.0 | 2.69664 | 2.96071 | 3.28963 | 3.52598 |
| | 0.4 | 2.70867 | 2.97499 | 3.30678 | 3.54523 |
| | 0.8 | 2.71978 | 2.98813 | 3.32252 | 3.56267 |
| | 1.2 | 2.72993 | 3.00011 | 3.33681 | 3.57885 |
| | 1.6 | 2.73911 | 3.01090 | 3.34966 | 3.59319 |
| | 2.0 | 2.74732 | 3.02053 | 3.36107 | 3.60590 |
| | 2.4 | 2.75458 | 3.02901 | 3.37109 | 3.61701 |
| | 2.8 | 2.76090 | 3.03637 | 3.37974 | 3.62660 |
| | 3.2 | 2.76632 | 3.04265 | 3.38710 | 3.63471 |
| | 3.6 | 2.77089 | 3.04792 | 3.39321 | 3.64142 |
| | 4.0 | 2.77465 | 3.05222 | 3.39817 | 3.64683 |
| | 6.0 | 2.78315 | 3.06151 | 3.40824 | 3.65732 |
| | 12.0 | 2.75454 | 3.02557 | 3.36225 | 3.60348 |
| INFINITY | | 2.57583 | 2.80703 | 3.09024 | 3.29051 |

FRACTILES $W_\alpha$ RHO = 0.40

| DELTAS | | $\alpha$:0.0005 | 0.0010 | 0.0025 | 0.0050 | 0.0075 |
|---|---|---|---|---|---|---|
| 0.0 | 0.0 | -3.42136 | -3.06372 | -2.59579 | -2.24647 | -2.04447 |
| 0.4 | 0.0 | -3.54080 | -3.17375 | -2.69168 | -2.33028 | -2.12065 |
| | 0.4 | -2.99623 | -2.69268 | -2.29561 | -1.99927 | -1.82797 |
| 0.8 | 0.0 | -3.59194 | -3.24070 | -2.77493 | -2.42158 | -2.21457 |
| | 0.4 | -2.88260 | -2.59955 | -2.22793 | -1.94949 | -1.78806 |
| | 0.8 | -2.41137 | -2.18783 | -1.89561 | -1.67775 | -1.55195 |
| 1.2 | 0.0 | -3.50529 | -3.18471 | -2.75680 | -2.42947 | -2.23633 |
| | 0.4 | -2.90005 | -2.63412 | -2.28149 | -2.01402 | -1.85734 |
| | 0.8 | -2.36631 | -2.15622 | -1.88061 | -1.67435 | -1.55492 |
| | 1.2 | -2.06983 | -1.90194 | -1.68275 | -1.51967 | -1.42569 |
| 1.6 | 0.0 | -3.38615 | -3.09445 | -2.70304 | -2.40171 | -2.22298 |
| | 0.4 | -2.89620 | -2.64896 | -2.31884 | -2.06621 | -1.91709 |
| | 0.8 | -2.44812 | -2.24470 | -1.97500 | -1.77047 | -1.65073 |
| | 1.2 | -2.08959 | -1.92756 | -1.71528 | -1.55677 | -1.46521 |
| | 1.6 | -1.91277 | -1.78105 | -1.60947 | -1.48227 | -1.40926 |
| 2.0 | 0.0 | -3.27978 | -3.01169 | -2.65031 | -2.37057 | -2.20388 |
| | 0.4 | -2.88250 | -2.65106 | -2.34029 | -2.10082 | -1.95866 |
| | 0.8 | -2.51619 | -2.32075 | -2.05965 | -1.85969 | -1.74161 |
| | 1.2 | -2.20419 | -2.04227 | -1.82763 | -1.66498 | -1.56983 |
| | 1.6 | -1.96531 | -1.83520 | -1.66515 | -1.53870 | -1.46601 |
| | 2.0 | -1.86460 | -1.75755 | -1.61868 | -1.51646 | -1.45829 |
| 2.4 | 0.0 | -3.19556 | -2.94612 | -2.60848 | -2.34581 | -2.18867 |
| | 0.4 | -2.86976 | -2.65100 | -2.35580 | -2.12694 | -1.99040 |
| | 0.8 | -2.56776 | -2.37925 | -2.12583 | -1.93026 | -1.81401 |
| | 1.2 | -2.30524 | -2.14518 | -1.93115 | -1.76709 | -1.67012 |
| | 1.6 | -2.08924 | -1.95562 | -1.77862 | -1.64468 | -1.56651 |
| | 2.0 | -1.93159 | -1.82390 | -1.68381 | -1.58054 | -1.52183 |
| | 2.4 | -1.88199 | -1.79293 | -1.67836 | -1.59550 | -1.54959 |
| 2.8 | 0.0 | -3.13186 | -2.89708 | -2.57807 | -2.32878 | -2.17909 |
| | 0.4 | -2.86121 | -2.65253 | -2.36965 | -2.14915 | -2.01701 |
| | 0.8 | -2.60941 | -2.42653 | -2.17935 | -1.98730 | -1.87249 |
| | 1.2 | -2.38764 | -2.22935 | -2.01623 | -1.85140 | -1.75322 |
| | 1.6 | -2.20139 | -2.06613 | -1.88506 | -1.74608 | -1.66386 |
| | 2.0 | -2.05326 | -1.93987 | -1.78990 | -1.67689 | -1.61136 |
| | 2.4 | -1.95410 | -1.86324 | -1.74627 | -1.66213 | -1.61650 |
| | 2.8 | -1.94155 | -1.86678 | -1.77294 | -1.71240 | -1.67563 |
| 3.2 | 0.0 | -3.08476 | -2.86156 | -2.55723 | -2.31842 | -2.17453 |
| | 0.4 | -2.85716 | -2.65648 | -2.38332 | -2.16934 | -2.04057 |
| | 0.8 | -2.64491 | -2.46655 | -2.22426 | -2.03487 | -1.92108 |
| | 1.2 | -2.45624 | -2.29935 | -2.08679 | -1.92112 | -1.82179 |
| | 1.6 | -2.29575 | -2.15920 | -1.97490 | -1.83189 | -1.74645 |
| | 2.0 | -2.16558 | -2.04830 | -1.89106 | -1.77019 | -1.69866 |
| | 2.4 | -2.06767 | -1.96947 | -1.84022 | -1.74436 | -1.69111 |
| | 2.8 | -2.01402 | -1.93709 | -1.84196 | -1.77750 | -1.73058 |
| | 3.2 | -2.03117 | -1.97043 | -1.89611 | -1.82251 | -1.77086 |

FRACTILES $W_\alpha$ RHO = 0.40

| DELTAS | | $\alpha$: 0.0005 | 0.0010 | 0.0025 | 0.0050 | 0.0075 |
|---|---|---|---|---|---|---|
| 3.6 | 0.0 | -3.05048 | -2.83651 | -2.54382 | -2.31324 | -2.17387 |
| | 0.4 | -2.85698 | -2.66268 | -2.39717 | -2.18824 | -2.06203 |
| | 0.8 | -2.67630 | -2.50152 | -2.26299 | -2.07550 | -1.96233 |
| | 1.2 | -2.51464 | -2.35875 | -2.14636 | -1.97967 | -1.87913 |
| | 1.6 | -2.37582 | -2.23799 | -2.05062 | -1.90389 | -1.81551 |
| | 2.0 | -2.26195 | -2.14134 | -1.97799 | -1.85060 | -1.77413 |
| | 2.4 | -2.17436 | -2.07046 | -1.93096 | -1.82385 | -1.76141 |
| | 2.8 | -2.11594 | -2.03000 | -1.92028 | -1.83911 | -1.78366 |
| | 3.2 | -2.10215 | -2.04241 | -1.95335 | -1.86967 | -1.81228 |
| | 3.6 | -2.14910 | -2.09017 | -1.99479 | -1.90527 | -1.84415 |
| 4.0 | 0.0 | -3.02598 | -2.81942 | -2.53599 | -2.31190 | -2.17607 |
| | 0.4 | -2.85989 | -2.67066 | -2.41119 | -2.20612 | -2.08182 |
| | 0.8 | -2.70472 | -2.53272 | -2.29701 | -2.11079 | -1.99791 |
| | 1.2 | -2.56524 | -2.40999 | -2.19739 | -2.02948 | -1.92768 |
| | 1.6 | -2.44461 | -2.30546 | -2.11508 | -1.96477 | -1.87361 |
| | 2.0 | -2.34481 | -2.22108 | -2.05203 | -1.91868 | -1.83778 |
| | 2.4 | -2.26718 | -2.15836 | -2.01016 | -1.89372 | -1.82355 |
| | 2.8 | -2.21326 | -2.11957 | -1.99465 | -1.89734 | -1.83379 |
| | 3.2 | -2.18915 | -2.11700 | -2.01139 | -1.91661 | -1.85308 |
| | 3.6 | -2.21296 | -2.14595 | -2.04005 | -1.94266 | -1.87700 |
| | 4.0 | -2.25704 | -2.18649 | -2.07453 | -1.97192 | -1.90305 |
| 6.0 | 0.0 | -2.98485 | -2.79896 | -2.54062 | -2.33330 | -2.20614 |
| | 0.4 | -2.89862 | -2.72289 | -2.47841 | -2.28190 | -2.16119 |
| | 0.8 | -2.81823 | -2.65254 | -2.42170 | -2.23577 | -2.12132 |
| | 1.2 | -2.74525 | -2.58936 | -2.37174 | -2.19597 | -2.08748 |
| | 1.6 | -2.68093 | -2.53447 | -2.32948 | -2.16328 | -2.06034 |
| | 2.0 | -2.62621 | -2.48871 | -2.29560 | -2.13821 | -2.04024 |
| | 2.4 | -2.58177 | -2.45270 | -2.27049 | -2.12091 | -2.02717 |
| | 2.8 | -2.54815 | -2.42682 | -2.25427 | -2.11109 | -2.02056 |
| | 3.2 | -2.52570 | -2.41121 | -2.24647 | -2.10783 | -2.01933 |
| | 3.6 | -2.51451 | -2.40536 | -2.24587 | -2.10981 | -2.02231 |
| | 4.0 | -2.51365 | -2.40777 | -2.25092 | -2.11577 | -2.02841 |
| | 6.0 | -2.59648 | -2.48514 | -2.31839 | -2.17414 | -2.08095 |
| 12.0 | 0.0 | -3.04881 | -2.87217 | -2.62222 | -2.41773 | -2.29054 |
| | 0.4 | -3.02629 | -2.85274 | -2.60684 | -2.40535 | -2.27988 |
| | 0.8 | -3.00524 | -2.83467 | -2.59263 | -2.39398 | -2.27014 |
| | 1.2 | -2.98578 | -2.81804 | -2.57965 | -2.38368 | -2.26135 |
| | 1.6 | -2.96799 | -2.80293 | -2.56797 | -2.37447 | -2.25353 |
| | 2.0 | -2.95195 | -2.78938 | -2.55758 | -2.36636 | -2.24668 |
| | 2.4 | -2.93769 | -2.77742 | -2.54851 | -2.35933 | -2.24079 |
| | 2.8 | -2.92520 | -2.76702 | -2.54072 | -2.35336 | -2.23582 |
| | 3.2 | -2.91448 | -2.75817 | -2.53418 | -2.34841 | -2.23174 |
| | 3.6 | -2.90547 | -2.75082 | -2.52883 | -2.34443 | -2.22849 |
| | 4.0 | -2.89810 | -2.74488 | -2.52460 | -2.34135 | -2.22601 |
| | 6.0 | -2.88273 | -2.73348 | -2.51765 | -2.33716 | -2.22319 |
| | 12.0 | -2.94470 | -2.78909 | -2.56432 | -2.37669 | -2.25840 |
| INFINITY | | -3.29051 | -3.09024 | -2.80703 | -2.57583 | -2.43238 |

FRACTILES $W_\alpha$ RHO = 0.40

| DELTAS | | α:0.0100 | 0.0175 | 0.0250 | 0.0375 | 0.0500 |
|---|---|---|---|---|---|---|
| 0.0 | 0.0 | -1.90241 | -1.62961 | -1.45871 | -1.26794 | -1.13537 |
| 0.4 | 0.0 | -1.97290 | -1.68842 | -1.50962 | -1.30948 | -1.17001 |
| | 0.4 | -1.70753 | -1.47635 | -1.33161 | -1.17018 | -1.05812 |
| 0.8 | 0.0 | -2.06759 | -1.78165 | -1.59956 | -1.39306 | -1.24715 |
| | 0.4 | -1.67434 | -1.45555 | -1.31821 | -1.16470 | -1.05794 |
| | 0.8 | -1.46356 | -1.29417 | -1.18836 | -1.07067 | -0.98928 |
| 1.2 | 0.0 | -2.09848 | -1.82817 | -1.65427 | -1.45488 | -1.31224 |
| | 0.4 | -1.74613 | -1.52984 | -1.39222 | -1.23630 | -1.12629 |
| | 0.8 | -1.47088 | -1.30949 | -1.20848 | -1.09601 | -1.01819 |
| | 1.2 | -1.35978 | -1.23386 | -1.15559 | -1.06911 | -1.00986 |
| 1.6 | 0.0 | -2.09491 | -1.84239 | -1.67876 | -1.48977 | -1.35347 |
| | 0.4 | -1.81063 | -1.60182 | -1.46742 | -1.31329 | -1.20301 |
| | 0.8 | -1.56576 | -1.40068 | -1.29582 | -1.17730 | -1.09400 |
| | 1.2 | -1.40093 | -1.27795 | -1.20146 | -1.11704 | -1.05940 |
| | 1.6 | -1.35825 | -1.26145 | -1.20196 | -1.13739 | -1.09444 |
| 2.0 | 0.0 | -2.08405 | -1.84666 | -1.69193 | -1.51216 | -1.38167 |
| | 0.4 | -1.85673 | -1.65556 | -1.52503 | -1.37408 | -1.26504 |
| | 0.8 | -1.65727 | -1.49176 | -1.38518 | -1.26293 | -1.17547 |
| | 1.2 | -1.50239 | -1.37163 | -1.28889 | -1.19599 | -1.13140 |
| | 1.6 | -1.41520 | -1.31886 | -1.25989 | -1.19657 | -1.15563 |
| | 2.0 | -1.41800 | -1.34296 | -1.29869 | -1.25569 | -1.22769 |
| 2.4 | 0.0 | -2.07537 | -1.84998 | -1.70231 | -1.52985 | -1.40397 |
| | 0.4 | -1.89214 | -1.69719 | -1.56985 | -1.42157 | -1.31366 |
| | 0.8 | -1.73058 | -1.56564 | -1.45842 | -1.33412 | -1.24412 |
| | 1.2 | -1.60083 | -1.46477 | -1.37713 | -1.27666 | -1.20498 |
| | 1.6 | -1.51124 | -1.40469 | -1.33809 | -1.26515 | -1.21755 |
| | 2.0 | -1.48129 | -1.40667 | -1.36485 | -1.32141 | -1.27585 |
| | 2.4 | -1.51894 | -1.46926 | -1.42763 | -1.36533 | -1.31087 |
| 2.8 | 0.0 | -2.07087 | -1.85479 | -1.71256 | -1.54566 | -1.42324 |
| | 0.4 | -1.92160 | -1.73141 | -1.60645 | -1.46005 | -1.35282 |
| | 0.8 | -1.78975 | -1.62517 | -1.51732 | -1.39126 | -1.29912 |
| | 1.2 | -1.68264 | -1.54279 | -1.45157 | -1.34547 | -1.26835 |
| | 1.6 | -1.60507 | -1.48965 | -1.41546 | -1.33117 | -1.27310 |
| | 2.0 | -1.56544 | -1.47918 | -1.43180 | -1.36926 | -1.31325 |
| | 2.4 | -1.58830 | -1.52650 | -1.47371 | -1.40099 | -1.33972 |
| | 2.8 | -1.64352 | -1.56756 | -1.50841 | -1.42876 | -1.36266 |
| 3.2 | 0.0 | -2.07025 | -1.86132 | -1.72321 | -1.56046 | -1.44054 |
| | 0.4 | -1.94732 | -1.76065 | -1.63735 | -1.49212 | -1.38512 |
| | 0.8 | -1.83875 | -1.67412 | -1.56548 | -1.43759 | -1.34336 |
| | 1.2 | -1.75002 | -1.60674 | -1.51236 | -1.40136 | -1.31962 |
| | 1.6 | -1.68489 | -1.56243 | -1.48216 | -1.38843 | -1.32059 |
| | 2.0 | -1.64757 | -1.54811 | -1.48825 | -1.41037 | -1.34562 |
| | 2.4 | -1.65741 | -1.57686 | -1.51507 | -1.43292 | -1.36533 |
| | 2.8 | -1.69275 | -1.60646 | -1.54100 | -1.45437 | -1.38343 |
| | 3.2 | -1.72963 | -1.63654 | -1.56664 | -1.47489 | -1.40031 |

FRACTILES $W_\alpha$ RHO = 0.40

| DELTAS | | $\alpha$:0.0100 | 0.0175 | 0.0250 | 0.0375 | 0.0500 |
|---|---|---|---|---|---|---|
| 3.6 | 0.0 | -2.07264 | -1.86918 | -1.73415 | -1.57442 | -1.45623 |
| | 0.4 | -1.97039 | -1.78622 | -1.66399 | -1.51932 | -1.41218 |
| | 0.8 | -1.88017 | -1.71509 | -1.60548 | -1.47564 | -1.37932 |
| | 1.2 | -1.80617 | -1.65958 | -1.56219 | -1.44665 | -1.36070 |
| | 1.6 | -1.75142 | -1.62269 | -1.53714 | -1.43548 | -1.35975 |
| | 2.0 | -1.71882 | -1.60834 | -1.53680 | -1.44662 | -1.37424 |
| | 2.4 | -1.71843 | -1.62302 | -1.55294 | -1.46197 | -1.38846 |
| | 2.8 | -1.74017 | -1.64345 | -1.57168 | -1.47816 | -1.40250 |
| | 3.2 | -1.76704 | -1.66626 | -1.59154 | -1.49441 | -1.41610 |
| | 3.6 | -1.79616 | -1.68988 | -1.61158 | -1.51036 | -1.42913 |
| 4.0 | 0.0 | -2.07719 | -1.87788 | -1.74512 | -1.58751 | -1.47047 |
| | 0.4 | -1.99134 | -1.80887 | -1.68723 | -1.54264 | -1.43507 |
| | 0.8 | -1.91572 | -1.74985 | -1.63912 | -1.50724 | -1.40885 |
| | 1.2 | -1.85352 | -1.70368 | -1.60344 | -1.48364 | -1.39382 |
| | 1.6 | -1.80715 | -1.67263 | -1.58229 | -1.47369 | -1.39157 |
| | 2.0 | -1.77876 | -1.65899 | -1.57848 | -1.47835 | -1.39946 |
| | 2.4 | -1.77293 | -1.66479 | -1.58725 | -1.48827 | -1.40935 |
| | 2.8 | -1.78484 | -1.67808 | -1.60026 | -1.50015 | -1.42002 |
| | 3.2 | -1.80361 | -1.69489 | -1.61531 | -1.51284 | -1.43084 |
| | 3.6 | -1.82581 | -1.71339 | -1.63125 | -1.52574 | -1.44154 |
| | 4.0 | -1.84953 | -1.73250 | -1.64740 | -1.53852 | -1.45194 |
| 6.0 | 0.0 | -2.11281 | -1.92250 | -1.79397 | -1.63937 | -1.52303 |
| | 0.4 | -2.07248 | -1.89124 | -1.76853 | -1.62052 | -1.50878 |
| | 0.8 | -2.03708 | -1.86454 | -1.74733 | -1.60544 | -1.49790 |
| | 1.2 | -2.00746 | -1.84303 | -1.73083 | -1.59439 | -1.49046 |
| | 1.6 | -1.98419 | -1.82703 | -1.71919 | -1.58728 | -1.48622 |
| | 2.0 | -1.96750 | -1.81650 | -1.71214 | -1.58366 | -1.48462 |
| | 2.4 | -1.95720 | -1.81092 | -1.70904 | -1.58280 | -1.48499 |
| | 2.8 | -1.95257 | -1.80942 | -1.70902 | -1.58401 | -1.48678 |
| | 3.2 | -1.95252 | -1.81105 | -1.71131 | -1.58671 | -1.48957 |
| | 3.6 | -1.95599 | -1.81501 | -1.71530 | -1.59049 | -1.49305 |
| | 4.0 | -1.96205 | -1.82066 | -1.72049 | -1.59499 | -1.49698 |
| | 6.0 | -2.01025 | -1.86006 | -1.75408 | -1.62183 | -1.51897 |
| 12.0 | 0.0 | -2.19633 | -2.00201 | -1.86913 | -1.70764 | -1.58502 |
| | 0.4 | -2.18687 | -1.99482 | -1.86334 | -1.70338 | -1.58179 |
| | 0.8 | -2.17826 | -1.98833 | -1.85815 | -1.69960 | -1.57895 |
| | 1.2 | -2.17051 | -1.98255 | -1.85355 | -1.69628 | -1.57649 |
| | 1.6 | -2.16365 | -1.97747 | -1.84955 | -1.69342 | -1.57440 |
| | 2.0 | -2.15767 | -1.97309 | -1.84612 | -1.69101 | -1.57265 |
| | 2.4 | -2.15255 | -1.96938 | -1.84325 | -1.68902 | -1.57124 |
| | 2.8 | -2.14826 | -1.96632 | -1.84091 | -1.68743 | -1.57013 |
| | 3.2 | -2.14476 | -1.96387 | -1.83906 | -1.68620 | -1.56930 |
| | 3.6 | -2.14200 | -1.96197 | -1.83767 | -1.68532 | -1.56873 |
| | 4.0 | -2.13992 | -1.96061 | -1.83669 | -1.68474 | -1.56840 |
| | 6.0 | -2.13796 | -1.96003 | -1.83680 | -1.68545 | -1.56941 |
| | 12.0 | -2.17005 | -1.98595 | -1.85874 | -1.70281 | -1.58352 |
| INFINITY | | -2.32635 | -2.10836 | -1.95996 | -1.78046 | -1.64485 |

FRACTILES $W_\alpha$ RHO = 0.40

| DELTAS | | $\alpha$:0.0750 | 0.1000 | 0.1500 | 0.2000 | 0.2500 |
|---|---|---|---|---|---|---|
| 0.0 | 0.0 | -0.95349 | -0.82894 | -0.66213 | -0.55253 | -0.47501 |
| 0.4 | 0.0 | -0.97808 | -0.84621 | -0.66886 | -0.55173 | -0.46834 |
| | 0.4 | -0.90462 | -0.79977 | -0.66000 | -0.56906 | -0.50585 |
| 0.8 | 0.0 | -1.04301 | -0.89996 | -0.70284 | -0.56854 | -0.46994 |
| | 0.4 | -0.91147 | -0.81133 | -0.67790 | -0.59146 | -0.53211 |
| | 0.8 | -0.87846 | -0.80351 | -0.70558 | -0.64492 | -0.60827 |
| 1.2 | 0.0 | -1.10946 | -0.96436 | -0.75843 | -0.61202 | -0.49935 |
| | 0.4 | -0.97280 | -0.86574 | -0.71962 | -0.62218 | -0.55357 |
| | 0.8 | -0.91233 | -0.84100 | -0.74895 | -0.69464 | -0.66330 |
| | 1.2 | -0.93050 | -0.87852 | -0.81681 | -0.77960 | -0.72292 |
| 1.6 | 0.0 | -1.15770 | -1.01574 | -0.81042 | -0.66029 | -0.54088 |
| | 0.4 | -1.04627 | -0.93421 | -0.77573 | -0.66423 | -0.58061 |
| | 0.8 | -0.97859 | -0.89916 | -0.79436 | -0.73231 | -0.68899 |
| | 1.2 | -0.98293 | -0.93420 | -0.88156 | -0.81756 | -0.74078 |
| | 1.6 | -1.04115 | -1.01259 | -0.93437 | -0.84553 | -0.75287 |
| 2.0 | 0.0 | -1.19275 | -1.05442 | -0.85168 | -0.70067 | -0.57811 |
| | 0.4 | -1.10812 | -0.99408 | -0.82868 | -0.70750 | -0.61146 |
| | 0.8 | -1.05134 | -0.96299 | -0.84013 | -0.76074 | -0.69999 |
| | 1.2 | -1.04405 | -0.98770 | -0.91761 | -0.83186 | -0.74030 |
| | 1.6 | -1.11008 | -1.06141 | -0.95705 | -0.85336 | -0.75076 |
| | 2.0 | -1.16090 | -1.09583 | -0.97417 | -0.86025 | -0.75052 |
| 2.4 | 0.0 | -1.22048 | -1.08501 | -0.88427 | -0.73256 | -0.60753 |
| | 0.4 | -1.15687 | -1.04152 | -0.87130 | -0.74324 | -0.63822 |
| | 0.8 | -1.11420 | -1.01945 | -0.88184 | -0.78365 | -0.70350 |
| | 1.2 | -1.10421 | -1.03515 | -0.93934 | -0.83658 | -0.73411 |
| | 1.6 | -1.15609 | -1.09132 | -0.96888 | -0.85439 | -0.74440 |
| | 2.0 | -1.19237 | -1.11745 | -0.98342 | -0.86158 | -0.74624 |
| | 2.4 | -1.21681 | -1.13500 | -0.99182 | -0.86388 | -0.74410 |
| 2.8 | 0.0 | -1.24372 | -1.11019 | -0.91044 | -0.75757 | -0.62999 |
| | 0.4 | -1.19574 | -1.07900 | -0.90431 | -0.77026 | -0.65780 |
| | 0.8 | -1.16444 | -1.06453 | -0.91527 | -0.80167 | -0.70382 |
| | 1.2 | -1.15666 | -1.07562 | -0.95444 | -0.83836 | -0.72739 |
| | 1.6 | -1.19076 | -1.11372 | -0.97684 | -0.85351 | -0.73740 |
| | 2.0 | -1.21710 | -1.13403 | -0.98945 | -0.86080 | -0.74065 |
| | 2.4 | -1.23657 | -1.14861 | -0.99721 | -0.86385 | -0.74019 |
| | 2.8 | -1.25276 | -1.16009 | -1.00227 | -0.86458 | -0.73777 |
| 3.2 | 0.0 | -1.26372 | -1.13134 | -0.93163 | -0.77710 | -0.64677 |
| | 0.4 | -1.22727 | -1.10892 | -0.92973 | -0.79007 | -0.67097 |
| | 0.8 | -1.20421 | -1.09962 | -0.94027 | -0.81452 | -0.70316 |
| | 1.2 | -1.19687 | -1.10812 | -0.96624 | -0.83955 | -0.72163 |
| | 1.6 | -1.21949 | -1.13238 | -0.98327 | -0.85234 | -0.73098 |
| | 2.0 | -1.23834 | -1.14801 | -0.99402 | -0.85933 | -0.73496 |
| | 2.4 | -1.25368 | -1.16000 | -1.00105 | -0.86279 | -0.73566 |
| | 2.8 | -1.26691 | -1.16971 | -1.00580 | -0.86410 | -0.73444 |
| | 3.2 | -1.27867 | -1.17791 | -1.00913 | -0.86417 | -0.73214 |

FRACTILES $W_\alpha$ RHO = 0.40

| DELTAS | | $\alpha$:0.0750 | 0.1000 | 0.1500 | 0.2000 | 0.2500 |
|---|---|---|---|---|---|---|
| 3.6 | 0.0 | -1.28111 | -1.14923 | -0.94878 | -0.79225 | -0.65906 |
| | 0.4 | -1.25314 | -1.13299 | -0.94931 | -0.80438 | -0.67940 |
| | 0.8 | -1.23589 | -1.12694 | -0.95861 | -0.82315 | -0.70233 |
| | 1.2 | -1.23199 | -1.13327 | -0.97596 | -0.84071 | -0.71700 |
| | 1.6 | -1.24394 | -1.14846 | -0.98889 | -0.85136 | -0.72543 |
| | 2.0 | -1.25709 | -1.16027 | -0.99785 | -0.85779 | -0.72967 |
| | 2.4 | -1.26891 | -1.16996 | -1.00409 | -0.86136 | -0.73111 |
| | 2.8 | -1.27959 | -1.17808 | -1.00846 | -0.86305 | -0.73077 |
| | 3.2 | -1.28934 | -1.18507 | -1.01160 | -0.86355 | -0.72933 |
| | 3.6 | -1.29832 | -1.19123 | -1.01393 | -0.86332 | -0.72727 |
| 4.0 | 0.0 | -1.29627 | -1.16439 | -0.96266 | -0.80390 | -0.66789 |
| | 0.4 | -1.27452 | -1.15245 | -0.96439 | -0.81466 | -0.68462 |
| | 0.8 | -1.26131 | -1.14833 | -0.97213 | -0.82899 | -0.70150 |
| | 1.2 | -1.25806 | -1.15273 | -0.98404 | -0.84185 | -0.71331 |
| | 1.6 | -1.26472 | -1.16235 | -0.99387 | -0.85065 | -0.72075 |
| | 2.0 | -1.27367 | -1.17112 | -1.00124 | -0.85640 | -0.72495 |
| | 2.4 | -1.28259 | -1.17882 | -1.00665 | -0.85989 | -0.72682 |
| | 2.8 | -1.29110 | -1.18554 | -1.01061 | -0.86178 | -0.72710 |
| | 3.2 | -1.29911 | -1.19147 | -1.01353 | -0.86258 | -0.72633 |
| | 3.6 | -1.30665 | -1.19677 | -1.01573 | -0.86268 | -0.72489 |
| | 4.0 | -1.31374 | -1.20157 | -1.01744 | -0.86233 | -0.72306 |
| 6.0 | 0.0 | -1.34724 | -1.21194 | -1.00113 | -0.83201 | -0.68517 |
| | 0.4 | -1.33929 | -1.20821 | -1.00272 | -0.83669 | -0.69166 |
| | 0.8 | -1.33397 | -1.20646 | -1.00519 | -0.84136 | -0.69748 |
| | 1.2 | -1.33111 | -1.20636 | -1.00809 | -0.84563 | -0.70234 |
| | 1.6 | -1.33028 | -1.20742 | -1.01100 | -0.84927 | -0.70617 |
| | 2.0 | -1.33091 | -1.20918 | -1.01371 | -0.85220 | -0.70902 |
| | 2.4 | -1.33251 | -1.21130 | -1.01611 | -0.85446 | -0.71101 |
| | 2.8 | -1.33474 | -1.21360 | -1.01816 | -0.85612 | -0.71225 |
| | 3.2 | -1.33738 | -1.21595 | -1.01989 | -0.85728 | -0.71290 |
| | 3.6 | -1.34025 | -1.21828 | -1.02133 | -0.85801 | -0.71306 |
| | 4.0 | -1.34326 | -1.22057 | -1.02253 | -0.85842 | -0.71285 |
| | 6.0 | -1.35841 | -1.23091 | -1.02626 | -0.85770 | -0.70890 |
| 12.0 | 0.0 | -1.39819 | -1.25337 | -1.02663 | -0.84443 | -0.68660 |
| | 0.4 | -1.39634 | -1.25243 | -1.02681 | -0.84530 | -0.68790 |
| | 0.8 | -1.39475 | -1.25164 | -1.02704 | -0.84613 | -0.68911 |
| | 1.2 | -1.39340 | -1.25102 | -1.02730 | -0.84691 | -0.69021 |
| | 1.6 | -1.39229 | -1.25054 | -1.02759 | -0.84763 | -0.69120 |
| | 2.0 | -1.39141 | -1.25019 | -1.02789 | -0.84830 | -0.69208 |
| | 2.4 | -1.39072 | -1.24996 | -1.02820 | -0.84890 | -0.69285 |
| | 2.8 | -1.39022 | -1.24984 | -1.02851 | -0.84944 | -0.69352 |
| | 3.2 | -1.38989 | -1.24980 | -1.02881 | -0.84992 | -0.69408 |
| | 3.6 | -1.38971 | -1.24985 | -1.02910 | -0.85033 | -0.69455 |
| | 4.0 | -1.38966 | -1.24997 | -1.02939 | -0.85068 | -0.69493 |
| | 6.0 | -1.39093 | -1.25129 | -1.03060 | -0.85169 | -0.69569 |
| | 12.0 | -1.40048 | -1.25765 | -1.03259 | -0.85075 | -0.69263 |
| INFINITY | | -1.43953 | -1.28155 | -1.03643 | -0.84162 | -0.67449 |

FRACTILES $W_\alpha$ RHO = 0.40

| DELTAS | | $\alpha$:0.3000 | 0.3500 | 0.4000 | 0.4500 | 0.5000 |
|---|---|---|---|---|---|---|
| 0.0 | 0.0 | -0.41905 | -0.38055 | -0.35477 | -0.31405 | -0.25994 |
| 0.4 | 0.0 | -0.40751 | -0.36445 | -0.33799 | -0.29939 | -0.24677 |
| | 0.4 | -0.46201 | -0.43616 | -0.40218 | -0.35472 | -0.29458 |
| 0.8 | 0.0 | -0.39538 | -0.33939 | -0.30117 | -0.26919 | -0.21912 |
| | 0.4 | -0.49260 | -0.46645 | -0.42517 | -0.37082 | -0.30352 |
| | 0.8 | -0.57973 | -0.53480 | -0.47748 | -0.40820 | -0.32646 |
| 1.2 | 0.0 | -0.40926 | -0.33645 | -0.27914 | -0.24115 | -0.19263 |
| | 0.4 | -0.50682 | -0.47514 | -0.42619 | -0.36318 | -0.28681 |
| | 0.8 | -0.61648 | -0.55619 | -0.48450 | -0.40192 | -0.30811 |
| | 1.2 | -0.65409 | -0.57586 | -0.48917 | -0.39408 | -0.28995 |
| 1.6 | 0.0 | -0.44135 | -0.35618 | -0.28281 | -0.22206 | -0.17200 |
| | 0.4 | -0.51836 | -0.47614 | -0.41694 | -0.34306 | -0.25646 |
| | 0.8 | -0.62588 | -0.55114 | -0.46704 | -0.37415 | -0.27209 |
| | 1.2 | -0.65656 | -0.56639 | -0.47045 | -0.36822 | -0.25874 |
| | 1.6 | -0.65741 | -0.55887 | -0.45653 | -0.34941 | -0.23624 |
| 2.0 | 0.0 | -0.47343 | -0.38095 | -0.29738 | -0.22128 | -0.15600 |
| | 0.4 | -0.53328 | -0.47450 | -0.40271 | -0.31763 | -0.22196 |
| | 0.8 | -0.62137 | -0.53454 | -0.44121 | -0.34140 | -0.23437 |
| | 1.2 | -0.64552 | -0.54760 | -0.44591 | -0.33951 | -0.22714 |
| | 1.6 | -0.64812 | -0.54424 | -0.43788 | -0.32778 | -0.21253 |
| | 2.0 | -0.64259 | -0.53463 | -0.42513 | -0.31264 | -0.19564 |
| 2.4 | 0.0 | -0.49893 | -0.40104 | -0.31019 | -0.22384 | -0.14233 |
| | 0.4 | -0.54789 | -0.47140 | -0.38679 | -0.29235 | -0.19001 |
| | 0.8 | -0.61184 | -0.51590 | -0.41605 | -0.31163 | -0.20148 |
| | 1.2 | -0.63172 | -0.52829 | -0.42260 | -0.31337 | -0.19918 |
| | 1.6 | -0.63639 | -0.52851 | -0.41921 | -0.30702 | -0.19041 |
| | 2.0 | -0.63412 | -0.52301 | -0.41113 | -0.29692 | -0.17878 |
| | 2.4 | -0.62863 | -0.51493 | -0.40108 | -0.28539 | -0.16623 |
| 2.8 | 0.0 | -0.51767 | -0.41490 | -0.31782 | -0.22340 | -0.12948 |
| | 0.4 | -0.55800 | -0.46701 | -0.37170 | -0.27040 | -0.16343 |
| | 0.8 | -0.60212 | -0.49938 | -0.39474 | -0.28694 | -0.17458 |
| | 1.2 | -0.61903 | -0.51126 | -0.40244 | -0.29103 | -0.17548 |
| | 1.6 | -0.62497 | -0.51383 | -0.40216 | -0.28833 | -0.17073 |
| | 2.0 | -0.62500 | -0.51127 | -0.39748 | -0.28193 | -0.16297 |
| | 2.4 | -0.62183 | -0.50597 | -0.39052 | -0.27370 | -0.15381 |
| | 2.8 | -0.61705 | -0.49938 | -0.38257 | -0.26476 | -0.14421 |
| 3.2 | 0.0 | -0.53078 | -0.42344 | -0.32083 | -0.21975 | -0.11724 |
| | 0.4 | -0.56331 | -0.46179 | -0.35876 | -0.25256 | -0.14227 |
| | 0.8 | -0.59387 | -0.48590 | -0.37749 | -0.26703 | -0.15289 |
| | 1.2 | -0.60833 | -0.49697 | -0.38553 | -0.27232 | -0.15566 |
| | 1.6 | -0.61474 | -0.50083 | -0.38716 | -0.27197 | -0.15359 |
| | 2.0 | -0.61624 | -0.50025 | -0.38484 | -0.26821 | -0.14862 |
| | 2.4 | -0.61477 | -0.49704 | -0.38023 | -0.26249 | -0.14206 |
| | 2.8 | -0.61160 | -0.49236 | -0.37439 | -0.25578 | -0.13474 |
| | 3.2 | -0.60754 | -0.48697 | -0.36801 | -0.24869 | -0.12720 |

FRACTILES $W_\alpha$ RHO = 0.40

| DELTAS | | $\alpha$:0.3000 | 0.3500 | 0.4000 | 0.4500 | 0.5000 |
|---|---|---|---|---|---|---|
| 3.6 | 0.0 | -0.53953 | -0.42803 | -0.32067 | -0.21433 | -0.10613 |
| | 0.4 | -0.56526 | -0.45645 | -0.34811 | -0.23833 | -0.12553 |
| | 0.8 | -0.58723 | -0.47509 | -0.36363 | -0.25097 | -0.13537 |
| | 1.2 | -0.59954 | -0.48512 | -0.37147 | -0.25672 | -0.13911 |
| | 1.6 | -0.60590 | -0.48957 | -0.37418 | -0.25783 | -0.13878 |
| | 2.0 | -0.60824 | -0.49028 | -0.37347 | -0.25591 | -0.13582 |
| | 2.4 | -0.60796 | -0.48858 | -0.37060 | -0.25208 | -0.13122 |
| | 2.8 | -0.60602 | -0.48538 | -0.36641 | -0.24712 | -0.12571 |
| | 3.2 | -0.60311 | -0.48134 | -0.36151 | -0.24160 | -0.11977 |
| | 3.6 | -0.59968 | -0.47688 | -0.35630 | -0.23586 | -0.11370 |
| 4.0 | 0.0 | -0.54509 | -0.42996 | -0.31872 | -0.20842 | -0.09649 |
| | 0.4 | -0.56532 | -0.45148 | -0.33938 | -0.22689 | -0.11213 |
| | 0.8 | -0.58186 | -0.46636 | -0.35239 | -0.23791 | -0.12108 |
| | 1.2 | -0.59232 | -0.47530 | -0.35976 | -0.24368 | -0.12525 |
| | 1.6 | -0.59834 | -0.47992 | -0.36302 | -0.24566 | -0.12604 |
| | 2.0 | -0.60112 | -0.48142 | -0.36338 | -0.24501 | -0.12448 |
| | 2.4 | -0.60164 | -0.48079 | -0.36177 | -0.24257 | -0.12136 |
| | 2.8 | -0.60062 | -0.47872 | -0.35886 | -0.23899 | -0.11728 |
| | 3.2 | -0.59862 | -0.47577 | -0.35516 | -0.23474 | -0.11263 |
| | 3.6 | -0.59603 | -0.47229 | -0.35103 | -0.23013 | -0.10772 |
| | 4.0 | -0.59311 | -0.46856 | -0.34671 | -0.22540 | -0.10275 |
| 6.0 | 0.0 | -0.55149 | -0.42574 | -0.30449 | -0.18516 | -0.06562 |
| | 0.4 | -0.55897 | -0.43364 | -0.31239 | -0.19276 | -0.07265 |
| | 0.8 | -0.56529 | -0.44005 | -0.31860 | -0.19857 | -0.07789 |
| | 1.2 | -0.57030 | -0.44495 | -0.32320 | -0.20273 | -0.08151 |
| | 1.6 | -0.57406 | -0.44846 | -0.32636 | -0.20546 | -0.08376 |
| | 2.0 | -0.57667 | -0.45075 | -0.32829 | -0.20699 | -0.08488 |
| | 2.4 | -0.57832 | -0.45204 | -0.32920 | -0.20754 | -0.08507 |
| | 2.8 | -0.57916 | -0.45249 | -0.32930 | -0.20731 | -0.08454 |
| | 3.2 | -0.57935 | -0.45229 | -0.32876 | -0.20647 | -0.08346 |
| | 3.6 | -0.57904 | -0.45159 | -0.32773 | -0.20518 | -0.08195 |
| | 4.0 | -0.57835 | -0.45050 | -0.32633 | -0.20354 | -0.08014 |
| | 6.0 | -0.57199 | -0.44234 | -0.31684 | -0.19312 | -0.06917 |
| 12.0 | 0.0 | -0.54357 | -0.40987 | -0.28192 | -0.15709 | -0.03320 |
| | 0.4 | -0.54515 | -0.41162 | -0.28376 | -0.15894 | -0.03501 |
| | 0.8 | -0.54659 | -0.41320 | -0.28541 | -0.16059 | -0.03660 |
| | 1.2 | -0.54789 | -0.41461 | -0.28686 | -0.16204 | -0.03799 |
| | 1.6 | -0.54904 | -0.41584 | -0.28812 | -0.16328 | -0.03918 |
| | 2.0 | -0.55004 | -0.41691 | -0.28920 | -0.16434 | -0.04018 |
| | 2.4 | -0.55091 | -0.41782 | -0.29011 | -0.16521 | -0.04100 |
| | 2.8 | -0.55164 | -0.41857 | -0.29085 | -0.16592 | -0.04165 |
| | 3.2 | -0.55224 | -0.41918 | -0.29144 | -0.16647 | -0.04215 |
| | 3.6 | -0.55272 | -0.41965 | -0.29189 | -0.16688 | -0.04250 |
| | 4.0 | -0.55310 | -0.42000 | -0.29220 | -0.16716 | -0.04273 |
| | 6.0 | -0.55361 | -0.42028 | -0.29225 | -0.16698 | -0.04235 |
| | 12.0 | -0.54899 | -0.41448 | -0.28560 | -0.15975 | -0.03479 |
| INFINITY | | -0.52440 | -0.38532 | -0.25335 | -0.12566 | 0.0 |

FRACTILES $w_\alpha$ RHO = 0.40

| DELTAS | | $\alpha$:0.5500 | 0.6000 | 0.6500 | 0.7000 | 0.7500 |
|---|---|---|---|---|---|---|
| 0.0 | 0.0 | -0.19209 | -0.10862 | -0.00636 | 0.11977 | 0.27787 |
| 0.4 | 0.0 | -0.18023 | -0.09805 | 0.00287 | 0.12749 | 0.28385 |
| | 0.4 | -0.22069 | -0.13092 | -0.02187 | 0.11168 | 0.27803 |
| 0.8 | 0.0 | -0.15409 | -0.07296 | 0.02710 | 0.15080 | 0.30587 |
| | 0.4 | -0.22215 | -0.12459 | -0.00757 | 0.13395 | 0.30793 |
| | 0.8 | -0.23102 | -0.11988 | 0.01005 | 0.16350 | 0.34803 |
| 1.2 | 0.0 | -0.12595 | -0.04196 | 0.06143 | 0.18842 | 0.34612 |
| | 0.4 | -0.19638 | -0.09019 | 0.03460 | 0.18248 | 0.36070 |
| | 0.8 | -0.20208 | -0.08200 | 0.05499 | 0.21330 | 0.40001 |
| | 1.2 | -0.17556 | -0.04896 | 0.09274 | 0.25390 | 0.44135 |
| 1.6 | 0.0 | -0.09931 | -0.00815 | 0.10221 | 0.23515 | 0.39691 |
| | 0.4 | -0.15701 | -0.04342 | 0.08679 | 0.23772 | 0.41602 |
| | 0.8 | -0.15977 | -0.03535 | 0.10397 | 0.26245 | 0.44679 |
| | 1.2 | -0.14055 | -0.01161 | 0.13101 | 0.29153 | 0.47651 |
| | 1.6 | -0.11540 | 0.01522 | 0.15853 | 0.31867 | 0.50199 |
| 2.0 | 0.0 | -0.07366 | 0.02666 | 0.14444 | 0.28249 | 0.44654 |
| | 0.4 | -0.11562 | 0.00272 | 0.13553 | 0.28675 | 0.46268 |
| | 0.8 | -0.11879 | 0.00730 | 0.14668 | 0.30350 | 0.48413 |
| | 1.2 | -0.10723 | 0.02235 | 0.16444 | 0.32318 | 0.50482 |
| | 1.6 | -0.09042 | 0.04067 | 0.18360 | 0.34242 | 0.52323 |
| | 2.0 | -0.07239 | 0.05925 | 0.20209 | 0.36009 | 0.53920 |
| 2.4 | 0.0 | -0.04908 | 0.05939 | 0.18263 | 0.32361 | 0.48791 |
| | 0.4 | -0.07915 | 0.04186 | 0.17559 | 0.32592 | 0.49891 |
| | 0.8 | -0.08409 | 0.04259 | 0.18134 | 0.33617 | 0.51316 |
| | 1.2 | -0.07834 | 0.05125 | 0.19243 | 0.34917 | 0.52751 |
| | 1.6 | -0.06764 | 0.06341 | 0.20557 | 0.36275 | 0.54088 |
| | 2.0 | -0.05493 | 0.07677 | 0.21908 | 0.37589 | 0.55297 |
| | 2.4 | -0.04178 | 0.09009 | 0.23212 | 0.38812 | 0.56374 |
| 2.8 | 0.0 | -0.02687 | 0.08735 | 0.21398 | 0.35632 | 0.51993 |
| | 0.4 | -0.04959 | 0.07300 | 0.20701 | 0.35625 | 0.52656 |
| | 0.8 | -0.05596 | 0.07099 | 0.20902 | 0.36203 | 0.53586 |
| | 1.2 | -0.05404 | 0.07540 | 0.21563 | 0.37051 | 0.54585 |
| | 1.6 | -0.04758 | 0.08325 | 0.22452 | 0.38007 | 0.55563 |
| | 2.0 | -0.03880 | 0.09272 | 0.23434 | 0.38983 | 0.56483 |
| | 2.4 | -0.02904 | 0.10276 | 0.24429 | 0.39929 | 0.57329 |
| | 2.8 | -0.01910 | 0.11272 | 0.25393 | 0.40820 | 0.58099 |
| 3.2 | 0.0 | -0.00811 | 0.10987 | 0.23857 | 0.38153 | 0.54424 |
| | 0.4 | -0.02630 | 0.09741 | 0.23152 | 0.37980 | 0.54787 |
| | 0.8 | -0.03332 | 0.09381 | 0.23122 | 0.38270 | 0.55388 |
| | 1.2 | -0.03376 | 0.09551 | 0.23488 | 0.38812 | 0.56085 |
| | 1.6 | -0.03019 | 0.10036 | 0.24078 | 0.39482 | 0.56804 |
| | 2.0 | -0.02427 | 0.10699 | 0.24786 | 0.40205 | 0.57507 |
| | 2.4 | -0.01712 | 0.11448 | 0.25542 | 0.40936 | 0.58173 |
| | 2.8 | -0.00944 | 0.12226 | 0.26302 | 0.41647 | 0.58795 |
| | 3.2 | -0.00169 | 0.12996 | 0.27041 | 0.42324 | 0.59371 |

FRACTILES $W_\alpha$ RHO = 0.40

| DELTAS | | α:0.5500 | 0.6000 | 0.6500 | 0.7000 | 0.7500 |
|---|---|---|---|---|---|---|
| 3.6 | 0.0 | 0.00716 | 0.12763 | 0.25771 | 0.40099 | 0.56286 |
| | 0.4 | -0.00790 | 0.11666 | 0.25084 | 0.39833 | 0.56459 |
| | 0.8 | -0.01499 | 0.11230 | 0.24921 | 0.39941 | 0.56840 |
| | 1.2 | -0.01682 | 0.11229 | 0.25092 | 0.40276 | 0.57326 |
| | 1.6 | -0.01519 | 0.11509 | 0.25473 | 0.40741 | 0.57856 |
| | 2.0 | -0.01135 | 0.11963 | 0.25978 | 0.41276 | 0.58394 |
| | 2.4 | -0.00618 | 0.12517 | 0.26549 | 0.41840 | 0.58920 |
| | 2.8 | -0.00031 | 0.13120 | 0.27146 | 0.42406 | 0.59424 |
| | 3.2 | 0.00585 | 0.13738 | 0.27745 | 0.42959 | 0.59900 |
| | 3.6 | 0.01205 | 0.14351 | 0.28329 | 0.43489 | 0.60345 |
| 4.0 | 0.0 | 0.01951 | 0.14172 | 0.27277 | 0.41624 | 0.57738 |
| | 0.4 | 0.00681 | 0.13206 | 0.26630 | 0.41315 | 0.57792 |
| | 0.8 | 0.0 | 0.12744 | 0.26394 | 0.41310 | 0.58027 |
| | 1.2 | -0.00261 | 0.12638 | 0.26439 | 0.41503 | 0.58363 |
| | 1.6 | -0.00227 | 0.12776 | 0.26672 | 0.41821 | 0.58753 |
| | 2.0 | 0.00008 | 0.13078 | 0.27027 | 0.42215 | 0.59166 |
| | 2.4 | 0.00373 | 0.13482 | 0.27455 | 0.42648 | 0.59583 |
| | 2.8 | 0.00817 | 0.13946 | 0.27922 | 0.43098 | 0.59991 |
| | 3.2 | 0.01304 | 0.14440 | 0.28405 | 0.43550 | 0.60384 |
| | 3.6 | 0.01810 | 0.14943 | 0.28888 | 0.43991 | 0.60759 |
| | 4.0 | 0.02315 | 0.15441 | 0.29361 | 0.44417 | 0.61113 |
| 6.0 | 0.0 | 0.05618 | 0.18239 | 0.31557 | 0.45907 | 0.61769 |
| | 0.4 | 0.04992 | 0.17711 | 0.31146 | 0.45634 | 0.61658 |
| | 0.8 | 0.04541 | 0.17344 | 0.30877 | 0.45475 | 0.61626 |
| | 1.2 | 0.04240 | 0.17114 | 0.30724 | 0.45409 | 0.61654 |
| | 1.6 | 0.04068 | 0.16997 | 0.30667 | 0.45414 | 0.61726 |
| | 2.0 | 0.03999 | 0.16972 | 0.30685 | 0.45476 | 0.61830 |
| | 2.4 | 0.04014 | 0.17019 | 0.30762 | 0.45580 | 0.61958 |
| | 2.8 | 0.04093 | 0.17121 | 0.30884 | 0.45715 | 0.62101 |
| | 3.2 | 0.04222 | 0.17266 | 0.31038 | 0.45873 | 0.62253 |
| | 3.6 | 0.04388 | 0.17441 | 0.31216 | 0.46047 | 0.62412 |
| | 4.0 | 0.04580 | 0.17638 | 0.31411 | 0.46230 | 0.62573 |
| | 6.0 | 0.05697 | 0.18738 | 0.32455 | 0.47170 | 0.63351 |
| 12.0 | 0.0 | 0.09176 | 0.21986 | 0.35350 | 0.49573 | 0.65087 |
| | 0.4 | 0.09006 | 0.21832 | 0.35218 | 0.49470 | 0.65022 |
| | 0.8 | 0.08856 | 0.21698 | 0.35104 | 0.49383 | 0.64968 |
| | 1.2 | 0.08727 | 0.21583 | 0.35008 | 0.49310 | 0.64925 |
| | 1.6 | 0.08618 | 0.21486 | 0.34928 | 0.49251 | 0.64893 |
| | 2.0 | 0.08527 | 0.21407 | 0.34864 | 0.49205 | 0.64869 |
| | 2.4 | 0.08453 | 0.21344 | 0.34813 | 0.49171 | 0.64854 |
| | 2.8 | 0.08395 | 0.21295 | 0.34776 | 0.49147 | 0.64845 |
| | 3.2 | 0.08352 | 0.21260 | 0.34751 | 0.49132 | 0.64844 |
| | 3.6 | 0.08323 | 0.21238 | 0.34736 | 0.49127 | 0.64848 |
| | 4.0 | 0.08305 | 0.21227 | 0.34731 | 0.49129 | 0.64858 |
| | 6.0 | 0.08362 | 0.21299 | 0.34818 | 0.49226 | 0.64962 |
| | 12.0 | 0.09127 | 0.22048 | 0.35523 | 0.49854 | 0.65471 |
| INFINITY | | 0.12566 | 0.25335 | 0.38532 | 0.52440 | 0.67449 |

FRACTILES $W_\alpha$ RHO = 0.40

| DELTAS | | $\alpha$: 0.8000 | 0.8500 | 0.9000 | 0.9250 | 0.9500 |
|---|---|---|---|---|---|---|
| 0.0 | 0.0 | 0.48185 | 0.75812 | 1.16676 | 1.46716 | 1.90195 |
| 0.4 | 0.0 | 0.48566 | 0.75903 | 1.16332 | 1.46043 | 1.89028 |
| | 0.4 | 0.49130 | 0.77822 | 1.19916 | 1.50637 | 1.94809 |
| 0.8 | 0.0 | 0.50556 | 0.77506 | 1.17138 | 1.46100 | 1.87775 |
| | 0.4 | 0.52792 | 0.81937 | 1.23963 | 1.54195 | 1.97149 |
| | 0.8 | 0.57644 | 0.87287 | 1.29150 | 1.58800 | 2.00440 |
| 1.2 | 0.0 | 0.54685 | 0.81402 | 1.20047 | 1.47894 | 1.87494 |
| | 0.4 | 0.58162 | 0.86861 | 1.27414 | 1.56146 | 1.96500 |
| | 0.8 | 0.62708 | 0.91698 | 1.32003 | 1.60227 | 1.99532 |
| | 1.2 | 0.66653 | 0.95078 | 1.34165 | 1.61315 | 1.98894 |
| 1.6 | 0.0 | 0.59861 | 0.86158 | 1.23401 | 1.49813 | 1.86927 |
| | 0.4 | 0.63306 | 0.91031 | 1.29582 | 1.56578 | 1.94170 |
| | 0.8 | 0.66821 | 0.94769 | 1.33194 | 1.59880 | 1.96812 |
| | 1.2 | 0.69684 | 0.97271 | 1.34894 | 1.60862 | 1.96625 |
| | 1.6 | 0.71896 | 0.98891 | 1.35467 | 1.60582 | 1.95029 |
| 2.0 | 0.0 | 0.64685 | 0.90310 | 1.25964 | 1.50932 | 1.85697 |
| | 0.4 | 0.67399 | 0.94061 | 1.30699 | 1.56134 | 1.91323 |
| | 0.8 | 0.69916 | 0.96826 | 1.33507 | 1.58815 | 1.93660 |
| | 1.2 | 0.71973 | 0.98706 | 1.34916 | 1.59774 | 1.93864 |
| | 1.6 | 0.73610 | 0.99956 | 1.35451 | 1.59714 | 1.92873 |
| | 2.0 | 0.74917 | 1.00791 | 1.35487 | 1.59113 | 1.91302 |
| 2.4 | 0.0 | 0.68531 | 0.93429 | 1.27619 | 1.51337 | 1.84130 |
| | 0.4 | 0.70463 | 0.96182 | 1.31202 | 1.55345 | 1.88566 |
| | 0.8 | 0.72237 | 0.98237 | 1.33425 | 1.57565 | 1.90656 |
| | 1.2 | 0.73732 | 0.99685 | 1.34630 | 1.58507 | 1.91129 |
| | 1.6 | 0.74963 | 1.00680 | 1.35156 | 1.58628 | 1.90601 |
| | 2.0 | 0.75977 | 1.01362 | 1.35260 | 1.58264 | 1.89517 |
| | 2.4 | 0.76820 | 1.01835 | 1.35119 | 1.57641 | 1.88165 |
| 2.8 | 0.0 | 0.71421 | 0.95670 | 1.28634 | 1.51331 | 1.82534 |
| | 0.4 | 0.72753 | 0.97690 | 1.31388 | 1.54482 | 1.86113 |
| | 0.8 | 0.74012 | 0.99246 | 1.33184 | 1.56353 | 1.87988 |
| | 1.2 | 0.75114 | 1.00383 | 1.34231 | 1.57262 | 1.88620 |
| | 1.6 | 0.76054 | 1.01194 | 1.34747 | 1.57508 | 1.88422 |
| | 2.0 | 0.76851 | 1.01765 | 1.34908 | 1.57330 | 1.87715 |
| | 2.4 | 0.77528 | 1.02168 | 1.34847 | 1.56900 | 1.86721 |
| | 2.8 | 0.78108 | 1.02455 | 1.34655 | 1.56333 | 1.85592 |
| 3.2 | 0.0 | 0.73577 | 0.97288 | 1.29256 | 1.51130 | 1.81053 |
| | 0.4 | 0.74495 | 0.98796 | 1.31419 | 1.53661 | 1.84000 |
| | 0.8 | 0.75400 | 0.99995 | 1.32895 | 1.55256 | 1.85681 |
| | 1.2 | 0.76224 | 1.00904 | 1.33811 | 1.56116 | 1.86394 |
| | 1.6 | 0.76950 | 1.01574 | 1.34308 | 1.56441 | 1.86423 |
| | 2.0 | 0.77582 | 1.02060 | 1.34510 | 1.56400 | 1.85996 |
| | 2.4 | 0.78131 | 1.02410 | 1.34514 | 1.56124 | 1.85287 |
| | 2.8 | 0.78610 | 1.02663 | 1.34391 | 1.55705 | 1.84420 |
| | 3.2 | 0.79030 | 1.02847 | 1.34193 | 1.55209 | 1.83478 |

FRACTILES $W_\alpha$ RHO = 0.40

| DELTAS | | α:0.8000 | 0.8500 | 0.9000 | 0.9250 | 0.9500 |
|---|---|---|---|---|---|---|
| 3.6 | 0.0 | 0.75210 | 0.98482 | 1.29642 | 1.50846 | 1.79731 |
| | 0.4 | 0.75849 | 0.99630 | 1.31374 | 1.52918 | 1.82198 |
| | 0.8 | 0.76507 | 1.00571 | 1.32603 | 1.54291 | 1.83704 |
| | 1.2 | 0.77130 | 1.01307 | 1.33407 | 1.55093 | 1.84449 |
| | 1.6 | 0.77696 | 1.01866 | 1.33880 | 1.55462 | 1.84626 |
| | 2.0 | 0.78201 | 1.02283 | 1.34106 | 1.55518 | 1.84404 |
| | 2.4 | 0.78649 | 1.02590 | 1.34159 | 1.55360 | 1.83917 |
| | 2.8 | 0.79047 | 1.02816 | 1.34094 | 1.55062 | 1.83263 |
| | 3.2 | 0.79402 | 1.02983 | 1.33952 | 1.54679 | 1.82517 |
| | 3.6 | 0.79719 | 1.03106 | 1.33764 | 1.54249 | 1.81726 |
| 4.0 | 0.0 | 0.76473 | 0.99384 | 1.29881 | 1.50537 | 1.78569 |
| | 0.4 | 0.76922 | 1.00275 | 1.31293 | 1.52260 | 1.80662 |
| | 0.8 | 0.77406 | 1.01024 | 1.32329 | 1.53449 | 1.82013 |
| | 1.2 | 0.77881 | 1.01627 | 1.33037 | 1.54191 | 1.82756 |
| | 1.6 | 0.78325 | 1.02097 | 1.33480 | 1.54580 | 1.83028 |
| | 2.0 | 0.78732 | 1.02457 | 1.33719 | 1.54702 | 1.82954 |
| | 2.4 | 0.79100 | 1.02729 | 1.33806 | 1.54633 | 1.82636 |
| | 2.8 | 0.79432 | 1.02932 | 1.33786 | 1.54431 | 1.82154 |
| | 3.2 | 0.79732 | 1.03085 | 1.33692 | 1.54142 | 1.81569 |
| | 3.6 | 0.80004 | 1.03198 | 1.33549 | 1.53800 | 1.80926 |
| | 4.0 | 0.80250 | 1.03284 | 1.33376 | 1.53427 | 1.80256 |
| 6.0 | 0.0 | 0.79910 | 1.01711 | 1.30176 | 1.49150 | 1.74559 |
| | 0.4 | 0.79991 | 1.02025 | 1.30792 | 1.49961 | 1.75623 |
| | 0.8 | 0.80104 | 1.02312 | 1.31296 | 1.50601 | 1.76431 |
| | 1.2 | 0.80238 | 1.02565 | 1.31692 | 1.51081 | 1.77011 |
| | 1.6 | 0.80380 | 1.02783 | 1.31990 | 1.51421 | 1.77390 |
| | 2.0 | 0.80526 | 1.02967 | 1.32202 | 1.51639 | 1.77602 |
| | 2.4 | 0.80670 | 1.03119 | 1.32342 | 1.51757 | 1.77674 |
| | 2.8 | 0.80812 | 1.03243 | 1.32422 | 1.51793 | 1.77635 |
| | 3.2 | 0.80948 | 1.03344 | 1.32453 | 1.51764 | 1.77508 |
| | 3.6 | 0.81078 | 1.03425 | 1.32445 | 1.51683 | 1.77313 |
| | 4.0 | 0.81202 | 1.03489 | 1.32408 | 1.51564 | 1.77069 |
| | 6.0 | 0.81736 | 1.03655 | 1.31980 | 1.50676 | 1.75492 |
| 12.0 | 0.0 | 0.82571 | 1.03236 | 1.29686 | 1.47007 | 1.69846 |
| | 0.4 | 0.82554 | 1.03284 | 1.29827 | 1.47213 | 1.70142 |
| | 0.8 | 0.82544 | 1.03331 | 1.29954 | 1.47397 | 1.70405 |
| | 1.2 | 0.82539 | 1.03375 | 1.30069 | 1.47560 | 1.70633 |
| | 1.6 | 0.82539 | 1.03418 | 1.30170 | 1.47701 | 1.70830 |
| | 2.0 | 0.82543 | 1.03458 | 1.30259 | 1.47823 | 1.70996 |
| | 2.4 | 0.82551 | 1.03495 | 1.30335 | 1.47926 | 1.71133 |
| | 2.8 | 0.82562 | 1.03529 | 1.30400 | 1.48010 | 1.71243 |
| | 3.2 | 0.82575 | 1.03561 | 1.30454 | 1.48078 | 1.71329 |
| | 3.6 | 0.82591 | 1.03589 | 1.30498 | 1.48131 | 1.71391 |
| | 4.0 | 0.82608 | 1.03615 | 1.30532 | 1.48169 | 1.71433 |
| | 6.0 | 0.82712 | 1.03707 | 1.30590 | 1.48193 | 1.71397 |
| | 12.0 | 0.83045 | 1.03776 | 1.30235 | 1.47509 | 1.70220 |
| INFINITY | | 0.84162 | 1.03643 | 1.28155 | 1.43953 | 1.64485 |

FRACTILES $W_\alpha$ RHO = 0.40

| DELTAS | | $\alpha$:0.9625 | 0.9750 | 0.9825 | 0.9900 | 0.9925 |
|---|---|---|---|---|---|---|
| 0.0 | 0.0 | 2.21702 | 2.66863 | 3.07197 | 3.71417 | 4.04805 |
| 0.4 | 0.0 | 2.20163 | 2.64767 | 3.04584 | 3.67937 | 4.00854 |
| | 0.4 | 2.26625 | 2.71971 | 3.12243 | 3.75962 | 4.08914 |
| 0.8 | 0.0 | 2.17807 | 2.60626 | 2.98662 | 3.58856 | 3.89989 |
| | 0.4 | 2.27775 | 2.71058 | 3.09191 | 3.69046 | 3.99805 |
| | 0.8 | 2.29856 | 2.71124 | 3.07241 | 3.63573 | 3.92379 |
| 1.2 | 0.0 | 2.15745 | 2.55685 | 2.90879 | 3.46132 | 3.74529 |
| | 0.4 | 2.25010 | 2.65008 | 3.00014 | 3.54611 | 3.82531 |
| | 0.8 | 2.27115 | 2.65608 | 2.99134 | 3.51177 | 3.77693 |
| | 1.2 | 2.25134 | 2.61603 | 2.93246 | 3.42181 | 3.67038 |
| 1.6 | 0.0 | 2.13153 | 2.49950 | 2.82155 | 3.32385 | 3.58071 |
| | 0.4 | 2.20548 | 2.57356 | 2.89412 | 3.39170 | 3.64519 |
| | 0.8 | 2.22599 | 2.58434 | 2.89526 | 3.37605 | 3.62026 |
| | 1.2 | 2.21495 | 2.55942 | 2.85733 | 3.31655 | 3.54919 |
| | 1.6 | 2.18900 | 2.51866 | 2.80299 | 3.24001 | 3.46090 |
| 2.0 | 0.0 | 2.10086 | 2.44114 | 2.73743 | 3.19724 | 3.43146 |
| | 0.4 | 2.15885 | 2.50011 | 2.79614 | 3.25380 | 3.48623 |
| | 0.8 | 2.17886 | 2.51434 | 2.80443 | 3.25153 | 3.47801 |
| | 1.2 | 2.17485 | 2.50103 | 2.78233 | 3.21465 | 3.43316 |
| | 1.6 | 2.15780 | 2.47334 | 2.74480 | 3.16099 | 3.37091 |
| | 2.0 | 2.13481 | 2.43962 | 2.70128 | 3.10154 | 3.30306 |
| 2.4 | 0.0 | 2.07009 | 2.38785 | 2.66338 | 3.08922 | 3.30543 |
| | 0.4 | 2.11652 | 2.43609 | 2.71235 | 3.13798 | 3.35353 |
| | 0.8 | 2.13577 | 2.45218 | 2.72498 | 3.14415 | 3.35597 |
| | 1.2 | 2.13660 | 2.44689 | 2.71379 | 3.12292 | 3.32926 |
| | 1.6 | 2.12629 | 2.42899 | 2.68883 | 3.08627 | 3.28635 |
| | 2.0 | 2.10999 | 2.40462 | 2.65705 | 3.04238 | 3.23606 |
| | 2.4 | 2.09102 | 2.37766 | 2.62281 | 2.99635 | 3.18382 |
| 2.8 | 0.0 | 2.04201 | 2.34179 | 2.60079 | 2.99963 | 3.20155 |
| | 0.4 | 2.08008 | 2.38219 | 2.64254 | 3.04242 | 3.24443 |
| | 0.8 | 2.09826 | 2.39890 | 2.65741 | 3.05354 | 3.25328 |
| | 1.2 | 2.10216 | 2.39885 | 2.65346 | 3.04281 | 3.23879 |
| | 1.6 | 2.09667 | 2.38800 | 2.63755 | 3.01844 | 3.20986 |
| | 2.0 | 2.08554 | 2.37082 | 2.61478 | 2.98650 | 3.17304 |
| | 2.4 | 2.07137 | 2.35041 | 2.58867 | 2.95111 | 3.13275 |
| | 2.8 | 2.05588 | 2.32879 | 2.56147 | 2.91490 | 3.09181 |
| 3.2 | 0.0 | 2.01747 | 2.30281 | 2.54855 | 2.92576 | 3.11623 |
| | 0.4 | 2.04928 | 2.33721 | 2.58467 | 2.96367 | 3.15469 |
| | 0.8 | 2.06621 | 2.35376 | 2.60042 | 2.97746 | 3.16718 |
| | 1.2 | 2.07193 | 2.35703 | 2.60116 | 2.97368 | 3.16084 |
| | 1.6 | 2.06979 | 2.35112 | 2.59166 | 2.95807 | 3.14192 |
| | 2.0 | 2.06253 | 2.33935 | 2.57569 | 2.93515 | 3.11528 |
| | 2.4 | 2.05217 | 2.32414 | 2.55602 | 2.90819 | 3.08446 |
| | 2.8 | 2.04014 | 2.30718 | 2.53457 | 2.87945 | 3.05188 |
| | 3.2 | 2.02741 | 2.28962 | 2.51261 | 2.85041 | 3.01912 |

FRACTILES $W_\alpha$ RHO = 0.40

| DELTAS | | $\alpha$:0.9625 | 0.9750 | 0.9825 | 0.9900 | 0.9925 |
|---|---|---|---|---|---|---|
| 3.6 | 0.0 | 1.99635 | 2.26997 | 2.50495 | 2.86458 | 3.04575 |
| | 0.4 | 2.02334 | 2.29964 | 2.53651 | 2.89837 | 3.08037 |
| | 0.8 | 2.03895 | 2.31557 | 2.55232 | 2.91340 | 3.09476 |
| | 1.2 | 2.04565 | 2.32085 | 2.55604 | 2.91419 | 3.09384 |
| | 1.6 | 2.04580 | 2.31839 | 2.55105 | 2.90482 | 3.08206 |
| | 2.0 | 2.04139 | 2.31064 | 2.54015 | 2.88866 | 3.06307 |
| | 2.4 | 2.03399 | 2.29948 | 2.52551 | 2.86830 | 3.03967 |
| | 2.8 | 2.02478 | 2.28632 | 2.50874 | 2.84564 | 3.01389 |
| | 3.2 | 2.01460 | 2.27216 | 2.49096 | 2.82200 | 2.98717 |
| | 3.6 | 2.00402 | 2.25768 | 2.47294 | 2.79828 | 2.96046 |
| 4.0 | 0.0 | 1.97824 | 2.24222 | 2.46834 | 2.81351 | 2.98702 |
| | 0.4 | 2.00140 | 2.26806 | 2.49614 | 2.84376 | 3.01827 |
| | 0.8 | 2.01572 | 2.28313 | 2.51154 | 2.85918 | 3.03348 |
| | 1.2 | 2.02288 | 2.28958 | 2.51711 | 2.86293 | 3.03613 |
| | 1.6 | 2.02455 | 2.28950 | 2.51527 | 2.85798 | 3.02945 |
| | 2.0 | 2.02222 | 2.28471 | 2.50813 | 2.84689 | 3.01620 |
| | 2.4 | 2.01711 | 2.27669 | 2.49741 | 2.83168 | 2.99859 |
| | 2.8 | 2.01016 | 2.26659 | 2.48440 | 2.81392 | 2.97831 |
| | 3.2 | 2.00210 | 2.25526 | 2.47010 | 2.79478 | 2.95661 |
| | 3.6 | 1.99343 | 2.24333 | 2.45519 | 2.77507 | 2.93438 |
| | 4.0 | 1.98453 | 2.23122 | 2.44018 | 2.75537 | 2.91223 |
| 6.0 | 0.0 | 1.91814 | 2.15237 | 2.35110 | 2.65145 | 2.80120 |
| | 0.4 | 1.93040 | 2.16674 | 2.36715 | 2.66987 | 2.82071 |
| | 0.8 | 1.93955 | 2.17721 | 2.37864 | 2.68270 | 2.83413 |
| | 1.2 | 1.94592 | 2.18424 | 2.38611 | 2.69066 | 2.84225 |
| | 1.6 | 1.94989 | 2.18832 | 2.39017 | 2.69450 | 2.84589 |
| | 2.0 | 1.95185 | 2.18995 | 2.39140 | 2.69494 | 2.84586 |
| | 2.4 | 1.95216 | 2.18957 | 2.39033 | 2.69264 | 2.84286 |
| | 2.8 | 1.95116 | 2.18761 | 2.38745 | 2.68818 | 2.83755 |
| | 3.2 | 1.94913 | 2.18441 | 2.38316 | 2.68207 | 2.83046 |
| | 3.6 | 1.94631 | 2.18029 | 2.37782 | 2.67474 | 2.82207 |
| | 4.0 | 1.94291 | 2.17548 | 2.37173 | 2.66653 | 2.81274 |
| | 6.0 | 1.92203 | 2.14713 | 2.33660 | 2.62046 | 2.76091 |
| 12.0 | 0.0 | 1.85138 | 2.05637 | 2.22809 | 2.48411 | 2.61028 |
| | 0.4 | 1.85497 | 2.06083 | 2.23329 | 2.49042 | 2.61714 |
| | 0.8 | 1.85813 | 2.06472 | 2.23781 | 2.49589 | 2.62307 |
| | 1.2 | 1.86087 | 2.06808 | 2.24169 | 2.50053 | 2.62810 |
| | 1.6 | 1.86321 | 2.07092 | 2.24495 | 2.50441 | 2.63228 |
| | 2.0 | 1.86517 | 2.07327 | 2.24763 | 2.50757 | 2.63567 |
| | 2.4 | 1.86677 | 2.07517 | 2.24977 | 2.51006 | 2.63831 |
| | 2.8 | 1.86803 | 2.07665 | 2.25141 | 2.51193 | 2.64028 |
| | 3.2 | 1.86900 | 2.07774 | 2.25260 | 2.51323 | 2.64163 |
| | 3.6 | 1.86968 | 2.07848 | 2.25337 | 2.51402 | 2.64242 |
| | 4.0 | 1.87010 | 2.07890 | 2.25376 | 2.51434 | 2.64269 |
| | 6.0 | 1.86925 | 2.07725 | 2.25134 | 2.51058 | 2.63818 |
| | 12.0 | 1.85381 | 2.05645 | 2.22567 | 2.47703 | 2.60049 |
| INFINITY | | 1.78046 | 1.95996 | 2.10836 | 2.32635 | 2.43238 |

FRACTILES $W_{\alpha}$ RHO = 0.40

| DELTAS | | $\alpha$:0.9950 | 0.9975 | 0.9990 | 0.9995 |
|---|---|---|---|---|---|
| 0.0 | 0.0 | 4.52228 | 5.34138 | 6.43723 | 7.27408 |
| 0.4 | 0.0 | 4.47584 | 5.28237 | 6.36021 | 7.18244 |
| | 0.4 | 4.55530 | 5.35578 | 6.41865 | 7.22495 |
| 0.8 | 0.0 | 4.34036 | 5.09678 | 6.10123 | 6.86325 |
| | 0.4 | 4.43127 | 5.17074 | 6.14565 | 6.88102 |
| | 0.8 | 4.32815 | 5.01526 | 5.91634 | 6.59313 |
| 1.2 | 0.0 | 4.14526 | 4.82802 | 5.72818 | 6.40717 |
| | 0.4 | 4.21721 | 4.88315 | 5.75645 | 6.41237 |
| | 0.8 | 4.14819 | 4.77683 | 5.59777 | 6.21224 |
| | 1.2 | 4.01766 | 4.60399 | 5.36689 | 5.93619 |
| 1.6 | 0.0 | 3.94124 | 4.55385 | 5.35715 | 5.96046 |
| | 0.4 | 4.00008 | 4.60099 | 5.38564 | 5.97291 |
| | 0.8 | 3.96144 | 4.53743 | 5.28684 | 5.84606 |
| | 1.2 | 3.87361 | 4.41990 | 5.12842 | 5.65569 |
| | 1.6 | 3.76842 | 4.28505 | 4.95318 | 5.44919 |
| 2.0 | 0.0 | 3.75934 | 4.31443 | 5.03911 | 5.58142 |
| | 0.4 | 3.81091 | 4.35898 | 5.07194 | 5.60389 |
| | 0.8 | 3.79380 | 4.32551 | 5.01503 | 5.52812 |
| | 1.2 | 3.73735 | 4.24835 | 4.90915 | 5.39970 |
| | 1.6 | 3.66272 | 4.15193 | 4.78293 | 5.25033 |
| | 2.0 | 3.58282 | 4.05094 | 4.65332 | 5.09862 |
| 2.4 | 0.0 | 3.60739 | 4.11699 | 4.77973 | 5.27410 |
| | 0.4 | 3.65404 | 4.15991 | 4.81576 | 5.30371 |
| | 0.8 | 3.65081 | 4.14603 | 4.78629 | 5.26153 |
| | 1.2 | 3.61606 | 4.09682 | 4.71684 | 5.17606 |
| | 1.6 | 3.56410 | 4.02885 | 4.62684 | 5.06888 |
| | 2.0 | 3.50460 | 3.95318 | 4.52915 | 4.95413 |
| | 2.4 | 3.44347 | 3.87653 | 4.43147 | 4.84025 |
| 2.8 | 0.0 | 3.48298 | 3.95657 | 4.57035 | 5.02686 |
| | 0.4 | 3.52555 | 3.99759 | 4.60770 | 5.06043 |
| | 0.8 | 3.53085 | 3.99604 | 4.59582 | 5.03996 |
| | 1.2 | 3.51080 | 3.96588 | 4.55131 | 4.98399 |
| | 1.6 | 3.47525 | 3.91852 | 4.48760 | 4.90746 |
| | 2.0 | 3.43139 | 3.86226 | 4.41437 | 4.82104 |
| | 2.4 | 3.38407 | 3.80263 | 4.33802 | 4.73176 |
| | 2.8 | 3.33635 | 3.74308 | 4.26248 | 4.64392 |
| 3.2 | 0.0 | 3.38120 | 3.82596 | 4.40054 | 4.82675 |
| | 0.4 | 3.42008 | 3.86471 | 4.43775 | 4.86196 |
| | 0.8 | 3.43045 | 3.87078 | 4.43706 | 4.85548 |
| | 1.2 | 3.42029 | 3.85354 | 4.40962 | 4.81981 |
| | 1.6 | 3.39651 | 3.82105 | 4.36494 | 4.76550 |
| | 2.0 | 3.36449 | 3.77950 | 4.31027 | 4.70059 |
| | 2.4 | 3.32811 | 3.73335 | 4.25081 | 4.63080 |
| | 2.8 | 3.29002 | 3.68563 | 4.19003 | 4.55995 |
| | 3.2 | 3.25194 | 3.63828 | 4.13016 | 4.49045 |

FRACTILES $W_\alpha$ RHO = 0.40

| DELTAS | | $\alpha$:0.9950 | 0.9975 | 0.9990 | 0.9995 |
|---|---|---|---|---|---|
| 3.6 | 0.0 | 3.29736 | 3.71870 | 4.26143 | 4.66303 |
| | 0.4 | 3.33286 | 3.75497 | 4.29758 | 4.69837 |
| | 0.8 | 3.34609 | 3.76565 | 4.30395 | 4.70091 |
| | 1.2 | 3.34257 | 3.75721 | 4.28828 | 4.67929 |
| | 1.6 | 3.32723 | 3.73543 | 4.25738 | 4.64114 |
| | 2.0 | 3.30413 | 3.70499 | 4.21677 | 4.59255 |
| | 2.4 | 3.27633 | 3.66944 | 4.17060 | 4.53811 |
| | 2.8 | 3.24608 | 3.63135 | 4.12184 | 4.48110 |
| | 3.2 | 3.21494 | 3.59250 | 4.07255 | 4.42375 |
| | 3.6 | 3.18395 | 3.55407 | 4.02407 | 4.36754 |
| 4.0 | 0.0 | 3.22763 | 3.62967 | 4.14616 | 4.52748 |
| | 0.4 | 3.26003 | 3.66343 | 4.18074 | 4.56205 |
| | 0.8 | 3.27475 | 3.67680 | 4.19150 | 4.57035 |
| | 1.2 | 3.27566 | 3.67435 | 4.18396 | 4.55854 |
| | 1.6 | 3.26639 | 3.66032 | 4.16312 | 4.53221 |
| | 2.0 | 3.24999 | 3.63827 | 4.13317 | 4.49602 |
| | 2.4 | 3.22891 | 3.61103 | 4.09743 | 4.45365 |
| | 2.8 | 3.20500 | 3.58073 | 4.05841 | 4.40786 |
| | 3.2 | 3.17964 | 3.54896 | 4.01792 | 4.36064 |
| | 3.6 | 3.15379 | 3.51681 | 3.97725 | 4.31339 |
| | 4.0 | 3.12813 | 3.48504 | 3.93725 | 4.26707 |
| 6.0 | 0.0 | 3.00763 | 3.34966 | 3.78441 | 4.10247 |
| | 0.4 | 3.02856 | 3.37272 | 3.80978 | 4.12927 |
| | 0.8 | 3.04269 | 3.38782 | 3.82574 | 4.14559 |
| | 1.2 | 3.05094 | 3.39607 | 3.83361 | 4.15295 |
| | 1.6 | 3.05424 | 3.39857 | 3.83475 | 4.15284 |
| | 2.0 | 3.05346 | 3.39637 | 3.83039 | 4.14668 |
| | 2.4 | 3.04943 | 3.39043 | 3.82168 | 4.13573 |
| | 2.8 | 3.04286 | 3.38157 | 3.80961 | 4.12110 |
| | 3.2 | 3.03434 | 3.37051 | 3.79501 | 4.10372 |
| | 3.6 | 3.02441 | 3.35785 | 3.77860 | 4.08438 |
| | 4.0 | 3.01346 | 3.34407 | 3.76094 | 4.06371 |
| | 6.0 | 2.95340 | 3.26965 | 3.66712 | 3.95494 |
| 12.0 | 0.0 | 2.78269 | 3.06480 | 3.41756 | 3.67190 |
| | 0.4 | 2.79032 | 3.07367 | 3.42797 | 3.68341 |
| | 0.8 | 2.79688 | 3.08126 | 3.43682 | 3.69315 |
| | 1.2 | 2.80242 | 3.08764 | 3.44420 | 3.70122 |
| | 1.6 | 2.80701 | 3.09287 | 3.45020 | 3.70774 |
| | 2.0 | 2.81070 | 3.09703 | 3.45490 | 3.71280 |
| | 2.4 | 2.81356 | 3.10020 | 3.45842 | 3.71653 |
| | 2.8 | 2.81565 | 3.10247 | 3.46085 | 3.71903 |
| | 3.2 | 2.81705 | 3.10392 | 3.46229 | 3.72042 |
| | 3.6 | 2.81782 | 3.10461 | 3.46283 | 3.72081 |
| | 4.0 | 2.81801 | 3.10464 | 3.46257 | 3.72030 |
| | 6.0 | 2.81240 | 3.09701 | 3.45205 | 3.70745 |
| | 12.0 | 2.76877 | 3.04300 | 3.38396 | 3.62849 |
| INFINITY | | 2.57583 | 2.80703 | 3.09024 | 3.29051 |

FRACTILES $W_\alpha$ RHO = 0.60

| DELTAS | | $\alpha$:0.0005 | 0.0010 | 0.0025 | 0.0050 | 0.0075 |
|---|---|---|---|---|---|---|
| 0.0 | 0.0 | -2.32672 | -2.10694 | -1.81955 | -1.60519 | -1.48134 |
| 0.4 | 0.0 | -2.48921 | -2.25497 | -1.94674 | -1.71517 | -1.58064 |
| | 0.4 | -2.07812 | -1.89095 | -1.64626 | -1.46381 | -1.35844 |
| 0.8 | 0.0 | -2.62835 | -2.39625 | -2.08744 | -1.85218 | -1.71386 |
| | 0.4 | -2.08324 | -1.90230 | -1.66428 | -1.48556 | -1.38179 |
| | 0.8 | -1.75138 | -1.61301 | -1.43227 | -1.29767 | -1.22003 |
| 1.2 | 0.0 | -2.65052 | -2.43197 | -2.13912 | -1.91404 | -1.78071 |
| | 0.4 | -2.17369 | -1.99888 | -1.76628 | -1.58905 | -1.48484 |
| | 0.8 | -1.78756 | -1.65307 | -1.47630 | -1.34373 | -1.26686 |
| | 1.2 | -1.58963 | -1.48555 | -1.34982 | -1.24898 | -1.19097 |
| 1.6 | 0.0 | -2.62968 | -2.42587 | -2.15117 | -1.93855 | -1.81186 |
| | 0.4 | -2.23292 | -2.06621 | -1.84268 | -1.67073 | -1.56880 |
| | 0.8 | -1.90544 | -1.77198 | -1.59436 | -1.45900 | -1.37940 |
| | 1.2 | -1.65892 | -1.55508 | -1.41879 | -1.31683 | -1.25787 |
| | 1.6 | -1.55133 | -1.46969 | -1.36351 | -1.28499 | -1.24004 |
| 2.0 | 0.0 | -2.60318 | -2.41225 | -2.15360 | -1.95219 | -1.83159 |
| | 0.4 | -2.27571 | -2.11617 | -1.90090 | -1.73402 | -1.63445 |
| | 0.8 | -2.00325 | -1.87204 | -1.69593 | -1.56027 | -1.47975 |
| | 1.2 | -1.78959 | -1.68331 | -1.54181 | -1.43396 | -1.37055 |
| | 1.6 | -1.63566 | -1.55220 | -1.44295 | -1.36161 | -1.31485 |
| | 2.0 | -1.58601 | -1.51977 | -1.43406 | -1.37127 | -1.33574 |
| 2.4 | 0.0 | -2.58301 | -2.40258 | -2.15706 | -1.96484 | -1.84924 |
| | 0.4 | -2.31177 | -2.15813 | -1.94965 | -1.78694 | -1.68931 |
| | 0.8 | -2.08437 | -1.95513 | -1.78042 | -1.64465 | -1.56347 |
| | 1.2 | -1.90327 | -1.79583 | -1.65139 | -1.53988 | -1.47357 |
| | 1.6 | -1.76639 | -1.67849 | -1.56144 | -1.47224 | -1.41984 |
| | 2.0 | -1.67488 | -1.60576 | -1.51578 | -1.44950 | -1.41195 |
| | 2.4 | -1.66476 | -1.60986 | -1.53964 | -1.48945 | -1.46216 |
| 2.8 | 0.0 | -2.57143 | -2.39938 | -2.16429 | -1.97931 | -1.86762 |
| | 0.4 | -2.34514 | -2.19618 | -1.99303 | -1.83351 | -1.73732 |
| | 0.8 | -2.15422 | -2.02649 | -1.85273 | -1.71666 | -1.63475 |
| | 1.2 | -2.00051 | -1.89203 | -1.74497 | -1.63027 | -1.56143 |
| | 1.6 | -1.88289 | -1.79185 | -1.66914 | -1.57410 | -1.51740 |
| | 2.0 | -1.79930 | -1.72440 | -1.62468 | -1.54884 | -1.50452 |
| | 2.4 | -1.75287 | -1.69461 | -1.61977 | -1.56650 | -1.53846 |
| | 2.8 | -1.77209 | -1.72646 | -1.67021 | -1.63651 | -1.61039 |
| 3.2 | 0.0 | -2.56766 | -2.40222 | -2.17530 | -1.99593 | -1.88721 |
| | 0.4 | -2.37734 | -2.23202 | -2.03291 | -1.87569 | -1.78044 |
| | 0.8 | -2.21602 | -2.08934 | -1.91603 | -1.77934 | -1.69658 |
| | 1.2 | -2.08517 | -1.97558 | -1.82593 | -1.70812 | -1.63686 |
| | 1.6 | -1.98428 | -1.89034 | -1.76247 | -1.66212 | -1.60155 |
| | 2.0 | -1.91207 | -1.83265 | -1.72521 | -1.64155 | -1.59140 |
| | 2.4 | -1.86763 | -1.80234 | -1.71567 | -1.65073 | -1.61479 |
| | 2.8 | -1.85681 | -1.80782 | -1.74871 | -1.70622 | -1.66954 |
| | 3.2 | -1.90040 | -1.86493 | -1.81751 | -1.75960 | -1.71636 |

FRACTILES $W_\alpha$ RHO = 0.60

| DELTAS | | α:0.0005 | 0.0010 | 0.0025 | 0.0050 | 0.0075 |
|---|---|---|---|---|---|---|
| 3.6 | 0.0 | -2.57017 | -2.40994 | -2.18935 | -2.01423 | -1.90772 |
| | 0.4 | -2.40887 | -2.26632 | -2.07017 | -1.91446 | -1.81972 |
| | 0.8 | -2.27173 | -2.14568 | -1.97229 | -1.83467 | -1.75089 |
| | 1.2 | -2.15994 | -2.04912 | -1.89679 | -1.77586 | -1.70220 |
| | 1.6 | -2.07336 | -1.97661 | -1.84374 | -1.73829 | -1.67399 |
| | 2.0 | -2.01128 | -1.92764 | -1.81303 | -1.72218 | -1.66677 |
| | 2.4 | -1.97306 | -1.90198 | -1.80524 | -1.72937 | -1.68417 |
| | 2.8 | -1.95941 | -1.90175 | -1.82806 | -1.76852 | -1.72345 |
| | 3.2 | -1.98111 | -1.94522 | -1.87907 | -1.81007 | -1.76066 |
| | 3.6 | -2.04869 | -2.00547 | -1.92825 | -1.85128 | -1.79716 |
| 4.0 | 0.0 | -2.57748 | -2.42132 | -2.20559 | -2.03362 | -1.92868 |
| | 0.4 | -2.43982 | -2.29932 | -2.10520 | -1.95034 | -1.85575 |
| | 0.8 | -2.32257 | -2.19675 | -2.02284 | -1.88396 | -1.79900 |
| | 1.2 | -2.22672 | -2.11453 | -1.95938 | -1.83528 | -1.75920 |
| | 1.6 | -2.15231 | -2.05277 | -1.91499 | -1.80454 | -1.73661 |
| | 2.0 | -2.09898 | -2.01124 | -1.88968 | -1.79190 | -1.73145 |
| | 2.4 | -2.06639 | -1.98986 | -1.88379 | -1.79814 | -1.74490 |
| | 2.8 | -2.05482 | -1.98955 | -1.90060 | -1.82544 | -1.77279 |
| | 3.2 | -2.06828 | -2.01957 | -1.93740 | -1.85766 | -1.80223 |
| | 3.6 | -2.11759 | -2.06521 | -1.97658 | -1.89119 | -1.83222 |
| | 4.0 | -2.17197 | -2.11355 | -2.01653 | -1.92459 | -1.86174 |
| 6.0 | 0.0 | -2.65005 | -2.50348 | -2.29794 | -2.13120 | -2.02801 |
| | 0.4 | -2.58236 | -2.44494 | -2.25179 | -2.09458 | -1.99699 |
| | 0.8 | -2.52483 | -2.39602 | -2.21441 | -2.06595 | -1.97344 |
| | 1.2 | -2.47782 | -2.35700 | -2.18595 | -2.04536 | -1.95731 |
| | 1.6 | -2.44142 | -2.32791 | -2.16635 | -2.03258 | -1.94826 |
| | 2.0 | -2.41552 | -2.30858 | -2.15528 | -2.02711 | -1.94564 |
| | 2.4 | -2.39989 | -2.29867 | -2.15212 | -2.02804 | -1.94838 |
| | 2.8 | -2.39412 | -2.29756 | -2.15587 | -2.03406 | -1.95513 |
| | 3.2 | -2.39754 | -2.30420 | -2.16498 | -2.04373 | -1.96467 |
| | 3.6 | -2.40882 | -2.31684 | -2.17789 | -2.05588 | -1.97605 |
| | 4.0 | -2.42589 | -2.33366 | -2.19331 | -2.06961 | -1.98860 |
| | 6.0 | -2.54535 | -2.44063 | -2.28249 | -2.14461 | -2.05510 |
| 12.0 | 0.0 | -2.87073 | -2.71720 | -2.49762 | -2.31584 | -2.20178 |
| | 0.4 | -2.85646 | -2.70528 | -2.48865 | -2.30897 | -2.19607 |
| | 0.8 | -2.84426 | -2.69519 | -2.48120 | -2.30336 | -2.19146 |
| | 1.2 | -2.83407 | -2.68687 | -2.47518 | -2.29892 | -2.18787 |
| | 1.6 | -2.82580 | -2.68024 | -2.47052 | -2.29558 | -2.18523 |
| | 2.0 | -2.81936 | -2.67519 | -2.46711 | -2.29325 | -2.18346 |
| | 2.4 | -2.81463 | -2.67161 | -2.46485 | -2.29184 | -2.18248 |
| | 2.8 | -2.81148 | -2.66937 | -2.46364 | -2.29125 | -2.18219 |
| | 3.2 | -2.80976 | -2.66836 | -2.46337 | -2.29140 | -2.18253 |
| | 3.6 | -2.80935 | -2.66844 | -2.46394 | -2.29220 | -2.18341 |
| | 4.0 | -2.81010 | -2.66949 | -2.46523 | -2.29357 | -2.18477 |
| | 6.0 | -2.82679 | -2.68542 | -2.47964 | -2.30642 | -2.19657 |
| | 12.0 | -2.92035 | -2.76788 | -2.54720 | -2.36262 | -2.24610 |
| INFINITY | | -3.29051 | -3.09024 | -2.80703 | -2.57583 | -2.43238 |

FRACTILES $W_\alpha$ RHO = 0.60

| DELTAS | | $\alpha$:0.0100 | 0.0175 | 0.0250 | 0.0375 | 0.0500 |
|---|---|---|---|---|---|---|
| 0.0 | 0.0 | -1.39430 | -1.22736 | -1.12297 | -1.00670 | -0.92613 |
| 0.4 | 0.0 | -1.48573 | -1.30277 | -1.18766 | -1.05873 | -0.96888 |
| | 0.4 | -1.28441 | -1.14250 | -1.05383 | -0.95517 | -0.88690 |
| 0.8 | 0.0 | -1.61539 | -1.42305 | -1.29995 | -1.15960 | -1.05985 |
| | 0.4 | -1.30862 | -1.16769 | -1.07915 | -0.98015 | -0.91134 |
| | 0.8 | -1.16554 | -1.06127 | -0.99631 | -0.92429 | -0.87470 |
| 1.2 | 0.0 | -1.68526 | -1.49730 | -1.37572 | -1.23554 | -1.13462 |
| | 0.4 | -1.41065 | -1.26576 | -1.17306 | -1.06742 | -0.99241 |
| | 0.8 | -1.21273 | -1.10869 | -1.04356 | -0.97107 | -0.92100 |
| | 1.2 | -1.15034 | -1.07292 | -1.02498 | -0.97231 | -0.93651 |
| 1.6 | 0.0 | -1.72079 | -1.54037 | -1.42278 | -1.28613 | -1.18693 |
| | 0.4 | -1.49578 | -1.35189 | -1.25871 | -1.15116 | -1.07364 |
| | 0.8 | -1.32274 | -1.21211 | -1.14139 | -1.06092 | -1.00395 |
| | 1.2 | -1.21646 | -1.13723 | -1.08801 | -1.03386 | -0.99713 |
| | 1.6 | -1.20872 | -1.14955 | -1.11349 | -1.07488 | -1.04985 |
| 2.0 | 0.0 | -1.74458 | -1.57133 | -1.45769 | -1.32478 | -1.22764 |
| | 0.4 | -1.56280 | -1.42062 | -1.32774 | -1.21954 | -1.14077 |
| | 0.8 | -1.42202 | -1.30809 | -1.23417 | -1.14868 | -1.08694 |
| | 1.2 | -1.32543 | -1.23743 | -1.18131 | -1.11780 | -1.07327 |
| | 1.6 | -1.28218 | -1.22038 | -1.18282 | -1.14310 | -1.11852 |
| | 2.0 | -1.31127 | -1.26630 | -1.24062 | -1.21848 | -1.19470 |
| 2.4 | 0.0 | -1.76556 | -1.59821 | -1.48781 | -1.35798 | -1.26250 |
| | 0.4 | -1.61877 | -1.47796 | -1.38529 | -1.27651 | -1.19665 |
| | 0.8 | -1.50494 | -1.38848 | -1.31211 | -1.22277 | -1.15739 |
| | 1.2 | -1.42595 | -1.33176 | -1.27047 | -1.19939 | -1.14793 |
| | 1.6 | -1.38261 | -1.31032 | -1.26472 | -1.21450 | -1.18222 |
| | 2.0 | -1.38614 | -1.33943 | -1.31528 | -1.28276 | -1.24524 |
| | 2.4 | -1.44449 | -1.41503 | -1.38287 | -1.33108 | -1.28362 |
| 2.8 | 0.0 | -1.78654 | -1.62371 | -1.51575 | -1.38812 | -1.29376 |
| | 0.4 | -1.66755 | -1.52756 | -1.43481 | -1.32520 | -1.24414 |
| | 0.8 | -1.57542 | -1.45653 | -1.37784 | -1.28490 | -1.21614 |
| | 1.2 | -1.51166 | -1.41218 | -1.34650 | -1.26908 | -1.21186 |
| | 1.6 | -1.47658 | -1.39557 | -1.34266 | -1.28132 | -1.23802 |
| | 2.0 | -1.47327 | -1.41435 | -1.38267 | -1.33244 | -1.28459 |
| | 2.4 | -1.52290 | -1.47593 | -1.43236 | -1.36934 | -1.31454 |
| | 2.8 | -1.58560 | -1.52276 | -1.47138 | -1.40014 | -1.33975 |
| 3.2 | 0.0 | -1.80808 | -1.64855 | -1.54227 | -1.41604 | -1.32222 |
| | 0.4 | -1.71112 | -1.57137 | -1.47821 | -1.36744 | -1.28496 |
| | 0.8 | -1.63637 | -1.51493 | -1.43391 | -1.33740 | -1.26532 |
| | 1.2 | -1.58503 | -1.48049 | -1.41065 | -1.32722 | -1.26462 |
| | 1.6 | -1.55755 | -1.46885 | -1.40958 | -1.33864 | -1.28541 |
| | 2.0 | -1.55520 | -1.48333 | -1.43871 | -1.37418 | -1.31774 |
| | 2.4 | -1.59182 | -1.52771 | -1.47509 | -1.40255 | -1.34134 |
| | 2.8 | -1.63845 | -1.56440 | -1.50626 | -1.42757 | -1.36202 |
| | 3.2 | -1.68086 | -1.59843 | -1.53502 | -1.45038 | -1.38066 |

FRACTILES $W_\alpha$ RHO = 0.60

| DELTAS | | $\alpha$:0.0100 | 0.0175 | 0.0250 | 0.0375 | 0.0500 |
|---|---|---|---|---|---|---|
| 3.6 | 0.0 | -1.82999 | -1.67273 | -1.56749 | -1.44193 | -1.34818 |
| | 0.4 | -1.75056 | -1.61050 | -1.51661 | -1.40433 | -1.32023 |
| | 0.8 | -1.68969 | -1.56556 | -1.48213 | -1.38200 | -1.30661 |
| | 1.2 | -1.64834 | -1.53883 | -1.46493 | -1.37569 | -1.30790 |
| | 1.6 | -1.62691 | -1.53086 | -1.46558 | -1.38585 | -1.32419 |
| | 2.0 | -1.62614 | -1.54296 | -1.48654 | -1.41019 | -1.34644 |
| | 2.4 | -1.65200 | -1.57369 | -1.51311 | -1.43204 | -1.36504 |
| | 2.8 | -1.68679 | -1.60241 | -1.53797 | -1.45233 | -1.38199 |
| | 3.2 | -1.72086 | -1.63021 | -1.56166 | -1.47130 | -1.39760 |
| | 3.6 | -1.75399 | -1.65679 | -1.58408 | -1.48901 | -1.41202 |
| 4.0 | 0.0 | -1.85191 | -1.69607 | -1.59133 | -1.46586 | -1.37176 |
| | 0.4 | -1.78649 | -1.64564 | -1.55074 | -1.43667 | -1.35075 |
| | 0.8 | -1.73671 | -1.60972 | -1.52381 | -1.42002 | -1.34133 |
| | 1.2 | -1.70331 | -1.58890 | -1.51102 | -1.41615 | -1.34342 |
| | 1.6 | -1.68653 | -1.58338 | -1.51238 | -1.42454 | -1.35573 |
| | 2.0 | -1.68663 | -1.59317 | -1.52742 | -1.44144 | -1.37144 |
| | 2.4 | -1.70501 | -1.61462 | -1.54699 | -1.45829 | -1.38610 |
| | 2.8 | -1.73099 | -1.63706 | -1.56677 | -1.47472 | -1.39998 |
| | 3.2 | -1.75826 | -1.65971 | -1.58627 | -1.49049 | -1.41306 |
| | 3.6 | -1.78565 | -1.68192 | -1.60512 | -1.50549 | -1.42533 |
| | 4.0 | -1.81242 | -1.70332 | -1.62311 | -1.51966 | -1.43682 |
| 6.0 | 0.0 | -1.95178 | -1.79488 | -1.68769 | -1.55726 | -1.45791 |
| | 0.4 | -1.92473 | -1.77547 | -1.67303 | -1.54777 | -1.45187 |
| | 0.8 | -1.90473 | -1.76216 | -1.66375 | -1.54271 | -1.44948 |
| | 1.2 | -1.89165 | -1.75466 | -1.65942 | -1.54148 | -1.45004 |
| | 1.6 | -1.88508 | -1.75232 | -1.65927 | -1.54322 | -1.45269 |
| | 2.0 | -1.88423 | -1.75418 | -1.66228 | -1.54700 | -1.45668 |
| | 2.4 | -1.88795 | -1.75908 | -1.66747 | -1.55210 | -1.46148 |
| | 2.8 | -1.89496 | -1.76606 | -1.67409 | -1.55803 | -1.46675 |
| | 3.2 | -1.90423 | -1.77441 | -1.68161 | -1.56444 | -1.47227 |
| | 3.6 | -1.91495 | -1.78360 | -1.68968 | -1.57111 | -1.47788 |
| | 4.0 | -1.92657 | -1.79327 | -1.69802 | -1.57786 | -1.48348 |
| | 6.0 | -1.98699 | -1.84173 | -1.73883 | -1.61001 | -1.50952 |
| 12.0 | 0.0 | -2.11679 | -1.94011 | -1.81822 | -1.66890 | -1.55466 |
| | 0.4 | -2.11187 | -1.93662 | -1.81558 | -1.66714 | -1.55346 |
| | 0.8 | -2.10793 | -1.93390 | -1.81356 | -1.66585 | -1.55263 |
| | 1.2 | -2.10491 | -1.93189 | -1.81212 | -1.66498 | -1.55212 |
| | 1.6 | -2.10273 | -1.93052 | -1.81120 | -1.66450 | -1.55191 |
| | 2.0 | -2.10133 | -1.92973 | -1.81074 | -1.66436 | -1.55195 |
| | 2.4 | -2.10061 | -1.92946 | -1.81069 | -1.66452 | -1.55221 |
| | 2.8 | -2.10051 | -1.92964 | -1.81101 | -1.66494 | -1.55267 |
| | 3.2 | -2.10095 | -1.93023 | -1.81164 | -1.66558 | -1.55329 |
| | 3.6 | -2.10187 | -1.93116 | -1.81255 | -1.66641 | -1.55405 |
| | 4.0 | -2.10320 | -1.93240 | -1.81369 | -1.66741 | -1.55493 |
| | 6.0 | -2.11420 | -1.94166 | -1.82176 | -1.67404 | -1.56048 |
| | 12.0 | -2.15898 | -1.97726 | -1.85153 | -1.69726 | -1.57912 |
| INFINITY | | -2.32635 | -2.10836 | -1.95996 | -1.78046 | -1.64485 |

FRACTILES $W_\alpha$ RHO = 0.60

| DELTAS | | $\alpha$:0.0750 | 0.1000 | 0.1500 | 0.2000 | 0.2500 |
|---|---|---|---|---|---|---|
| 0.0 | 0.0 | -0.81610 | -0.74129 | -0.64245 | -0.57942 | -0.53743 |
| 0.4 | 0.0 | -0.84538 | -0.76077 | -0.64788 | -0.57473 | -0.52469 |
| | 0.4 | -0.79387 | -0.73084 | -0.64819 | -0.59641 | -0.56361 |
| 0.8 | 0.0 | -0.91930 | -0.81996 | -0.68167 | -0.58647 | -0.51638 |
| | 0.4 | -0.81711 | -0.75299 | -0.66856 | -0.61565 | -0.58259 |
| | 0.8 | -0.80768 | -0.76294 | -0.70622 | -0.67478 | -0.65205 |
| 1.2 | 0.0 | -0.99002 | -0.88548 | -0.73499 | -0.62577 | -0.53980 |
| | 0.4 | -0.88693 | -0.81267 | -0.71030 | -0.64165 | -0.59431 |
| | 0.8 | -0.85323 | -0.80805 | -0.75161 | -0.72354 | -0.68896 |
| | 1.2 | -0.88931 | -0.85946 | -0.82824 | -0.78824 | -0.73374 |
| 1.6 | 0.0 | -1.04328 | -0.93804 | -0.78369 | -0.66859 | -0.57504 |
| | 0.4 | -0.96240 | -0.88186 | -0.76582 | -0.68189 | -0.61699 |
| | 0.8 | -0.92432 | -0.86901 | -0.79586 | -0.75627 | -0.71023 |
| | 1.2 | -0.94920 | -0.92025 | -0.88132 | -0.82173 | -0.75034 |
| | 1.6 | -1.02224 | -0.99813 | -0.92841 | -0.84734 | -0.76045 |
| 2.0 | 0.0 | -1.08577 | -0.98075 | -0.82459 | -0.70593 | -0.60756 |
| | 0.4 | -1.02626 | -0.94193 | -0.81729 | -0.72328 | -0.64602 |
| | 0.8 | -0.99816 | -0.93381 | -0.84174 | -0.78130 | -0.72188 |
| | 1.2 | -1.01269 | -0.97422 | -0.91614 | -0.83795 | -0.75221 |
| | 1.6 | -1.08751 | -1.04567 | -0.95250 | -0.85618 | -0.75868 |
| | 2.0 | -1.13869 | -1.08133 | -0.96971 | -0.86218 | -0.75690 |
| 2.4 | 0.0 | -1.12203 | -1.01712 | -0.85926 | -0.73742 | -0.63476 |
| | 0.4 | -1.07933 | -0.99176 | -0.85989 | -0.75766 | -0.67075 |
| | 0.8 | -1.06165 | -0.99042 | -0.88356 | -0.80237 | -0.72682 |
| | 1.2 | -1.07402 | -1.02183 | -0.93969 | -0.84510 | -0.74811 |
| | 1.6 | -1.13407 | -1.07781 | -0.96630 | -0.85859 | -0.75320 |
| | 2.0 | -1.17253 | -1.10466 | -0.97974 | -0.86370 | -0.75241 |
| | 2.4 | -1.19881 | -1.12317 | -0.98806 | -0.86535 | -0.74924 |
| 2.8 | 0.0 | -1.15399 | -1.04876 | -0.88871 | -0.76344 | -0.65641 |
| | 0.4 | -1.12393 | -1.03318 | -0.89432 | -0.78428 | -0.68846 |
| | 0.8 | -1.11400 | -1.03656 | -0.91677 | -0.81962 | -0.72752 |
| | 1.2 | -1.12689 | -1.06263 | -0.95616 | -0.84827 | -0.74256 |
| | 1.6 | -1.17016 | -1.10185 | -0.97567 | -0.85886 | -0.74713 |
| | 2.0 | -1.19888 | -1.12258 | -0.98670 | -0.86357 | -0.74723 |
| | 2.4 | -1.22000 | -1.13778 | -0.99392 | -0.86543 | -0.74518 |
| | 2.8 | -1.23754 | -1.15005 | -0.99906 | -0.86581 | -0.74213 |
| 3.2 | 0.0 | -1.18242 | -1.07636 | -0.91350 | -0.78444 | -0.67282 |
| | 0.4 | -1.16164 | -1.06759 | -0.92171 | -0.80403 | -0.69979 |
| | 0.8 | -1.15696 | -1.07351 | -0.94158 | -0.83101 | -0.72616 |
| | 1.2 | -1.16954 | -1.09494 | -0.96832 | -0.84994 | -0.73727 |
| | 1.6 | -1.19943 | -1.12125 | -0.98296 | -0.85852 | -0.74147 |
| | 2.0 | -1.22103 | -1.13749 | -0.99213 | -0.86284 | -0.74215 |
| | 2.4 | -1.23814 | -1.15007 | -0.99841 | -0.86481 | -0.74093 |
| | 2.8 | -1.25275 | -1.16043 | -1.00296 | -0.86543 | -0.73869 |
| | 3.2 | -1.26561 | -1.16930 | -1.00642 | -0.86529 | -0.73597 |

FRACTILES $W_\alpha$ RHO = 0.60

| DELTAS | | α:0.0750 | 0.1000 | 0.1500 | 0.2000 | 0.2500 |
|---|---|---|---|---|---|---|
| 3.6 | 0.0 | -1.20767 | -1.10035 | -0.93410 | -0.80092 | -0.68460 |
| | 0.4 | -1.19355 | -1.09607 | -0.94315 | -0.81812 | -0.70621 |
| | 0.8 | -1.19212 | -1.10287 | -0.95964 | -0.83794 | -0.72409 |
| | 1.2 | -1.20333 | -1.11966 | -0.97779 | -0.85103 | -0.73265 |
| | 1.6 | -1.22380 | -1.13752 | -0.98901 | -0.85806 | -0.73644 |
| | 2.0 | -1.24018 | -1.15031 | -0.99663 | -0.86194 | -0.73745 |
| | 2.4 | -1.25404 | -1.16070 | -1.00206 | -0.86390 | -0.73680 |
| | 2.8 | -1.26621 | -1.16946 | -1.00608 | -0.86468 | -0.73520 |
| | 3.2 | -1.27710 | -1.17705 | -1.00917 | -0.86474 | -0.73307 |
| | 3.6 | -1.28696 | -1.18376 | -1.01163 | -0.86439 | -0.73070 |
| 4.0 | 0.0 | -1.22998 | -1.12102 | -0.95096 | -0.81351 | -0.69260 |
| | 0.4 | -1.22051 | -1.11952 | -0.95972 | -0.82787 | -0.70934 |
| | 0.8 | -1.22086 | -1.12611 | -0.97277 | -0.84224 | -0.72189 |
| | 1.2 | -1.23014 | -1.13875 | -0.98542 | -0.85182 | -0.72869 |
| | 1.6 | -1.24434 | -1.15135 | -0.99415 | -0.85761 | -0.73205 |
| | 2.0 | -1.25689 | -1.16146 | -1.00046 | -0.86104 | -0.73320 |
| | 2.4 | -1.26810 | -1.17003 | -1.00513 | -0.86291 | -0.73295 |
| | 2.8 | -1.27823 | -1.17743 | -1.00867 | -0.86378 | -0.73182 |
| | 3.2 | -1.28746 | -1.18394 | -1.01144 | -0.86399 | -0.73018 |
| | 3.6 | -1.29593 | -1.18976 | -1.01365 | -0.86379 | -0.72825 |
| | 4.0 | -1.30372 | -1.19500 | -1.01547 | -0.86335 | -0.72619 |
| 6.0 | 0.0 | -1.30562 | -1.18643 | -0.99691 | -0.84130 | -0.70356 |
| | 0.4 | -1.30396 | -1.18739 | -1.00062 | -0.84609 | -0.70862 |
| | 0.8 | -1.30469 | -1.18978 | -1.00439 | -0.85012 | -0.71245 |
| | 1.2 | -1.30708 | -1.19292 | -1.00785 | -0.85330 | -0.71515 |
| | 1.6 | -1.31042 | -1.19635 | -1.01087 | -0.85571 | -0.71691 |
| | 2.0 | -1.31426 | -1.19981 | -1.01345 | -0.85744 | -0.71789 |
| | 2.4 | -1.31834 | -1.20319 | -1.01562 | -0.85864 | -0.71828 |
| | 2.8 | -1.32250 | -1.20643 | -1.01745 | -0.85941 | -0.71822 |
| | 3.2 | -1.32663 | -1.20952 | -1.01899 | -0.85984 | -0.71781 |
| | 3.6 | -1.33069 | -1.21244 | -1.02030 | -0.86003 | -0.71715 |
| | 4.0 | -1.33463 | -1.21521 | -1.02142 | -0.86002 | -0.71632 |
| | 6.0 | -1.35221 | -1.22692 | -1.02522 | -0.85857 | -0.71112 |
| 12.0 | 0.0 | -1.37918 | -1.24201 | -1.02524 | -0.84937 | -0.69585 |
| | 0.4 | -1.37869 | -1.24194 | -1.02566 | -0.85005 | -0.69665 |
| | 0.8 | -1.37842 | -1.24200 | -1.02608 | -0.85064 | -0.69732 |
| | 1.2 | -1.37834 | -1.24217 | -1.02650 | -0.85116 | -0.69788 |
| | 1.6 | -1.37843 | -1.24243 | -1.02691 | -0.85161 | -0.69832 |
| | 2.0 | -1.37867 | -1.24275 | -1.02730 | -0.85199 | -0.69866 |
| | 2.4 | -1.37903 | -1.24314 | -1.02767 | -0.85231 | -0.69891 |
| | 2.8 | -1.37949 | -1.24358 | -1.02802 | -0.85256 | -0.69907 |
| | 3.2 | -1.38005 | -1.24406 | -1.02835 | -0.85277 | -0.69917 |
| | 3.6 | -1.38067 | -1.24457 | -1.02866 | -0.85292 | -0.69919 |
| | 4.0 | -1.38136 | -1.24510 | -1.02896 | -0.85303 | -0.69916 |
| | 6.0 | -1.38534 | -1.24794 | -1.03017 | -0.85310 | -0.69836 |
| | 12.0 | -1.39764 | -1.25588 | -1.03224 | -0.85131 | -0.69383 |
| INFINITY | | -1.43953 | -1.28155 | -1.03643 | -0.84162 | -0.67449 |

FRACTILES $W_\alpha$ RHO = 0.60

| DELTAS | | $\alpha$:0.3000 | 0.3500 | 0.4000 | 0.4500 | 0.5000 |
|---|---|---|---|---|---|---|
| 0.0 | 0.0 | -0.51302 | -0.48586 | -0.44619 | -0.39541 | -0.33287 |
| 0.4 | 0.0 | -0.49193 | -0.46861 | -0.43117 | -0.38198 | -0.32077 |
| | 0.4 | -0.54319 | -0.51081 | -0.46749 | -0.41334 | -0.34745 |
| 0.8 | 0.0 | -0.46414 | -0.42811 | -0.39909 | -0.35357 | -0.29461 |
| | 0.4 | -0.56022 | -0.52415 | -0.47669 | -0.41792 | -0.34696 |
| | 0.8 | -0.61513 | -0.56711 | -0.50833 | -0.43834 | -0.35620 |
| 1.2 | 0.0 | -0.46935 | -0.41105 | -0.36546 | -0.32686 | -0.26861 |
| | 0.4 | -0.56574 | -0.52632 | -0.47317 | -0.40758 | -0.32920 |
| | 0.8 | -0.64059 | -0.58086 | -0.51039 | -0.42909 | -0.33632 |
| | 1.2 | -0.66830 | -0.59340 | -0.50956 | -0.41668 | -0.31409 |
| 1.6 | 0.0 | -0.49504 | -0.42434 | -0.36074 | -0.30550 | -0.24729 |
| | 0.4 | -0.56823 | -0.52541 | -0.46372 | -0.38871 | -0.30106 |
| | 0.8 | -0.64897 | -0.57673 | -0.49485 | -0.40356 | -0.30239 |
| | 1.2 | -0.67069 | -0.58407 | -0.49076 | -0.39035 | -0.28196 |
| | 1.6 | -0.66924 | -0.57382 | -0.47374 | -0.36816 | -0.25590 |
| 2.0 | 0.0 | -0.52149 | -0.44319 | -0.36967 | -0.29876 | -0.22906 |
| | 0.4 | -0.57982 | -0.52312 | -0.45027 | -0.36482 | -0.26828 |
| | 0.8 | -0.64684 | -0.56295 | -0.47164 | -0.37295 | -0.26618 |
| | 1.2 | -0.66164 | -0.56671 | -0.46708 | -0.36196 | -0.25020 |
| | 1.6 | -0.65970 | -0.55847 | -0.45400 | -0.34517 | -0.23062 |
| | 2.0 | -0.65220 | -0.54662 | -0.43883 | -0.32751 | -0.21120 |
| 2.4 | 0.0 | -0.54335 | -0.45853 | -0.37692 | -0.29549 | -0.21128 |
| | 0.4 | -0.59224 | -0.51878 | -0.43412 | -0.33952 | -0.23616 |
| | 0.8 | -0.63922 | -0.54603 | -0.44779 | -0.34402 | -0.23370 |
| | 1.2 | -0.64946 | -0.54857 | -0.44453 | -0.33622 | -0.22231 |
| | 1.6 | -0.64847 | -0.54293 | -0.43525 | -0.32410 | -0.20800 |
| | 2.0 | -0.64323 | -0.53427 | -0.42394 | -0.31076 | -0.19321 |
| | 2.4 | -0.63645 | -0.52473 | -0.41230 | -0.29759 | -0.17898 |
| 2.8 | 0.0 | -0.55970 | -0.46853 | -0.37929 | -0.28865 | -0.19286 |
| | 0.4 | -0.59939 | -0.51196 | -0.41734 | -0.31600 | -0.20798 |
| | 0.8 | -0.62997 | -0.52982 | -0.42658 | -0.31923 | -0.20648 |
| | 1.2 | -0.63771 | -0.53227 | -0.42488 | -0.31418 | -0.19871 |
| | 1.6 | -0.63776 | -0.52877 | -0.41853 | -0.30556 | -0.18831 |
| | 2.0 | -0.63433 | -0.52257 | -0.41017 | -0.29552 | -0.17703 |
| | 2.4 | -0.62931 | -0.51527 | -0.40112 | -0.28517 | -0.16577 |
| | 2.8 | -0.62368 | -0.50770 | -0.39209 | -0.27509 | -0.15501 |
| 3.2 | 0.0 | -0.57075 | -0.47339 | -0.37710 | -0.27862 | -0.17470 |
| | 0.4 | -0.60136 | -0.50356 | -0.40180 | -0.29578 | -0.18455 |
| | 0.8 | -0.62114 | -0.51576 | -0.40873 | -0.29867 | -0.18410 |
| | 1.2 | -0.62743 | -0.51831 | -0.40821 | -0.29560 | -0.17889 |
| | 1.6 | -0.62819 | -0.51630 | -0.40394 | -0.28948 | -0.17131 |
| | 2.0 | -0.62604 | -0.51190 | -0.39776 | -0.28190 | -0.16267 |
| | 2.4 | -0.62237 | -0.50634 | -0.39072 | -0.27376 | -0.15374 |
| | 2.8 | -0.61798 | -0.50030 | -0.38345 | -0.26558 | -0.14496 |
| | 3.2 | -0.61334 | -0.49421 | -0.37629 | -0.25766 | -0.13657 |

FRACTILES $w_\alpha$ RHO = 0.60

| DELTAS | | $\alpha$:0.3000 | 0.3500 | 0.4000 | 0.4500 | 0.5000 |
|---|---|---|---|---|---|---|
| 3.6 | 0.0 | -0.57730 | -0.47422 | -0.37186 | -0.26735 | -0.15820 |
| | 0.4 | -0.60000 | -0.49495 | -0.38830 | -0.27892 | -0.16538 |
| | 0.8 | -0.61339 | -0.50393 | -0.39390 | -0.28169 | -0.16568 |
| | 1.2 | -0.61867 | -0.50648 | -0.39413 | -0.27993 | -0.16221 |
| | 1.6 | -0.61982 | -0.50546 | -0.39129 | -0.27558 | -0.15664 |
| | 2.0 | -0.61855 | -0.50235 | -0.38671 | -0.26983 | -0.14998 |
| | 2.4 | -0.61589 | -0.49811 | -0.38123 | -0.26340 | -0.14287 |
| | 2.8 | -0.61247 | -0.49330 | -0.37536 | -0.25675 | -0.13569 |
| | 3.2 | -0.60871 | -0.48830 | -0.36943 | -0.25016 | -0.12868 |
| | 3.6 | -0.60484 | -0.48332 | -0.36363 | -0.24380 | -0.12197 |
| 4.0 | 0.0 | -0.58045 | -0.47240 | -0.36523 | -0.25642 | -0.14393 |
| | 0.4 | -0.59704 | -0.48699 | -0.37684 | -0.26493 | -0.14961 |
| | 0.8 | -0.60674 | -0.49400 | -0.38153 | -0.26755 | -0.15035 |
| | 1.2 | -0.61122 | -0.49644 | -0.38219 | -0.26664 | -0.14808 |
| | 1.6 | -0.61254 | -0.49604 | -0.38033 | -0.26355 | -0.14396 |
| | 2.0 | -0.61186 | -0.49387 | -0.37693 | -0.25916 | -0.13880 |
| | 2.4 | -0.60995 | -0.49062 | -0.37264 | -0.25406 | -0.13311 |
| | 2.8 | -0.60729 | -0.48679 | -0.36790 | -0.24864 | -0.12722 |
| | 3.2 | -0.60425 | -0.48268 | -0.36298 | -0.24315 | -0.12135 |
| | 3.6 | -0.60103 | -0.47849 | -0.35808 | -0.23774 | -0.11563 |
| | 4.0 | -0.59778 | -0.47435 | -0.35330 | -0.23252 | -0.11015 |
| 6.0 | 0.0 | -0.57607 | -0.45444 | -0.33570 | -0.21761 | -0.09816 |
| | 0.4 | -0.58097 | -0.45893 | -0.33965 | -0.22091 | -0.10076 |
| | 0.8 | -0.58438 | -0.46182 | -0.34196 | -0.22262 | -0.10186 |
| | 1.2 | -0.58654 | -0.46340 | -0.34298 | -0.22309 | -0.10181 |
| | 1.6 | -0.58766 | -0.46395 | -0.34298 | -0.22260 | -0.10088 |
| | 2.0 | -0.58799 | -0.46370 | -0.34223 | -0.22141 | -0.09932 |
| | 2.4 | -0.58771 | -0.46285 | -0.34091 | -0.21970 | -0.09731 |
| | 2.8 | -0.58697 | -0.46157 | -0.33919 | -0.21764 | -0.09498 |
| | 3.2 | -0.58590 | -0.45998 | -0.33719 | -0.21533 | -0.09245 |
| | 3.6 | -0.58460 | -0.45818 | -0.33501 | -0.21287 | -0.08980 |
| | 4.0 | -0.58315 | -0.45625 | -0.33273 | -0.21033 | -0.08710 |
| | 6.0 | -0.57518 | -0.44624 | -0.32123 | -0.19784 | -0.07404 |
| 12.0 | 0.0 | -0.55583 | -0.42420 | -0.29759 | -0.17348 | -0.04977 |
| | 0.4 | -0.55668 | -0.42506 | -0.29842 | -0.17426 | -0.05046 |
| | 0.8 | -0.55737 | -0.42574 | -0.29906 | -0.17484 | -0.05096 |
| | 1.2 | -0.55792 | -0.42625 | -0.29952 | -0.17524 | -0.05129 |
| | 1.6 | -0.55833 | -0.42661 | -0.29983 | -0.17547 | -0.05146 |
| | 2.0 | -0.55862 | -0.42684 | -0.29999 | -0.17557 | -0.05149 |
| | 2.4 | -0.55880 | -0.42695 | -0.30003 | -0.17554 | -0.05139 |
| | 2.8 | -0.55888 | -0.42695 | -0.29995 | -0.17540 | -0.05119 |
| | 3.2 | -0.55887 | -0.42685 | -0.29978 | -0.17517 | -0.05090 |
| | 3.6 | -0.55878 | -0.42668 | -0.29953 | -0.17485 | -0.05053 |
| | 4.0 | -0.55863 | -0.42643 | -0.29921 | -0.17446 | -0.05010 |
| | 6.0 | -0.55718 | -0.42446 | -0.29683 | -0.17179 | -0.04723 |
| | 12.0 | -0.55065 | -0.41647 | -0.28781 | -0.16210 | -0.03721 |
| INFINITY | | -0.52440 | -0.38532 | -0.25335 | -0.12566 | 0.0 |

FRACTILES $w_\alpha$ RHO = 0.60

| DELTAS | | $\alpha$:0.5500 | 0.6000 | 0.6500 | 0.7000 | 0.7500 |
|---|---|---|---|---|---|---|
| 0.0 | 0.0 | -0.25705 | -0.16554 | -0.05473 | 0.08086 | 0.24991 |
| 0.4 | 0.0 | -0.24616 | -0.15583 | -0.04624 | 0.08804 | 0.25559 |
| | 0.4 | -0.26816 | -0.17298 | -0.05824 | 0.08160 | 0.25525 |
| 0.8 | 0.0 | -0.22155 | -0.13238 | -0.02377 | 0.10951 | 0.27578 |
| | 0.4 | -0.26222 | -0.16127 | -0.04057 | 0.10523 | 0.28454 |
| | 0.8 | -0.26037 | -0.14861 | -0.01757 | 0.13774 | 0.32524 |
| 1.2 | 0.0 | -0.19427 | -0.10285 | 0.00846 | 0.14446 | 0.31295 |
| | 0.4 | -0.23680 | -0.12835 | -0.00068 | 0.15105 | 0.33457 |
| | 0.8 | -0.23084 | -0.11067 | 0.02719 | 0.18738 | 0.37725 |
| | 1.2 | -0.20053 | -0.07403 | 0.06837 | 0.23115 | 0.42132 |
| 1.6 | 0.0 | -0.16752 | -0.06973 | 0.04786 | 0.18934 | 0.36172 |
| | 0.4 | -0.20015 | -0.08437 | 0.04905 | 0.20449 | 0.38902 |
| | 0.8 | -0.19019 | -0.06508 | 0.07584 | 0.23697 | 0.42525 |
| | 1.2 | -0.16417 | -0.03494 | 0.10868 | 0.27102 | 0.45884 |
| | 1.6 | -0.13541 | -0.00456 | 0.13958 | 0.30125 | 0.48696 |
| 2.0 | 0.0 | -0.14033 | -0.03406 | 0.09055 | 0.23704 | 0.41180 |
| | 0.4 | -0.16025 | -0.03922 | 0.09743 | 0.25388 | 0.43681 |
| | 0.8 | -0.15006 | -0.02262 | 0.11901 | 0.27910 | 0.46427 |
| | 1.2 | -0.13026 | -0.00004 | 0.14338 | 0.30422 | 0.48894 |
| | 1.6 | -0.10872 | 0.02269 | 0.16650 | 0.32683 | 0.50996 |
| | 2.0 | -0.08818 | 0.04368 | 0.18724 | 0.34654 | 0.52765 |
| 2.4 | 0.0 | -0.11251 | 0.00136 | 0.13120 | 0.28049 | 0.45538 |
| | 0.4 | -0.12330 | 0.00076 | 0.13869 | 0.29459 | 0.47486 |
| | 0.8 | -0.11534 | 0.01310 | 0.15447 | 0.31293 | 0.49481 |
| | 1.2 | -0.10115 | 0.02938 | 0.17215 | 0.33126 | 0.51292 |
| | 1.6 | -0.08526 | 0.04627 | 0.18944 | 0.34826 | 0.52880 |
| | 2.0 | -0.06952 | 0.06244 | 0.20548 | 0.36355 | 0.54257 |
| | 2.4 | -0.05471 | 0.07737 | 0.22003 | 0.37714 | 0.55448 |
| 2.8 | 0.0 | -0.08584 | 0.03326 | 0.16611 | 0.31638 | 0.49013 |
| | 0.4 | -0.09206 | 0.03366 | 0.17193 | 0.32675 | 0.50430 |
| | 0.8 | -0.08670 | 0.04219 | 0.18303 | 0.33984 | 0.51873 |
| | 1.2 | -0.07675 | 0.05384 | 0.19587 | 0.35332 | 0.53220 |
| | 1.6 | -0.06502 | 0.06645 | 0.20890 | 0.36624 | 0.54437 |
| | 2.0 | -0.05291 | 0.07898 | 0.22140 | 0.37822 | 0.55521 |
| | 2.4 | -0.04113 | 0.09090 | 0.23306 | 0.38915 | 0.56481 |
| | 2.8 | -0.03003 | 0.10199 | 0.24376 | 0.39902 | 0.57331 |
| 3.2 | 0.0 | -0.06225 | 0.06003 | 0.19446 | 0.34481 | 0.51702 |
| | 0.4 | -0.06660 | 0.06011 | 0.19835 | 0.35205 | 0.52717 |
| | 0.8 | -0.06331 | 0.06583 | 0.20611 | 0.36144 | 0.53774 |
| | 1.2 | -0.05634 | 0.07421 | 0.21552 | 0.37148 | 0.54791 |
| | 1.6 | -0.04762 | 0.08371 | 0.22544 | 0.38142 | 0.55736 |
| | 2.0 | -0.03825 | 0.09348 | 0.23527 | 0.39090 | 0.56599 |
| | 2.4 | -0.02885 | 0.10305 | 0.24467 | 0.39974 | 0.57380 |
| | 2.8 | -0.01976 | 0.11216 | 0.25349 | 0.40790 | 0.58084 |
| | 3.2 | -0.01116 | 0.12069 | 0.26166 | 0.41537 | 0.58718 |

FRACTILES $W_\alpha$ RHO = 0.60

| DELTAS | | $\alpha$:0.5500 | 0.6000 | 0.6500 | 0.7000 | 0.7500 |
|---|---|---|---|---|---|---|
| 3.6 | 0.0 | -0.04239 | 0.08184 | 0.21710 | 0.36720 | 0.53790 |
| | 0.4 | -0.04597 | 0.08141 | 0.21952 | 0.37219 | 0.54521 |
| | 0.8 | -0.04409 | 0.08521 | 0.22496 | 0.37901 | 0.55308 |
| | 1.2 | -0.03919 | 0.09128 | 0.23194 | 0.38659 | 0.56088 |
| | 1.6 | -0.03266 | 0.09851 | 0.23958 | 0.39432 | 0.56831 |
| | 2.0 | -0.02535 | 0.10619 | 0.24737 | 0.40189 | 0.57525 |
| | 2.4 | -0.01781 | 0.11392 | 0.25500 | 0.40910 | 0.58165 |
| | 2.8 | -0.01034 | 0.12143 | 0.26229 | 0.41587 | 0.58752 |
| | 3.2 | -0.00314 | 0.12860 | 0.26918 | 0.42218 | 0.59289 |
| | 3.6 | 0.00370 | 0.13535 | 0.27561 | 0.42801 | 0.59779 |
| 4.0 | 0.0 | -0.02594 | 0.09957 | 0.23530 | 0.38502 | 0.55435 |
| | 0.4 | -0.02909 | 0.09877 | 0.23670 | 0.38848 | 0.55970 |
| | 0.8 | -0.02813 | 0.10127 | 0.24056 | 0.39349 | 0.56565 |
| | 1.2 | -0.02467 | 0.10572 | 0.24579 | 0.39928 | 0.57171 |
| | 1.6 | -0.01974 | 0.11126 | 0.25173 | 0.40537 | 0.57763 |
| | 2.0 | -0.01401 | 0.11735 | 0.25795 | 0.41146 | 0.58326 |
| | 2.4 | -0.00792 | 0.12362 | 0.26418 | 0.41738 | 0.58855 |
| | 2.8 | -0.00176 | 0.12984 | 0.27025 | 0.42304 | 0.59347 |
| | 3.2 | 0.00429 | 0.13589 | 0.27607 | 0.42838 | 0.59804 |
| | 3.6 | 0.01013 | 0.14166 | 0.28158 | 0.43340 | 0.60225 |
| | 4.0 | 0.01570 | 0.14714 | 0.28677 | 0.43807 | 0.60614 |
| 6.0 | 0.0 | 0.02457 | 0.15275 | 0.28899 | 0.43677 | 0.60120 |
| | 0.4 | 0.02272 | 0.15168 | 0.28871 | 0.43731 | 0.60256 |
| | 0.8 | 0.02223 | 0.15179 | 0.28940 | 0.43856 | 0.60433 |
| | 1.2 | 0.02277 | 0.15278 | 0.29080 | 0.44031 | 0.60634 |
| | 1.6 | 0.02408 | 0.15441 | 0.29270 | 0.44239 | 0.60850 |
| | 2.0 | 0.02594 | 0.15651 | 0.29494 | 0.44468 | 0.61073 |
| | 2.4 | 0.02819 | 0.15891 | 0.29740 | 0.44710 | 0.61297 |
| | 2.8 | 0.03070 | 0.16151 | 0.30000 | 0.44958 | 0.61519 |
| | 3.2 | 0.03337 | 0.16422 | 0.30265 | 0.45206 | 0.61736 |
| | 3.6 | 0.03611 | 0.16697 | 0.30532 | 0.45453 | 0.61947 |
| | 4.0 | 0.03889 | 0.16973 | 0.30797 | 0.45694 | 0.62150 |
| | 6.0 | 0.05210 | 0.18266 | 0.32014 | 0.46782 | 0.63041 |
| 12.0 | 0.0 | 0.07555 | 0.20454 | 0.33964 | 0.48401 | 0.64213 |
| | 0.4 | 0.07495 | 0.20406 | 0.33930 | 0.48383 | 0.64213 |
| | 0.8 | 0.07454 | 0.20375 | 0.33911 | 0.48376 | 0.64221 |
| | 1.2 | 0.07430 | 0.20360 | 0.33905 | 0.48381 | 0.64236 |
| | 1.6 | 0.07420 | 0.20358 | 0.33911 | 0.48395 | 0.64258 |
| | 2.0 | 0.07424 | 0.20368 | 0.33927 | 0.48416 | 0.64284 |
| | 2.4 | 0.07439 | 0.20388 | 0.33952 | 0.48445 | 0.64315 |
| | 2.8 | 0.07464 | 0.20417 | 0.33984 | 0.48479 | 0.64350 |
| | 3.2 | 0.07497 | 0.20454 | 0.34023 | 0.48518 | 0.64387 |
| | 3.6 | 0.07538 | 0.20498 | 0.34067 | 0.48562 | 0.64427 |
| | 4.0 | 0.07585 | 0.20546 | 0.34116 | 0.48608 | 0.64469 |
| | 6.0 | 0.07882 | 0.20844 | 0.34403 | 0.48872 | 0.64692 |
| | 12.0 | 0.08887 | 0.21818 | 0.35310 | 0.49670 | 0.65328 |
| INFINITY | | 0.12566 | 0.25335 | 0.36532 | 0.52440 | 0.67449 |

FRACTILES $w_\alpha$ RHO = 0.60

| DELTAS | | $\alpha$:0.8000 | 0.8500 | 0.9000 | 0.9250 | 0.9500 |
|---|---|---|---|---|---|---|
| 0.0 | 0.0 | 0.46716 | 0.76058 | 1.19362 | 1.51158 | 1.97143 |
| 0.4 | 0.0 | 0.47101 | 0.76202 | 1.19150 | 1.50678 | 1.96262 |
| | 0.4 | 0.47747 | 0.77614 | 1.21429 | 1.53420 | 1.99447 |
| 0.8 | 0.0 | 0.48927 | 0.77688 | 1.19954 | 1.50845 | 1.95314 |
| | 0.4 | 0.51153 | 0.81289 | 1.24864 | 1.56289 | 2.01034 |
| | 0.8 | 0.55829 | 0.86203 | 1.29288 | 1.59908 | 2.03019 |
| 1.2 | 0.0 | 0.52733 | 0.81295 | 1.22695 | 1.52591 | 1.95189 |
| | 0.4 | 0.56295 | 0.86085 | 1.28366 | 1.58422 | 2.00744 |
| | 0.8 | 0.60924 | 0.90674 | 1.32214 | 1.61397 | 2.02135 |
| | 1.2 | 0.65072 | 0.94143 | 1.34274 | 1.62231 | 2.01015 |
| 1.6 | 0.0 | 0.57726 | 0.85925 | 1.26023 | 1.54553 | 1.94745 |
| | 0.4 | 0.61472 | 0.90432 | 1.30881 | 1.59300 | 1.98972 |
| | 0.8 | 0.65238 | 0.94021 | 1.33751 | 1.61428 | 1.99820 |
| | 1.2 | 0.68335 | 0.96547 | 1.35162 | 1.61891 | 1.98783 |
| | 1.6 | 0.70748 | 0.98276 | 1.35703 | 1.61473 | 1.96894 |
| 2.0 | 0.0 | 0.62614 | 0.90160 | 1.28664 | 1.55723 | 1.93499 |
| | 0.4 | 0.65753 | 0.93724 | 1.32328 | 1.59215 | 1.96504 |
| | 0.8 | 0.68557 | 0.96358 | 1.34401 | 1.60727 | 1.97059 |
| | 1.2 | 0.70825 | 0.98197 | 1.35406 | 1.61022 | 1.96229 |
| | 1.6 | 0.72623 | 0.99475 | 1.35773 | 1.60651 | 1.94723 |
| | 2.0 | 0.74059 | 1.00377 | 1.35781 | 1.59952 | 1.92950 |
| 2.4 | 0.0 | 0.66658 | 0.93426 | 1.30364 | 1.56082 | 1.91742 |
| | 0.4 | 0.69022 | 0.96062 | 1.33042 | 1.58622 | 1.93912 |
| | 0.8 | 0.71063 | 0.97987 | 1.34568 | 1.59742 | 1.94333 |
| | 1.2 | 0.72738 | 0.99355 | 1.35324 | 1.59970 | 1.93718 |
| | 1.6 | 0.74103 | 1.00328 | 1.35602 | 1.59681 | 1.92551 |
| | 2.0 | 0.75221 | 1.01029 | 1.35597 | 1.59115 | 1.91131 |
| | 2.4 | 0.76148 | 1.01543 | 1.35432 | 1.58418 | 1.89632 |
| 2.8 | 0.0 | 0.69757 | 0.95785 | 1.31352 | 1.55937 | 1.89837 |
| | 0.4 | 0.71479 | 0.97716 | 1.33333 | 1.57828 | 1.91470 |
| | 0.8 | 0.72976 | 0.99148 | 1.34492 | 1.58697 | 1.91829 |
| | 1.2 | 0.74236 | 1.00192 | 1.35087 | 1.58899 | 1.91395 |
| | 1.6 | 0.75290 | 1.00953 | 1.35317 | 1.58690 | 1.90504 |
| | 2.0 | 0.76176 | 1.01512 | 1.35318 | 1.58245 | 1.89377 |
| | 2.4 | 0.76924 | 1.01927 | 1.35182 | 1.57675 | 1.88151 |
| | 2.8 | 0.77563 | 1.02242 | 1.34969 | 1.57051 | 1.86911 |
| 3.2 | 0.0 | 0.72092 | 0.97476 | 1.31891 | 1.55538 | 1.87994 |
| | 0.4 | 0.73350 | 0.98912 | 1.33393 | 1.56990 | 1.89271 |
| | 0.8 | 0.74466 | 1.00002 | 1.34301 | 1.57692 | 1.89599 |
| | 1.2 | 0.75432 | 1.00816 | 1.34786 | 1.57881 | 1.89306 |
| | 1.6 | 0.76260 | 1.01423 | 1.34985 | 1.57740 | 1.88629 |
| | 2.0 | 0.76970 | 1.01878 | 1.34998 | 1.57395 | 1.87737 |
| | 2.4 | 0.77582 | 1.02221 | 1.34892 | 1.56934 | 1.86738 |
| | 2.8 | 0.78113 | 1.02483 | 1.34715 | 1.56415 | 1.85702 |
| | 3.2 | 0.78576 | 1.02686 | 1.34499 | 1.55873 | 1.84674 |

FRACTILES $W_\alpha$ RHO = 0.60

| DELTAS | | $\alpha$:0.8000 | 0.8500 | 0.9000 | 0.9250 | 0.9500 |
|---|---|---|---|---|---|---|
| 3.6 | 0.0 | 0.73869 | 0.98708 | 1.32164 | 1.55034 | 1.86297 |
| | 0.4 | 0.74802 | 0.99798 | 1.33332 | 1.56181 | 1.87329 |
| | 0.8 | 0.75651 | 1.00645 | 1.34061 | 1.56763 | 1.87636 |
| | 1.2 | 0.76403 | 1.01293 | 1.34466 | 1.56944 | 1.87447 |
| | 1.6 | 0.77062 | 1.01786 | 1.34644 | 1.56856 | 1.86937 |
| | 2.0 | 0.77639 | 1.02162 | 1.34667 | 1.56592 | 1.86230 |
| | 2.4 | 0.78145 | 1.02450 | 1.34589 | 1.56222 | 1.85416 |
| | 2.8 | 0.78589 | 1.02672 | 1.34445 | 1.55791 | 1.84552 |
| | 3.2 | 0.78982 | 1.02846 | 1.34263 | 1.55332 | 1.83678 |
| | 3.6 | 0.79331 | 1.02983 | 1.34059 | 1.54866 | 1.82819 |
| 4.0 | 0.0 | 0.75248 | 0.99626 | 1.32279 | 1.54501 | 1.84769 |
| | 0.4 | 0.75953 | 1.00471 | 1.33208 | 1.55428 | 1.85625 |
| | 0.8 | 0.76610 | 1.01142 | 1.33805 | 1.55920 | 1.85914 |
| | 1.2 | 0.77204 | 1.01666 | 1.34149 | 1.56095 | 1.85802 |
| | 1.6 | 0.77736 | 1.02072 | 1.34311 | 1.56046 | 1.85417 |
| | 2.0 | 0.78209 | 1.02387 | 1.34343 | 1.55846 | 1.84858 |
| | 2.4 | 0.78630 | 1.02632 | 1.34287 | 1.55550 | 1.84193 |
| | 2.8 | 0.79005 | 1.02822 | 1.34172 | 1.55194 | 1.83472 |
| | 3.2 | 0.79341 | 1.02973 | 1.34020 | 1.54806 | 1.82730 |
| | 3.6 | 0.79642 | 1.03092 | 1.33846 | 1.54404 | 1.81988 |
| | 4.0 | 0.79914 | 1.03187 | 1.33659 | 1.54001 | 1.81262 |
| 6.0 | 0.0 | 0.79043 | 1.01928 | 1.32010 | 1.52168 | 1.79276 |
| | 0.4 | 0.79262 | 1.02229 | 1.32385 | 1.52571 | 1.79693 |
| | 0.8 | 0.79484 | 1.02485 | 1.32653 | 1.52826 | 1.79905 |
| | 1.2 | 0.79702 | 1.02701 | 1.32832 | 1.52959 | 1.79953 |
| | 1.6 | 0.79911 | 1.02880 | 1.32938 | 1.52997 | 1.79873 |
| | 2.0 | 0.80110 | 1.03029 | 1.32987 | 1.52959 | 1.79696 |
| | 2.4 | 0.80297 | 1.03151 | 1.32991 | 1.52864 | 1.79447 |
| | 2.8 | 0.80474 | 1.03252 | 1.32960 | 1.52727 | 1.79146 |
| | 3.2 | 0.80639 | 1.03335 | 1.32903 | 1.52560 | 1.78809 |
| | 3.6 | 0.80794 | 1.03402 | 1.32827 | 1.52371 | 1.78449 |
| | 4.0 | 0.80939 | 1.03458 | 1.32738 | 1.52168 | 1.78075 |
| | 6.0 | 0.81538 | 1.03621 | 1.32205 | 1.51100 | 1.76209 |
| 12.0 | 0.0 | 0.82108 | 1.03356 | 1.30693 | 1.48674 | 1.72471 |
| | 0.4 | 0.82129 | 1.03401 | 1.30770 | 1.48769 | 1.72588 |
| | 0.8 | 0.82153 | 1.03444 | 1.30832 | 1.48843 | 1.72675 |
| | 1.2 | 0.82180 | 1.03482 | 1.30883 | 1.48898 | 1.72734 |
| | 1.6 | 0.82209 | 1.03517 | 1.30922 | 1.48937 | 1.72768 |
| | 2.0 | 0.82239 | 1.03549 | 1.30950 | 1.48960 | 1.72780 |
| | 2.4 | 0.82270 | 1.03577 | 1.30970 | 1.48970 | 1.72772 |
| | 2.8 | 0.82302 | 1.03603 | 1.30981 | 1.48968 | 1.72748 |
| | 3.2 | 0.82335 | 1.03626 | 1.30985 | 1.48955 | 1.72709 |
| | 3.6 | 0.82368 | 1.03646 | 1.30982 | 1.48934 | 1.72657 |
| | 4.0 | 0.82401 | 1.03664 | 1.30974 | 1.48904 | 1.72593 |
| | 6.0 | 0.82562 | 1.03727 | 1.30873 | 1.48673 | 1.72165 |
| | 12.0 | 0.82960 | 1.03776 | 1.30368 | 1.47743 | 1.70603 |
| INFINITY | | 0.84162 | 1.03643 | 1.28155 | 1.43953 | 1.64485 |

FRACTILES $w_\alpha$ RHO = 0.60

| DELTAS | | $\alpha$:0.9625 | 0.9750 | 0.9825 | 0.9900 | 0.9925 |
|---|---|---|---|---|---|---|
| 0.0 | 0.0 | 2.30450 | 2.78173 | 3.20787 | 3.88620 | 4.23881 |
| 0.4 | 0.0 | 2.29267 | 2.76541 | 3.16734 | 3.85865 | 4.20745 |
| | 0.4 | 2.32622 | 2.79941 | 3.21995 | 3.88591 | 4.23055 |
| 0.8 | 0.0 | 2.27379 | 2.73125 | 3.13789 | 3.78193 | 4.11526 |
| | 0.4 | 2.32995 | 2.78232 | 3.18144 | 3.80878 | 4.13152 |
| | 0.8 | 2.33536 | 2.76419 | 3.14005 | 3.72712 | 4.02766 |
| 1.2 | 0.0 | 2.25630 | 2.68727 | 3.06758 | 3.66545 | 3.97306 |
| | 0.4 | 2.30706 | 2.72809 | 3.09713 | 3.67354 | 3.96863 |
| | 0.8 | 2.30780 | 2.70815 | 3.05735 | 3.60021 | 3.87710 |
| | 1.2 | 2.28146 | 2.65912 | 2.98727 | 3.49548 | 3.75393 |
| 1.6 | 0.0 | 2.23205 | 2.63203 | 2.98264 | 3.53028 | 3.81065 |
| | 0.4 | 2.26866 | 2.65852 | 2.99855 | 3.52712 | 3.79672 |
| | 0.8 | 2.26676 | 2.64058 | 2.96538 | 3.46838 | 3.72417 |
| | 1.2 | 2.24486 | 2.60140 | 2.91021 | 3.38693 | 3.62873 |
| | 1.6 | 2.21485 | 2.55498 | 2.84876 | 3.30098 | 3.52983 |
| 2.0 | 0.0 | 2.20058 | 2.57176 | 2.89548 | 3.39865 | 3.65528 |
| | 0.4 | 2.22586 | 2.58884 | 2.90419 | 3.39250 | 3.64080 |
| | 0.8 | 2.22367 | 2.57471 | 2.87873 | 3.34799 | 3.58600 |
| | 1.2 | 2.20670 | 2.54473 | 2.83668 | 3.28606 | 3.51347 |
| | 1.6 | 2.18304 | 2.50836 | 2.78864 | 3.21900 | 3.43633 |
| | 2.0 | 2.15727 | 2.47078 | 2.74029 | 3.15318 | 3.36131 |
| 2.4 | 0.0 | 2.16677 | 2.51372 | 2.81508 | 3.28164 | 3.51885 |
| | 0.4 | 2.18488 | 2.52570 | 2.82080 | 3.27622 | 3.50717 |
| | 0.8 | 2.18340 | 2.51538 | 2.80207 | 3.24328 | 3.46653 |
| | 1.2 | 2.17071 | 2.49285 | 2.77037 | 3.19644 | 3.41159 |
| | 1.6 | 2.15238 | 2.46462 | 2.73305 | 3.14423 | 3.35150 |
| | 2.0 | 2.13177 | 2.43457 | 2.69437 | 3.09156 | 3.29144 |
| | 2.4 | 2.11080 | 2.40485 | 2.65668 | 3.04099 | 3.23409 |
| 2.8 | 0.0 | 2.13436 | 2.46151 | 2.74470 | 3.18161 | 3.40313 |
| | 0.4 | 2.14810 | 2.47076 | 2.74932 | 3.17793 | 3.39476 |
| | 0.8 | 2.14750 | 2.46360 | 2.73585 | 3.15375 | 3.36475 |
| | 1.2 | 2.13819 | 2.44677 | 2.71200 | 3.11823 | 3.32298 |
| | 1.6 | 2.12407 | 2.42489 | 2.68297 | 3.07747 | 3.27599 |
| | 2.0 | 2.10766 | 2.40090 | 2.65203 | 3.03524 | 3.22779 |
| | 2.4 | 2.09050 | 2.37655 | 2.62113 | 2.99373 | 3.18069 |
| | 2.8 | 2.07351 | 2.35286 | 2.59136 | 2.95414 | 3.13594 |
| 3.2 | 0.0 | 2.10499 | 2.41598 | 2.68436 | 3.09716 | 3.30593 |
| | 0.4 | 2.11593 | 2.42365 | 2.68861 | 3.09520 | 3.30043 |
| | 0.8 | 2.11609 | 2.41888 | 2.67907 | 3.07749 | 3.27826 |
| | 1.2 | 2.10935 | 2.40634 | 2.66107 | 3.05040 | 3.24629 |
| | 1.6 | 2.09848 | 2.38934 | 2.63840 | 3.01839 | 3.20930 |
| | 2.0 | 2.08541 | 2.37014 | 2.61358 | 2.98439 | 3.17044 |
| | 2.4 | 2.07139 | 2.35018 | 2.58821 | 2.95024 | 3.13166 |
| | 2.8 | 2.05719 | 2.33036 | 2.56327 | 2.91704 | 3.09411 |
| | 3.2 | 2.04329 | 2.31119 | 2.53933 | 2.88539 | 3.05842 |

FRACTILES $W_\alpha$ RHO = 0.60

| DELTAS | | $\alpha$:0.9625 | 0.9750 | 0.9825 | 0.9900 | 0.9925 |
|---|---|---|---|---|---|---|
| 3.6 | 0.0 | 2.07901 | 2.37669 | 2.63290 | 3.02589 | 3.22420 |
| | 0.4 | 2.08804 | 2.38335 | 2.63702 | 3.02531 | 3.22093 |
| | 0.8 | 2.08878 | 2.38034 | 2.63036 | 3.01236 | 3.20451 |
| | 1.2 | 2.08394 | 2.37098 | 2.61672 | 2.99157 | 3.17986 |
| | 1.6 | 2.07557 | 2.35773 | 2.59892 | 2.96627 | 3.15055 |
| | 2.0 | 2.06515 | 2.34231 | 2.57892 | 2.93875 | 3.11905 |
| | 2.4 | 2.05366 | 2.32591 | 2.55802 | 2.91054 | 3.08698 |
| | 2.8 | 2.04179 | 2.30930 | 2.53710 | 2.88263 | 3.05539 |
| | 3.2 | 2.02997 | 2.29297 | 2.51668 | 2.85561 | 3.02490 |
| | 3.6 | 2.01846 | 2.27722 | 2.49709 | 2.82983 | 2.99587 |
| 4.0 | 0.0 | 2.05623 | 2.34285 | 2.58893 | 2.96544 | 3.15506 |
| | 0.4 | 2.06388 | 2.34877 | 2.59295 | 2.96590 | 3.15345 |
| | 0.8 | 2.06501 | 2.34704 | 2.58840 | 2.95644 | 3.14127 |
| | 1.2 | 2.06159 | 2.34006 | 2.57803 | 2.94039 | 3.12213 |
| | 1.6 | 2.05514 | 2.32967 | 2.56398 | 2.92025 | 3.09874 |
| | 2.0 | 2.04679 | 2.31724 | 2.54777 | 2.89785 | 3.07304 |
| | 2.4 | 2.03737 | 2.30372 | 2.53050 | 2.87445 | 3.04641 |
| | 2.8 | 2.02743 | 2.28977 | 2.51289 | 2.85092 | 3.01974 |
| | 3.2 | 2.01736 | 2.27584 | 2.49545 | 2.82780 | 2.99364 |
| | 3.6 | 2.00742 | 2.26221 | 2.47849 | 2.80545 | 2.96847 |
| | 4.0 | 1.99776 | 2.24907 | 2.46220 | 2.78408 | 2.94444 |
| 6.0 | 0.0 | 1.97749 | 2.22901 | 2.44300 | 2.76733 | 2.92939 |
| | 0.4 | 1.98159 | 2.23280 | 2.44635 | 2.76973 | 2.93120 |
| | 0.8 | 1.98325 | 2.23366 | 2.44635 | 2.76815 | 2.92870 |
| | 1.2 | 1.98300 | 2.23221 | 2.44373 | 2.76348 | 2.92290 |
| | 1.6 | 1.98125 | 2.22900 | 2.43912 | 2.75650 | 2.91463 |
| | 2.0 | 1.97839 | 2.22449 | 2.43305 | 2.74784 | 2.90458 |
| | 2.4 | 1.97472 | 2.21903 | 2.42594 | 2.73801 | 2.89329 |
| | 2.8 | 1.97046 | 2.21292 | 2.41813 | 2.72740 | 2.88120 |
| | 3.2 | 1.96582 | 2.20640 | 2.40988 | 2.71634 | 2.86865 |
| | 3.6 | 1.96093 | 2.19963 | 2.40140 | 2.70507 | 2.85590 |
| | 4.0 | 1.95592 | 2.19276 | 2.39283 | 2.69375 | 2.84315 |
| | 6.0 | 1.93135 | 2.15958 | 2.35187 | 2.64026 | 2.78307 |
| 12.0 | 0.0 | 1.88457 | 2.09947 | 2.28001 | 2.54995 | 2.68331 |
| | 0.4 | 1.88587 | 2.10093 | 2.28156 | 2.55162 | 2.68501 |
| | 0.8 | 1.88680 | 2.10191 | 2.28256 | 2.55260 | 2.68596 |
| | 1.2 | 1.88739 | 2.10246 | 2.28306 | 2.55297 | 2.68624 |
| | 1.6 | 1.88767 | 2.10264 | 2.28312 | 2.55279 | 2.68592 |
| | 2.0 | 1.88769 | 2.10248 | 2.28278 | 2.55213 | 2.68508 |
| | 2.4 | 1.88747 | 2.10204 | 2.28211 | 2.55106 | 2.68380 |
| | 2.8 | 1.88705 | 2.10133 | 2.28114 | 2.54963 | 2.68212 |
| | 3.2 | 1.88644 | 2.10041 | 2.27991 | 2.54789 | 2.68010 |
| | 3.6 | 1.88569 | 2.09929 | 2.27846 | 2.54589 | 2.67780 |
| | 4.0 | 1.88480 | 2.09802 | 2.27683 | 2.54367 | 2.67526 |
| | 6.0 | 1.87903 | 2.09005 | 2.26684 | 2.53039 | 2.66023 |
| | 12.0 | 1.85874 | 2.06297 | 2.23362 | 2.48727 | 2.61193 |
| INFINITY | | 1.78046 | 1.95996 | 2.10836 | 2.32635 | 2.43238 |

FRACTILES $W_\alpha$ RHO = 0.60

| DELTAS | | $\alpha$:0.9950 | 0.9975 | 0.9990 | 0.9995 |
|---|---|---|---|---|---|
| 0.0 | 0.0 | 4.73960 | 5.60449 | 6.76149 | 7.64498 |
| 0.4 | 0.0 | 4.70263 | 5.55730 | 6.69964 | 7.57118 |
| | 0.4 | 4.71836 | 5.55665 | 6.67077 | 7.51661 |
| 0.8 | 0.0 | 4.58709 | 5.39797 | 6.47572 | 7.29399 |
| | 0.4 | 4.58643 | 5.36371 | 6.38972 | 7.16437 |
| | 0.8 | 4.44987 | 5.16805 | 6.11105 | 6.82004 |
| 1.2 | 0.0 | 4.40665 | 5.14756 | 6.12558 | 6.86402 |
| | 0.4 | 4.38317 | 5.08831 | 6.01419 | 6.71030 |
| | 0.8 | 4.26508 | 4.92276 | 5.78276 | 6.42715 |
| | 1.2 | 4.11530 | 4.72611 | 5.52197 | 6.11656 |
| 1.6 | 0.0 | 4.20451 | 4.87445 | 5.75409 | 6.41542 |
| | 0.4 | 4.17447 | 4.81479 | 5.65203 | 6.27936 |
| | 0.8 | 4.08182 | 4.68631 | 5.47392 | 6.06233 |
| | 1.2 | 3.96620 | 4.53513 | 5.27409 | 5.82471 |
| | 1.6 | 3.84871 | 4.38507 | 5.07975 | 5.59612 |
| 2.0 | 0.0 | 4.01482 | 4.62424 | 5.42100 | 6.01794 |
| | 0.4 | 3.98795 | 4.57466 | 5.33902 | 5.91002 |
| | 0.8 | 3.91816 | 4.47810 | 5.20534 | 5.74718 |
| | 1.2 | 3.83033 | 4.36327 | 5.05350 | 5.56655 |
| | 1.6 | 3.73871 | 4.24627 | 4.90194 | 5.38824 |
| | 2.0 | 3.65050 | 4.13500 | 4.75942 | 5.22163 |
| 2.4 | 0.0 | 3.85045 | 4.41080 | 5.14071 | 5.68590 |
| | 0.4 | 3.82946 | 4.37269 | 5.07811 | 5.60364 |
| | 0.8 | 3.77756 | 4.30066 | 4.97806 | 5.48153 |
| | 1.2 | 3.71092 | 4.21332 | 4.86228 | 5.34358 |
| | 1.6 | 3.63948 | 4.12195 | 4.74372 | 5.20395 |
| | 2.0 | 3.56882 | 4.03274 | 4.62935 | 5.07015 |
| | 2.4 | 3.50177 | 3.94877 | 4.52249 | 4.94566 |
| 2.8 | 0.0 | 3.71220 | 4.23306 | 4.90928 | 5.41297 |
| | 0.4 | 3.69682 | 4.20473 | 4.86233 | 5.35101 |
| | 0.8 | 3.65827 | 4.15085 | 4.78702 | 5.25878 |
| | 1.2 | 3.60744 | 4.08396 | 4.69799 | 5.15245 |
| | 1.6 | 3.55146 | 4.01217 | 4.60460 | 5.04228 |
| | 2.0 | 3.49469 | 3.94038 | 4.51239 | 4.93429 |
| | 2.4 | 3.43959 | 3.87131 | 4.42438 | 4.83169 |
| | 2.8 | 3.38746 | 3.80632 | 4.34203 | 4.73597 |
| 3.2 | 0.0 | 3.59672 | 4.08555 | 4.71829 | 5.18839 |
| | 0.4 | 3.58589 | 4.06484 | 4.68326 | 5.14178 |
| | 0.8 | 3.55715 | 4.02427 | 4.62607 | 5.07142 |
| | 1.2 | 3.51808 | 3.97258 | 4.55693 | 4.98859 |
| | 1.6 | 3.47392 | 3.91575 | 4.48275 | 4.90091 |
| | 2.0 | 3.42808 | 3.85765 | 4.40795 | 4.81318 |
| | 2.4 | 3.38265 | 3.80062 | 4.33516 | 4.72824 |
| | 2.8 | 3.33888 | 3.74599 | 4.26586 | 4.64763 |
| | 3.2 | 3.29742 | 3.69447 | 4.20075 | 4.57208 |

FRACTILES $W_\alpha$ RHO = 0.60

| DELTAS | | $\alpha$:0.9950 | 0.9975 | 0.9990 | 0.9995 |
|---|---|---|---|---|---|
| 3.6 | 0.0 | 3.49997 | 3.96254 | 4.55961 | 5.00218 |
| | 0.4 | 3.49262 | 3.94756 | 4.53351 | 4.96703 |
| | 0.8 | 3.47109 | 3.91678 | 4.48967 | 4.91279 |
| | 1.2 | 3.44082 | 3.87647 | 4.43542 | 4.84760 |
| | 1.6 | 3.40571 | 3.83112 | 4.37599 | 4.77719 |
| | 2.0 | 3.36847 | 3.78379 | 4.31490 | 4.70542 |
| | 2.4 | 3.33086 | 3.73648 | 4.25440 | 4.63472 |
| | 2.8 | 3.29400 | 3.69041 | 4.19587 | 4.56658 |
| | 3.2 | 3.25856 | 3.64632 | 4.14010 | 4.50182 |
| | 3.6 | 3.22489 | 3.60458 | 4.08747 | 4.44083 |
| 4.0 | 0.0 | 3.41835 | 3.85909 | 4.42654 | 4.84625 |
| | 0.4 | 3.41359 | 3.84838 | 4.40710 | 4.81965 |
| | 0.8 | 3.39739 | 3.82486 | 4.37319 | 4.77744 |
| | 1.2 | 3.37374 | 3.79315 | 4.33023 | 4.72562 |
| | 1.6 | 3.34563 | 3.75667 | 4.28222 | 4.66860 |
| | 2.0 | 3.31519 | 3.71787 | 4.23197 | 4.60946 |
| | 2.4 | 3.28390 | 3.67842 | 4.18141 | 4.55029 |
| | 2.8 | 3.25275 | 3.63943 | 4.13179 | 4.49246 |
| | 3.2 | 3.22238 | 3.60160 | 4.08388 | 4.43678 |
| | 3.6 | 3.19317 | 3.56536 | 4.03813 | 4.38373 |
| | 4.0 | 3.16534 | 3.53092 | 3.99479 | 4.33354 |
| 6.0 | 0.0 | 3.15317 | 3.52477 | 3.99843 | 4.34576 |
| | 0.4 | 3.15401 | 3.52371 | 3.99440 | 4.33920 |
| | 0.8 | 3.15014 | 3.51725 | 3.98414 | 4.32583 |
| | 1.2 | 3.14265 | 3.50669 | 3.96920 | 4.30737 |
| | 1.6 | 3.13249 | 3.49312 | 3.95085 | 4.28524 |
| | 2.0 | 3.12041 | 3.47742 | 3.93014 | 4.26059 |
| | 2.4 | 3.10701 | 3.46030 | 3.90788 | 4.23432 |
| | 2.8 | 3.09279 | 3.44231 | 3.88473 | 4.20716 |
| | 3.2 | 3.07810 | 3.42387 | 3.86118 | 4.17964 |
| | 3.6 | 3.06324 | 3.40530 | 3.83759 | 4.15216 |
| | 4.0 | 3.04841 | 3.38686 | 3.81424 | 4.12503 |
| | 6.0 | 2.97894 | 3.30105 | 3.70639 | 4.00024 |
| 12.0 | 0.0 | 2.86588 | 3.16538 | 3.54110 | 3.81277 |
| | 0.4 | 2.86761 | 3.16709 | 3.54269 | 3.81420 |
| | 0.8 | 2.86849 | 3.16781 | 3.54309 | 3.81430 |
| | 1.2 | 2.86862 | 3.16764 | 3.54244 | 3.81323 |
| | 1.6 | 2.86809 | 3.16669 | 3.54087 | 3.81113 |
| | 2.0 | 2.86698 | 3.16507 | 3.53849 | 3.80814 |
| | 2.4 | 2.86537 | 3.16286 | 3.53543 | 3.80438 |
| | 2.8 | 2.86332 | 3.16014 | 3.53176 | 3.79996 |
| | 3.2 | 2.86090 | 3.15700 | 3.52760 | 3.79499 |
| | 3.6 | 2.85816 | 3.15348 | 3.52301 | 3.78954 |
| | 4.0 | 2.85516 | 3.14967 | 3.51806 | 3.78371 |
| | 6.0 | 2.83762 | 3.12770 | 3.49003 | 3.75097 |
| | 12.0 | 2.78191 | 3.05910 | 3.40405 | 3.65162 |
| INFINITY | | 2.57583 | 2.80703 | 3.09024 | 3.29051 |

FRACTILES $W_{\alpha}$ RHO = 0.80

| DELTAS | | $\alpha$:0.0005 | 0.0010 | 0.0025 | 0.0050 | 0.0075 |
|---|---|---|---|---|---|---|
| 0.0 | 0.0 | -1.39921 | -1.29952 | -1.16933 | -1.07239 | -1.01649 |
| 0.4 | 0.0 | -1.55890 | -1.44489 | -1.29413 | -1.18018 | -1.11365 |
| | 0.4 | -1.30118 | -1.21551 | -1.10365 | -1.02041 | -0.97242 |
| 0.8 | 0.0 | -1.71327 | -1.59391 | -1.43405 | -1.31125 | -1.23854 |
| | 0.4 | -1.38320 | -1.29417 | -1.17645 | -1.08751 | -1.03561 |
| | 0.8 | -1.19457 | -1.13056 | -1.04705 | -0.98501 | -0.94931 |
| 1.2 | 0.0 | -1.79545 | -1.67736 | -1.51788 | -1.39413 | -1.32024 |
| | 0.4 | -1.50175 | -1.41144 | -1.29042 | -1.19739 | -1.14228 |
| | 0.8 | -1.28528 | -1.21855 | -1.13035 | -1.06377 | -1.02496 |
| | 1.2 | -1.18605 | -1.13766 | -1.07467 | -1.02801 | -1.00126 |
| 1.6 | 0.0 | -1.84355 | -1.72907 | -1.57341 | -1.45167 | -1.37849 |
| | 0.4 | -1.59460 | -1.50468 | -1.38309 | -1.28859 | -1.23206 |
| | 0.8 | -1.41183 | -1.34255 | -1.24963 | -1.17810 | -1.13568 |
| | 1.2 | -1.28803 | -1.23620 | -1.16775 | -1.11619 | -1.08620 |
| | 1.6 | -1.24899 | -1.21098 | -1.16169 | -1.12541 | -1.10476 |
| 2.0 | 0.0 | -1.88005 | -1.76946 | -1.61826 | -1.49920 | -1.42726 |
| | 0.4 | -1.67306 | -1.58393 | -1.46252 | -1.36729 | -1.30991 |
| | 0.8 | -1.52110 | -1.45017 | -1.35405 | -1.27912 | -1.23418 |
| | 1.2 | -1.41711 | -1.36162 | -1.28707 | -1.22957 | -1.19540 |
| | 1.6 | -1.35408 | -1.31219 | -1.25699 | -1.21557 | -1.19163 |
| | 2.0 | -1.35432 | -1.32353 | -1.28389 | -1.25511 | -1.23902 |
| 2.4 | 0.0 | -1.91437 | -1.80723 | -1.66002 | -1.54341 | -1.47261 |
| | 0.4 | -1.74340 | -1.65500 | -1.53382 | -1.43803 | -1.37995 |
| | 0.8 | -1.61777 | -1.54547 | -1.44665 | -1.36880 | -1.32170 |
| | 1.2 | -1.53179 | -1.47337 | -1.39391 | -1.33167 | -1.29417 |
| | 1.6 | -1.47974 | -1.43349 | -1.37119 | -1.32297 | -1.29422 |
| | 2.0 | -1.45769 | -1.42279 | -1.37702 | -1.34307 | -1.32380 |
| | 2.4 | -1.48537 | -1.46000 | -1.42790 | -1.40553 | -1.39398 |
| 2.8 | 0.0 | -1.94974 | -1.84540 | -1.70137 | -1.58666 | -1.51670 |
| | 0.4 | -1.80869 | -1.72080 | -1.59961 | -1.50314 | -1.44431 |
| | 0.8 | -1.70515 | -1.63156 | -1.53022 | -1.44964 | -1.40050 |
| | 1.2 | -1.63472 | -1.57362 | -1.48967 | -1.42306 | -1.38246 |
| | 1.6 | -1.59311 | -1.54310 | -1.47470 | -1.42068 | -1.38786 |
| | 2.0 | -1.57684 | -1.53697 | -1.48304 | -1.44107 | -1.41594 |
| | 2.4 | -1.58455 | -1.55492 | -1.51666 | -1.48963 | -1.47638 |
| | 2.8 | -1.63325 | -1.61252 | -1.58800 | -1.57251 | -1.55376 |
| 3.2 | 0.0 | -1.98685 | -1.88466 | -1.74297 | -1.62953 | -1.56005 |
| | 0.4 | -1.87043 | -1.78276 | -1.66121 | -1.56382 | -1.50409 |
| | 0.8 | -1.78533 | -1.71041 | -1.60653 | -1.52322 | -1.47206 |
| | 1.2 | -1.72817 | -1.66450 | -1.57624 | -1.50538 | -1.46177 |
| | 1.6 | -1.69572 | -1.64213 | -1.56790 | -1.50826 | -1.47146 |
| | 2.0 | -1.68538 | -1.64104 | -1.57982 | -1.53071 | -1.50036 |
| | 2.4 | -1.69538 | -1.65996 | -1.61174 | -1.57406 | -1.55237 |
| | 2.8 | -1.72687 | -1.70190 | -1.67305 | -1.64472 | -1.61621 |
| | 3.2 | -1.79386 | -1.77955 | -1.74958 | -1.70436 | -1.66815 |

FRACTILES $w_\alpha$ RHO = 0.80

| DELTAS | | α:0.0005 | 0.0010 | 0.0025 | 0.0050 | 0.0075 |
|---|---|---|---|---|---|---|
| 3.6 | 0.0 | -2.02551 | -1.92485 | -1.78472 | -1.67195 | -1.60258 |
| | 0.4 | -1.92936 | -1.84160 | -1.71929 | -1.62067 | -1.55987 |
| | 0.8 | -1.85963 | -1.78329 | -1.67673 | -1.59059 | -1.53731 |
| | 1.2 | -1.81372 | -1.74752 | -1.65496 | -1.57985 | -1.53319 |
| | 1.6 | -1.78918 | -1.73210 | -1.65211 | -1.58685 | -1.54602 |
| | 2.0 | -1.78402 | -1.73529 | -1.66683 | -1.61055 | -1.57491 |
| | 2.4 | -1.79688 | -1.75611 | -1.69873 | -1.65109 | -1.62039 |
| | 2.8 | -1.82753 | -1.79511 | -1.75194 | -1.70783 | -1.67108 |
| | 3.2 | -1.88163 | -1.86309 | -1.81453 | -1.75757 | -1.71486 |
| | 3.6 | -1.96298 | -1.93194 | -1.86955 | -1.80317 | -1.75505 |
| 4.0 | 0.0 | -2.06521 | -1.96557 | -1.82627 | -1.71361 | -1.64403 |
| | 0.4 | -1.98580 | -1.89766 | -1.77419 | -1.67402 | -1.61195 |
| | 0.8 | -1.92892 | -1.85100 | -1.74158 | -1.65242 | -1.59693 |
| | 1.2 | -1.89253 | -1.82374 | -1.72679 | -1.64732 | -1.59752 |
| | 1.6 | -1.87472 | -1.81414 | -1.72832 | -1.65733 | -1.61237 |
| | 2.0 | -1.87393 | -1.82081 | -1.74500 | -1.68133 | -1.64024 |
| | 2.4 | -1.88908 | -1.84290 | -1.77616 | -1.71843 | -1.67943 |
| | 2.8 | -1.91986 | -1.88045 | -1.82243 | -1.76398 | -1.72003 |
| | 3.2 | -1.96826 | -1.93725 | -1.87336 | -1.80590 | -1.75724 |
| | 3.6 | -2.03557 | -1.99479 | -1.92039 | -1.84520 | -1.79201 |
| | 4.0 | -2.09737 | -2.04896 | -1.96462 | -1.88193 | -1.82437 |
| 6.0 | 0.0 | -2.26340 | -2.16286 | -2.01947 | -1.90072 | -1.82594 |
| | 0.4 | -2.23397 | -2.13995 | -2.00517 | -1.89273 | -1.82147 |
| | 0.8 | -2.21575 | -2.12747 | -2.00003 | -1.89274 | -1.82419 |
| | 1.2 | -2.20789 | -2.12453 | -2.00311 | -1.89968 | -1.83293 |
| | 1.6 | -2.20950 | -2.13020 | -2.01331 | -1.91225 | -1.84628 |
| | 2.0 | -2.21966 | -2.14340 | -2.02930 | -1.92893 | -1.86264 |
| | 2.4 | -2.23730 | -2.16288 | -2.04947 | -1.94803 | -1.88053 |
| | 2.8 | -2.26110 | -2.18697 | -2.07203 | -1.96823 | -1.89900 |
| | 3.2 | -2.28924 | -2.21373 | -2.09564 | -1.98875 | -1.91751 |
| | 3.6 | -2.31964 | -2.24163 | -2.11949 | -2.00911 | -1.93571 |
| | 4.0 | -2.35087 | -2.26976 | -2.14306 | -2.02901 | -1.95339 |
| | 6.0 | -2.49834 | -2.39964 | -2.24934 | -2.11731 | -2.03117 |
| 12.0 | 0.0 | -2.67646 | -2.54939 | -2.36417 | -2.20773 | -2.10815 |
| | 0.4 | -2.67718 | -2.55077 | -2.36616 | -2.20997 | -2.11045 |
| | 0.8 | -2.67954 | -2.55352 | -2.36918 | -2.21299 | -2.11338 |
| | 1.2 | -2.68330 | -2.55743 | -2.37304 | -2.21665 | -2.11685 |
| | 1.6 | -2.68826 | -2.56231 | -2.37762 | -2.22083 | -2.12073 |
| | 2.0 | -2.69422 | -2.56800 | -2.38276 | -2.22542 | -2.12495 |
| | 2.4 | -2.70101 | -2.57435 | -2.38837 | -2.23035 | -2.12943 |
| | 2.8 | -2.70845 | -2.58122 | -2.39433 | -2.23552 | -2.13411 |
| | 3.2 | -2.71643 | -2.58849 | -2.40056 | -2.24087 | -2.13891 |
| | 3.6 | -2.72481 | -2.59607 | -2.40698 | -2.24635 | -2.14381 |
| | 4.0 | -2.73348 | -2.60386 | -2.41353 | -2.25190 | -2.14876 |
| | 6.0 | -2.77861 | -2.64392 | -2.44666 | -2.27966 | -2.17334 |
| | 12.0 | -2.89756 | -2.74802 | -2.53117 | -2.34944 | -2.23456 |
| INFINITY | | -3.29051 | -3.09024 | -2.80703 | -2.57583 | -2.43238 |

FRACTILES $W_\alpha$ RHO = 0.80

| DELTAS | | $\alpha$:0.0100 | 0.0175 | 0.0250 | 0.0375 | 0.0500 |
|---|---|---|---|---|---|---|
| 0.0 | 0.0 | -0.97726 | -0.90223 | -0.85549 | -0.80372 | -0.76809 |
| 0.4 | 0.0 | -1.06656 | -0.97535 | -0.91766 | -0.85274 | -0.80733 |
| | 0.4 | -0.93877 | -0.87444 | -0.83442 | -0.79015 | -0.75976 |
| 0.8 | 0.0 | -1.18651 | -1.08411 | -1.01790 | -0.94162 | -0.88675 |
| | 0.4 | -0.99888 | -0.92781 | -0.88291 | -0.83250 | -0.79734 |
| | 0.8 | -0.92430 | -0.87662 | -0.84709 | -0.81460 | -0.79247 |
| 1.2 | 0.0 | -1.26703 | -1.16137 | -1.09229 | -1.01175 | -0.95307 |
| | 0.4 | -1.10282 | -1.02509 | -0.97480 | -0.91681 | -0.87505 |
| | 0.8 | -0.99753 | -0.94456 | -0.91122 | -0.87399 | -0.84824 |
| | 1.2 | -0.98258 | -0.94717 | -0.92545 | -0.90190 | -0.88625 |
| 1.6 | 0.0 | -1.32556 | -1.21974 | -1.14997 | -1.06796 | -1.00766 |
| | 0.4 | -1.19132 | -1.11025 | -1.05710 | -0.99496 | -0.94952 |
| | 0.8 | -1.10528 | -1.04532 | -1.00647 | -0.96158 | -0.92922 |
| | 1.2 | -1.06506 | -1.02443 | -0.99910 | -0.97124 | -0.95251 |
| | 1.6 | -1.09044 | -1.06368 | -1.04770 | -1.03126 | -1.02173 |
| 2.0 | 0.0 | -1.37501 | -1.26997 | -1.20024 | -1.11771 | -1.05658 |
| | 0.4 | -1.26832 | -1.18492 | -1.12970 | -1.06447 | -1.01624 |
| | 0.8 | -1.20171 | -1.13689 | -1.09419 | -1.04397 | -1.00699 |
| | 1.2 | -1.17087 | -1.12240 | -1.09091 | -1.05446 | -1.02821 |
| | 1.6 | -1.17485 | -1.14308 | -1.12388 | -1.10424 | -1.09410 |
| | 2.0 | -1.22809 | -1.20869 | -1.19906 | -1.18513 | -1.16479 |
| 2.4 | 0.0 | -1.42100 | -1.31676 | -1.24713 | -1.16419 | -1.10237 |
| | 0.4 | -1.33764 | -1.25225 | -1.19522 | -1.12727 | -1.07653 |
| | 0.8 | -1.28744 | -1.21837 | -1.17228 | -1.11734 | -1.07626 |
| | 1.2 | -1.26697 | -1.21233 | -1.17599 | -1.13277 | -1.10047 |
| | 1.6 | -1.27354 | -1.23256 | -1.20590 | -1.17541 | -1.15587 |
| | 2.0 | -1.31061 | -1.28749 | -1.27592 | -1.24339 | -1.21713 |
| | 2.4 | -1.38740 | -1.36821 | -1.34300 | -1.29991 | -1.25851 |
| 2.8 | 0.0 | -1.46554 | -1.36171 | -1.29196 | -1.20840 | -1.14570 |
| | 0.4 | -1.40127 | -1.31387 | -1.25505 | -1.18438 | -1.13116 |
| | 0.8 | -1.36455 | -1.29145 | -1.24213 | -1.18263 | -1.13754 |
| | 1.2 | -1.35276 | -1.29230 | -1.25137 | -1.20169 | -1.16362 |
| | 1.6 | -1.36388 | -1.31509 | -1.28200 | -1.24156 | -1.21034 |
| | 2.0 | -1.39784 | -1.36241 | -1.34102 | -1.29976 | -1.25858 |
| | 2.4 | -1.46758 | -1.43143 | -1.39546 | -1.34071 | -1.29152 |
| | 2.8 | -1.53475 | -1.48287 | -1.43816 | -1.37420 | -1.31879 |
| 3.2 | 0.0 | -1.50908 | -1.40518 | -1.33498 | -1.25041 | -1.18657 |
| | 0.4 | -1.46023 | -1.37063 | -1.30988 | -1.23633 | -1.18048 |
| | 0.8 | -1.43442 | -1.35727 | -1.30468 | -1.24055 | -1.19134 |
| | 1.2 | -1.42961 | -1.36338 | -1.31784 | -1.26158 | -1.21759 |
| | 1.6 | -1.44424 | -1.38772 | -1.34824 | -1.29813 | -1.25700 |
| | 2.0 | -1.47784 | -1.43057 | -1.39678 | -1.34237 | -1.29273 |
| | 2.4 | -1.53612 | -1.48470 | -1.43966 | -1.37523 | -1.31949 |
| | 2.8 | -1.59068 | -1.52693 | -1.47507 | -1.40325 | -1.34242 |
| | 3.2 | -1.63750 | -1.56421 | -1.50643 | -1.42803 | -1.36262 |

FRACTILES $w_\alpha$ RHO = 0.80

| DELTAS | | α:0.0100 | 0.0175 | 0.0250 | 0.0375 | 0.0500 |
|---|---|---|---|---|---|---|
| 3.6 | 0.0 | -1.55155 | -1.44705 | -1.37606 | -1.29004 | -1.22471 |
| | 0.4 | -1.51505 | -1.42297 | -1.36008 | -1.28339 | -1.22467 |
| | 0.8 | -1.49793 | -1.41659 | -1.36060 | -1.29163 | -1.23812 |
| | 1.2 | -1.49854 | -1.42641 | -1.37610 | -1.31304 | -1.26292 |
| | 1.6 | -1.51549 | -1.45103 | -1.40494 | -1.34499 | -1.29471 |
| | 2.0 | -1.54790 | -1.48902 | -1.44395 | -1.37837 | -1.32167 |
| | 2.4 | -1.59571 | -1.53080 | -1.47801 | -1.40523 | -1.34379 |
| | 2.8 | -1.64001 | -1.56595 | -1.50774 | -1.42891 | -1.36323 |
| | 3.2 | -1.67968 | -1.59777 | -1.53461 | -1.45020 | -1.38061 |
| | 3.6 | -1.71607 | -1.62682 | -1.55904 | -1.46945 | -1.39625 |
| 4.0 | 0.0 | -1.59267 | -1.48707 | -1.41493 | -1.32704 | -1.25989 |
| | 0.4 | -1.56602 | -1.47113 | -1.40587 | -1.32573 | -1.26390 |
| | 0.8 | -1.55570 | -1.46994 | -1.41038 | -1.33632 | -1.27831 |
| | 1.2 | -1.56030 | -1.48206 | -1.42683 | -1.35677 | -1.30047 |
| | 1.6 | -1.57843 | -1.50578 | -1.45299 | -1.38351 | -1.32537 |
| | 2.0 | -1.60861 | -1.53827 | -1.48398 | -1.40929 | -1.34660 |
| | 2.4 | -1.64754 | -1.57121 | -1.51171 | -1.43158 | -1.36510 |
| | 2.8 | -1.68406 | -1.60078 | -1.53688 | -1.45175 | -1.38170 |
| | 3.2 | -1.71793 | -1.62811 | -1.56002 | -1.47012 | -1.39673 |
| | 3.6 | -1.74946 | -1.65337 | -1.58130 | -1.48692 | -1.41039 |
| | 4.0 | -1.77874 | -1.67669 | -1.60088 | -1.50231 | -1.42285 |
| 6.0 | 0.0 | -1.76999 | -1.65268 | -1.57066 | -1.46854 | -1.38885 |
| | 0.4 | -1.76789 | -1.65475 | -1.57494 | -1.47472 | -1.39586 |
| | 0.8 | -1.77234 | -1.66190 | -1.58321 | -1.48352 | -1.40451 |
| | 1.2 | -1.78207 | -1.67273 | -1.59406 | -1.49368 | -1.41374 |
| | 1.6 | -1.79563 | -1.68578 | -1.60618 | -1.50425 | -1.42297 |
| | 2.0 | -1.81143 | -1.69983 | -1.61872 | -1.51480 | -1.43196 |
| | 2.4 | -1.82826 | -1.71417 | -1.63126 | -1.52512 | -1.44061 |
| | 2.8 | -1.84538 | -1.72842 | -1.64354 | -1.53507 | -1.44887 |
| | 3.2 | -1.86238 | -1.74236 | -1.65544 | -1.54461 | -1.45672 |
| | 3.6 | -1.87900 | -1.75583 | -1.66687 | -1.55370 | -1.46414 |
| | 4.0 | -1.89508 | -1.76877 | -1.67780 | -1.56234 | -1.47116 |
| | 6.0 | -1.96542 | -1.82471 | -1.72465 | -1.59899 | -1.50070 |
| 12.0 | 0.0 | -2.03325 | -1.87568 | -1.76556 | -1.62914 | -1.52369 |
| | 0.4 | -2.03555 | -1.87787 | -1.76761 | -1.63098 | -1.52533 |
| | 0.8 | -2.03839 | -1.88045 | -1.76996 | -1.63301 | -1.52711 |
| | 1.2 | -2.04168 | -1.88334 | -1.77254 | -1.63519 | -1.52898 |
| | 1.6 | -2.04533 | -1.88646 | -1.77529 | -1.63749 | -1.53092 |
| | 2.0 | -2.04927 | -1.88978 | -1.77818 | -1.63986 | -1.53291 |
| | 2.4 | -2.05341 | -1.89323 | -1.78116 | -1.64228 | -1.53493 |
| | 2.8 | -2.05771 | -1.89678 | -1.78421 | -1.64474 | -1.53696 |
| | 3.2 | -2.06212 | -1.90039 | -1.78729 | -1.64721 | -1.53899 |
| | 3.6 | -2.06660 | -1.90403 | -1.79039 | -1.64968 | -1.54101 |
| | 4.0 | -2.07112 | -1.90768 | -1.79348 | -1.65214 | -1.54301 |
| | 6.0 | -2.09342 | -1.92554 | -1.80850 | -1.66393 | -1.55254 |
| | 12.0 | -2.14861 | -1.96911 | -1.84477 | -1.69205 | -1.57498 |
| INFINITY | | -2.32635 | -2.10836 | -1.95996 | -1.78046 | -1.64485 |

FRACTILES $w_\alpha$ RHO = 0.80

| DELTAS | | $\alpha$:0.0750 | 0.1000 | 0.1500 | 0.2000 | 0.2500 |
|---|---|---|---|---|---|---|
| 0.0 | 0.0 | -0.72001 | -0.68798 | -0.64758 | -0.62576 | -0.60823 |
| 0.4 | 0.0 | -0.74476 | -0.70195 | -0.64559 | -0.61109 | -0.59220 |
| | 0.4 | -0.71889 | -0.69186 | -0.65844 | -0.64206 | -0.62112 |
| 0.8 | 0.0 | -0.80822 | -0.75157 | -0.67040 | -0.61224 | -0.56784 |
| | 0.4 | -0.74915 | -0.71651 | -0.67468 | -0.65214 | -0.63192 |
| | 0.8 | -0.76317 | -0.74441 | -0.72452 | -0.70473 | -0.67407 |
| 1.2 | 0.0 | -0.86768 | -0.80474 | -0.71160 | -0.64123 | -0.58318 |
| | 0.4 | -0.81525 | -0.77210 | -0.71048 | -0.66721 | -0.63856 |
| | 0.8 | -0.81355 | -0.79092 | -0.76654 | -0.74120 | -0.70304 |
| | 1.2 | -0.86666 | -0.85680 | -0.83067 | -0.78975 | -0.73741 |
| 1.6 | 0.0 | -0.91894 | -0.85266 | -0.75278 | -0.67542 | -0.60990 |
| | 0.4 | -0.88308 | -0.83377 | -0.76002 | -0.70335 | -0.65576 |
| | 0.8 | -0.88286 | -0.84934 | -0.80358 | -0.77215 | -0.72451 |
| | 1.2 | -0.92906 | -0.91689 | -0.87711 | -0.82124 | -0.75440 |
| | 1.6 | -1.00675 | -0.98264 | -0.91976 | -0.84540 | -0.76384 |
| 2.0 | 0.0 | -0.96583 | -0.89728 | -0.79249 | -0.70976 | -0.63830 |
| | 0.4 | -0.94469 | -0.89061 | -0.80759 | -0.74123 | -0.68254 |
| | 0.8 | -0.95237 | -0.91121 | -0.84810 | -0.79914 | -0.73900 |
| | 1.2 | -0.99108 | -0.96767 | -0.91258 | -0.84029 | -0.75962 |
| | 1.6 | -1.06631 | -1.02962 | -0.94584 | -0.85620 | -0.76347 |
| | 2.0 | -1.11720 | -1.06647 | -0.96391 | -0.86225 | -0.76111 |
| 2.4 | 0.0 | -1.00983 | -0.93924 | -0.82988 | -0.74203 | -0.66477 |
| | 0.4 | -1.00036 | -0.94190 | -0.85019 | -0.77463 | -0.70549 |
| | 0.8 | -1.01429 | -0.96624 | -0.88893 | -0.82107 | -0.74628 |
| | 1.2 | -1.05160 | -1.01359 | -0.93854 | -0.85048 | -0.75829 |
| | 1.6 | -1.11381 | -1.06423 | -0.96215 | -0.86054 | -0.75939 |
| | 2.0 | -1.15358 | -1.09194 | -0.97517 | -0.86440 | -0.75682 |
| | 2.4 | -1.18185 | -1.11168 | -0.98380 | -0.86581 | -0.75306 |
| 2.8 | 0.0 | -1.05115 | -0.97831 | -0.86404 | -0.77067 | -0.68712 |
| | 0.4 | -1.05035 | -0.98746 | -0.88684 | -0.80169 | -0.72171 |
| | 0.8 | -1.06829 | -1.01332 | -0.92170 | -0.83721 | -0.74779 |
| | 1.2 | -1.10381 | -1.05401 | -0.95665 | -0.85544 | -0.75433 |
| | 1.6 | -1.15148 | -1.09021 | -0.97321 | -0.86217 | -0.75443 |
| | 2.0 | -1.18177 | -1.11138 | -0.98320 | -0.86503 | -0.75218 |
| | 2.4 | -1.20450 | -1.12737 | -0.99024 | -0.86614 | -0.74902 |
| | 2.8 | -1.22335 | -1.14047 | -0.99561 | -0.86638 | -0.74557 |
| 3.2 | 0.0 | -1.08953 | -1.01409 | -0.89423 | -0.79467 | -0.70420 |
| | 0.4 | -1.09476 | -1.02715 | -0.91705 | -0.82178 | -0.73086 |
| | 0.8 | -1.11457 | -1.05239 | -0.94607 | -0.84659 | -0.74607 |
| | 1.2 | -1.14653 | -1.08535 | -0.96936 | -0.85786 | -0.74977 |
| | 1.6 | -1.18146 | -1.11055 | -0.98146 | -0.86274 | -0.74957 |
| | 2.0 | -1.20492 | -1.12725 | -0.98947 | -0.86502 | -0.74769 |
| | 2.4 | -1.22364 | -1.14050 | -0.99535 | -0.86596 | -0.74505 |
| | 2.8 | -1.23956 | -1.15159 | -0.99991 | -0.86617 | -0.74211 |
| | 3.2 | -1.25344 | -1.16113 | -1.00358 | -0.86596 | -0.73912 |

FRACTILES $W_{\alpha}$ RHO = 0.80

| DELTAS | | α:0.0750 | 0.1000 | 0.1500 | 0.2000 | 0.2500 |
|---|---|---|---|---|---|---|
| 3.6 | 0.0 | -1.12467 | -1.04617 | -0.91996 | -0.81359 | -0.71574 |
| | 0.4 | -1.13366 | -1.06099 | -0.94086 | -0.83539 | -0.73443 |
| | 0.8 | -1.15347 | -1.08383 | -0.96323 | -0.85161 | -0.74312 |
| | 1.2 | -1.18051 | -1.10893 | -0.97876 | -0.85912 | -0.74540 |
| | 1.6 | -1.20586 | -1.12713 | -0.98803 | -0.86285 | -0.74510 |
| | 2.0 | -1.22456 | -1.14065 | -0.99460 | -0.86472 | -0.74352 |
| | 2.4 | -1.24018 | -1.15176 | -0.99955 | -0.86552 | -0.74130 |
| | 2.8 | -1.25373 | -1.16122 | -1.00346 | -0.86570 | -0.73879 |
| | 3.2 | -1.26570 | -1.16945 | -1.00663 | -0.86552 | -0.73619 |
| | 3.6 | -1.27636 | -1.17669 | -1.00927 | -0.86511 | -0.73361 |
| 4.0 | 0.0 | -1.15631 | -1.07432 | -0.94111 | -0.82757 | -0.72241 |
| | 0.4 | -1.16719 | -1.08918 | -0.95887 | -0.84386 | -0.73466 |
| | 0.8 | -1.18557 | -1.10855 | -0.97520 | -0.85441 | -0.73990 |
| | 1.2 | -1.20732 | -1.12713 | -0.98611 | -0.85981 | -0.74139 |
| | 1.6 | -1.22624 | -1.14102 | -0.99345 | -0.86274 | -0.74106 |
| | 2.0 | -1.24147 | -1.15214 | -0.99888 | -0.86427 | -0.73971 |
| | 2.4 | -1.25463 | -1.16153 | -1.00308 | -0.86496 | -0.73782 |
| | 2.8 | -1.26624 | -1.16966 | -1.00644 | -0.86511 | -0.73566 |
| | 3.2 | -1.27660 | -1.17679 | -1.00920 | -0.86495 | -0.73339 |
| | 3.6 | -1.28593 | -1.18313 | -1.01150 | -0.86459 | -0.73112 |
| | 4.0 | -1.29436 | -1.18880 | -1.01347 | -0.86411 | -0.72890 |
| 6.0 | 0.0 | -1.26319 | -1.16171 | -0.99463 | -0.85247 | -0.72338 |
| | 0.4 | -1.27048 | -1.16851 | -0.99977 | -0.85592 | -0.72532 |
| | 0.8 | -1.27813 | -1.17495 | -1.00399 | -0.85831 | -0.72620 |
| | 1.2 | -1.28557 | -1.18084 | -1.00744 | -0.85991 | -0.72633 |
| | 1.6 | -1.29263 | -1.18621 | -1.01029 | -0.86092 | -0.72594 |
| | 2.0 | -1.29927 | -1.19110 | -1.01266 | -0.86151 | -0.72519 |
| | 2.4 | -1.30549 | -1.19557 | -1.01465 | -0.86179 | -0.72421 |
| | 2.8 | -1.31132 | -1.19967 | -1.01635 | -0.86185 | -0.72307 |
| | 3.2 | -1.31677 | -1.20344 | -1.01780 | -0.86175 | -0.72183 |
| | 3.6 | -1.32187 | -1.20691 | -1.01907 | -0.86154 | -0.72056 |
| | 4.0 | -1.32665 | -1.21013 | -1.02018 | -0.86125 | -0.71926 |
| | 6.0 | -1.34639 | -1.22316 | -1.02419 | -0.85932 | -0.71312 |
| 12.0 | 0.0 | -1.36000 | -1.23067 | -1.02405 | -0.85453 | -0.70529 |
| | 0.4 | -1.36131 | -1.23171 | -1.02465 | -0.85479 | -0.70529 |
| | 0.8 | -1.36268 | -1.23275 | -1.02520 | -0.85499 | -0.70519 |
| | 1.2 | -1.36408 | -1.23380 | -1.02572 | -0.85512 | -0.70503 |
| | 1.6 | -1.36550 | -1.23483 | -1.02620 | -0.85520 | -0.70481 |
| | 2.0 | -1.36693 | -1.23586 | -1.02664 | -0.85523 | -0.70454 |
| | 2.4 | -1.36836 | -1.23686 | -1.02705 | -0.85523 | -0.70423 |
| | 2.8 | -1.36978 | -1.23785 | -1.02743 | -0.85519 | -0.70389 |
| | 3.2 | -1.37118 | -1.23881 | -1.02778 | -0.85513 | -0.70352 |
| | 3.6 | -1.37257 | -1.23975 | -1.02811 | -0.85504 | -0.70313 |
| | 4.0 | -1.37393 | -1.24067 | -1.02841 | -0.85494 | -0.70273 |
| | 6.0 | -1.38032 | -1.24488 | -1.02967 | -0.85424 | -0.70063 |
| | 12.0 | -1.39497 | -1.25420 | -1.03189 | -0.85182 | -0.69495 |
| INFINITY | | -1.43953 | -1.28155 | -1.03643 | -0.84162 | -0.67449 |

FRACTILES $W_\alpha$ RHO = 0.80

| DELTAS | | $\alpha$:0.3000 | 0.3500 | 0.4000 | 0.4500 | 0.5000 |
|---|---|---|---|---|---|---|
| 0.0 | 0.0 | -0.57982 | -0.54254 | -0.49621 | -0.44005 | -0.37276 |
| 0.4 | 0.0 | -0.56715 | -0.53177 | -0.48680 | -0.43170 | -0.36529 |
| | 0.4 | -0.59079 | -0.55180 | -0.50377 | -0.44585 | -0.37669 |
| 0.8 | 0.0 | -0.53495 | -0.50819 | -0.46698 | -0.41399 | -0.34883 |
| | 0.4 | -0.60015 | -0.55888 | -0.50793 | -0.44656 | -0.37349 |
| | 0.8 | -0.63402 | -0.58461 | -0.52535 | -0.45542 | -0.37361 |
| 1.2 | 0.0 | -0.53271 | -0.48739 | -0.44832 | -0.39769 | -0.33182 |
| | 0.4 | -0.60736 | -0.56269 | -0.50673 | -0.43936 | -0.35966 |
| | 0.8 | -0.65427 | -0.59536 | -0.52625 | -0.44645 | -0.35507 |
| | 1.2 | -0.67501 | -0.60322 | -0.52224 | -0.43178 | -0.33110 |
| 1.6 | 0.0 | -0.55106 | -0.49572 | -0.44132 | -0.38576 | -0.31614 |
| | 0.4 | -0.61568 | -0.56570 | -0.50165 | -0.42545 | -0.33693 |
| | 0.8 | -0.66468 | -0.59454 | -0.51473 | -0.42519 | -0.32528 |
| | 1.2 | -0.67888 | -0.59572 | -0.50513 | -0.40678 | -0.29982 |
| | 1.6 | -0.67672 | -0.58442 | -0.48668 | -0.38280 | -0.27170 |
| 2.0 | 0.0 | -0.57275 | -0.50963 | -0.44585 | -0.37775 | -0.29911 |
| | 0.4 | -0.62639 | -0.56630 | -0.49150 | -0.40525 | -0.30788 |
| | 0.8 | -0.66643 | -0.58500 | -0.49561 | -0.39822 | -0.29212 |
| | 1.2 | -0.67297 | -0.58099 | -0.48351 | -0.37987 | -0.26900 |
| | 1.6 | -0.66799 | -0.56936 | -0.46681 | -0.35933 | -0.24565 |
| | 2.0 | -0.65945 | -0.55614 | -0.45003 | -0.33991 | -0.22437 |
| 2.4 | 0.0 | -0.59253 | -0.52153 | -0.44828 | -0.36881 | -0.27828 |
| | 0.4 | -0.63674 | -0.56223 | -0.47635 | -0.38123 | -0.27695 |
| | 0.8 | -0.66161 | -0.57075 | -0.47409 | -0.37117 | -0.26103 |
| | 1.2 | -0.66310 | -0.56469 | -0.46237 | -0.35514 | -0.24175 |
| | 1.6 | -0.65774 | -0.55445 | -0.44840 | -0.33835 | -0.22290 |
| | 2.0 | -0.65038 | -0.54346 | -0.43462 | -0.32250 | -0.20560 |
| | 2.4 | -0.64275 | -0.53288 | -0.42183 | -0.30807 | -0.19007 |
| 2.8 | 0.0 | -0.60763 | -0.52815 | -0.44505 | -0.35478 | -0.25443 |
| | 0.4 | -0.64056 | -0.55341 | -0.45861 | -0.35702 | -0.24809 |
| | 0.8 | -0.65344 | -0.55555 | -0.45368 | -0.34695 | -0.23412 |
| | 1.2 | -0.65269 | -0.54947 | -0.44356 | -0.33372 | -0.21855 |
| | 1.6 | -0.64790 | -0.54096 | -0.43216 | -0.32012 | -0.20335 |
| | 2.0 | -0.64181 | -0.53192 | -0.42087 | -0.30714 | -0.18918 |
| | 2.4 | -0.63544 | -0.52311 | -0.41021 | -0.29513 | -0.17626 |
| | 2.8 | -0.62924 | -0.51482 | -0.40036 | -0.28416 | -0.16457 |
| 3.2 | 0.0 | -0.61689 | -0.52868 | -0.43636 | -0.33747 | -0.23058 |
| | 0.4 | -0.63858 | -0.54213 | -0.44107 | -0.33508 | -0.22307 |
| | 0.8 | -0.64453 | -0.54144 | -0.43574 | -0.32622 | -0.21149 |
| | 1.2 | -0.64307 | -0.53608 | -0.42735 | -0.31546 | -0.19893 |
| | 1.6 | -0.63903 | -0.52908 | -0.41803 | -0.30438 | -0.18655 |
| | 2.0 | -0.63400 | -0.52161 | -0.40870 | -0.29365 | -0.17485 |
| | 2.4 | -0.62866 | -0.51422 | -0.39975 | -0.28357 | -0.16400 |
| | 2.8 | -0.62338 | -0.50716 | -0.39137 | -0.27423 | -0.15404 |
| | 3.2 | -0.61831 | -0.50054 | -0.38360 | -0.26565 | -0.14496 |

FRACTILES $w_\alpha$ RHO = 0.80

| DELTAS | | $\alpha$:0.3000 | 0.3500 | 0.4000 | 0.4500 | 0.5000 |
|---|---|---|---|---|---|---|
| 3.6 | 0.0 | -0.62057 | -0.52438 | -0.42468 | -0.31992 | -0.20901 |
| | 0.4 | -0.63345 | -0.53067 | -0.42527 | -0.31616 | -0.20199 |
| | 0.8 | -0.63613 | -0.52904 | -0.42037 | -0.30869 | -0.19250 |
| | 1.2 | -0.63456 | -0.52447 | -0.41343 | -0.29989 | -0.18226 |
| | 1.6 | -0.63117 | -0.51867 | -0.40575 | -0.29075 | -0.17207 |
| | 2.0 | -0.62697 | -0.51244 | -0.39795 | -0.28179 | -0.16229 |
| | 2.4 | -0.62247 | -0.50619 | -0.39038 | -0.27325 | -0.15310 |
| | 2.8 | -0.61794 | -0.50014 | -0.38319 | -0.26525 | -0.14457 |
| | 3.2 | -0.61355 | -0.49440 | -0.37646 | -0.25781 | -0.13669 |
| | 3.6 | -0.60936 | -0.48902 | -0.37019 | -0.25093 | -0.12945 |
| 4.0 | 0.0 | -0.62010 | -0.51744 | -0.41250 | -0.30392 | -0.19040 |
| | 0.4 | -0.62739 | -0.52015 | -0.41153 | -0.30008 | -0.18428 |
| | 0.8 | -0.62860 | -0.51830 | -0.40723 | -0.29382 | -0.17645 |
| | 1.2 | -0.62709 | -0.51441 | -0.40144 | -0.28652 | -0.16800 |
| | 1.6 | -0.62422 | -0.50955 | -0.39502 | -0.27889 | -0.15950 |
| | 2.0 | -0.62067 | -0.50429 | -0.38844 | -0.27133 | -0.15125 |
| | 2.4 | -0.61683 | -0.49896 | -0.38198 | -0.26404 | -0.14340 |
| | 2.8 | -0.61293 | -0.49374 | -0.37578 | -0.25713 | -0.13604 |
| | 3.2 | -0.60910 | -0.48874 | -0.36990 | -0.25065 | -0.12917 |
| | 3.6 | -0.60542 | -0.48400 | -0.36439 | -0.24460 | -0.12279 |
| | 4.0 | -0.60191 | -0.47954 | -0.35924 | -0.23897 | -0.11689 |
| 6.0 | 0.0 | -0.60158 | -0.48362 | -0.36705 | -0.24992 | -0.13040 |
| | 0.4 | -0.60220 | -0.48310 | -0.36556 | -0.24759 | -0.12737 |
| | 0.8 | -0.60183 | -0.48169 | -0.36328 | -0.24459 | -0.12381 |
| | 1.2 | -0.60079 | -0.47969 | -0.36050 | -0.24120 | -0.11995 |
| | 1.6 | -0.59929 | -0.47731 | -0.35743 | -0.23760 | -0.11595 |
| | 2.0 | -0.59751 | -0.47472 | -0.35421 | -0.23391 | -0.11193 |
| | 2.4 | -0.59555 | -0.47202 | -0.35094 | -0.23022 | -0.10796 |
| | 2.8 | -0.59351 | -0.46928 | -0.34768 | -0.22659 | -0.10409 |
| | 3.2 | -0.59143 | -0.46656 | -0.34449 | -0.22306 | -0.10035 |
| | 3.6 | -0.58936 | -0.46390 | -0.34139 | -0.21965 | -0.09676 |
| | 4.0 | -0.58732 | -0.46131 | -0.33840 | -0.21638 | -0.09333 |
| | 6.0 | -0.57810 | -0.44982 | -0.32528 | -0.20219 | -0.07855 |
| 12.0 | 0.0 | -0.56823 | -0.43863 | -0.31331 | -0.18988 | -0.06629 |
| | 0.4 | -0.56800 | -0.43821 | -0.31273 | -0.18918 | -0.06550 |
| | 0.8 | -0.56768 | -0.43770 | -0.31207 | -0.18840 | -0.06463 |
| | 1.2 | -0.56728 | -0.43712 | -0.31134 | -0.18756 | -0.06370 |
| | 1.6 | -0.56682 | -0.43647 | -0.31056 | -0.18666 | -0.06273 |
| | 2.0 | -0.56632 | -0.43578 | -0.30973 | -0.18573 | -0.06173 |
| | 2.4 | -0.56577 | -0.43506 | -0.30887 | -0.18477 | -0.06070 |
| | 2.8 | -0.56520 | -0.43431 | -0.30798 | -0.18379 | -0.05966 |
| | 3.2 | -0.56461 | -0.43354 | -0.30709 | -0.18280 | -0.05862 |
| | 3.6 | -0.56399 | -0.43276 | -0.30618 | -0.18181 | -0.05757 |
| | 4.0 | -0.56337 | -0.43197 | -0.30527 | -0.18082 | -0.05654 |
| | 6.0 | -0.56024 | -0.42808 | -0.30083 | -0.17601 | -0.05153 |
| | 12.0 | -0.55219 | -0.41832 | -0.28988 | -0.16431 | -0.03946 |
| INFINITY | | -0.52440 | -0.38532 | -0.25335 | -0.12566 | 0.0 |

FRACTILES $W_\alpha$ RHO = 0.80

| DELTAS | | $\alpha$:0.5500 | 0.6000 | 0.6500 | 0.7000 | 0.7500 |
|---|---|---|---|---|---|---|
| 0.0 | 0.0 | -0.29250 | -0.19660 | -0.08126 | 0.05921 | 0.23377 |
| 0.4 | 0.0 | -0.28580 | -0.19062 | -0.07597 | 0.06378 | 0.23755 |
| | 0.4 | -0.29442 | -0.19636 | -0.07868 | 0.06431 | 0.24153 |
| 0.8 | 0.0 | -0.27005 | -0.17524 | -0.06075 | 0.07895 | 0.25260 |
| | 0.4 | -0.28690 | -0.18420 | -0.06167 | 0.08624 | 0.26817 |
| | 0.8 | -0.27821 | -0.16679 | -0.03586 | 0.11979 | 0.30830 |
| 1.2 | 0.0 | -0.25105 | -0.15351 | -0.03589 | 0.10705 | 0.28365 |
| | 0.4 | -0.26612 | -0.15643 | -0.02719 | 0.12671 | 0.31333 |
| | 0.8 | -0.25072 | -0.13126 | 0.00645 | 0.16721 | 0.35859 |
| | 1.2 | -0.21889 | -0.09316 | 0.04912 | 0.21251 | 0.40419 |
| 1.6 | 0.0 | -0.22942 | -0.12543 | -0.00146 | 0.14720 | 0.32826 |
| | 0.4 | -0.23503 | -0.11784 | 0.01765 | 0.17609 | 0.36493 |
| | 0.8 | -0.21378 | -0.08873 | 0.05284 | 0.21546 | 0.40627 |
| | 1.2 | -0.18287 | -0.05391 | 0.09006 | 0.25345 | 0.44314 |
| | 1.6 | -0.15185 | -0.02114 | 0.12338 | 0.28603 | 0.47344 |
| 2.0 | 0.0 | -0.20343 | -0.09123 | 0.03968 | 0.19354 | 0.37744 |
| | 0.4 | -0.19859 | -0.07562 | 0.06383 | 0.22419 | 0.41245 |
| | 0.8 | -0.17600 | -0.04789 | 0.09515 | 0.25750 | 0.44598 |
| | 1.2 | -0.14940 | -0.01898 | 0.12522 | 0.28749 | 0.47446 |
| | 1.6 | -0.12416 | 0.00729 | 0.15161 | 0.31300 | 0.49788 |
| | 2.0 | -0.10173 | 0.03016 | 0.17418 | 0.33442 | 0.51710 |
| 2.4 | 0.0 | -0.17347 | -0.05422 | 0.08171 | 0.23838 | 0.42257 |
| | 0.4 | -0.16252 | -0.03611 | 0.10512 | 0.26545 | 0.45159 |
| | 0.8 | -0.14219 | -0.01260 | 0.13066 | 0.29184 | 0.47750 |
| | 1.2 | -0.12060 | 0.01046 | 0.15432 | 0.31517 | 0.49939 |
| | 1.6 | -0.10038 | 0.03137 | 0.17522 | 0.33525 | 0.51768 |
| | 2.0 | -0.08219 | 0.04986 | 0.19341 | 0.35245 | 0.53301 |
| | 2.4 | -0.06606 | 0.06610 | 0.20921 | 0.36720 | 0.54596 |
| 2.8 | 0.0 | -0.14283 | -0.01899 | 0.11960 | 0.27703 | 0.45983 |
| | 0.4 | -0.13049 | -0.00225 | 0.13948 | 0.29891 | 0.48249 |
| | 0.8 | -0.11361 | 0.01668 | 0.15965 | 0.31944 | 0.50238 |
| | 1.2 | -0.09637 | 0.03496 | 0.17830 | 0.33772 | 0.51940 |
| | 1.6 | -0.08011 | 0.05173 | 0.19502 | 0.35374 | 0.53391 |
| | 2.0 | -0.06524 | 0.06684 | 0.20985 | 0.36772 | 0.54631 |
| | 2.4 | -0.05182 | 0.08033 | 0.22297 | 0.37993 | 0.55698 |
| | 2.8 | -0.03977 | 0.09235 | 0.23456 | 0.39063 | 0.56621 |
| 3.2 | 0.0 | -0.11475 | 0.01167 | 0.15139 | 0.30849 | 0.48932 |
| | 0.4 | -0.10349 | 0.02572 | 0.16741 | 0.32569 | 0.50679 |
| | 0.8 | -0.08986 | 0.04080 | 0.18332 | 0.34176 | 0.52224 |
| | 1.2 | -0.07601 | 0.05542 | 0.19819 | 0.35629 | 0.53569 |
| | 1.6 | -0.06279 | 0.06905 | 0.21176 | 0.36925 | 0.54738 |
| | 2.0 | -0.05050 | 0.08152 | 0.22400 | 0.38076 | 0.55756 |
| | 2.4 | -0.03924 | 0.09284 | 0.23498 | 0.39097 | 0.56644 |
| | 2.8 | -0.02899 | 0.10307 | 0.24484 | 0.40005 | 0.57424 |
| | 3.2 | -0.01969 | 0.11229 | 0.25368 | 0.40814 | 0.58112 |

FRACTILES $w_{\alpha}$ RHO = 0.80

| DELTAS | | α:0.5500 | 0.6000 | 0.6500 | 0.7000 | 0.7500 |
|---|---|---|---|---|---|---|
| 3.6 | 0.0 | -0.09061 | 0.03725 | 0.17735 | 0.33373 | 0.51253 |
| | 0.4 | -0.08107 | 0.04869 | 0.19014 | 0.34727 | 0.52613 |
| | 0.8 | -0.07004 | 0.06080 | 0.20284 | 0.36004 | 0.53833 |
| | 1.2 | -0.05880 | 0.07265 | 0.21488 | 0.37177 | 0.54915 |
| | 1.6 | -0.04791 | 0.08387 | 0.22603 | 0.38240 | 0.55871 |
| | 2.0 | -0.03764 | 0.09429 | 0.23624 | 0.39199 | 0.56715 |
| | 2.4 | -0.02811 | 0.10387 | 0.24553 | 0.40061 | 0.57463 |
| | 2.8 | -0.01932 | 0.11263 | 0.25397 | 0.40837 | 0.58128 |
| | 3.2 | -0.01125 | 0.12063 | 0.26162 | 0.41536 | 0.58721 |
| | 3.6 | -0.00387 | 0.12792 | 0.26857 | 0.42167 | 0.59251 |
| 4.0 | 0.0 | -0.07033 | 0.05839 | 0.19854 | 0.35409 | 0.53101 |
| | 0.4 | -0.06239 | 0.06771 | 0.20884 | 0.36490 | 0.54178 |
| | 0.8 | -0.05337 | 0.07757 | 0.21914 | 0.37521 | 0.55158 |
| | 1.2 | -0.04410 | 0.08732 | 0.22902 | 0.38482 | 0.56041 |
| | 1.6 | -0.03503 | 0.09666 | 0.23830 | 0.39366 | 0.56833 |
| | 2.0 | -0.02636 | 0.10545 | 0.24691 | 0.40173 | 0.57542 |
| | 2.4 | -0.01822 | 0.11363 | 0.25484 | 0.40908 | 0.58178 |
| | 2.8 | -0.01064 | 0.12119 | 0.26211 | 0.41576 | 0.58749 |
| | 3.2 | -0.00361 | 0.12816 | 0.26877 | 0.42184 | 0.59263 |
| | 3.6 | 0.00289 | 0.13457 | 0.27488 | 0.42738 | 0.59728 |
| | 4.0 | 0.00888 | 0.14047 | 0.28047 | 0.43243 | 0.60148 |
| 6.0 | 0.0 | -0.00662 | 0.12354 | 0.26276 | 0.41469 | 0.58467 |
| | 0.4 | -0.00304 | 0.12754 | 0.26703 | 0.41903 | 0.58886 |
| | 0.8 | 0.00095 | 0.13181 | 0.27142 | 0.42336 | 0.59289 |
| | 1.2 | 0.00514 | 0.13618 | 0.27581 | 0.42758 | 0.59671 |
| | 1.6 | 0.00939 | 0.14053 | 0.28011 | 0.43165 | 0.60031 |
| | 2.0 | 0.01360 | 0.14480 | 0.28428 | 0.43553 | 0.60369 |
| | 2.4 | 0.01772 | 0.14893 | 0.28827 | 0.43922 | 0.60685 |
| | 2.8 | 0.02170 | 0.15290 | 0.29208 | 0.44271 | 0.60981 |
| | 3.2 | 0.02553 | 0.15669 | 0.29570 | 0.44600 | 0.61257 |
| | 3.6 | 0.02918 | 0.16029 | 0.29913 | 0.44910 | 0.61515 |
| | 4.0 | 0.03267 | 0.16372 | 0.30237 | 0.45202 | 0.61757 |
| | 6.0 | 0.04757 | 0.17827 | 0.31605 | 0.46420 | 0.62750 |
| 12.0 | 0.0 | 0.05940 | 0.18930 | 0.32587 | 0.47237 | 0.63342 |
| | 0.4 | 0.06027 | 0.19020 | 0.32678 | 0.47324 | 0.63422 |
| | 0.8 | 0.06120 | 0.19116 | 0.32774 | 0.47415 | 0.63502 |
| | 1.2 | 0.06217 | 0.19216 | 0.32871 | 0.47507 | 0.63583 |
| | 1.6 | 0.06319 | 0.19318 | 0.32971 | 0.47599 | 0.63662 |
| | 2.0 | 0.06423 | 0.19423 | 0.33072 | 0.47692 | 0.63741 |
| | 2.4 | 0.06529 | 0.19528 | 0.33173 | 0.47785 | 0.63819 |
| | 2.8 | 0.06635 | 0.19634 | 0.33274 | 0.47876 | 0.63896 |
| | 3.2 | 0.06742 | 0.19739 | 0.33374 | 0.47967 | 0.63971 |
| | 3.6 | 0.06848 | 0.19843 | 0.33474 | 0.48056 | 0.64045 |
| | 4.0 | 0.06953 | 0.19947 | 0.33571 | 0.48144 | 0.64116 |
| | 6.0 | 0.07458 | 0.20439 | 0.34033 | 0.48554 | 0.64448 |
| | 12.0 | 0.08662 | 0.21602 | 0.35111 | 0.49497 | 0.65193 |
| INFINITY | | 0.12566 | 0.25335 | 0.38532 | 0.52440 | 0.67449 |

FRACTILES $W_\alpha$ RHO = 0.80

| DELTAS | | $\alpha$:0.8000 | 0.8500 | 0.9000 | 0.9250 | 0.9500 |
|---|---|---|---|---|---|---|
| 0.0 | 0.0 | 0.45756 | 0.75927 | 1.20397 | 1.53024 | 2.00190 |
| 0.4 | 0.0 | 0.46039 | 0.76088 | 1.20374 | 1.52861 | 1.99813 |
| | 0.4 | 0.46807 | 0.77239 | 1.21885 | 1.54496 | 2.01442 |
| 0.8 | 0.0 | 0.47500 | 0.77421 | 1.21362 | 1.53477 | 1.99723 |
| | 0.4 | 0.49871 | 0.80529 | 1.24960 | 1.57068 | 2.02865 |
| | 0.8 | 0.54342 | 0.85099 | 1.28896 | 1.60114 | 2.04165 |
| 1.2 | 0.0 | 0.50811 | 0.80722 | 1.24133 | 1.55529 | 2.00329 |
| | 0.4 | 0.54628 | 0.85117 | 1.28548 | 1.59510 | 2.03205 |
| | 0.8 | 0.59339 | 0.89570 | 1.31945 | 1.61801 | 2.03569 |
| | 1.2 | 0.63628 | 0.93145 | 1.34033 | 1.62594 | 2.02292 |
| 1.6 | 0.0 | 0.55489 | 0.85204 | 1.27584 | 1.57815 | 2.00489 |
| | 0.4 | 0.59678 | 0.89542 | 1.31415 | 1.60919 | 2.02196 |
| | 0.8 | 0.63732 | 0.93119 | 1.33827 | 1.62260 | 2.01782 |
| | 1.2 | 0.67065 | 0.95744 | 1.35125 | 1.62450 | 2.00241 |
| | 1.6 | 0.69664 | 0.97609 | 1.35716 | 1.62016 | 1.98235 |
| 2.0 | 0.0 | 0.60364 | 0.89528 | 1.30444 | 1.59281 | 1.99628 |
| | 0.4 | 0.64048 | 0.93054 | 1.33235 | 1.61299 | 2.00305 |
| | 0.8 | 0.67203 | 0.95695 | 1.34815 | 1.61958 | 1.99493 |
| | 1.2 | 0.69712 | 0.97587 | 1.35596 | 1.61827 | 1.97950 |
| | 1.6 | 0.71682 | 0.98941 | 1.35898 | 1.61287 | 1.96123 |
| | 2.0 | 0.73244 | 0.99928 | 1.35925 | 1.60556 | 1.94245 |
| 2.4 | 0.0 | 0.64581 | 0.92984 | 1.32333 | 1.59814 | 1.98006 |
| | 0.4 | 0.67481 | 0.95612 | 1.34229 | 1.61017 | 1.98057 |
| | 0.8 | 0.69857 | 0.97526 | 1.35247 | 1.61275 | 1.97113 |
| | 1.2 | 0.71752 | 0.98906 | 1.35714 | 1.60998 | 1.95687 |
| | 1.6 | 0.73270 | 0.99913 | 1.35852 | 1.60442 | 1.94073 |
| | 2.0 | 0.74499 | 1.00661 | 1.35799 | 1.59757 | 1.92432 |
| | 2.4 | 0.75509 | 1.01227 | 1.35638 | 1.59027 | 1.90844 |
| 2.8 | 0.0 | 0.67904 | 0.95523 | 1.33424 | 1.59710 | 1.96047 |
| | 0.4 | 0.70100 | 0.97439 | 1.34697 | 1.60398 | 1.95780 |
| | 0.8 | 0.71893 | 0.98841 | 1.35359 | 1.60438 | 1.94840 |
| | 1.2 | 0.73349 | 0.99871 | 1.35638 | 1.60106 | 1.93567 |
| | 1.6 | 0.74540 | 1.00637 | 1.35684 | 1.59577 | 1.92160 |
| | 2.0 | 0.75525 | 1.01217 | 1.35591 | 1.58954 | 1.90734 |
| | 2.4 | 0.76348 | 1.01663 | 1.35418 | 1.58297 | 1.89349 |
| | 2.8 | 0.77044 | 1.02012 | 1.35202 | 1.57641 | 1.88034 |
| 3.2 | 0.0 | 0.70444 | 0.97347 | 1.33991 | 1.59259 | 1.94034 |
| | 0.4 | 0.72106 | 0.98758 | 1.34857 | 1.59640 | 1.93627 |
| | 0.8 | 0.73482 | 0.99806 | 1.35295 | 1.59565 | 1.92748 |
| | 1.2 | 0.74622 | 1.00592 | 1.35459 | 1.59224 | 1.91628 |
| | 1.6 | 0.75573 | 1.01189 | 1.35452 | 1.58737 | 1.90408 |
| | 2.0 | 0.76374 | 1.01647 | 1.35342 | 1.58178 | 1.89169 |
| | 2.4 | 0.77055 | 1.02005 | 1.35170 | 1.57592 | 1.87960 |
| | 2.8 | 0.77638 | 1.02287 | 1.34964 | 1.57006 | 1.86804 |
| | 3.2 | 0.78143 | 1.02514 | 1.34742 | 1.56436 | 1.85714 |

FRACTILES $W_\alpha$ RHO = 0.80

| DELTAS | | $\alpha$:0.8000 | 0.8500 | 0.9000 | 0.9250 | 0.9500 |
|---|---|---|---|---|---|---|
| 3.6 | 0.0 | 0.72393 | 0.98671 | 1.34241 | 1.58649 | 1.92111 |
| | 0.4 | 0.73669 | 0.99729 | 1.34842 | 1.58846 | 1.91655 |
| | 0.8 | 0.74746 | 1.00531 | 1.35135 | 1.58715 | 1.90857 |
| | 1.2 | 0.75655 | 1.01144 | 1.35228 | 1.58385 | 1.89877 |
| | 1.6 | 0.76427 | 1.01617 | 1.35193 | 1.57944 | 1.88816 |
| | 2.0 | 0.77087 | 1.01985 | 1.35076 | 1.57445 | 1.87739 |
| | 2.4 | 0.77656 | 1.02276 | 1.34912 | 1.56924 | 1.86680 |
| | 2.8 | 0.78150 | 1.02509 | 1.34719 | 1.56401 | 1.85661 |
| | 3.2 | 0.78582 | 1.02697 | 1.34513 | 1.55890 | 1.84694 |
| | 3.6 | 0.78962 | 1.02850 | 1.34302 | 1.55399 | 1.83783 |
| 4.0 | 0.0 | 0.73912 | 0.99650 | 1.34304 | 1.57982 | 1.90334 |
| | 0.4 | 0.74911 | 1.00461 | 1.34729 | 1.58069 | 1.89875 |
| | 0.8 | 0.75769 | 1.01088 | 1.34928 | 1.57911 | 1.89158 |
| | 1.2 | 0.76505 | 1.01574 | 1.34975 | 1.57603 | 1.88300 |
| | 1.6 | 0.77141 | 1.01954 | 1.34924 | 1.57206 | 1.87377 |
| | 2.0 | 0.77692 | 1.02255 | 1.34808 | 1.56762 | 1.86435 |
| | 2.4 | 0.78172 | 1.02495 | 1.34653 | 1.56298 | 1.85505 |
| | 2.8 | 0.78594 | 1.02689 | 1.34476 | 1.55832 | 1.84605 |
| | 3.2 | 0.78967 | 1.02846 | 1.34286 | 1.55373 | 1.83744 |
| | 3.6 | 0.79298 | 1.02976 | 1.34093 | 1.54930 | 1.82928 |
| | 4.0 | 0.79593 | 1.03083 | 1.33899 | 1.54505 | 1.82158 |
| 6.0 | 0.0 | 0.78136 | 1.02033 | 1.33671 | 1.54953 | 1.83676 |
| | 0.4 | 0.78508 | 1.02329 | 1.33760 | 1.54882 | 1.83349 |
| | 0.8 | 0.78848 | 1.02556 | 1.33782 | 1.54735 | 1.82938 |
| | 1.2 | 0.79157 | 1.02743 | 1.33756 | 1.54536 | 1.82471 |
| | 1.6 | 0.79439 | 1.02897 | 1.33694 | 1.54301 | 1.81973 |
| | 2.0 | 0.79695 | 1.03025 | 1.33608 | 1.54045 | 1.81459 |
| | 2.4 | 0.79930 | 1.03132 | 1.33504 | 1.53776 | 1.80941 |
| | 2.8 | 0.80144 | 1.03221 | 1.33389 | 1.53502 | 1.80427 |
| | 3.2 | 0.80340 | 1.03296 | 1.33267 | 1.53228 | 1.79924 |
| | 3.6 | 0.80521 | 1.03359 | 1.33142 | 1.52956 | 1.79434 |
| | 4.0 | 0.80687 | 1.03412 | 1.33015 | 1.52690 | 1.78960 |
| | 6.0 | 0.81350 | 1.03584 | 1.32407 | 1.51484 | 1.76864 |
| 12.0 | 0.0 | 0.81642 | 1.03459 | 1.31662 | 1.50285 | 1.75012 |
| | 0.4 | 0.81706 | 1.03497 | 1.31652 | 1.50236 | 1.74902 |
| | 0.8 | 0.81769 | 1.03531 | 1.31635 | 1.50177 | 1.74780 |
| | 1.2 | 0.81830 | 1.03560 | 1.31611 | 1.50111 | 1.74647 |
| | 1.6 | 0.81889 | 1.03587 | 1.31582 | 1.50038 | 1.74507 |
| | 2.0 | 0.81946 | 1.03610 | 1.31549 | 1.49960 | 1.74361 |
| | 2.4 | 0.82002 | 1.03630 | 1.31512 | 1.49878 | 1.74211 |
| | 2.8 | 0.82055 | 1.03649 | 1.31473 | 1.49793 | 1.74057 |
| | 3.2 | 0.82107 | 1.03665 | 1.31431 | 1.49706 | 1.73902 |
| | 3.6 | 0.82157 | 1.03679 | 1.31388 | 1.49618 | 1.73747 |
| | 4.0 | 0.82205 | 1.03692 | 1.31343 | 1.49529 | 1.73591 |
| | 6.0 | 0.82421 | 1.03735 | 1.31112 | 1.49086 | 1.72833 |
| | 12.0 | 0.82880 | 1.03775 | 1.30490 | 1.47960 | 1.70960 |
| INFINITY | | 0.84162 | 1.03643 | 1.28155 | 1.43953 | 1.64485 |

FRACTILES $W_\alpha$ RHO = 0.80

| DELTAS | | $\alpha$:0.9625 | 0.9750 | 0.9825 | 0.9900 | 0.9925 |
|---|---|---|---|---|---|---|
| 0.0 | 0.0 | 2.34342 | 2.83268 | 3.26948 | 3.96470 | 4.32606 |
| 0.4 | 0.0 | 2.33800 | 2.82474 | 3.25914 | 3.95027 | 4.30936 |
| | 0.4 | 2.35298 | 2.83618 | 3.26587 | 3.94679 | 4.29937 |
| 0.8 | 0.0 | 2.33083 | 2.80699 | 3.23048 | 3.90162 | 4.24916 |
| | 0.4 | 2.35628 | 2.82060 | 3.23074 | 3.87621 | 4.20858 |
| | 0.8 | 2.35405 | 2.79367 | 3.17949 | 3.78289 | 4.09210 |
| 1.2 | 0.0 | 2.32388 | 2.77829 | 3.17973 | 3.81155 | 4.13691 |
| | 0.4 | 2.34194 | 2.77804 | 3.16078 | 3.75936 | 4.06610 |
| | 0.8 | 2.32987 | 2.74161 | 3.10120 | 3.66089 | 3.94665 |
| | 1.2 | 2.30109 | 2.68881 | 3.02612 | 3.54916 | 3.81542 |
| 1.6 | 0.0 | 2.30759 | 2.73362 | 3.10754 | 3.69234 | 3.99203 |
| | 0.4 | 2.31269 | 2.71960 | 3.07497 | 3.62808 | 3.91047 |
| | 0.8 | 2.29475 | 2.68072 | 3.01651 | 3.53719 | 3.80224 |
| | 1.2 | 2.26611 | 2.63241 | 2.95007 | 3.44106 | 3.69035 |
| | 1.6 | 2.23420 | 2.58301 | 2.88466 | 3.34960 | 3.58513 |
| 2.0 | 0.0 | 2.28047 | 2.67821 | 3.02557 | 3.56620 | 3.84221 |
| | 0.4 | 2.27634 | 2.65724 | 2.98860 | 3.50238 | 3.76390 |
| | 0.8 | 2.25683 | 2.62061 | 2.93607 | 3.42364 | 3.67120 |
| | 1.2 | 2.23067 | 2.57853 | 2.87935 | 3.34300 | 3.57788 |
| | 1.6 | 2.20271 | 2.53629 | 2.82406 | 3.26649 | 3.49015 |
| | 2.0 | 2.17534 | 2.49632 | 2.77261 | 3.19642 | 3.41028 |
| 2.4 | 0.0 | 2.24763 | 2.62052 | 2.94488 | 3.44777 | 3.70373 |
| | 0.4 | 2.23899 | 2.59790 | 2.90911 | 3.39007 | 3.63426 |
| | 0.8 | 2.22030 | 2.56536 | 2.86374 | 3.32360 | 3.55654 |
| | 1.2 | 2.19732 | 2.52945 | 2.81596 | 3.25643 | 3.47910 |
| | 1.6 | 2.17322 | 2.49363 | 2.76942 | 3.19246 | 3.40594 |
| | 2.0 | 2.14965 | 2.45956 | 2.72579 | 3.13333 | 3.33864 |
| | 2.4 | 2.12739 | 2.42795 | 2.68568 | 3.07948 | 3.27756 |
| 2.8 | 0.0 | 2.21393 | 2.56592 | 2.87109 | 3.34266 | 3.58204 |
| | 0.4 | 2.20375 | 2.54431 | 2.83877 | 3.29253 | 3.52236 |
| | 0.8 | 2.18682 | 2.51613 | 2.80018 | 3.23681 | 3.45753 |
| | 1.2 | 2.16695 | 2.48568 | 2.76000 | 3.18077 | 3.39308 |
| | 1.6 | 2.14629 | 2.45530 | 2.72075 | 3.12707 | 3.33176 |
| | 2.0 | 2.12602 | 2.42622 | 2.68363 | 3.07694 | 3.27477 |
| | 2.4 | 2.10673 | 2.39898 | 2.64917 | 3.03080 | 3.22249 |
| | 2.8 | 2.08869 | 2.37378 | 2.61749 | 2.98864 | 3.17482 |
| 3.2 | 0.0 | 2.18202 | 2.51660 | 2.80584 | 3.25149 | 3.47719 |
| | 0.4 | 2.17178 | 2.49700 | 2.77748 | 3.20857 | 3.42647 |
| | 0.8 | 2.15681 | 2.47281 | 2.74475 | 3.16181 | 3.37223 |
| | 1.2 | 2.13971 | 2.44695 | 2.71085 | 3.11477 | 3.31824 |
| | 1.6 | 2.12199 | 2.42110 | 2.67757 | 3.06940 | 3.26649 |
| | 2.0 | 2.10452 | 2.39617 | 2.64584 | 3.02666 | 3.21793 |
| | 2.4 | 2.08777 | 2.37262 | 2.61610 | 2.98691 | 3.17293 |
| | 2.8 | 2.07197 | 2.35063 | 2.58850 | 2.95025 | 3.13150 |
| | 3.2 | 2.05721 | 2.33025 | 2.56302 | 2.91655 | 3.09349 |

FRACTILES $W_\alpha$ RHO = 0.80

| DELTAS | | $\alpha$:0.9625 | 0.9750 | 0.9825 | 0.9900 | 0.9925 |
|---|---|---|---|---|---|---|
| 3.6 | 0.0 | 2.15291 | 2.47295 | 2.74890 | 3.17297 | 3.38728 |
| | 0.4 | 2.14325 | 2.45555 | 2.72427 | 3.13633 | 3.34421 |
| | 0.8 | 2.13014 | 2.43480 | 2.69644 | 3.09686 | 3.29854 |
| | 1.2 | 2.11541 | 2.41274 | 2.66766 | 3.05710 | 3.25295 |
| | 1.6 | 2.10015 | 2.39062 | 2.63926 | 3.01848 | 3.20895 |
| | 2.0 | 2.08503 | 2.36914 | 2.61198 | 2.98179 | 3.16730 |
| | 2.4 | 2.07044 | 2.34868 | 2.58619 | 2.94737 | 3.12834 |
| | 2.8 | 2.05657 | 2.32942 | 2.56204 | 2.91533 | 3.09215 |
| | 3.2 | 2.04350 | 2.31142 | 2.53957 | 2.88564 | 3.05867 |
| | 3.6 | 2.03128 | 2.29467 | 2.51872 | 2.85819 | 3.02776 |
| 4.0 | 0.0 | 2.12681 | 2.43459 | 2.69937 | 3.10529 | 3.31004 |
| | 0.4 | 2.11795 | 2.41928 | 2.67803 | 3.07393 | 3.27332 |
| | 0.8 | 2.10650 | 2.40142 | 2.65423 | 3.04038 | 3.23457 |
| | 1.2 | 2.09375 | 2.38249 | 2.62963 | 3.00650 | 3.19576 |
| | 1.6 | 2.08055 | 2.36344 | 2.60523 | 2.97339 | 3.15805 |
| | 2.0 | 2.06740 | 2.34482 | 2.58162 | 2.94168 | 3.12208 |
| | 2.4 | 2.05463 | 2.32696 | 2.55913 | 2.91170 | 3.08815 |
| | 2.8 | 2.04240 | 2.31002 | 2.53790 | 2.88357 | 3.05639 |
| | 3.2 | 2.03081 | 2.29407 | 2.51800 | 2.85729 | 3.02677 |
| | 3.6 | 2.01988 | 2.27912 | 2.49941 | 2.83283 | 2.99923 |
| | 4.0 | 2.00963 | 2.26515 | 2.48208 | 2.81008 | 2.97364 |
| 6.0 | 0.0 | 2.03309 | 2.30105 | 2.52958 | 2.87677 | 3.05060 |
| | 0.4 | 2.02783 | 2.29280 | 2.51851 | 2.86105 | 3.03239 |
| | 0.8 | 2.02168 | 2.28361 | 2.50651 | 2.84442 | 3.01328 |
| | 1.2 | 2.01499 | 2.27391 | 2.49404 | 2.82740 | 2.99386 |
| | 1.6 | 2.00801 | 2.26399 | 2.48142 | 2.81039 | 2.97452 |
| | 2.0 | 2.00094 | 2.25407 | 2.46890 | 2.79364 | 2.95554 |
| | 2.4 | 1.99389 | 2.24429 | 2.45663 | 2.77733 | 2.93710 |
| | 2.8 | 1.98697 | 2.23475 | 2.44472 | 2.76158 | 2.91932 |
| | 3.2 | 1.98023 | 2.22552 | 2.43323 | 2.74643 | 2.90226 |
| | 3.6 | 1.97371 | 2.21664 | 2.42220 | 2.73195 | 2.88595 |
| | 4.0 | 1.96743 | 2.20812 | 2.41165 | 2.71812 | 2.87041 |
| | 6.0 | 1.93990 | 2.17103 | 2.36593 | 2.65851 | 2.80352 |
| 12.0 | 0.0 | 1.91671 | 2.14122 | 2.33028 | 2.61371 | 2.75402 |
| | 0.4 | 1.91514 | 2.13894 | 2.32735 | 2.60968 | 2.74942 |
| | 0.8 | 1.91342 | 2.13650 | 2.32423 | 2.60546 | 2.74459 |
| | 1.2 | 1.91160 | 2.13393 | 2.32097 | 2.60108 | 2.73962 |
| | 1.6 | 1.90969 | 2.13127 | 2.31762 | 2.59659 | 2.73454 |
| | 2.0 | 1.90771 | 2.12854 | 2.31420 | 2.59205 | 2.72939 |
| | 2.4 | 1.90570 | 2.12577 | 2.31074 | 2.58746 | 2.72421 |
| | 2.8 | 1.90366 | 2.12297 | 2.30726 | 2.58287 | 2.71904 |
| | 3.2 | 1.90160 | 2.12018 | 2.30379 | 2.57830 | 2.71388 |
| | 3.6 | 1.89954 | 2.11738 | 2.30033 | 2.57376 | 2.70877 |
| | 4.0 | 1.89749 | 2.11461 | 2.29690 | 2.56926 | 2.70372 |
| | 6.0 | 1.88757 | 2.10127 | 2.28047 | 2.54785 | 2.67969 |
| | 12.0 | 1.86333 | 2.06905 | 2.24103 | 2.49684 | 2.62261 |
| INFINITY | | 1.78046 | 1.95996 | 2.10836 | 2.32635 | 2.43238 |

FRACTILES $w_\alpha$ RHO = 0.80

| DELTAS | | $\alpha$:0.9950 | 0.9975 | 0.9990 | 0.9995 |
|---|---|---|---|---|---|
| 0.0 | 0.0 | 4.83925 | 5.72551 | 6.91102 | 7.81620 |
| 0.4 | 0.0 | 4.81916 | 5.69913 | 6.87538 | 7.77287 |
| | 0.4 | 4.79865 | 5.65716 | 6.79904 | 7.66653 |
| 0.8 | 0.0 | 4.74131 | 5.58760 | 6.71326 | 7.56845 |
| | 0.4 | 4.67738 | 5.47912 | 6.53853 | 7.33906 |
| | 0.8 | 4.52677 | 5.26683 | 6.23960 | 6.97162 |
| 1.2 | 0.0 | 4.59584 | 5.38073 | 6.41789 | 7.20162 |
| | 0.4 | 4.49731 | 5.23147 | 6.19649 | 6.92267 |
| | 0.8 | 4.34733 | 5.02717 | 5.91714 | 6.58461 |
| | 1.2 | 4.18797 | 4.81827 | 5.64051 | 6.25541 |
| 1.6 | 0.0 | 4.41333 | 5.13061 | 6.07344 | 6.78292 |
| | 0.4 | 4.30643 | 4.97825 | 5.85770 | 6.51727 |
| | 0.8 | 4.17309 | 4.80052 | 5.61901 | 6.23110 |
| | 1.2 | 4.03854 | 4.62613 | 5.39029 | 5.96026 |
| | 1.6 | 3.91356 | 4.46656 | 5.18371 | 5.71737 |
| 2.0 | 0.0 | 4.22922 | 4.88583 | 5.74532 | 6.38991 |
| | 0.4 | 4.12982 | 4.74887 | 5.55640 | 6.16026 |
| | 0.8 | 4.01694 | 4.60040 | 5.35915 | 5.92509 |
| | 1.2 | 3.90539 | 4.45683 | 5.17196 | 5.70410 |
| | 1.6 | 3.80158 | 4.32488 | 5.00177 | 5.50439 |
| | 2.0 | 3.70766 | 4.20642 | 4.85008 | 5.32707 |
| 2.4 | 0.0 | 4.06186 | 4.66769 | 5.45791 | 6.04880 |
| | 0.4 | 3.97528 | 4.55073 | 5.29903 | 5.85713 |
| | 0.8 | 3.88135 | 4.42821 | 5.13736 | 5.66503 |
| | 1.2 | 3.78913 | 4.31007 | 4.98388 | 5.48420 |
| | 1.6 | 3.70277 | 4.20060 | 4.84305 | 5.31914 |
| | 2.0 | 3.62377 | 4.10118 | 4.71595 | 5.17071 |
| | 2.4 | 3.55236 | 4.01174 | 4.60214 | 5.03813 |
| 2.8 | 0.0 | 3.91635 | 4.48043 | 5.21384 | 5.76082 |
| | 0.4 | 3.84281 | 4.38230 | 5.08182 | 5.60230 |
| | 0.8 | 3.76483 | 4.28115 | 4.94894 | 5.44475 |
| | 1.2 | 3.68828 | 4.18337 | 4.82224 | 5.29566 |
| | 1.6 | 3.61602 | 4.09196 | 4.70483 | 5.15816 |
| | 2.0 | 3.54921 | 4.00800 | 4.59762 | 5.03303 |
| | 2.4 | 3.48814 | 3.93159 | 4.50048 | 4.91993 |
| | 2.8 | 3.43261 | 3.86236 | 4.41274 | 4.81795 |
| 3.2 | 0.0 | 3.79185 | 4.32153 | 5.00825 | 5.51916 |
| | 0.4 | 3.72981 | 4.23944 | 4.89850 | 5.38780 |
| | 0.8 | 3.66479 | 4.15541 | 4.78846 | 5.25753 |
| | 1.2 | 3.60079 | 4.07384 | 4.68294 | 5.13346 |
| | 1.6 | 3.53988 | 3.99689 | 4.58421 | 5.01790 |
| | 2.0 | 3.48301 | 3.92549 | 4.49311 | 4.91161 |
| | 2.4 | 3.43048 | 3.85982 | 4.40967 | 4.81449 |
| | 2.8 | 3.38224 | 3.79971 | 4.33354 | 4.72603 |
| | 3.2 | 3.33807 | 3.74482 | 4.26417 | 4.64553 |

FRACTILES $W_\alpha$ RHO = 0.80

| DELTAS | | $\alpha$:0.9950 | 0.9975 | 0.9990 | 0.9995 |
|---|---|---|---|---|---|
| 3.6 | 0.0 | 3.68561 | 4.18674 | 4.83472 | 5.31574 |
| | 0.4 | 3.63320 | 4.11779 | 4.74297 | 5.20617 |
| | 0.8 | 3.57860 | 4.04742 | 4.65101 | 5.09742 |
| | 1.2 | 3.52463 | 3.97875 | 4.56229 | 4.99317 |
| | 1.6 | 3.47289 | 3.91344 | 4.47856 | 4.89519 |
| | 2.0 | 3.42413 | 3.85226 | 4.40055 | 4.80419 |
| | 2.4 | 3.37867 | 3.79547 | 4.32841 | 4.72024 |
| | 2.8 | 3.33655 | 3.74302 | 4.26199 | 4.64307 |
| | 3.2 | 3.29766 | 3.69470 | 4.20095 | 4.57225 |
| | 3.6 | 3.26180 | 3.65023 | 4.14488 | 4.50727 |
| 4.0 | 0.0 | 3.59466 | 4.07183 | 4.68736 | 5.14334 |
| | 0.4 | 3.55016 | 4.01354 | 4.61004 | 5.05116 |
| | 0.8 | 3.50391 | 3.95407 | 4.53247 | 4.95950 |
| | 1.2 | 3.45803 | 3.89575 | 4.45718 | 4.87106 |
| | 1.6 | 3.41371 | 3.83986 | 4.38556 | 4.78727 |
| | 2.0 | 3.37162 | 3.78707 | 4.31826 | 4.70878 |
| | 2.4 | 3.33205 | 3.73765 | 4.25551 | 4.63575 |
| | 2.8 | 3.29509 | 3.69163 | 4.19724 | 4.56807 |
| | 3.2 | 3.26069 | 3.64891 | 4.14328 | 4.50546 |
| | 3.6 | 3.22875 | 3.60931 | 4.09336 | 4.44761 |
| | 4.0 | 3.19911 | 3.57263 | 4.04718 | 4.39414 |
| 6.0 | 0.0 | 3.29094 | 3.69083 | 4.20174 | 4.57712 |
| | 0.4 | 3.26911 | 3.66258 | 4.16462 | 4.53306 |
| | 0.8 | 3.24644 | 3.63360 | 4.12698 | 4.48868 |
| | 1.2 | 3.22354 | 3.60459 | 4.08962 | 4.44482 |
| | 1.6 | 3.20085 | 3.57603 | 4.05305 | 4.40205 |
| | 2.0 | 3.17867 | 3.54823 | 4.01763 | 4.36073 |
| | 2.4 | 3.15717 | 3.52140 | 3.98356 | 4.32108 |
| | 2.8 | 3.13649 | 3.49566 | 3.95097 | 4.28320 |
| | 3.2 | 3.11669 | 3.47107 | 3.91990 | 4.24715 |
| | 3.6 | 3.09779 | 3.44764 | 3.89036 | 4.21291 |
| | 4.0 | 3.07979 | 3.42537 | 3.86232 | 4.18044 |
| | 6.0 | 3.00251 | 3.33005 | 3.74269 | 4.04214 |
| 12.0 | 0.0 | 2.94642 | 3.26273 | 3.66066 | 3.94910 |
| | 0.4 | 2.94097 | 3.25578 | 3.65164 | 3.93844 |
| | 0.8 | 2.93529 | 3.24859 | 3.64236 | 3.92753 |
| | 1.2 | 2.92945 | 3.24122 | 3.63290 | 3.91645 |
| | 1.6 | 2.92350 | 3.23376 | 3.62336 | 3.90528 |
| | 2.0 | 2.91750 | 3.22624 | 3.61378 | 3.89410 |
| | 2.4 | 2.91147 | 3.21872 | 3.60421 | 3.88295 |
| | 2.8 | 2.90545 | 3.21122 | 3.59471 | 3.87190 |
| | 3.2 | 2.89947 | 3.20378 | 3.58530 | 3.86096 |
| | 3.6 | 2.89354 | 3.19643 | 3.57601 | 3.85018 |
| | 4.0 | 2.88769 | 3.18918 | 3.56687 | 3.83957 |
| | 6.0 | 2.85992 | 3.15489 | 3.52376 | 3.78968 |
| | 12.0 | 2.79419 | 3.07415 | 3.42282 | 3.67324 |
| INFINITY | | 2.57583 | 2.80703 | 3.09024 | 3.29051 |

FRACTILES $W_\alpha$ RHO = 0.90

| DELTAS | | α:0.0005 | 0.0010 | 0.0025 | 0.0050 | 0.0075 |
|---|---|---|---|---|---|---|
| 0.0 | 0.0 | -1.01369 | -0.96646 | -0.90487 | -0.85911 | -0.83279 |
| 0.4 | 0.0 | -1.14392 | -1.08520 | -1.00703 | -0.94746 | -0.91243 |
| | 0.4 | -0.97518 | -0.93436 | -0.88115 | -0.84165 | -0.81893 |
| 0.8 | 0.0 | -1.26936 | -1.20427 | -1.11638 | -1.04819 | -1.00747 |
| | 0.4 | -1.06825 | -1.02201 | -0.96047 | -0.91356 | -0.88597 |
| | 0.8 | -0.95856 | -0.92785 | -0.88788 | -0.85825 | -0.84126 |
| 1.2 | 0.0 | -1.35392 | -1.28660 | -1.19483 | -1.12282 | -1.07942 |
| | 0.4 | -1.17663 | -1.12709 | -1.06011 | -1.00807 | -0.97695 |
| | 0.8 | -1.05857 | -1.02367 | -0.97719 | -0.94176 | -0.92093 |
| | 1.2 | -1.01446 | -0.99118 | -0.96095 | -0.93864 | -0.92590 |
| 1.6 | 0.0 | -1.41797 | -1.35044 | -1.25773 | -1.18437 | -1.13986 |
| | 0.4 | -1.26990 | -1.21845 | -1.14819 | -1.09291 | -1.05950 |
| | 0.8 | -1.17397 | -1.13573 | -1.08394 | -1.04358 | -1.01938 |
| | 1.2 | -1.12100 | -1.09372 | -1.05737 | -1.02967 | -1.01339 |
| | 1.6 | -1.12063 | -1.10234 | -1.07870 | -1.06141 | -1.05163 |
| 2.0 | 0.0 | -1.47445 | -1.40745 | -1.31495 | -1.24124 | -1.19628 |
| | 0.4 | -1.35446 | -1.30179 | -1.22927 | -1.17164 | -1.13655 |
| | 0.8 | -1.27859 | -1.23777 | -1.18183 | -1.13759 | -1.11074 |
| | 1.2 | -1.23872 | -1.20786 | -1.16592 | -1.13308 | -1.11331 |
| | 1.6 | -1.22810 | -1.20588 | -1.17628 | -1.15374 | -1.14053 |
| | 2.0 | -1.25638 | -1.24160 | -1.22268 | -1.20911 | -1.20166 |
| 2.4 | 0.0 | -1.52917 | -1.46285 | -1.37079 | -1.29700 | -1.25175 |
| | 0.4 | -1.43377 | -1.38014 | -1.30580 | -1.24623 | -1.20971 |
| | 0.8 | -1.37523 | -1.33224 | -1.27277 | -1.22519 | -1.19602 |
| | 1.2 | -1.34697 | -1.31305 | -1.26629 | -1.22901 | -1.20620 |
| | 1.6 | -1.34369 | -1.31770 | -1.28217 | -1.25412 | -1.23708 |
| | 2.0 | -1.36135 | -1.34284 | -1.31794 | -1.29908 | -1.28818 |
| | 2.4 | -1.41003 | -1.39794 | -1.38284 | -1.37275 | -1.36825 |
| 2.8 | 0.0 | -1.58420 | -1.51842 | -1.42667 | -1.35269 | -1.30711 |
| | 0.4 | -1.50953 | -1.45501 | -1.37896 | -1.31757 | -1.27968 |
| | 0.8 | -1.46569 | -1.42074 | -1.35803 | -1.30733 | -1.27598 |
| | 1.2 | -1.44751 | -1.41080 | -1.35961 | -1.31817 | -1.29248 |
| | 1.6 | -1.45085 | -1.42142 | -1.38045 | -1.34731 | -1.32671 |
| | 2.0 | -1.47267 | -1.44990 | -1.41847 | -1.39323 | -1.37761 |
| | 2.4 | -1.51117 | -1.49501 | -1.47370 | -1.45832 | -1.45122 |
| | 2.8 | -1.57537 | -1.56570 | -1.55515 | -1.54371 | -1.52781 |
| 3.2 | 0.0 | -1.64000 | -1.57452 | -1.48276 | -1.40835 | -1.36228 |
| | 0.4 | -1.58256 | -1.52711 | -1.44929 | -1.38600 | -1.34670 |
| | 0.8 | -1.55110 | -1.50425 | -1.43838 | -1.38461 | -1.35107 |
| | 1.2 | -1.54155 | -1.50219 | -1.44670 | -1.40117 | -1.37260 |
| | 1.6 | -1.55068 | -1.51796 | -1.47170 | -1.43348 | -1.40926 |
| | 2.0 | -1.57611 | -1.54946 | -1.51171 | -1.48023 | -1.45996 |
| | 2.4 | -1.61633 | -1.59556 | -1.56625 | -1.54175 | -1.52596 |
| | 2.8 | -1.67125 | -1.65725 | -1.64114 | -1.61676 | -1.59173 |
| | 3.2 | -1.74961 | -1.74349 | -1.71857 | -1.67892 | -1.64588 |

FRACTILES $W_\alpha$ RHO = 0.90

| DELTAS | | $\alpha$:0.0005 | 0.0010 | 0.0025 | 0.0050 | 0.0075 |
|---|---|---|---|---|---|---|
| 3.6 | 0.0 | -1.69639 | -1.63094 | -1.53879 | -1.46362 | -1.41686 |
| | 0.4 | -1.65323 | -1.59673 | -1.51696 | -1.45161 | -1.41077 |
| | 0.8 | -1.63215 | -1.58339 | -1.51429 | -1.45734 | -1.42152 |
| | 1.2 | -1.62993 | -1.58793 | -1.52812 | -1.47839 | -1.44681 |
| | 1.6 | -1.64403 | -1.60803 | -1.55640 | -1.51291 | -1.48485 |
| | 2.0 | -1.67254 | -1.64198 | -1.59771 | -1.55957 | -1.53422 |
| | 2.4 | -1.71424 | -1.68882 | -1.65137 | -1.61757 | -1.59304 |
| | 2.8 | -1.76881 | -1.74870 | -1.71890 | -1.68037 | -1.64718 |
| | 3.2 | -1.83930 | -1.82637 | -1.78504 | -1.73340 | -1.69368 |
| | 3.6 | -1.92423 | -1.89832 | -1.84249 | -1.78086 | -1.73546 |
| 4.0 | 0.0 | -1.75300 | -1.68730 | -1.59435 | -1.51807 | -1.47037 |
| | 0.4 | -1.72168 | -1.66397 | -1.58201 | -1.51434 | -1.47178 |
| | 0.8 | -1.70928 | -1.65851 | -1.58602 | -1.52567 | -1.48740 |
| | 1.2 | -1.71322 | -1.66852 | -1.60422 | -1.55004 | -1.51522 |
| | 1.6 | -1.73147 | -1.69211 | -1.63487 | -1.58573 | -1.55348 |
| | 2.0 | -1.76246 | -1.72785 | -1.67664 | -1.63120 | -1.60015 |
| | 2.4 | -1.80514 | -1.77479 | -1.72839 | -1.68416 | -1.65088 |
| | 2.8 | -1.85906 | -1.83253 | -1.78831 | -1.73630 | -1.69607 |
| | 3.2 | -1.92508 | -1.90049 | -1.84422 | -1.78213 | -1.73649 |
| | 3.6 | -1.99834 | -1.96256 | -1.89446 | -1.82385 | -1.77328 |
| | 4.0 | -2.06310 | -2.01912 | -1.94048 | -1.86201 | -1.80687 |
| 6.0 | 0.0 | -2.02626 | -1.95511 | -1.85174 | -1.76406 | -1.70770 |
| | 0.4 | -2.02963 | -1.96268 | -1.86447 | -1.78011 | -1.72527 |
| | 0.8 | -2.04231 | -1.97874 | -1.88434 | -1.80191 | -1.74758 |
| | 1.2 | -2.06311 | -2.00205 | -1.90989 | -1.82779 | -1.77285 |
| | 1.6 | -2.09083 | -2.03126 | -1.93950 | -1.85589 | -1.79917 |
| | 2.0 | -2.12422 | -2.06487 | -1.97128 | -1.88437 | -1.82510 |
| | 2.4 | -2.16173 | -2.10100 | -2.00334 | -1.91210 | -1.85001 |
| | 2.8 | -2.20139 | -2.13762 | -2.03456 | -1.93866 | -1.87370 |
| | 3.2 | -2.24114 | -2.17341 | -2.06452 | -1.96391 | -1.89612 |
| | 3.6 | -2.27981 | -2.20785 | -2.09304 | -1.98780 | -1.91725 |
| | 4.0 | -2.31697 | -2.24071 | -2.12005 | -2.01031 | -1.93712 |
| | 6.0 | -2.47609 | -2.38022 | -2.23362 | -2.10434 | -2.01980 |
| 12.0 | 0.0 | -2.57015 | -2.45860 | -2.29309 | -2.15083 | -2.05922 |
| | 0.4 | -2.58216 | -2.46973 | -2.30280 | -2.15929 | -2.06687 |
| | 0.8 | -2.59473 | -2.48126 | -2.31271 | -2.16783 | -2.07456 |
| | 1.2 | -2.60768 | -2.49302 | -2.32272 | -2.17639 | -2.08223 |
| | 1.6 | -2.62085 | -2.50489 | -2.33274 | -2.18490 | -2.08982 |
| | 2.0 | -2.63410 | -2.51677 | -2.34269 | -2.19331 | -2.09731 |
| | 2.4 | -2.64733 | -2.52858 | -2.35252 | -2.20159 | -2.10466 |
| | 2.8 | -2.66046 | -2.54025 | -2.36220 | -2.20971 | -2.11185 |
| | 3.2 | -2.67343 | -2.55175 | -2.37169 | -2.21765 | -2.11887 |
| | 3.6 | -2.68618 | -2.56302 | -2.38096 | -2.22539 | -2.12571 |
| | 4.0 | -2.69868 | -2.57405 | -2.39001 | -2.23293 | -2.13235 |
| | 6.0 | -2.75675 | -2.62505 | -2.43163 | -2.26743 | -2.16270 |
| | 12.0 | -2.88666 | -2.73851 | -2.52349 | -2.34313 | -2.22904 |
| INFINITY | | -3.29051 | -3.09024 | -2.80703 | -2.57583 | -2.43238 |

FRACTILES $W_\alpha$ RHO = 0.90

| DELTAS | | $\alpha$:0.0100 | 0.0175 | 0.0250 | 0.0375 | 0.0500 |
|---|---|---|---|---|---|---|
| 0.0 | 0.0 | -0.81435 | -0.77922 | -0.75747 | -0.73356 | -0.71730 |
| 0.4 | 0.0 | -0.88749 | -0.83881 | -0.80769 | -0.77229 | -0.74723 |
| | 0.4 | -0.80304 | -0.77278 | -0.75407 | -0.73356 | -0.71966 |
| 0.8 | 0.0 | -0.97815 | -0.91990 | -0.88179 | -0.83733 | -0.80489 |
| | 0.4 | -0.86634 | -0.82801 | -0.80353 | -0.77571 | -0.75607 |
| | 0.8 | -0.82939 | -0.80688 | -0.79304 | -0.77801 | -0.76798 |
| 1.2 | 0.0 | -1.04796 | -0.98487 | -0.94310 | -0.89378 | -0.85734 |
| | 0.4 | -0.95452 | -0.90988 | -0.88061 | -0.84637 | -0.82131 |
| | 0.8 | -0.90611 | -0.87720 | -0.85876 | -0.83791 | -0.82329 |
| | 1.2 | -0.91705 | -0.90039 | -0.89031 | -0.87965 | -0.87291 |
| 1.6 | 0.0 | -1.10744 | -1.04197 | -0.99827 | -0.94623 | -0.90744 |
| | 0.4 | -1.03524 | -0.98642 | -0.95395 | -0.91541 | -0.88675 |
| | 0.8 | -1.00191 | -0.96701 | -0.94401 | -0.91696 | -0.89701 |
| | 1.2 | -1.00182 | -0.97931 | -0.96505 | -0.94916 | -0.93839 |
| | 1.6 | -1.04491 | -1.03255 | -1.02546 | -1.01900 | -1.01433 |
| 2.0 | 0.0 | -1.16338 | -1.09661 | -1.05171 | -0.99788 | -0.95746 |
| | 0.4 | -1.11091 | -1.05888 | -1.02392 | -0.98197 | -0.95040 |
| | 0.8 | -1.09117 | -1.05154 | -1.02495 | -0.99305 | -0.96901 |
| | 1.2 | -1.09897 | -1.07015 | -1.05097 | -1.02814 | -1.01103 |
| | 1.6 | -1.13117 | -1.11318 | -1.10217 | -1.09137 | -1.08314 |
| | 2.0 | -1.19671 | -1.18870 | -1.18368 | -1.16917 | -1.15043 |
| 2.4 | 0.0 | -1.21853 | -1.15074 | -1.10487 | -1.04952 | -1.00766 |
| | 0.4 | -1.18289 | -1.12808 | -1.09090 | -1.04587 | -1.01163 |
| | 0.8 | -1.17459 | -1.13074 | -1.10090 | -1.06455 | -1.03668 |
| | 1.2 | -1.18945 | -1.15515 | -1.13172 | -1.10296 | -1.08057 |
| | 1.6 | -1.22463 | -1.19926 | -1.18201 | -1.16085 | -1.14499 |
| | 2.0 | -1.28065 | -1.26797 | -1.25755 | -1.23235 | -1.20373 |
| | 2.4 | -1.36477 | -1.34669 | -1.32434 | -1.28510 | -1.24649 |
| 2.8 | 0.0 | -1.27353 | -1.20467 | -1.15777 | -1.10081 | -1.05742 |
| | 0.4 | -1.25173 | -1.19421 | -1.15486 | -1.10674 | -1.06977 |
| | 0.8 | -1.25279 | -1.20488 | -1.17184 | -1.13104 | -1.09923 |
| | 1.2 | -1.27342 | -1.23378 | -1.20613 | -1.17133 | -1.14340 |
| | 1.6 | -1.31137 | -1.27914 | -1.25617 | -1.22598 | -1.19955 |
| | 2.0 | -1.36598 | -1.34147 | -1.32260 | -1.28472 | -1.24638 |
| | 2.4 | -1.44274 | -1.41089 | -1.37824 | -1.32725 | -1.28063 |
| | 2.8 | -1.51125 | -1.46430 | -1.42264 | -1.36202 | -1.30890 |
| 3.2 | 0.0 | -1.32823 | -1.25802 | -1.20990 | -1.15105 | -1.10588 |
| | 0.4 | -1.31757 | -1.25722 | -1.21555 | -1.16413 | -1.12419 |
| | 0.8 | -1.32611 | -1.27403 | -1.23766 | -1.19211 | -1.15602 |
| | 1.2 | -1.35120 | -1.30604 | -1.27390 | -1.23252 | -1.19842 |
| | 1.6 | -1.39094 | -1.35145 | -1.32222 | -1.28203 | -1.24529 |
| | 2.0 | -1.44432 | -1.40875 | -1.37836 | -1.32785 | -1.28112 |
| | 2.4 | -1.51133 | -1.46501 | -1.42328 | -1.36249 | -1.30923 |
| | 2.8 | -1.56866 | -1.50955 | -1.46053 | -1.39185 | -1.33319 |
| | 3.2 | -1.61742 | -1.54827 | -1.49308 | -1.41754 | -1.35411 |

FRACTILES $w_\alpha$ RHO = 0.90

| DELTAS | | $\alpha$:0.0100 | 0.0175 | 0.0250 | 0.0375 | 0.0500 |
|---|---|---|---|---|---|---|
| 3.6 | 0.0 | -1.38217 | -1.31025 | -1.26063 | -1.19949 | -1.15219 |
| | 0.4 | -1.38036 | -1.31689 | -1.27267 | -1.21755 | -1.17426 |
| | 0.8 | -1.39469 | -1.33815 | -1.29815 | -1.24735 | -1.20646 |
| | 1.2 | -1.42294 | -1.37183 | -1.33475 | -1.28599 | -1.24494 |
| | 1.6 | -1.46331 | -1.41579 | -1.37949 | -1.32832 | -1.28193 |
| | 2.0 | -1.51410 | -1.46628 | -1.42510 | -1.36389 | -1.31023 |
| | 2.4 | -1.57071 | -1.51129 | -1.46189 | -1.39278 | -1.33384 |
| | 2.8 | -1.61856 | -1.54908 | -1.49369 | -1.41796 | -1.35440 |
| | 3.2 | -1.66059 | -1.58264 | -1.52195 | -1.44028 | -1.37258 |
| | 3.6 | -1.69839 | -1.61278 | -1.54728 | -1.46023 | -1.38879 |
| 4.0 | 0.0 | -1.43485 | -1.36080 | -1.30933 | -1.24543 | -1.19557 |
| | 0.4 | -1.43993 | -1.37297 | -1.32586 | -1.26654 | -1.21944 |
| | 0.8 | -1.45853 | -1.39710 | -1.35308 | -1.29640 | -1.25014 |
| | 1.2 | -1.48867 | -1.43101 | -1.38843 | -1.33153 | -1.28313 |
| | 1.6 | -1.52839 | -1.47193 | -1.42797 | -1.36608 | -1.31190 |
| | 2.0 | -1.57503 | -1.51457 | -1.46461 | -1.39467 | -1.33516 |
| | 2.4 | -1.62196 | -1.55153 | -1.49555 | -1.41922 | -1.35529 |
| | 2.8 | -1.66263 | -1.58406 | -1.52303 | -1.44102 | -1.37311 |
| | 3.2 | -1.69926 | -1.61339 | -1.54775 | -1.46055 | -1.38902 |
| | 3.6 | -1.73257 | -1.63998 | -1.57010 | -1.47816 | -1.40331 |
| | 4.0 | -1.76295 | -1.66416 | -1.59039 | -1.49410 | -1.41622 |
| 6.0 | 0.0 | -1.66486 | -1.57297 | -1.50686 | -1.42214 | -1.35410 |
| | 0.4 | -1.68323 | -1.59200 | -1.52546 | -1.43926 | -1.36946 |
| | 0.8 | -1.70554 | -1.61315 | -1.54496 | -1.45605 | -1.38389 |
| | 1.2 | -1.72992 | -1.63467 | -1.56400 | -1.47181 | -1.39715 |
| | 1.6 | -1.75460 | -1.65548 | -1.58201 | -1.48646 | -1.40935 |
| | 2.0 | -1.77853 | -1.67521 | -1.59892 | -1.50009 | -1.42060 |
| | 2.4 | -1.80135 | -1.69381 | -1.61476 | -1.51274 | -1.43098 |
| | 2.8 | -1.82296 | -1.71129 | -1.62957 | -1.52450 | -1.44058 |
| | 3.2 | -1.84334 | -1.72768 | -1.64340 | -1.53543 | -1.44947 |
| | 3.6 | -1.86251 | -1.74302 | -1.65631 | -1.54559 | -1.45771 |
| | 4.0 | -1.88051 | -1.75738 | -1.66836 | -1.55506 | -1.46536 |
| | 6.0 | -1.95517 | -1.81660 | -1.71789 | -1.59374 | -1.49649 |
| 12.0 | 0.0 | -1.98981 | -1.84250 | -1.73861 | -1.60895 | -1.50806 |
| | 0.4 | -1.99686 | -1.84833 | -1.74362 | -1.61299 | -1.51140 |
| | 0.8 | -2.00392 | -1.85412 | -1.74856 | -1.61697 | -1.51467 |
| | 1.2 | -2.01094 | -1.85983 | -1.75343 | -1.62085 | -1.51785 |
| | 1.6 | -2.01787 | -1.86545 | -1.75820 | -1.62464 | -1.52095 |
| | 2.0 | -2.02469 | -1.87096 | -1.76285 | -1.62833 | -1.52395 |
| | 2.4 | -2.03138 | -1.87633 | -1.76739 | -1.63190 | -1.52685 |
| | 2.8 | -2.03791 | -1.88157 | -1.77179 | -1.63537 | -1.52966 |
| | 3.2 | -2.04427 | -1.88665 | -1.77607 | -1.63873 | -1.53237 |
| | 3.6 | -2.05047 | -1.89159 | -1.78021 | -1.64198 | -1.53499 |
| | 4.0 | -2.05648 | -1.89638 | -1.78423 | -1.64512 | -1.53751 |
| | 6.0 | -2.08389 | -1.91812 | -1.80239 | -1.65926 | -1.54886 |
| | 12.0 | -2.14364 | -1.96521 | -1.84153 | -1.68955 | -1.57300 |
| INFINITY | | -2.32635 | -2.10836 | -1.95996 | -1.78046 | -1.64485 |

FRACTILES $w_\alpha$ RHO = 0.90

| DELTAS | | $\alpha$:0.0750 | 0.1000 | 0.1500 | 0.2000 | 0.2500 |
|---|---|---|---|---|---|---|
| 0.0 | 0.0 | -0.69581 | -0.68212 | -0.66787 | -0.65205 | -0.62787 |
| 0.4 | 0.0 | -0.71225 | -0.68800 | -0.65611 | -0.63917 | -0.61861 |
| | 0.4 | -0.70143 | -0.69003 | -0.67808 | -0.66071 | -0.63544 |
| 0.8 | 0.0 | -0.75760 | -0.72265 | -0.67067 | -0.63111 | -0.59835 |
| | 0.4 | -0.72878 | -0.71011 | -0.68686 | -0.67159 | -0.64606 |
| | 0.8 | -0.75526 | -0.74826 | -0.73458 | -0.71087 | -0.67867 |
| 1.2 | 0.0 | -0.80335 | -0.76264 | -0.70045 | -0.65122 | -0.60836 |
| | 0.4 | -0.78460 | -0.75728 | -0.71624 | -0.68461 | -0.65906 |
| | 0.8 | -0.80339 | -0.79061 | -0.77375 | -0.74524 | -0.70632 |
| | 1.2 | -0.86592 | -0.85709 | -0.82844 | -0.78786 | -0.73673 |
| 1.6 | 0.0 | -0.84935 | -0.80499 | -0.73609 | -0.68038 | -0.63087 |
| | 0.4 | -0.84390 | -0.81115 | -0.75993 | -0.71764 | -0.67834 |
| | 0.8 | -0.86749 | -0.84518 | -0.81109 | -0.77798 | -0.72968 |
| | 1.2 | -0.92607 | -0.91373 | -0.87354 | -0.81938 | -0.75463 |
| | 1.6 | -0.99745 | -0.97412 | -0.91439 | -0.84313 | -0.76414 |
| 2.0 | 0.0 | -0.89639 | -0.84923 | -0.77492 | -0.71370 | -0.65825 |
| | 0.4 | -0.90249 | -0.86519 | -0.80529 | -0.75396 | -0.70437 |
| | 0.8 | -0.93230 | -0.90329 | -0.85477 | -0.80752 | -0.74645 |
| | 1.2 | -0.98509 | -0.96551 | -0.91030 | -0.84038 | -0.76195 |
| | 1.6 | -1.05588 | -1.02137 | -0.94177 | -0.85528 | -0.76485 |
| | 2.0 | -1.10649 | -1.05881 | -0.96052 | -0.86167 | -0.76253 |
| 2.4 | 0.0 | -0.94387 | -0.89407 | -0.81445 | -0.74752 | -0.68562 |
| | 0.4 | -0.95896 | -0.91724 | -0.84859 | -0.78771 | -0.72683 |
| | 0.8 | -0.99310 | -0.95756 | -0.89508 | -0.83077 | -0.75509 |
| | 1.2 | -1.04432 | -1.01222 | -0.93777 | -0.85231 | -0.76225 |
| | 1.6 | -1.10434 | -1.05747 | -0.95959 | -0.86082 | -0.76169 |
| | 2.0 | -1.14429 | -1.08554 | -0.97260 | -0.86431 | -0.75850 |
| | 2.4 | -1.17364 | -1.10602 | -0.98153 | -0.86575 | -0.75458 |
| 2.8 | 0.0 | -0.99070 | -0.93803 | -0.85244 | -0.77891 | -0.70935 |
| | 0.4 | -1.01215 | -0.96569 | -0.88732 | -0.81544 | -0.74170 |
| | 0.8 | -1.04841 | -1.00572 | -0.92756 | -0.84582 | -0.75697 |
| | 1.2 | -1.09617 | -1.05192 | -0.95670 | -0.85826 | -0.75920 |
| | 1.6 | -1.14293 | -1.08455 | -0.97162 | -0.86324 | -0.75735 |
| | 2.0 | -1.17355 | -1.10586 | -0.98124 | -0.86538 | -0.75416 |
| | 2.4 | -1.19709 | -1.12230 | -0.98830 | -0.86626 | -0.75060 |
| | 2.8 | -1.21659 | -1.13584 | -0.99383 | -0.86649 | -0.74702 |
| 3.2 | 0.0 | -1.03575 | -0.97969 | -0.88703 | -0.80559 | -0.72694 |
| | 0.4 | -1.06106 | -1.00924 | -0.91964 | -0.83523 | -0.74835 |
| | 0.8 | -1.09710 | -1.04627 | -0.95088 | -0.85389 | -0.75508 |
| | 1.2 | -1.13878 | -1.08204 | -0.96964 | -0.86111 | -0.75508 |
| | 1.6 | -1.17323 | -1.10538 | -0.98039 | -0.86428 | -0.75292 |
| | 2.0 | -1.19726 | -1.12224 | -0.98794 | -0.86573 | -0.74997 |
| | 2.4 | -1.21673 | -1.13586 | -0.99372 | -0.86630 | -0.74678 |
| | 2.8 | -1.23328 | -1.14733 | -0.99834 | -0.86637 | -0.74358 |
| | 3.2 | -1.24765 | -1.15720 | -1.00215 | -0.86616 | -0.74050 |

FRACTILES $W_\alpha$ RHO = 0.90

| DELTAS | | α:0.0750 | 0.1000 | 0.1500 | 0.2000 | 0.2500 |
|---|---|---|---|---|---|---|
| 3.6 | 0.0 | -1.07797 | -1.01788 | -0.91681 | -0.82613 | -0.73741 |
| | 0.4 | -1.10485 | -1.04686 | -0.94456 | -0.84727 | -0.74924 |
| | 0.8 | -1.13838 | -1.07853 | -0.96647 | -0.85796 | -0.75175 |
| | 1.2 | -1.17214 | -1.10452 | -0.97902 | -0.86250 | -0.75089 |
| | 1.6 | -1.19763 | -1.12214 | -0.98725 | -0.86470 | -0.74874 |
| | 2.0 | -1.21716 | -1.13593 | -0.99337 | -0.86572 | -0.74606 |
| | 2.4 | -1.23358 | -1.14742 | -0.99818 | -0.86608 | -0.74321 |
| | 2.8 | -1.24779 | -1.15725 | -1.00209 | -0.86605 | -0.74034 |
| | 3.2 | -1.26026 | -1.16579 | -1.00534 | -0.86578 | -0.73756 |
| | 3.6 | -1.27131 | -1.17329 | -1.00808 | -0.86538 | -0.73491 |
| 4.0 | 0.0 | -1.11650 | -1.05166 | -0.94088 | -0.84018 | -0.74172 |
| | 0.4 | -1.14288 | -1.07799 | -0.96241 | -0.85373 | -0.74742 |
| | 0.8 | -1.17209 | -1.10315 | -0.97714 | -0.86009 | -0.74812 |
| | 1.2 | -1.19810 | -1.12204 | -0.98626 | -0.86319 | -0.74693 |
| | 1.6 | -1.21791 | -1.13609 | -0.99281 | -0.86477 | -0.74488 |
| | 2.0 | -1.23420 | -1.14761 | -0.99787 | -0.86549 | -0.74245 |
| | 2.4 | -1.24823 | -1.15741 | -1.00192 | -0.86570 | -0.73989 |
| | 2.8 | -1.26053 | -1.16590 | -1.00526 | -0.86560 | -0.73732 |
| | 3.2 | -1.27143 | -1.17334 | -1.00805 | -0.86530 | -0.73481 |
| | 3.6 | -1.28116 | -1.17994 | -1.01043 | -0.86489 | -0.73241 |
| | 4.0 | -1.28990 | -1.18582 | -1.01247 | -0.86441 | -0.73013 |
| 6.0 | 0.0 | -1.24338 | -1.15117 | -0.99503 | -0.85910 | -0.73393 |
| | 0.4 | -1.25540 | -1.16038 | -1.00000 | -0.86104 | -0.73357 |
| | 0.8 | -1.26610 | -1.16831 | -1.00396 | -0.86221 | -0.73265 |
| | 1.2 | -1.27571 | -1.17527 | -1.00718 | -0.86286 | -0.73139 |
| | 1.6 | -1.28439 | -1.18143 | -1.00986 | -0.86315 | -0.72993 |
| | 2.0 | -1.29227 | -1.18695 | -1.01211 | -0.86319 | -0.72837 |
| | 2.4 | -1.29947 | -1.19192 | -1.01403 | -0.86307 | -0.72677 |
| | 2.8 | -1.30606 | -1.19642 | -1.01569 | -0.86283 | -0.72517 |
| | 3.2 | -1.31212 | -1.20051 | -1.01713 | -0.86252 | -0.72359 |
| | 3.6 | -1.31770 | -1.20426 | -1.01841 | -0.86215 | -0.72206 |
| | 4.0 | -1.32285 | -1.20770 | -1.01953 | -0.86176 | -0.72058 |
| | 6.0 | -1.34361 | -1.22135 | -1.02368 | -0.85966 | -0.71406 |
| 12.0 | 0.0 | -1.35042 | -1.22508 | -1.02356 | -0.85721 | -0.71010 |
| | 0.4 | -1.35277 | -1.22672 | -1.02420 | -0.85718 | -0.70956 |
| | 0.8 | -1.35504 | -1.22828 | -1.02479 | -0.85710 | -0.70900 |
| | 1.2 | -1.35723 | -1.22978 | -1.02533 | -0.85699 | -0.70843 |
| | 1.6 | -1.35935 | -1.23121 | -1.02583 | -0.85686 | -0.70784 |
| | 2.0 | -1.36139 | -1.23258 | -1.02628 | -0.85671 | -0.70724 |
| | 2.4 | -1.36335 | -1.23389 | -1.02670 | -0.85654 | -0.70665 |
| | 2.8 | -1.36524 | -1.23514 | -1.02709 | -0.85635 | -0.70606 |
| | 3.2 | -1.36705 | -1.23634 | -1.02746 | -0.85616 | -0.70547 |
| | 3.6 | -1.36880 | -1.23749 | -1.02779 | -0.85596 | -0.70489 |
| | 4.0 | -1.37048 | -1.23859 | -1.02811 | -0.85576 | -0.70432 |
| | 6.0 | -1.37797 | -1.24343 | -1.02941 | -0.85473 | -0.70165 |
| | 12.0 | -1.39369 | -1.25340 | -1.03172 | -0.85206 | -0.69548 |
| INFINITY | | -1.43953 | -1.28155 | -1.03643 | -0.84162 | -0.67449 |

FRACTILES $W_{\alpha}$ RHO = 0.90

| DELTAS | | $\alpha$:0.3000 | 0.3500 | 0.4000 | 0.4500 | 0.5000 |
|---|---|---|---|---|---|---|
| 0.0 | 0.0 | -0.59611 | -0.55651 | -0.50840 | -0.45078 | -0.38224 |
| 0.4 | 0.0 | -0.58872 | -0.55033 | -0.50310 | -0.44613 | -0.37811 |
| | 0.4 | -0.60264 | -0.56193 | -0.51263 | -0.45370 | -0.38371 |
| 0.8 | 0.0 | -0.57178 | -0.53805 | -0.49281 | -0.43681 | -0.36914 |
| | 0.4 | -0.61187 | -0.56905 | -0.51704 | -0.45491 | -0.38128 |
| | 0.8 | -0.63797 | -0.58839 | -0.52926 | -0.45966 | -0.37830 |
| 1.2 | 0.0 | -0.56839 | -0.52864 | -0.48563 | -0.42926 | -0.35960 |
| | 0.4 | -0.62332 | -0.57659 | -0.51942 | -0.45129 | -0.37111 |
| | 0.8 | -0.65765 | -0.59929 | -0.53093 | -0.45197 | -0.36142 |
| | 1.2 | -0.67584 | -0.60560 | -0.52607 | -0.43689 | -0.33728 |
| 1.6 | 0.0 | -0.58380 | -0.53621 | -0.48470 | -0.42414 | -0.34877 |
| | 0.4 | -0.63726 | -0.58367 | -0.51813 | -0.44116 | -0.35215 |
| | 0.8 | -0.67034 | -0.60108 | -0.52222 | -0.43355 | -0.33436 |
| | 1.2 | -0.68107 | -0.59960 | -0.51040 | -0.41315 | -0.30702 |
| | 1.6 | -0.67904 | -0.58831 | -0.49178 | -0.38881 | -0.27836 |
| 2.0 | 0.0 | -0.60449 | -0.54910 | -0.48838 | -0.41769 | -0.33304 |
| | 0.4 | -0.65085 | -0.58664 | -0.51035 | -0.42342 | -0.32556 |
| | 0.8 | -0.67461 | -0.59413 | -0.50561 | -0.40888 | -0.30319 |
| | 1.2 | -0.67715 | -0.58661 | -0.49022 | -0.38737 | -0.27704 |
| | 1.6 | -0.67109 | -0.57375 | -0.47217 | -0.36539 | -0.25220 |
| | 2.0 | -0.66234 | -0.56013 | -0.45486 | -0.34535 | -0.23022 |
| 2.4 | 0.0 | -0.62428 | -0.55973 | -0.48808 | -0.40563 | -0.31068 |
| | 0.4 | -0.65991 | -0.58273 | -0.49590 | -0.40033 | -0.29558 |
| | 0.8 | -0.67139 | -0.58148 | -0.48555 | -0.38308 | -0.27311 |
| | 1.2 | -0.66868 | -0.57147 | -0.47001 | -0.36336 | -0.25031 |
| | 1.6 | -0.66151 | -0.55932 | -0.45408 | -0.34460 | -0.22951 |
| | 2.0 | -0.65337 | -0.54742 | -0.43932 | -0.32772 | -0.21116 |
| | 2.4 | -0.64545 | -0.53647 | -0.42607 | -0.31280 | -0.19510 |
| 2.8 | 0.0 | -0.63891 | -0.56366 | -0.48036 | -0.38730 | -0.28420 |
| | 0.4 | -0.66117 | -0.57284 | -0.47778 | -0.37599 | -0.26664 |
| | 0.8 | -0.66387 | -0.56695 | -0.46573 | -0.35933 | -0.24656 |
| | 1.2 | -0.65905 | -0.55691 | -0.45174 | -0.34236 | -0.22740 |
| | 1.6 | -0.65218 | -0.54623 | -0.43814 | -0.32658 | -0.21008 |
| | 2.0 | -0.64500 | -0.53601 | -0.42562 | -0.31235 | -0.19468 |
| | 2.4 | -0.63811 | -0.52659 | -0.41430 | -0.29964 | -0.18104 |
| | 2.8 | -0.63169 | -0.51802 | -0.40412 | -0.28831 | -0.16896 |
| 3.2 | 0.0 | -0.64615 | -0.55997 | -0.46676 | -0.36615 | -0.25772 |
| | 0.4 | -0.65656 | -0.56022 | -0.45940 | -0.35339 | -0.24103 |
| | 0.8 | -0.65501 | -0.55295 | -0.44789 | -0.33867 | -0.22393 |
| | 1.2 | -0.64984 | -0.54386 | -0.43581 | -0.32432 | -0.20794 |
| | 1.6 | -0.64368 | -0.53465 | -0.42426 | -0.31103 | -0.19342 |
| | 2.0 | -0.63743 | -0.52588 | -0.41358 | -0.29894 | -0.18037 |
| | 2.4 | -0.63142 | -0.51774 | -0.40383 | -0.28803 | -0.16869 |
| | 2.8 | -0.62578 | -0.51026 | -0.39498 | -0.27820 | -0.15823 |
| | 3.2 | -0.62056 | -0.50343 | -0.38696 | -0.26934 | -0.14885 |

FRACTILES $W_\alpha$ RHO = 0.90

| DELTAS | | α:0.3000 | 0.3500 | 0.4000 | 0.4500 | 0.5000 |
|---|---|---|---|---|---|---|
| 3.6 | 0.0 | -0.64641 | -0.55119 | -0.45109 | -0.34563 | -0.23381 |
| | 0.4 | -0.64943 | -0.54751 | -0.44262 | -0.33362 | -0.21916 |
| | 0.8 | -0.64639 | -0.54039 | -0.43237 | -0.32099 | -0.20477 |
| | 1.2 | -0.64151 | -0.53242 | -0.42203 | -0.30885 | -0.19133 |
| | 1.6 | -0.63608 | -0.52448 | -0.41217 | -0.29755 | -0.17903 |
| | 2.0 | -0.63062 | -0.51690 | -0.40298 | -0.28718 | -0.16788 |
| | 2.4 | -0.62536 | -0.50981 | -0.39452 | -0.27774 | -0.15779 |
| | 2.8 | -0.62038 | -0.50324 | -0.38677 | -0.26915 | -0.14867 |
| | 3.2 | -0.61574 | -0.49720 | -0.37969 | -0.26135 | -0.14042 |
| | 3.6 | -0.61142 | -0.49164 | -0.37323 | -0.25426 | -0.13294 |
| 4.0 | 0.0 | -0.64230 | -0.54067 | -0.43601 | -0.32729 | -0.21319 |
| | 0.4 | -0.64192 | -0.53587 | -0.42790 | -0.31664 | -0.20063 |
| | 0.8 | -0.63854 | -0.52936 | -0.41896 | -0.30585 | -0.18846 |
| | 1.2 | -0.63409 | -0.52241 | -0.41007 | -0.29549 | -0.17705 |
| | 1.6 | -0.62931 | -0.51551 | -0.40157 | -0.28579 | -0.16653 |
| | 2.0 | -0.62451 | -0.50890 | -0.39359 | -0.27682 | -0.15690 |
| | 2.4 | -0.61986 | -0.50269 | -0.38620 | -0.26858 | -0.14811 |
| | 2.8 | -0.61545 | -0.49689 | -0.37938 | -0.26103 | -0.14011 |
| | 3.2 | -0.61130 | -0.49152 | -0.37310 | -0.25412 | -0.13281 |
| | 3.6 | -0.60743 | -0.48654 | -0.36732 | -0.24779 | -0.12614 |
| | 4.0 | -0.60382 | -0.48195 | -0.36201 | -0.24199 | -0.12005 |
| 6.0 | 0.0 | -0.61469 | -0.49836 | -0.38273 | -0.26598 | -0.14637 |
| | 0.4 | -0.61253 | -0.49477 | -0.37802 | -0.26039 | -0.14014 |
| | 0.8 | -0.60999 | -0.49097 | -0.37324 | -0.25488 | -0.13410 |
| | 1.2 | -0.60727 | -0.48713 | -0.36853 | -0.24953 | -0.12832 |
| | 1.6 | -0.60449 | -0.48333 | -0.36398 | -0.24442 | -0.12284 |
| | 2.0 | -0.60173 | -0.47965 | -0.35961 | -0.23957 | -0.11767 |
| | 2.4 | -0.59902 | -0.47611 | -0.35546 | -0.23498 | -0.11281 |
| | 2.8 | -0.59641 | -0.47274 | -0.35152 | -0.23066 | -0.10826 |
| | 3.2 | -0.59390 | -0.46954 | -0.34781 | -0.22659 | -0.10399 |
| | 3.6 | -0.59150 | -0.46650 | -0.34431 | -0.22278 | -0.09998 |
| | 4.0 | -0.58922 | -0.46363 | -0.34101 | -0.21919 | -0.09624 |
| | 6.0 | -0.57947 | -0.45151 | -0.32720 | -0.20425 | -0.08069 |
| 12.0 | 0.0 | -0.57449 | -0.44587 | -0.32118 | -0.19807 | -0.07453 |
| | 0.4 | -0.57358 | -0.44467 | -0.31976 | -0.19649 | -0.07286 |
| | 0.8 | -0.57266 | -0.44347 | -0.31835 | -0.19494 | -0.07122 |
| | 1.2 | -0.57173 | -0.44228 | -0.31696 | -0.19341 | -0.06961 |
| | 1.6 | -0.57080 | -0.44110 | -0.31559 | -0.19192 | -0.06803 |
| | 2.0 | -0.56988 | -0.43993 | -0.31425 | -0.19045 | -0.06650 |
| | 2.4 | -0.56897 | -0.43879 | -0.31294 | -0.18903 | -0.06501 |
| | 2.8 | -0.56808 | -0.43767 | -0.31166 | -0.18764 | -0.06356 |
| | 3.2 | -0.56720 | -0.43658 | -0.31042 | -0.18629 | -0.06215 |
| | 3.6 | -0.56634 | -0.43552 | -0.30920 | -0.18498 | -0.06079 |
| | 4.0 | -0.56550 | -0.43448 | -0.30803 | -0.18371 | -0.05948 |
| | 6.0 | -0.56163 | -0.42973 | -0.30266 | -0.17795 | -0.05350 |
| | 12.0 | -0.55293 | -0.41921 | -0.29087 | -0.16536 | -0.04054 |
| INFINITY | | -0.52440 | -0.38532 | -0.25335 | -0.12566 | 0.0 |

FRACTILES $W_\alpha$ RHO = 0.90

| DELTAS | | α:0.5500 | 0.6000 | 0.6500 | 0.7000 | 0.7500 |
|---|---|---|---|---|---|---|
| 0.0 | 0.0 | -0.30084 | -0.20387 | -0.08746 | 0.05412 | 0.22988 |
| 0.4 | 0.0 | -0.29713 | -0.20053 | -0.08446 | 0.05679 | 0.23221 |
| | 0.4 | -0.30072 | -0.20200 | -0.08370 | 0.05993 | 0.23786 |
| 0.8 | 0.0 | -0.28810 | -0.19114 | -0.07451 | 0.06744 | 0.24359 |
| | 0.4 | -0.29425 | -0.19118 | -0.06830 | 0.08001 | 0.26248 |
| | 0.8 | -0.28338 | -0.17244 | -0.04193 | 0.11344 | 0.30190 |
| 1.2 | 0.0 | -0.27573 | -0.17541 | -0.05507 | 0.09071 | 0.27052 |
| | 0.4 | -0.27720 | -0.16717 | -0.03752 | 0.11697 | 0.30450 |
| | 0.8 | -0.25779 | -0.13892 | -0.00157 | 0.15909 | 0.35075 |
| | 1.2 | -0.22591 | -0.10076 | 0.04121 | 0.20461 | 0.39668 |
| 1.6 | 0.0 | -0.25837 | -0.15128 | -0.02432 | 0.12753 | 0.31228 |
| | 0.4 | -0.24983 | -0.13212 | 0.00410 | 0.16361 | 0.35403 |
| | 0.8 | -0.22336 | -0.09855 | 0.04307 | 0.20609 | 0.39775 |
| | 1.2 | -0.19063 | -0.06195 | 0.08199 | 0.24567 | 0.43601 |
| | 1.6 | -0.15891 | -0.02838 | 0.11620 | 0.27917 | 0.46722 |
| 2.0 | 0.0 | -0.23372 | -0.11852 | 0.01538 | 0.17259 | 0.36054 |
| | 0.4 | -0.21570 | -0.09195 | 0.04861 | 0.21052 | 0.40093 |
| | 0.8 | -0.18722 | -0.05897 | 0.08452 | 0.24769 | 0.43746 |
| | 1.2 | -0.15772 | -0.02733 | 0.11710 | 0.27989 | 0.46774 |
| | 1.6 | -0.13098 | 0.00041 | 0.14488 | 0.30668 | 0.49226 |
| | 2.0 | -0.10780 | 0.02405 | 0.16822 | 0.32884 | 0.51217 |
| 2.4 | 0.0 | -0.20294 | -0.08104 | 0.05775 | 0.21781 | 0.40618 |
| | 0.4 | -0.18047 | -0.05307 | 0.08953 | 0.25171 | 0.44033 |
| | 0.8 | -0.15417 | -0.02418 | 0.11979 | 0.28206 | 0.46928 |
| | 1.2 | -0.12924 | 0.00195 | 0.14621 | 0.30775 | 0.49302 |
| | 1.6 | -0.10715 | 0.02463 | 0.16872 | 0.32924 | 0.51246 |
| | 2.0 | -0.08793 | 0.04413 | 0.18786 | 0.34729 | 0.52852 |
| | 2.4 | -0.07125 | 0.06092 | 0.20421 | 0.36257 | 0.54194 |
| 2.8 | 0.0 | -0.17045 | -0.04434 | 0.09693 | 0.25765 | 0.44460 |
| | 0.4 | -0.14832 | -0.01900 | 0.12422 | 0.28562 | 0.47183 |
| | 0.8 | -0.12581 | 0.00502 | 0.14884 | 0.30987 | 0.49453 |
| | 1.2 | -0.10520 | 0.02638 | 0.17023 | 0.33046 | 0.51332 |
| | 1.6 | -0.08692 | 0.04503 | 0.18864 | 0.34792 | 0.52897 |
| | 2.0 | -0.07085 | 0.06128 | 0.20452 | 0.36282 | 0.54211 |
| | 2.4 | -0.05673 | 0.07545 | 0.21828 | 0.37562 | 0.55328 |
| | 2.8 | -0.04428 | 0.08788 | 0.23027 | 0.38670 | 0.56284 |
| 3.2 | 0.0 | -0.14028 | -0.01189 | 0.13030 | 0.29054 | 0.47536 |
| | 0.4 | -0.12075 | 0.00954 | 0.15274 | 0.31302 | 0.49678 |
| | 0.8 | -0.10200 | 0.02926 | 0.17271 | 0.33247 | 0.51475 |
| | 1.2 | -0.08493 | 0.04683 | 0.19019 | 0.34918 | 0.52986 |
| | 1.6 | -0.06967 | 0.06234 | 0.20544 | 0.36357 | 0.54265 |
| | 2.0 | -0.05610 | 0.07603 | 0.21877 | 0.37602 | 0.55356 |
| | 2.4 | -0.04402 | 0.08812 | 0.23047 | 0.38686 | 0.56296 |
| | 2.8 | -0.03327 | 0.09883 | 0.24079 | 0.39636 | 0.57111 |
| | 3.2 | -0.02366 | 0.10837 | 0.24993 | 0.40472 | 0.57823 |

FRACTILES $W_\alpha$ RHO = 0.90

| DELTAS | | $\alpha$:0.5500 | 0.6000 | 0.6500 | 0.7000 | 0.7500 |
|---|---|---|---|---|---|---|
| 3.6 | 0.0 | -0.11414 | 0.01545 | 0.15783 | 0.31715 | 0.49974 |
| | 0.4 | -0.09759 | 0.03322 | 0.17614 | 0.33525 | 0.51674 |
| | 0.8 | -0.08199 | 0.04948 | 0.19250 | 0.35105 | 0.53119 |
| | 1.2 | -0.06773 | 0.06410 | 0.20697 | 0.36481 | 0.54353 |
| | 1.6 | -0.05484 | 0.07716 | 0.21976 | 0.37683 | 0.55413 |
| | 2.0 | -0.04326 | 0.08881 | 0.23108 | 0.38736 | 0.56331 |
| | 2.4 | -0.03285 | 0.09922 | 0.24112 | 0.39663 | 0.57130 |
| | 2.8 | -0.02348 | 0.10853 | 0.25007 | 0.40484 | 0.57831 |
| | 3.2 | -0.01504 | 0.11689 | 0.25807 | 0.41214 | 0.58450 |
| | 3.6 | -0.00742 | 0.12442 | 0.26525 | 0.41866 | 0.59000 |
| 4.0 | 0.0 | -0.09206 | 0.03820 | 0.18046 | 0.33876 | 0.51925 |
| | 0.4 | -0.07814 | 0.05297 | 0.19552 | 0.35351 | 0.53294 |
| | 0.8 | -0.06505 | 0.06654 | 0.20909 | 0.36653 | 0.54475 |
| | 1.2 | -0.05299 | 0.07885 | 0.22123 | 0.37802 | 0.55498 |
| | 1.6 | -0.04199 | 0.08996 | 0.23209 | 0.38817 | 0.56389 |
| | 2.0 | -0.03201 | 0.09998 | 0.24179 | 0.39717 | 0.57169 |
| | 2.4 | -0.02296 | 0.10901 | 0.25049 | 0.40518 | 0.57856 |
| | 2.8 | -0.01475 | 0.11716 | 0.25830 | 0.41233 | 0.58464 |
| | 3.2 | -0.00729 | 0.12454 | 0.26535 | 0.41875 | 0.59006 |
| | 3.6 | -0.00051 | 0.13123 | 0.27172 | 0.42452 | 0.59490 |
| | 4.0 | 0.00568 | 0.13732 | 0.27750 | 0.42975 | 0.59926 |
| 6.0 | 0.0 | -0.02205 | 0.10910 | 0.24979 | 0.40372 | 0.57638 |
| | 0.4 | -0.01541 | 0.11593 | 0.25656 | 0.41015 | 0.58212 |
| | 0.8 | -0.00906 | 0.12239 | 0.26289 | 0.41609 | 0.58733 |
| | 1.2 | -0.00304 | 0.12845 | 0.26879 | 0.42157 | 0.59207 |
| | 1.6 | 0.00263 | 0.13412 | 0.27426 | 0.42661 | 0.59640 |
| | 2.0 | 0.00795 | 0.13941 | 0.27934 | 0.43126 | 0.60034 |
| | 2.4 | 0.01292 | 0.14434 | 0.28404 | 0.43554 | 0.60395 |
| | 2.8 | 0.01757 | 0.14892 | 0.28841 | 0.43950 | 0.60726 |
| | 3.2 | 0.02191 | 0.15320 | 0.29246 | 0.44316 | 0.61030 |
| | 3.6 | 0.02597 | 0.15718 | 0.29623 | 0.44655 | 0.61310 |
| | 4.0 | 0.02976 | 0.16090 | 0.29974 | 0.44969 | 0.61569 |
| | 6.0 | 0.04543 | 0.17618 | 0.31409 | 0.46248 | 0.62610 |
| 12.0 | 0.0 | 0.05136 | 0.18172 | 0.31902 | 0.46657 | 0.62908 |
| | 0.4 | 0.05308 | 0.18342 | 0.32066 | 0.46806 | 0.63033 |
| | 0.8 | 0.05476 | 0.18508 | 0.32224 | 0.46949 | 0.63152 |
| | 1.2 | 0.05640 | 0.18670 | 0.32378 | 0.47088 | 0.63267 |
| | 1.6 | 0.05800 | 0.18828 | 0.32527 | 0.47222 | 0.63378 |
| | 2.0 | 0.05956 | 0.18980 | 0.32671 | 0.47351 | 0.63484 |
| | 2.4 | 0.06107 | 0.19128 | 0.32810 | 0.47475 | 0.63585 |
| | 2.8 | 0.06253 | 0.19271 | 0.32945 | 0.47595 | 0.63683 |
| | 3.2 | 0.06395 | 0.19409 | 0.33074 | 0.47710 | 0.63776 |
| | 3.6 | 0.06532 | 0.19543 | 0.33199 | 0.47821 | 0.63866 |
| | 4.0 | 0.06664 | 0.19672 | 0.33320 | 0.47928 | 0.63952 |
| | 6.0 | 0.07263 | 0.20252 | 0.33862 | 0.48406 | 0.64333 |
| | 12.0 | 0.08555 | 0.21499 | 0.35016 | 0.49414 | 0.65128 |
| INFINITY | | 0.12566 | 0.25335 | 0.38532 | 0.52440 | 0.67449 |

FRACTILES $w_\alpha$ RHO = 0.90

| DELTAS | | $\alpha$:0.8000 | 0.8500 | 0.9000 | 0.9250 | 0.9500 |
|---|---|---|---|---|---|---|
| 0.0 | 0.0 | 0.45507 | 0.75851 | 1.20561 | 1.53356 | 2.00761 |
| 0.4 | 0.0 | 0.45699 | 0.75989 | 1.20612 | 1.53338 | 2.00630 |
| | 0.4 | 0.46525 | 0.77070 | 1.21889 | 1.54633 | 2.01784 |
| 0.8 | 0.0 | 0.46897 | 0.77196 | 1.21680 | 1.54191 | 2.01014 |
| | 0.4 | 0.49381 | 0.80167 | 1.24828 | 1.57133 | 2.03246 |
| | 0.8 | 0.53734 | 0.84584 | 1.28591 | 1.60001 | 2.04370 |
| 1.2 | 0.0 | 0.49886 | 0.80310 | 1.24481 | 1.56444 | 2.02081 |
| | 0.4 | 0.53889 | 0.84610 | 1.28443 | 1.59732 | 2.03932 |
| | 0.8 | 0.58635 | 0.89024 | 1.31700 | 1.61807 | 2.03969 |
| | 1.2 | 0.62966 | 0.92647 | 1.33830 | 1.62631 | 2.02701 |
| 1.6 | 0.0 | 0.54354 | 0.84696 | 1.28017 | 1.58950 | 2.02655 |
| | 0.4 | 0.58819 | 0.89032 | 1.31466 | 1.61406 | 2.03333 |
| | 0.8 | 0.63023 | 0.92642 | 1.33739 | 1.62479 | 2.02465 |
| | 1.2 | 0.66466 | 0.95331 | 1.35027 | 1.62603 | 2.00773 |
| | 1.6 | 0.69150 | 0.97267 | 1.35664 | 1.62192 | 1.98757 |
| 2.0 | 0.0 | 0.59190 | 0.89057 | 1.31020 | 1.60632 | 2.02102 |
| | 0.4 | 0.63195 | 0.92629 | 1.33472 | 1.62034 | 2.01771 |
| | 0.8 | 0.66541 | 0.95317 | 1.34889 | 1.62377 | 2.00425 |
| | 1.2 | 0.69176 | 0.97260 | 1.35610 | 1.62106 | 1.98625 |
| | 1.6 | 0.71232 | 0.98665 | 1.35908 | 1.61520 | 1.96693 |
| | 2.0 | 0.72853 | 0.99697 | 1.35957 | 1.60793 | 1.94790 |
| 2.4 | 0.0 | 0.63481 | 0.92616 | 1.33046 | 1.61319 | 2.00652 |
| | 0.4 | 0.66691 | 0.95292 | 1.34619 | 1.61935 | 1.99742 |
| | 0.8 | 0.69254 | 0.97240 | 1.35451 | 1.61850 | 1.98232 |
| | 1.2 | 0.71269 | 0.98653 | 1.35824 | 1.61387 | 1.96490 |
| | 1.6 | 0.72867 | 0.99692 | 1.35924 | 1.60740 | 1.94710 |
| | 2.0 | 0.74152 | 1.00470 | 1.35863 | 1.60019 | 1.92991 |
| | 2.4 | 0.75202 | 1.01065 | 1.35711 | 1.59283 | 1.91375 |
| 2.8 | 0.0 | 0.66916 | 0.95261 | 1.34231 | 1.61296 | 1.98752 |
| | 0.4 | 0.69384 | 0.97209 | 1.35194 | 1.61433 | 1.97592 |
| | 0.8 | 0.71345 | 0.98629 | 1.35661 | 1.61125 | 1.96089 |
| | 1.2 | 0.72909 | 0.99676 | 1.35826 | 1.60584 | 1.94473 |
| | 1.6 | 0.74174 | 1.00461 | 1.35811 | 1.59936 | 1.92864 |
| | 2.0 | 0.75210 | 1.01061 | 1.35690 | 1.59249 | 1.91324 |
| | 2.4 | 0.76071 | 1.01526 | 1.35508 | 1.58563 | 1.89878 |
| | 2.8 | 0.76794 | 1.01893 | 1.35295 | 1.57899 | 1.88537 |
| 3.2 | 0.0 | 0.69566 | 0.97172 | 1.34849 | 1.60869 | 1.96724 |
| | 0.4 | 0.71457 | 0.98596 | 1.35423 | 1.60741 | 1.95501 |
| | 0.8 | 0.72980 | 0.99651 | 1.35666 | 1.60329 | 1.94083 |
| | 1.2 | 0.74217 | 1.00444 | 1.35707 | 1.59771 | 1.92613 |
| | 1.6 | 0.75236 | 1.01050 | 1.35626 | 1.59149 | 1.91171 |
| | 2.0 | 0.76084 | 1.01520 | 1.35474 | 1.58508 | 1.89795 |
| | 2.4 | 0.76800 | 1.01890 | 1.35281 | 1.57876 | 1.88503 |
| | 2.8 | 0.77409 | 1.02186 | 1.35067 | 1.57267 | 1.87300 |
| | 3.2 | 0.77934 | 1.02424 | 1.34844 | 1.56687 | 1.86186 |

FRACTILES $W_\alpha$ RHO = 0.90

| DELTAS | | $\alpha$:0.8000 | 0.8500 | 0.9000 | 0.9250 | 0.9500 |
|---|---|---|---|---|---|---|
| 3.6 | 0.0 | 0.71608 | 0.98558 | 1.35118 | 1.60247 | 1.94742 |
| | 0.4 | 0.73078 | 0.99619 | 1.35449 | 1.59980 | 1.93549 |
| | 0.8 | 0.74282 | 1.00419 | 1.35555 | 1.59528 | 1.92243 |
| | 1.2 | 0.75278 | 1.01032 | 1.35522 | 1.58983 | 1.90919 |
| | 1.6 | 0.76112 | 1.01508 | 1.35405 | 1.58399 | 1.89629 |
| | 2.0 | 0.76817 | 1.01883 | 1.35238 | 1.57808 | 1.88399 |
| | 2.4 | 0.77419 | 1.02181 | 1.35043 | 1.57229 | 1.87242 |
| | 2.8 | 0.77938 | 1.02422 | 1.34834 | 1.56671 | 1.86162 |
| | 3.2 | 0.78388 | 1.02619 | 1.34621 | 1.56141 | 1.85157 |
| | 3.6 | 0.78783 | 1.02781 | 1.34409 | 1.55640 | 1.84225 |
| 4.0 | 0.0 | 0.73205 | 0.99583 | 1.35180 | 1.59546 | 1.92883 |
| | 0.4 | 0.74369 | 1.00389 | 1.35358 | 1.59213 | 1.91760 |
| | 0.8 | 0.75338 | 1.01009 | 1.35380 | 1.58756 | 1.90572 |
| | 1.2 | 0.76153 | 1.01490 | 1.35303 | 1.58238 | 1.89383 |
| | 1.6 | 0.76844 | 1.01870 | 1.35167 | 1.57696 | 1.88228 |
| | 2.0 | 0.77437 | 1.02173 | 1.34995 | 1.57153 | 1.87127 |
| | 2.4 | 0.77949 | 1.02417 | 1.34804 | 1.56623 | 1.86089 |
| | 2.8 | 0.78395 | 1.02616 | 1.34604 | 1.56114 | 1.85116 |
| | 3.2 | 0.78786 | 1.02780 | 1.34402 | 1.55628 | 1.84207 |
| | 3.6 | 0.79132 | 1.02915 | 1.34201 | 1.55169 | 1.83361 |
| | 4.0 | 0.79438 | 1.03029 | 1.34006 | 1.54735 | 1.82572 |
| 6.0 | 0.0 | 0.77666 | 1.02080 | 1.34436 | 1.56259 | 1.85759 |
| | 0.4 | 0.78122 | 1.02343 | 1.34372 | 1.55934 | 1.85035 |
| | 0.8 | 0.78525 | 1.02558 | 1.34273 | 1.55586 | 1.84310 |
| | 1.2 | 0.78883 | 1.02735 | 1.34151 | 1.55229 | 1.83597 |
| | 1.6 | 0.79203 | 1.02882 | 1.34015 | 1.54871 | 1.82906 |
| | 2.0 | 0.79490 | 1.03005 | 1.33870 | 1.54519 | 1.82241 |
| | 2.4 | 0.79749 | 1.03107 | 1.33722 | 1.54176 | 1.81606 |
| | 2.8 | 0.79982 | 1.03194 | 1.33573 | 1.53844 | 1.81001 |
| | 3.2 | 0.80195 | 1.03268 | 1.33425 | 1.53525 | 1.80427 |
| | 3.6 | 0.80388 | 1.03331 | 1.33280 | 1.53219 | 1.79882 |
| | 4.0 | 0.80565 | 1.03386 | 1.33138 | 1.52928 | 1.79366 |
| | 6.0 | 0.81259 | 1.03565 | 1.32501 | 1.51665 | 1.77173 |
| 12.0 | 0.0 | 0.81406 | 1.03505 | 1.32131 | 1.51068 | 1.76252 |
| | 0.4 | 0.81494 | 1.03537 | 1.32072 | 1.50937 | 1.76012 |
| | 0.8 | 0.81578 | 1.03565 | 1.32010 | 1.50805 | 1.75775 |
| | 1.2 | 0.81657 | 1.03589 | 1.31947 | 1.50674 | 1.75541 |
| | 1.6 | 0.81732 | 1.03611 | 1.31883 | 1.50544 | 1.75311 |
| | 2.0 | 0.81803 | 1.03631 | 1.31819 | 1.50415 | 1.75086 |
| | 2.4 | 0.81871 | 1.03648 | 1.31755 | 1.50289 | 1.74866 |
| | 2.8 | 0.81936 | 1.03663 | 1.31692 | 1.50164 | 1.74652 |
| | 3.2 | 0.81997 | 1.03676 | 1.31629 | 1.50043 | 1.74443 |
| | 3.6 | 0.82055 | 1.03688 | 1.31567 | 1.49924 | 1.74239 |
| | 4.0 | 0.82111 | 1.03699 | 1.31506 | 1.49808 | 1.74042 |
| | 6.0 | 0.82354 | 1.03736 | 1.31219 | 1.49273 | 1.73138 |
| | 12.0 | 0.82841 | 1.03774 | 1.30549 | 1.48064 | 1.71130 |
| INFINITY | | 0.84162 | 1.03643 | 1.28155 | 1.43953 | 1.64485 |

FRACTILES $W_\alpha$ RHO = 0.90

| DELTAS | | $\alpha$:0.9625 | 0.9750 | 0.9825 | 0.9900 | 0.9925 |
|---|---|---|---|---|---|---|
| 0.0 | 0.0 | 2.35084 | 2.84251 | 3.28144 | 3.98005 | 4.34316 |
| 0.4 | 0.0 | 2.34862 | 2.83884 | 3.27634 | 3.97239 | 4.33405 |
| | 0.4 | 2.35798 | 2.84354 | 3.27546 | 3.96010 | 4.31471 |
| 0.8 | 0.0 | 2.34795 | 2.83022 | 3.25924 | 3.93931 | 4.29156 |
| | 0.4 | 2.36257 | 2.83068 | 3.24440 | 3.89584 | 4.23144 |
| | 0.8 | 2.35862 | 2.80208 | 3.19151 | 3.80091 | 4.11333 |
| 1.2 | 0.0 | 2.34756 | 2.81095 | 3.22052 | 3.86546 | 4.19772 |
| | 0.4 | 2.35305 | 2.79485 | 3.18282 | 3.78994 | 4.10120 |
| | 0.8 | 2.33689 | 2.75312 | 3.11683 | 3.68327 | 3.97260 |
| | 1.2 | 2.30799 | 2.69986 | 3.04098 | 3.57023 | 3.83976 |
| 1.6 | 0.0 | 2.33680 | 2.77370 | 3.15740 | 3.75784 | 4.06568 |
| | 0.4 | 2.32888 | 2.74280 | 3.10450 | 3.66779 | 3.95551 |
| | 0.8 | 2.30504 | 2.69609 | 3.03648 | 3.56461 | 3.83356 |
| | 1.2 | 2.27427 | 2.64474 | 2.96620 | 3.46334 | 3.71588 |
| | 1.6 | 2.24201 | 2.59461 | 2.89972 | 3.37026 | 3.60874 |
| 2.0 | 0.0 | 2.31335 | 2.72277 | 3.08053 | 3.63768 | 3.92227 |
| | 0.4 | 2.29635 | 2.68495 | 3.02321 | 3.54800 | 3.81525 |
| | 0.8 | 2.26993 | 2.63919 | 2.95960 | 3.45511 | 3.70682 |
| | 1.2 | 2.24036 | 2.59251 | 2.89722 | 3.36715 | 3.60532 |
| | 1.6 | 2.21092 | 2.54817 | 2.83928 | 3.28709 | 3.51359 |
| | 2.0 | 2.18309 | 2.50743 | 2.78677 | 3.21552 | 3.43198 |
| 2.4 | 0.0 | 2.28232 | 2.66694 | 3.00172 | 3.52111 | 3.78560 |
| | 0.4 | 2.26141 | 2.62831 | 2.94665 | 3.43894 | 3.68901 |
| | 0.8 | 2.23547 | 2.58628 | 2.88982 | 3.35792 | 3.59516 |
| | 1.2 | 2.20840 | 2.54496 | 2.83546 | 3.28234 | 3.50836 |
| | 1.6 | 2.18210 | 2.50617 | 2.78528 | 3.21366 | 3.42993 |
| | 2.0 | 2.15745 | 2.47058 | 2.73973 | 3.15198 | 3.35977 |
| | 2.4 | 2.13473 | 2.43827 | 2.69870 | 3.09686 | 3.29723 |
| 2.8 | 0.0 | 2.24903 | 2.61247 | 2.92778 | 3.41538 | 3.66304 |
| | 0.4 | 2.22750 | 2.57610 | 2.87771 | 3.34282 | 3.57854 |
| | 0.8 | 2.20341 | 2.53861 | 2.82792 | 3.27295 | 3.49803 |
| | 1.2 | 2.17915 | 2.50241 | 2.78082 | 3.20810 | 3.42382 |
| | 1.6 | 2.15587 | 2.46858 | 2.73735 | 3.14902 | 3.35651 |
| | 2.0 | 2.13410 | 2.43746 | 2.69774 | 3.09567 | 3.29592 |
| | 2.4 | 2.11399 | 2.40908 | 2.66185 | 3.04764 | 3.24151 |
| | 2.8 | 2.09555 | 2.38330 | 2.62940 | 3.00443 | 3.19265 |
| 3.2 | 0.0 | 2.21667 | 2.56226 | 2.86125 | 3.32228 | 3.55591 |
| | 0.4 | 2.19610 | 2.52928 | 2.81684 | 3.25913 | 3.48282 |
| | 0.8 | 2.17430 | 2.49624 | 2.77348 | 3.19897 | 3.41376 |
| | 1.2 | 2.15276 | 2.46461 | 2.73264 | 3.14316 | 3.35006 |
| | 1.6 | 2.13220 | 2.43504 | 2.69487 | 3.09209 | 3.29199 |
| | 2.0 | 2.11295 | 2.40776 | 2.66028 | 3.04568 | 3.23936 |
| | 2.4 | 2.09511 | 2.38275 | 2.62875 | 3.00362 | 3.19176 |
| | 2.8 | 2.07868 | 2.35988 | 2.60006 | 2.96551 | 3.14871 |
| | 3.2 | 2.06358 | 2.33900 | 2.57394 | 2.93095 | 3.10972 |

FRACTILES $W_\alpha$ RHO = 0.90

| DELTAS | | α:0.9625 | 0.9750 | 0.9825 | 0.9900 | 0.9925 |
|---|---|---|---|---|---|---|
| 3.6 | 0.0 | 2.18663 | 2.51719 | 2.80246 | 3.24120 | 3.46308 |
| | 0.4 | 2.16766 | 2.48777 | 2.76342 | 3.18643 | 3.39996 |
| | 0.8 | 2.14815 | 2.45874 | 2.72568 | 3.13448 | 3.34051 |
| | 1.2 | 2.12907 | 2.43106 | 2.69014 | 3.08620 | 3.28550 |
| | 1.6 | 2.11089 | 2.40513 | 2.65716 | 3.04180 | 3.23508 |
| | 2.0 | 2.09383 | 2.38111 | 2.62681 | 3.00120 | 3.18909 |
| | 2.4 | 2.07796 | 2.35897 | 2.59897 | 2.96415 | 3.14721 |
| | 2.8 | 2.06328 | 2.33862 | 2.57348 | 2.93038 | 3.10909 |
| | 3.2 | 2.04971 | 2.31993 | 2.55015 | 2.89955 | 3.07433 |
| | 3.6 | 2.03720 | 2.30275 | 2.52876 | 2.87138 | 3.04260 |
| 4.0 | 0.0 | 2.15936 | 2.47718 | 2.75083 | 3.17072 | 3.38267 |
| | 0.4 | 2.14215 | 2.45109 | 2.71658 | 3.12314 | 3.32802 |
| | 0.8 | 2.12475 | 2.42556 | 2.68361 | 3.07806 | 3.27654 |
| | 1.2 | 2.10783 | 2.40124 | 2.65253 | 3.03603 | 3.22873 |
| | 1.6 | 2.09170 | 2.37840 | 2.62359 | 2.99718 | 3.18467 |
| | 2.0 | 2.07653 | 2.35715 | 2.59680 | 2.96145 | 3.14424 |
| | 2.4 | 2.06237 | 2.33746 | 2.57210 | 2.92866 | 3.10719 |
| | 2.8 | 2.04920 | 2.31927 | 2.54936 | 2.89857 | 3.07325 |
| | 3.2 | 2.03697 | 2.30247 | 2.52842 | 2.87096 | 3.04214 |
| | 3.6 | 2.02564 | 2.28696 | 2.50914 | 2.84559 | 3.01357 |
| | 4.0 | 2.01513 | 2.27263 | 2.49135 | 2.82223 | 2.98730 |
| 6.0 | 0.0 | 2.05950 | 2.33540 | 2.57093 | 2.92915 | 3.10866 |
| | 0.4 | 2.04925 | 2.32071 | 2.55217 | 2.90378 | 3.07979 |
| | 0.8 | 2.03916 | 2.30645 | 2.53411 | 2.87953 | 3.05228 |
| | 1.2 | 2.02938 | 2.29277 | 2.51688 | 2.85654 | 3.02625 |
| | 1.6 | 2.01998 | 2.27973 | 2.50052 | 2.83482 | 3.00171 |
| | 2.0 | 2.01101 | 2.26736 | 2.48507 | 2.81439 | 2.97865 |
| | 2.4 | 2.00248 | 2.25567 | 2.47051 | 2.79518 | 2.95701 |
| | 2.8 | 1.99441 | 2.24464 | 2.45680 | 2.77715 | 2.93671 |
| | 3.2 | 1.98677 | 2.23424 | 2.44390 | 2.76023 | 2.91768 |
| | 3.6 | 1.97955 | 2.22445 | 2.43178 | 2.74435 | 2.89983 |
| | 4.0 | 1.97274 | 2.21522 | 2.42038 | 2.72944 | 2.88308 |
| | 6.0 | 1.94393 | 2.17643 | 2.37257 | 2.66714 | 2.81319 |
| 12.0 | 0.0 | 1.93239 | 2.16160 | 2.35483 | 2.64484 | 2.78855 |
| | 0.4 | 1.92919 | 2.15721 | 2.34936 | 2.63762 | 2.78041 |
| | 0.8 | 1.92603 | 2.15290 | 2.34401 | 2.63057 | 2.77247 |
| | 1.2 | 1.92293 | 2.14869 | 2.33878 | 2.62370 | 2.76473 |
| | 1.6 | 1.91990 | 2.14457 | 2.33368 | 2.61701 | 2.75721 |
| | 2.0 | 1.91693 | 2.14055 | 2.32871 | 2.61052 | 2.74991 |
| | 2.4 | 1.91403 | 2.13665 | 2.32389 | 2.60422 | 2.74283 |
| | 2.8 | 1.91122 | 2.13286 | 2.31922 | 2.59811 | 2.73598 |
| | 3.2 | 1.90848 | 2.12918 | 2.31468 | 2.59220 | 2.72934 |
| | 3.6 | 1.90582 | 2.12560 | 2.31029 | 2.58648 | 2.72292 |
| | 4.0 | 1.90324 | 2.12215 | 2.30603 | 2.58094 | 2.71672 |
| | 6.0 | 1.89147 | 2.10642 | 2.28673 | 2.55590 | 2.68866 |
| | 12.0 | 1.86552 | 2.07194 | 2.24457 | 2.50140 | 2.62771 |
| INFINITY | | 1.78046 | 1.95996 | 2.10836 | 2.32635 | 2.43238 |

FRACTILES $W_\alpha$ RHO = 0.90

| DELTAS | | $\alpha$:0.9950 | 0.9975 | 0.9990 | 0.9995 |
|---|---|---|---|---|---|
| 0.0 | 0.0 | 4.85883 | 5.74936 | 6.94056 | 7.85012 |
| 0.4 | 0.0 | 4.84750 | 5.73379 | 6.91855 | 7.82260 |
| | 0.4 | 4.81694 | 5.68076 | 6.83011 | 7.70352 |
| 0.8 | 0.0 | 4.79046 | 5.64858 | 6.79035 | 7.65802 |
| | 0.4 | 4.70493 | 5.51502 | 6.58597 | 7.39554 |
| | 0.8 | 4.55266 | 5.30096 | 6.28506 | 7.02589 |
| 1.2 | 0.0 | 4.66651 | 5.46857 | 6.52891 | 7.33048 |
| | 0.4 | 4.53889 | 5.28440 | 6.26483 | 7.00289 |
| | 0.8 | 4.37841 | 5.06726 | 5.96947 | 6.64640 |
| | 1.2 | 4.21701 | 4.85556 | 5.68899 | 6.31254 |
| 1.6 | 0.0 | 4.49855 | 5.23588 | 6.20553 | 6.93549 |
| | 0.4 | 4.35907 | 5.04407 | 5.94124 | 6.61439 |
| | 0.8 | 4.21001 | 4.84719 | 5.67883 | 6.30104 |
| | 1.2 | 4.06871 | 4.66442 | 5.43955 | 6.01799 |
| | 1.6 | 3.94139 | 4.50176 | 5.22889 | 5.77025 |
| 2.0 | 0.0 | 4.32142 | 4.99894 | 5.88629 | 6.55206 |
| | 0.4 | 4.18932 | 4.82244 | 5.64879 | 6.26702 |
| | 0.8 | 4.05848 | 4.65220 | 5.42475 | 6.00125 |
| | 1.2 | 3.93753 | 4.49716 | 5.22332 | 5.76395 |
| | 1.6 | 3.82907 | 4.35944 | 5.04588 | 5.55585 |
| | 2.0 | 3.73307 | 4.23830 | 4.89071 | 5.37443 |
| 2.4 | 0.0 | 4.15579 | 4.78234 | 5.60008 | 6.21186 |
| | 0.4 | 4.03837 | 4.62819 | 5.39563 | 5.96832 |
| | 0.8 | 3.92607 | 4.48349 | 5.20677 | 5.74524 |
| | 1.2 | 3.82318 | 4.35241 | 5.03738 | 5.54624 |
| | 1.6 | 3.73076 | 4.23555 | 4.88737 | 5.37067 |
| | 2.0 | 3.64844 | 4.13201 | 4.75511 | 5.21626 |
| | 2.4 | 3.57530 | 4.04038 | 4.63847 | 5.08038 |
| 2.8 | 0.0 | 4.00904 | 4.59316 | 5.35315 | 5.92024 |
| | 0.4 | 3.90731 | 4.46111 | 5.17967 | 5.71461 |
| | 0.8 | 3.81152 | 4.33851 | 5.02056 | 5.52724 |
| | 1.2 | 3.72386 | 4.22733 | 4.87744 | 5.35945 |
| | 1.6 | 3.64477 | 4.12763 | 4.74982 | 5.21029 |
| | 2.0 | 3.57382 | 4.03861 | 4.63633 | 5.07796 |
| | 2.4 | 3.51028 | 3.95916 | 4.53537 | 4.96044 |
| | 2.8 | 3.45335 | 3.88815 | 4.44536 | 4.85581 |
| 3.2 | 0.0 | 3.88177 | 4.43063 | 5.14273 | 5.67284 |
| | 0.4 | 3.79436 | 4.31805 | 4.99579 | 5.49925 |
| | 0.8 | 3.71253 | 4.21382 | 4.86109 | 5.34098 |
| | 1.2 | 3.63749 | 4.11896 | 4.73933 | 5.19844 |
| | 1.6 | 3.56938 | 4.03333 | 4.62994 | 5.07074 |
| | 2.0 | 3.50785 | 3.95627 | 4.53187 | 4.95649 |
| | 2.4 | 3.45234 | 3.88695 | 4.44391 | 4.85417 |
| | 2.8 | 3.40222 | 3.82453 | 4.36487 | 4.76235 |
| | 3.2 | 3.35690 | 3.76819 | 4.29365 | 4.67971 |

FRACTILES $W_\alpha$ RHO = 0.90

| DELTAS | | $\alpha$:0.9950 | 0.9975 | 0.9990 | 0.9995 |
|---|---|---|---|---|---|
| 3.6 | 0.0 | 3.77208 | 4.29147 | 4.96360 | 5.46286 |
| | 0.4 | 3.69696 | 4.19526 | 4.83862 | 5.31559 |
| | 0.8 | 3.62672 | 4.10612 | 4.72380 | 5.18090 |
| | 1.2 | 3.56206 | 4.02460 | 4.61939 | 5.05882 |
| | 1.6 | 3.50303 | 3.95052 | 4.52491 | 4.94863 |
| | 2.0 | 3.44933 | 3.88336 | 4.43956 | 4.84927 |
| | 2.4 | 3.40054 | 3.82252 | 4.36243 | 4.75960 |
| | 2.8 | 3.35619 | 3.76734 | 4.29262 | 4.67854 |
| | 3.2 | 3.31582 | 3.71720 | 4.22930 | 4.60509 |
| | 3.6 | 3.27901 | 3.67154 | 4.17172 | 4.53835 |
| 4.0 | 0.0 | 3.67744 | 4.17198 | 4.81044 | 5.28374 |
| | 0.4 | 3.61263 | 4.08933 | 4.70347 | 5.15793 |
| | 0.8 | 3.55196 | 4.01256 | 4.60481 | 5.04236 |
| | 1.2 | 3.49586 | 3.94198 | 4.51458 | 4.93696 |
| | 1.6 | 3.44435 | 3.87742 | 4.43238 | 4.84115 |
| | 2.0 | 3.39718 | 3.81852 | 4.35759 | 4.75413 |
| | 2.4 | 3.35405 | 3.76479 | 4.28954 | 4.67506 |
| | 2.8 | 3.31460 | 3.71575 | 4.22754 | 4.60310 |
| | 3.2 | 3.27848 | 3.67092 | 4.17096 | 4.53749 |
| | 3.6 | 3.24536 | 3.62986 | 4.11921 | 4.47753 |
| | 4.0 | 3.21492 | 3.59218 | 4.07178 | 4.42260 |
| 6.0 | 0.0 | 3.35701 | 3.77057 | 4.29950 | 4.68848 |
| | 0.4 | 3.32310 | 3.72785 | 4.24478 | 4.62447 |
| | 0.8 | 3.29091 | 3.68746 | 4.19327 | 4.56436 |
| | 1.2 | 3.26053 | 3.64948 | 4.14498 | 4.50811 |
| | 1.6 | 3.23196 | 3.61385 | 4.09981 | 4.45558 |
| | 2.0 | 3.20515 | 3.58050 | 4.05761 | 4.40657 |
| | 2.4 | 3.18002 | 3.54930 | 4.01820 | 4.36086 |
| | 2.8 | 3.15648 | 3.52012 | 3.98141 | 4.31820 |
| | 3.2 | 3.13444 | 3.49283 | 3.94702 | 4.27837 |
| | 3.6 | 3.11377 | 3.46727 | 3.91487 | 4.24115 |
| | 4.0 | 3.09440 | 3.44333 | 3.88478 | 4.20633 |
| | 6.0 | 3.01368 | 3.34379 | 3.75990 | 4.06201 |
| 12.0 | 0.0 | 2.98575 | 3.31027 | 3.71905 | 4.01567 |
| | 0.4 | 2.97628 | 3.29849 | 3.70413 | 3.99831 |
| | 0.8 | 2.96706 | 3.28703 | 3.68962 | 3.98147 |
| | 1.2 | 2.95808 | 3.27589 | 3.67556 | 3.96515 |
| | 1.6 | 2.94937 | 3.26509 | 3.66194 | 3.94935 |
| | 2.0 | 2.94091 | 3.25463 | 3.64876 | 3.93408 |
| | 2.4 | 2.93273 | 3.24451 | 3.63602 | 3.91932 |
| | 2.8 | 2.92480 | 3.23471 | 3.62371 | 3.90507 |
| | 3.2 | 2.91713 | 3.22525 | 3.61182 | 3.89131 |
| | 3.6 | 2.90972 | 3.21611 | 3.60034 | 3.87804 |
| | 4.0 | 2.90256 | 3.20728 | 3.58926 | 3.86523 |
| | 6.0 | 2.87020 | 3.16745 | 3.53936 | 3.80760 |
| | 12.0 | 2.80005 | 3.08134 | 3.43178 | 3.68357 |
| INFINITY | | 2.57583 | 2.80703 | 3.09024 | 3.29051 |

FRACTILES $W_\alpha$ RHO = 0.95

| DELTAS | | $\alpha$:0.0005 | 0.0010 | 0.0025 | 0.0050 | 0.0075 |
|---|---|---|---|---|---|---|
| 0.0 | 0.0 | -0.84529 | -0.82238 | -0.79255 | -0.77045 | -0.75778 |
| 0.4 | 0.0 | -0.94431 | -0.91264 | -0.87016 | -0.83745 | -0.81805 |
| | 0.4 | -0.83187 | -0.81201 | -0.78617 | -0.76705 | -0.75609 |
| 0.8 | 0.0 | -1.03904 | -1.00172 | -0.95083 | -0.91087 | -0.88676 |
| | 0.4 | -0.91553 | -0.89047 | -0.85683 | -0.83091 | -0.81553 |
| | 0.8 | -0.85410 | -0.83911 | -0.81963 | -0.80526 | -0.79704 |
| 1.2 | 0.0 | -1.11212 | -1.07187 | -1.01646 | -0.97244 | -0.94565 |
| | 0.4 | -1.00742 | -0.97887 | -0.93986 | -0.90916 | -0.89061 |
| | 0.8 | -0.94825 | -0.92922 | -0.90365 | -0.88392 | -0.87220 |
| | 1.2 | -0.93838 | -0.92701 | -0.91227 | -0.90146 | -0.89532 |
| 1.6 | 0.0 | -1.17539 | -1.13379 | -1.07611 | -1.02992 | -1.00162 |
| | 0.4 | -1.09275 | -1.06187 | -1.01922 | -0.98521 | -0.96443 |
| | 0.8 | -1.05024 | -1.02806 | -0.99767 | -0.97365 | -0.95906 |
| | 1.2 | -1.03984 | -1.02487 | -1.00471 | -0.98913 | -0.97984 |
| | 1.6 | -1.06383 | -1.05490 | -1.04341 | -1.03507 | -1.03040 |
| 2.0 | 0.0 | -1.23631 | -1.19409 | -1.13521 | -1.08775 | -1.05851 |
| | 0.4 | -1.17460 | -1.14203 | -1.09668 | -1.06016 | -1.03766 |
| | 0.8 | -1.14692 | -1.12225 | -1.08802 | -1.06053 | -1.04362 |
| | 1.2 | -1.14627 | -1.12822 | -1.10335 | -1.08354 | -1.07141 |
| | 1.6 | -1.16722 | -1.15492 | -1.13832 | -1.12542 | -1.11770 |
| | 2.0 | -1.21324 | -1.20604 | -1.19691 | -1.19047 | -1.18704 |
| 2.4 | 0.0 | -1.29793 | -1.25536 | -1.19569 | -1.14730 | -1.11734 |
| | 0.4 | -1.25445 | -1.22051 | -1.17294 | -1.13430 | -1.11034 |
| | 0.8 | -1.23950 | -1.21272 | -1.17519 | -1.14467 | -1.12571 |
| | 1.2 | -1.24735 | -1.22664 | -1.19766 | -1.17411 | -1.15945 |
| | 1.6 | -1.27371 | -1.25832 | -1.23691 | -1.21960 | -1.20883 |
| | 2.0 | -1.31557 | -1.30508 | -1.29083 | -1.27969 | -1.27300 |
| | 2.4 | -1.37706 | -1.37123 | -1.36409 | -1.35973 | -1.35733 |
| 2.8 | 0.0 | -1.36105 | -1.31818 | -1.25778 | -1.20851 | -1.17785 |
| | 0.4 | -1.33291 | -1.29775 | -1.24813 | -1.20752 | -1.18216 |
| | 0.8 | -1.32869 | -1.30000 | -1.25944 | -1.22610 | -1.20519 |
| | 1.2 | -1.34385 | -1.32071 | -1.28793 | -1.26084 | -1.24373 |
| | 1.6 | -1.37500 | -1.35676 | -1.33087 | -1.30934 | -1.29561 |
| | 2.0 | -1.41976 | -1.40602 | -1.38655 | -1.37031 | -1.35982 |
| | 2.4 | -1.47668 | -1.46739 | -1.45470 | -1.44487 | -1.43910 |
| | 2.8 | -1.55023 | -1.54571 | -1.54105 | -1.52985 | -1.51526 |
| 3.2 | 0.0 | -1.42557 | -1.38231 | -1.32107 | -1.27079 | -1.23935 |
| | 0.4 | -1.41014 | -1.37378 | -1.32214 | -1.27953 | -1.25274 |
| | 0.8 | -1.41491 | -1.38440 | -1.34089 | -1.30473 | -1.28182 |
| | 1.2 | -1.43629 | -1.41082 | -1.37431 | -1.34367 | -1.32404 |
| | 1.6 | -1.47159 | -1.45059 | -1.42025 | -1.39441 | -1.37754 |
| | 2.0 | -1.51890 | -1.50198 | -1.47728 | -1.45573 | -1.44114 |
| | 2.4 | -1.57705 | -1.56407 | -1.54492 | -1.52737 | -1.51402 |
| | 2.8 | -1.64580 | -1.63730 | -1.62590 | -1.60335 | -1.57993 |
| | 3.2 | -1.73046 | -1.72610 | -1.70364 | -1.66664 | -1.63512 |

FRACTILES $W_\alpha$ RHO = 0.95

| DELTAS | | $\alpha$:0.0005 | 0.0010 | 0.0025 | 0.0050 | 0.0075 |
|---|---|---|---|---|---|---|
| 3.6 | 0.0 | -1.49104 | -1.44725 | -1.38495 | -1.33346 | -1.30108 |
| | 0.4 | -1.48612 | -1.44851 | -1.39473 | -1.34998 | -1.32163 |
| | 0.8 | -1.49839 | -1.46604 | -1.41950 | -1.38038 | -1.35535 |
| | 1.2 | -1.52500 | -1.49720 | -1.45686 | -1.42246 | -1.40009 |
| | 1.6 | -1.56379 | -1.54000 | -1.50502 | -1.47451 | -1.45412 |
| | 2.0 | -1.61322 | -1.59302 | -1.56272 | -1.53516 | -1.51568 |
| | 2.4 | -1.67230 | -1.65544 | -1.62915 | -1.60267 | -1.58047 |
| | 2.8 | -1.74071 | -1.72717 | -1.70351 | -1.66728 | -1.63572 |
| | 3.2 | -1.82021 | -1.80892 | -1.77087 | -1.72175 | -1.68347 |
| | 3.6 | -1.90566 | -1.88215 | -1.82944 | -1.77009 | -1.72599 |
| 4.0 | 0.0 | -1.55697 | -1.51248 | -1.44883 | -1.39585 | -1.36232 |
| | 0.4 | -1.56073 | -1.52175 | -1.46562 | -1.41848 | -1.38836 |
| | 0.8 | -1.57923 | -1.54495 | -1.49518 | -1.45282 | -1.42540 |
| | 1.2 | -1.61017 | -1.57993 | -1.53551 | -1.49696 | -1.47149 |
| | 1.6 | -1.65180 | -1.62506 | -1.58505 | -1.54924 | -1.52471 |
| | 2.0 | -1.70282 | -1.67910 | -1.64249 | -1.60778 | -1.58227 |
| | 2.4 | -1.76233 | -1.74113 | -1.70635 | -1.66877 | -1.63748 |
| | 2.8 | -1.82983 | -1.81039 | -1.77237 | -1.72313 | -1.68461 |
| | 3.2 | -1.90532 | -1.88308 | -1.83026 | -1.77070 | -1.72649 |
| | 3.6 | -1.98050 | -1.94707 | -1.88195 | -1.81353 | -1.76422 |
| | 4.0 | -2.04660 | -2.00472 | -1.92880 | -1.85235 | -1.79839 |
| 6.0 | 0.0 | -1.87835 | -1.82683 | -1.75039 | -1.68371 | -1.63977 |
| | 0.4 | -1.90654 | -1.85746 | -1.78353 | -1.71769 | -1.67350 |
| | 0.8 | -1.94171 | -1.89427 | -1.82133 | -1.75466 | -1.70896 |
| | 1.2 | -1.98263 | -1.93586 | -1.86204 | -1.79247 | -1.74393 |
| | 1.6 | -2.02801 | -1.98065 | -1.90354 | -1.82905 | -1.77682 |
| | 2.0 | -2.07630 | -2.02668 | -1.94374 | -1.86325 | -1.80728 |
| | 2.4 | -2.12546 | -2.07170 | -1.98151 | -1.89505 | -1.83549 |
| | 2.8 | -2.17323 | -2.11431 | -2.01679 | -1.92460 | -1.86165 |
| | 3.2 | -2.21842 | -2.15431 | -2.04971 | -1.95206 | -1.88589 |
| | 3.6 | -2.26091 | -2.19177 | -2.08040 | -1.97758 | -1.90839 |
| | 4.0 | -2.30077 | -2.22680 | -2.10900 | -2.00131 | -1.92928 |
| | 6.0 | -2.46525 | -2.37075 | -2.22595 | -2.09801 | -2.01424 |
| 12.0 | 0.0 | -2.51372 | -2.41092 | -2.25622 | -2.12158 | -2.03418 |
| | 0.4 | -2.53277 | -2.42798 | -2.27049 | -2.13362 | -2.04488 |
| | 0.8 | -2.55146 | -2.44464 | -2.28433 | -2.14526 | -2.05520 |
| | 1.2 | -2.56969 | -2.46083 | -2.29773 | -2.15648 | -2.06514 |
| | 1.6 | -2.58741 | -2.47653 | -2.31067 | -2.16729 | -2.07469 |
| | 2.0 | -2.60460 | -2.49170 | -2.32314 | -2.17769 | -2.08387 |
| | 2.4 | -2.62122 | -2.50636 | -2.33516 | -2.18768 | -2.09268 |
| | 2.8 | -2.63727 | -2.52049 | -2.34671 | -2.19728 | -2.10113 |
| | 3.2 | -2.65276 | -2.53409 | -2.35782 | -2.20650 | -2.10924 |
| | 3.6 | -2.66768 | -2.54720 | -2.36850 | -2.21535 | -2.11703 |
| | 4.0 | -2.68206 | -2.55980 | -2.37877 | -2.22384 | -2.12449 |
| | 6.0 | -2.74629 | -2.61602 | -2.42442 | -2.26155 | -2.15759 |
| | 12.0 | -2.88132 | -2.73386 | -2.51974 | -2.34004 | -2.22634 |
| INFINITY | | -3.29051 | -3.09024 | -2.80703 | -2.57583 | -2.43238 |

FRACTILES $w_\alpha$ RHO = 0.95

| DELTAS | | $\alpha$:0.0100 | 0.0175 | 0.0250 | 0.0375 | 0.0500 |
|---|---|---|---|---|---|---|
| 0.0 | 0.0 | -0.74893 | -0.73216 | -0.72186 | -0.71068 | -0.70323 |
| 0.4 | 0.0 | -0.80414 | -0.77671 | -0.75893 | -0.73839 | -0.72357 |
| | 0.4 | -0.74844 | -0.73396 | -0.72510 | -0.71552 | -0.70918 |
| 0.8 | 0.0 | -0.86928 | -0.83416 | -0.81086 | -0.78328 | -0.76282 |
| | 0.4 | -0.80450 | -0.78273 | -0.76861 | -0.75228 | -0.74050 |
| | 0.8 | -0.79133 | -0.78057 | -0.77404 | -0.76710 | -0.76266 |
| 1.2 | 0.0 | -0.92608 | -0.88642 | -0.85980 | -0.82794 | -0.80403 |
| | 0.4 | -0.87712 | -0.84996 | -0.83186 | -0.81033 | -0.79426 |
| | 0.8 | -0.86379 | -0.84715 | -0.83634 | -0.82383 | -0.81481 |
| | 1.2 | -0.89107 | -0.88318 | -0.87853 | -0.87387 | -0.87151 |
| 1.6 | 0.0 | -0.98085 | -0.93850 | -0.90985 | -0.87529 | -0.84915 |
| | 0.4 | -0.94921 | -0.91822 | -0.89728 | -0.87201 | -0.85286 |
| | 0.8 | -0.94842 | -0.92687 | -0.91238 | -0.89496 | -0.88176 |
| | 1.2 | -0.97317 | -0.95994 | -0.95132 | -0.94137 | -0.93433 |
| | 1.6 | -1.02723 | -1.02158 | -1.01867 | -1.01527 | -1.00940 |
| 2.0 | 0.0 | -1.03697 | -0.99281 | -0.96274 | -0.92621 | -0.89839 |
| | 0.4 | -1.02109 | -0.98707 | -0.96384 | -0.93551 | -0.91379 |
| | 0.8 | -1.03116 | -1.00557 | -0.98805 | -0.96655 | -0.94991 |
| | 1.2 | -1.06250 | -1.04424 | -1.03173 | -1.01631 | -1.00419 |
| | 1.6 | -1.11214 | -1.10111 | -1.09402 | -1.08680 | -1.07750 |
| | 2.0 | -1.18488 | -1.18128 | -1.17537 | -1.16116 | -1.14324 |
| 2.4 | 0.0 | -1.09519 | -1.04953 | -1.01825 | -0.97999 | -0.95063 |
| | 0.4 | -1.09259 | -1.05588 | -1.03058 | -0.99942 | -0.97525 |
| | 0.8 | -1.11164 | -1.08238 | -1.06205 | -1.03669 | -1.01669 |
| | 1.2 | -1.14853 | -1.12569 | -1.10961 | -1.08911 | -1.07235 |
| | 1.6 | -1.20081 | -1.18390 | -1.17174 | -1.15543 | -1.14010 |
| | 2.0 | -1.26821 | -1.25955 | -1.24868 | -1.22454 | -1.19714 |
| | 2.4 | -1.35334 | -1.33620 | -1.31520 | -1.27782 | -1.24057 |
| 2.8 | 0.0 | -1.15510 | -1.10795 | -1.07542 | -1.03534 | -1.00433 |
| | 0.4 | -1.16329 | -1.12395 | -1.09657 | -1.06248 | -1.03573 |
| | 0.8 | -1.18955 | -1.15668 | -1.13349 | -1.10410 | -1.08045 |
| | 1.2 | -1.23084 | -1.20340 | -1.18360 | -1.15761 | -1.13556 |
| | 1.6 | -1.28516 | -1.26234 | -1.24503 | -1.22008 | -1.19498 |
| | 2.0 | -1.35168 | -1.33269 | -1.31406 | -1.27752 | -1.24046 |
| | 2.4 | -1.43090 | -1.40096 | -1.36987 | -1.32069 | -1.27531 |
| | 2.8 | -1.49986 | -1.45528 | -1.41509 | -1.35609 | -1.30409 |
| 3.2 | 0.0 | -1.21592 | -1.16709 | -1.13314 | -1.09096 | -1.05802 |
| | 0.4 | -1.23269 | -1.19057 | -1.16093 | -1.12359 | -1.09387 |
| | 0.8 | -1.26457 | -1.22788 | -1.20158 | -1.16763 | -1.13971 |
| | 1.2 | -1.30909 | -1.27667 | -1.25266 | -1.22015 | -1.19155 |
| | 1.6 | -1.36446 | -1.33493 | -1.31142 | -1.27574 | -1.23992 |
| | 2.0 | -1.42932 | -1.39940 | -1.36980 | -1.32092 | -1.27552 |
| | 2.4 | -1.49969 | -1.45557 | -1.41539 | -1.35632 | -1.30425 |
| | 2.8 | -1.55804 | -1.50115 | -1.45349 | -1.38632 | -1.32869 |
| | 3.2 | -1.60771 | -1.54056 | -1.48660 | -1.41244 | -1.34996 |

FRACTILES $W_\alpha$ RHO = 0.95

| DELTAS | | $\alpha$:0.0100 | 0.0175 | 0.0250 | 0.0375 | 0.0500 |
|---|---|---|---|---|---|---|
| 3.6 | 0.0 | -1.27685 | -1.22601 | -1.19034 | -1.14562 | -1.11031 |
| | 0.4 | -1.30029 | -1.25504 | -1.22283 | -1.18167 | -1.14840 |
| | 0.8 | -1.33633 | -1.29537 | -1.26549 | -1.22612 | -1.19297 |
| | 1.2 | -1.38285 | -1.34471 | -1.31567 | -1.27514 | -1.23852 |
| | 1.6 | -1.43798 | -1.40034 | -1.36909 | -1.32113 | -1.27590 |
| | 2.0 | -1.49929 | -1.45612 | -1.41622 | -1.35698 | -1.30472 |
| | 2.4 | -1.55894 | -1.50197 | -1.45414 | -1.38677 | -1.32901 |
| | 2.8 | -1.60825 | -1.54095 | -1.48690 | -1.41264 | -1.35011 |
| | 3.2 | -1.65137 | -1.57532 | -1.51581 | -1.43546 | -1.36867 |
| | 3.6 | -1.68983 | -1.60598 | -1.54157 | -1.45574 | -1.38515 |
| 4.0 | 0.0 | -1.33711 | -1.28381 | -1.24606 | -1.19820 | -1.15995 |
| | 0.4 | -1.36555 | -1.31669 | -1.28142 | -1.23569 | -1.19808 |
| | 0.8 | -1.40439 | -1.35848 | -1.32434 | -1.27841 | -1.23893 |
| | 1.2 | -1.45160 | -1.40665 | -1.37152 | -1.32146 | -1.27623 |
| | 1.6 | -1.50490 | -1.45738 | -1.41731 | -1.35804 | -1.30551 |
| | 2.0 | -1.56032 | -1.50355 | -1.45544 | -1.38768 | -1.32965 |
| | 2.4 | -1.60986 | -1.54213 | -1.48780 | -1.41326 | -1.35054 |
| | 2.8 | -1.65236 | -1.57601 | -1.51634 | -1.43582 | -1.36893 |
| | 3.2 | -1.69026 | -1.60627 | -1.54180 | -1.45590 | -1.38526 |
| | 3.6 | -1.72439 | -1.63349 | -1.56466 | -1.47389 | -1.39986 |
| | 4.0 | -1.75528 | -1.65808 | -1.58529 | -1.49010 | -1.41298 |
| 6.0 | 0.0 | -1.60570 | -1.53046 | -1.47431 | -1.39983 | -1.33808 |
| | 0.4 | -1.63880 | -1.56082 | -1.50170 | -1.42272 | -1.35728 |
| | 0.8 | -1.67260 | -1.58991 | -1.52692 | -1.44310 | -1.37416 |
| | 1.2 | -1.70505 | -1.61663 | -1.54972 | -1.46140 | -1.38923 |
| | 1.6 | -1.73513 | -1.64105 | -1.57048 | -1.47796 | -1.40282 |
| | 2.0 | -1.76289 | -1.66347 | -1.58946 | -1.49303 | -1.41513 |
| | 2.4 | -1.78854 | -1.68409 | -1.60687 | -1.50680 | -1.42634 |
| | 2.8 | -1.81226 | -1.70309 | -1.62287 | -1.51941 | -1.43658 |
| | 3.2 | -1.83423 | -1.72063 | -1.63760 | -1.53099 | -1.44596 |
| | 3.6 | -1.85458 | -1.73684 | -1.65120 | -1.54167 | -1.45459 |
| | 4.0 | -1.87347 | -1.75186 | -1.66379 | -1.55152 | -1.46254 |
| | 6.0 | -1.95016 | -1.81264 | -1.71459 | -1.59117 | -1.49443 |
| 12.0 | 0.0 | -1.96766 | -1.82569 | -1.72501 | -1.59880 | -1.50023 |
| | 0.4 | -1.97740 | -1.83353 | -1.73163 | -1.60404 | -1.50448 |
| | 0.8 | -1.98677 | -1.84106 | -1.73797 | -1.60904 | -1.50854 |
| | 1.2 | -1.99579 | -1.84828 | -1.74405 | -1.61382 | -1.51240 |
| | 1.6 | -2.00444 | -1.85519 | -1.74985 | -1.61837 | -1.51608 |
| | 2.0 | -2.01275 | -1.86181 | -1.75541 | -1.62273 | -1.51959 |
| | 2.4 | -2.02072 | -1.86816 | -1.76072 | -1.62688 | -1.52294 |
| | 2.8 | -2.02837 | -1.87423 | -1.76581 | -1.63085 | -1.52613 |
| | 3.2 | -2.03570 | -1.88005 | -1.77067 | -1.63464 | -1.52918 |
| | 3.6 | -2.04272 | -1.88562 | -1.77532 | -1.63827 | -1.53209 |
| | 4.0 | -2.04947 | -1.89096 | -1.77978 | -1.64174 | -1.53487 |
| | 6.0 | -2.07932 | -1.91455 | -1.79944 | -1.65701 | -1.54708 |
| | 12.0 | -2.14121 | -1.96330 | -1.83995 | -1.68833 | -1.57202 |
| INFINITY | | -2.32635 | -2.10836 | -1.95996 | -1.78046 | -1.64485 |

FRACTILES $w_\alpha$ RHO = 0.95

| DELTAS | | $\alpha$:0.0750 | 0.1000 | 0.1500 | 0.2000 | 0.2500 |
|---|---|---|---|---|---|---|
| 0.0 | 0.0 | -0.69385 | -0.68885 | -0.67781 | -0.65895 | -0.63325 |
| 0.4 | 0.0 | -0.70234 | -0.68709 | -0.66614 | -0.65155 | -0.62797 |
| | 0.4 | -0.70135 | -0.69751 | -0.68530 | -0.66573 | -0.63930 |
| 0.8 | 0.0 | -0.73234 | -0.70915 | -0.67313 | -0.64369 | -0.61670 |
| | 0.4 | -0.72363 | -0.71160 | -0.69654 | -0.67817 | -0.65106 |
| | 0.8 | -0.75796 | -0.75321 | -0.73659 | -0.71194 | -0.67935 |
| 1.2 | 0.0 | -0.76791 | -0.73999 | -0.69576 | -0.65884 | -0.62458 |
| | 0.4 | -0.77010 | -0.75145 | -0.72168 | -0.69589 | -0.66866 |
| | 0.8 | -0.80202 | -0.79356 | -0.77575 | -0.74624 | -0.70708 |
| | 1.2 | -0.86536 | -0.85543 | -0.82636 | -0.78617 | -0.73572 |
| 1.6 | 0.0 | -0.80929 | -0.77813 | -0.72810 | -0.68567 | -0.64581 |
| | 0.4 | -0.82352 | -0.80033 | -0.76214 | -0.72779 | -0.69187 |
| | 0.8 | -0.86146 | -0.84515 | -0.81654 | -0.78078 | -0.73196 |
| | 1.2 | -0.92515 | -0.91155 | -0.87137 | -0.81806 | -0.75431 |
| | 1.6 | -0.99232 | -0.96952 | -0.91138 | -0.84167 | -0.76395 |
| 2.0 | 0.0 | -0.85560 | -0.82178 | -0.76671 | -0.71910 | -0.67352 |
| | 0.4 | -0.88002 | -0.85283 | -0.80689 | -0.76412 | -0.71823 |
| | 0.8 | -0.92351 | -0.90143 | -0.86034 | -0.81193 | -0.75008 |
| | 1.2 | -0.98411 | -0.96471 | -0.90915 | -0.84022 | -0.76283 |
| | 1.6 | -1.05065 | -1.01715 | -0.93955 | -0.85459 | -0.76531 |
| | 2.0 | -1.10107 | -1.05487 | -0.95869 | -0.86124 | -0.76308 |
| 2.4 | 0.0 | -0.90504 | -0.86858 | -0.80818 | -0.75469 | -0.70210 |
| | 0.4 | -0.93711 | -0.90581 | -0.85133 | -0.79851 | -0.74007 |
| | 0.8 | -0.98412 | -0.95590 | -0.90060 | -0.83548 | -0.75935 |
| | 1.2 | -1.04281 | -1.01247 | -0.93745 | -0.85307 | -0.76400 |
| | 1.6 | -1.09983 | -1.05413 | -0.95821 | -0.86081 | -0.76266 |
| | 2.0 | -1.13966 | -1.08231 | -0.97124 | -0.86417 | -0.75922 |
| | 2.4 | -1.16957 | -1.10320 | -0.98035 | -0.86565 | -0.75526 |
| 2.8 | 0.0 | -0.95566 | -0.91618 | -0.84940 | -0.78845 | -0.72667 |
| | 0.4 | -0.99278 | -0.95674 | -0.89184 | -0.82619 | -0.75296 |
| | 0.8 | -1.04085 | -1.00519 | -0.93228 | -0.84988 | -0.76135 |
| | 1.2 | -1.09464 | -1.05144 | -0.95673 | -0.85952 | -0.76143 |
| | 1.6 | -1.13888 | -1.08178 | -0.97075 | -0.86364 | -0.75866 |
| | 2.0 | -1.16951 | -1.10312 | -0.98021 | -0.86547 | -0.75505 |
| | 2.4 | -1.19345 | -1.11979 | -0.98730 | -0.86626 | -0.75132 |
| | 2.8 | -1.21328 | -1.13356 | -0.99294 | -0.86651 | -0.74769 |
| 3.2 | 0.0 | -1.00564 | -0.96245 | -0.88755 | -0.81689 | -0.74347 |
| | 0.4 | -1.04524 | -1.00335 | -0.92526 | -0.84440 | -0.75724 |
| | 0.8 | -1.09148 | -1.04649 | -0.95399 | -0.85735 | -0.75937 |
| | 1.2 | -1.13663 | -1.08058 | -0.96975 | -0.86258 | -0.75754 |
| | 1.6 | -1.16931 | -1.10286 | -0.97979 | -0.86494 | -0.75445 |
| | 2.0 | -1.19352 | -1.11976 | -0.98713 | -0.86601 | -0.75101 |
| | 2.4 | -1.21335 | -1.13357 | -0.99288 | -0.86641 | -0.74758 |
| | 2.8 | -1.23021 | -1.14523 | -0.99755 | -0.86644 | -0.74427 |
| | 3.2 | -1.24482 | -1.15527 | -1.00142 | -0.86624 | -0.74114 |

FRACTILES $W_\alpha$ RHO = 0.95

| DELTAS | | α:0.0750 | 0.1000 | 0.1500 | 0.2000 | 0.2500 |
|---|---|---|---|---|---|---|
| 3.6 | 0.0 | -1.05333 | -1.00544 | -0.92015 | -0.83747 | -0.75135 |
| | 0.4 | -1.09277 | -1.04359 | -0.94972 | -0.85416 | -0.75648 |
| | 0.8 | -1.13411 | -1.07827 | -0.96822 | -0.86100 | -0.75587 |
| | 1.2 | -1.16881 | -1.10242 | -0.97910 | -0.86405 | -0.75344 |
| | 1.6 | -1.19368 | -1.11971 | -0.98679 | -0.86550 | -0.75041 |
| | 2.0 | -1.21356 | -1.13361 | -0.99271 | -0.86613 | -0.74723 |
| | 2.4 | -1.23036 | -1.14528 | -0.99747 | -0.86630 | -0.74409 |
| | 2.8 | -1.24489 | -1.15529 | -1.00140 | -0.86618 | -0.74107 |
| | 3.2 | -1.25761 | -1.16399 | -1.00469 | -0.86589 | -0.73821 |
| | 3.6 | -1.26884 | -1.17162 | -1.00748 | -0.86549 | -0.73553 |
| 4.0 | 0.0 | -1.09721 | -1.04340 | -0.94545 | -0.84985 | -0.75271 |
| | 0.4 | -1.13388 | -1.07603 | -0.96600 | -0.85890 | -0.75368 |
| | 0.8 | -1.16792 | -1.10168 | -0.97815 | -0.86283 | -0.75206 |
| | 1.2 | -1.19391 | -1.11963 | -0.98629 | -0.86475 | -0.74951 |
| | 1.6 | -1.21391 | -1.13367 | -0.99244 | -0.86567 | -0.74665 |
| | 2.0 | -1.23066 | -1.14537 | -0.99733 | -0.86601 | -0.74372 |
| | 2.4 | -1.24510 | -1.15537 | -1.00132 | -0.86602 | -0.74085 |
| | 2.8 | -1.25774 | -1.16405 | -1.00465 | -0.86580 | -0.73809 |
| | 3.2 | -1.26890 | -1.17165 | -1.00747 | -0.86546 | -0.73548 |
| | 3.6 | -1.27883 | -1.17837 | -1.00989 | -0.86503 | -0.73302 |
| | 4.0 | -1.28772 | -1.18436 | -1.01197 | -0.86455 | -0.73071 |
| 6.0 | 0.0 | -1.23475 | -1.14685 | -0.99572 | -0.86270 | -0.73937 |
| | 0.4 | -1.24849 | -1.15673 | -1.00029 | -0.86365 | -0.73768 |
| | 0.8 | -1.26043 | -1.16520 | -1.00399 | -0.86411 | -0.73578 |
| | 1.2 | -1.27099 | -1.17259 | -1.00704 | -0.86425 | -0.73380 |
| | 1.6 | -1.28041 | -1.17911 | -1.00961 | -0.86417 | -0.73181 |
| | 2.0 | -1.28889 | -1.18492 | -1.01180 | -0.86396 | -0.72986 |
| | 2.4 | -1.29654 | -1.19012 | -1.01369 | -0.86364 | -0.72797 |
| | 2.8 | -1.30350 | -1.19482 | -1.01534 | -0.86327 | -0.72616 |
| | 3.2 | -1.30985 | -1.19908 | -1.01679 | -0.86286 | -0.72442 |
| | 3.6 | -1.31566 | -1.20296 | -1.01807 | -0.86243 | -0.72277 |
| | 4.0 | -1.32100 | -1.20650 | -1.01921 | -0.86199 | -0.72121 |
| | 6.0 | -1.34224 | -1.22046 | -1.02342 | -0.85982 | -0.71452 |
| 12.0 | 0.0 | -1.34565 | -1.22232 | -1.02335 | -0.85858 | -0.71252 |
| | 0.4 | -1.34854 | -1.22426 | -1.02400 | -0.85837 | -0.71169 |
| | 0.8 | -1.35128 | -1.22609 | -1.02459 | -0.85815 | -0.71088 |
| | 1.2 | -1.35388 | -1.22781 | -1.02514 | -0.85791 | -0.71008 |
| | 1.6 | -1.35635 | -1.22944 | -1.02564 | -0.85766 | -0.70930 |
| | 2.0 | -1.35869 | -1.23098 | -1.02610 | -0.85741 | -0.70854 |
| | 2.4 | -1.36092 | -1.23244 | -1.02653 | -0.85715 | -0.70781 |
| | 2.8 | -1.36304 | -1.23383 | -1.02692 | -0.85690 | -0.70709 |
| | 3.2 | -1.36506 | -1.23514 | -1.02729 | -0.85665 | -0.70640 |
| | 3.6 | -1.36698 | -1.23639 | -1.02763 | -0.85639 | -0.70573 |
| | 4.0 | -1.36882 | -1.23757 | -1.02795 | -0.85615 | -0.70508 |
| | 6.0 | -1.37684 | -1.24273 | -1.02927 | -0.85497 | -0.70214 |
| | 12.0 | -1.39306 | -1.25300 | -1.03164 | -0.85218 | -0.69574 |
| INFINITY | | -1.43953 | -1.28155 | -1.03643 | -0.84162 | -0.67449 |

FRACTILES $w_\alpha$ RHO = 0.95

| DELTAS | | $\alpha$:0.3000 | 0.3500 | 0.4000 | 0.4500 | 0.5000 |
|---|---|---|---|---|---|---|
| 0.0 | 0.0 | -0.60050 | -0.56019 | -0.51155 | -0.45351 | -0.38461 |
| 0.4 | 0.0 | -0.59641 | -0.55686 | -0.50874 | -0.45107 | -0.38244 |
| | 0.4 | -0.60575 | -0.56454 | -0.51487 | -0.45566 | -0.38546 |
| 0.8 | 0.0 | -0.58904 | -0.55172 | -0.50439 | -0.44684 | -0.37793 |
| | 0.4 | -0.61594 | -0.57252 | -0.52013 | -0.45773 | -0.38393 |
| | 0.8 | -0.63856 | -0.58905 | -0.53011 | -0.46077 | -0.37970 |
| 1.2 | 0.0 | -0.58992 | -0.55144 | -0.50405 | -0.44457 | -0.37284 |
| | 0.4 | -0.63069 | -0.58282 | -0.52498 | -0.45646 | -0.37603 |
| | 0.8 | -0.65853 | -0.60047 | -0.53252 | -0.45400 | -0.36390 |
| | 1.2 | -0.67563 | -0.60619 | -0.52740 | -0.43889 | -0.33984 |
| 1.6 | 0.0 | -0.60526 | -0.56069 | -0.50787 | -0.44291 | -0.36475 |
| | 0.4 | -0.64809 | -0.59243 | -0.52597 | -0.44850 | -0.35920 |
| | 0.8 | -0.67275 | -0.60386 | -0.52544 | -0.43720 | -0.33837 |
| | 1.2 | -0.68173 | -0.60109 | -0.51260 | -0.41592 | -0.31022 |
| | 1.6 | -0.67986 | -0.58993 | -0.49402 | -0.39152 | -0.28140 |
| 2.0 | 0.0 | -0.62627 | -0.57364 | -0.51152 | -0.43704 | -0.34964 |
| | 0.4 | -0.66327 | -0.59659 | -0.51945 | -0.43209 | -0.33394 |
| | 0.8 | -0.67841 | -0.59832 | -0.51019 | -0.41378 | -0.30831 |
| | 1.2 | -0.67891 | -0.58908 | -0.49324 | -0.39080 | -0.28075 |
| | 1.6 | -0.67240 | -0.57571 | -0.47461 | -0.36820 | -0.25525 |
| | 2.0 | -0.66361 | -0.56196 | -0.45711 | -0.34789 | -0.23297 |
| 2.4 | 0.0 | -0.64615 | -0.58279 | -0.50882 | -0.42337 | -0.32651 |
| | 0.4 | -0.67135 | -0.59264 | -0.50533 | -0.40950 | -0.30450 |
| | 0.8 | -0.67601 | -0.58651 | -0.49091 | -0.38867 | -0.27880 |
| | 1.2 | -0.67120 | -0.57458 | -0.47355 | -0.36720 | -0.25432 |
| | 1.6 | -0.66321 | -0.56156 | -0.45672 | -0.34753 | -0.23263 |
| | 2.0 | -0.65473 | -0.54926 | -0.44152 | -0.33019 | -0.21380 |
| | 2.4 | -0.64670 | -0.53815 | -0.42809 | -0.31504 | -0.19751 |
| 2.8 | 0.0 | -0.65942 | -0.58345 | -0.49784 | -0.40303 | -0.29879 |
| | 0.4 | -0.67103 | -0.58220 | -0.48703 | -0.38512 | -0.27556 |
| | 0.8 | -0.66880 | -0.57233 | -0.47142 | -0.36520 | -0.25246 |
| | 1.2 | -0.66199 | -0.56038 | -0.45557 | -0.34642 | -0.23159 |
| | 1.6 | -0.65415 | -0.54868 | -0.44095 | -0.32963 | -0.21327 |
| | 2.0 | -0.64648 | -0.53793 | -0.42786 | -0.31483 | -0.19730 |
| | 2.4 | -0.63936 | -0.52824 | -0.41624 | -0.30180 | -0.18334 |
| | 2.8 | -0.63285 | -0.51955 | -0.40591 | -0.29030 | -0.17108 |
| 3.2 | 0.0 | -0.66349 | -0.57602 | -0.48158 | -0.38019 | -0.27107 |
| | 0.4 | -0.66518 | -0.56896 | -0.46826 | -0.36223 | -0.24970 |
| | 0.8 | -0.65998 | -0.55841 | -0.45366 | -0.34460 | -0.22987 |
| | 1.2 | -0.65299 | -0.54752 | -0.43980 | -0.32852 | -0.21222 |
| | 1.6 | -0.64583 | -0.53727 | -0.42720 | -0.31418 | -0.19668 |
| | 2.0 | -0.63903 | -0.52790 | -0.41589 | -0.30146 | -0.18301 |
| | 2.4 | -0.63272 | -0.51941 | -0.40577 | -0.29016 | -0.17094 |
| | 2.8 | -0.62692 | -0.51174 | -0.39671 | -0.28011 | -0.16025 |
| | 3.2 | -0.62163 | -0.50481 | -0.38858 | -0.27112 | -0.15073 |

FRACTILES $W_\alpha$ RHO = 0.95

| DELTAS | | $\alpha$:0.3000 | 0.3500 | 0.4000 | 0.4500 | 0.5000 |
|---|---|---|---|---|---|---|
| 3.6 | 0.0 | -0.66024 | -0.56449 | -0.46407 | -0.35831 | -0.24605 |
| | 0.4 | -0.65718 | -0.55568 | -0.45102 | -0.34207 | -0.22748 |
| | 0.8 | -0.65128 | -0.54579 | -0.43809 | -0.32686 | -0.21065 |
| | 1.2 | -0.64476 | -0.53617 | -0.42610 | -0.31310 | -0.19565 |
| | 1.6 | -0.63837 | -0.52721 | -0.41520 | -0.30078 | -0.18235 |
| | 2.0 | -0.63233 | -0.51900 | -0.40536 | -0.28975 | -0.17054 |
| | 2.4 | -0.62672 | -0.51152 | -0.39649 | -0.27988 | -0.16003 |
| | 2.8 | -0.62154 | -0.50472 | -0.38849 | -0.27103 | -0.15064 |
| | 3.2 | -0.61678 | -0.49854 | -0.38125 | -0.26306 | -0.14222 |
| | 3.6 | -0.61241 | -0.49291 | -0.37470 | -0.25586 | -0.13463 |
| 4.0 | 0.0 | -0.65362 | -0.55221 | -0.44766 | -0.33885 | -0.22445 |
| | 0.4 | -0.64902 | -0.54351 | -0.43585 | -0.32468 | -0.20857 |
| | 0.8 | -0.64328 | -0.53464 | -0.42458 | -0.31161 | -0.19422 |
| | 1.2 | -0.63739 | -0.52619 | -0.41417 | -0.29976 | -0.18137 |
| | 1.6 | -0.63168 | -0.51832 | -0.40467 | -0.28907 | -0.16988 |
| | 2.0 | -0.62630 | -0.51108 | -0.39604 | -0.27944 | -0.15960 |
| | 2.4 | -0.62129 | -0.50445 | -0.38821 | -0.27075 | -0.15037 |
| | 2.8 | -0.61664 | -0.49839 | -0.38110 | -0.26291 | -0.14206 |
| | 3.2 | -0.61235 | -0.49285 | -0.37463 | -0.25580 | -0.13456 |
| | 3.6 | -0.60840 | -0.48777 | -0.36874 | -0.24934 | -0.12777 |
| | 4.0 | -0.60474 | -0.48311 | -0.36335 | -0.24346 | -0.12159 |
| 6.0 | 0.0 | -0.62133 | -0.50576 | -0.39057 | -0.27399 | -0.15431 |
| | 0.4 | -0.61762 | -0.50050 | -0.38413 | -0.26667 | -0.14640 |
| | 0.8 | -0.61394 | -0.49546 | -0.37806 | -0.25986 | -0.13909 |
| | 1.2 | -0.61037 | -0.49069 | -0.37239 | -0.25354 | -0.13235 |
| | 1.6 | -0.60696 | -0.48620 | -0.36710 | -0.24769 | -0.12614 |
| | 2.0 | -0.60372 | -0.48199 | -0.36218 | -0.24227 | -0.12042 |
| | 2.4 | -0.60066 | -0.47806 | -0.35761 | -0.23725 | -0.11514 |
| | 2.8 | -0.59778 | -0.47439 | -0.35336 | -0.23260 | -0.11025 |
| | 3.2 | -0.59507 | -0.47095 | -0.34940 | -0.22829 | -0.10573 |
| | 3.6 | -0.59252 | -0.46775 | -0.34571 | -0.22428 | -0.10154 |
| | 4.0 | -0.59013 | -0.46475 | -0.34228 | -0.22055 | -0.09764 |
| | 6.0 | -0.58014 | -0.45234 | -0.32814 | -0.20526 | -0.08173 |
| 12.0 | 0.0 | -0.57764 | -0.44951 | -0.32511 | -0.20216 | -0.07865 |
| | 0.4 | -0.57636 | -0.44788 | -0.32324 | -0.20012 | -0.07651 |
| | 0.8 | -0.57511 | -0.44631 | -0.32144 | -0.19815 | -0.07445 |
| | 1.2 | -0.57390 | -0.44479 | -0.31970 | -0.19627 | -0.07249 |
| | 1.6 | -0.57273 | -0.44334 | -0.31804 | -0.19447 | -0.07061 |
| | 2.0 | -0.57160 | -0.44193 | -0.31644 | -0.19274 | -0.06881 |
| | 2.4 | -0.57050 | -0.44058 | -0.31490 | -0.19108 | -0.06708 |
| | 2.8 | -0.56945 | -0.43928 | -0.31342 | -0.18949 | -0.06543 |
| | 3.2 | -0.56844 | -0.43803 | -0.31201 | -0.18796 | -0.06385 |
| | 3.6 | -0.56746 | -0.43683 | -0.31065 | -0.18650 | -0.06234 |
| | 4.0 | -0.56652 | -0.43568 | -0.30934 | -0.18510 | -0.06089 |
| | 6.0 | -0.56230 | -0.43053 | -0.30354 | -0.17888 | -0.05446 |
| | 12.0 | -0.55329 | -0.41964 | -0.29136 | -0.16587 | -0.04107 |
| INFINITY | | -0.52440 | -0.38532 | -0.25335 | -0.12566 | 0.0 |

FRACTILES $W_\alpha$ RHO = 0.95

| DELTAS | | $\alpha$:0.5500 | 0.6000 | 0.6500 | 0.7000 | 0.7500 |
|---|---|---|---|---|---|---|
| 0.0 | 0.0 | -0.30290 | -0.20565 | -0.08898 | 0.05287 | 0.22893 |
| 0.4 | 0.0 | -0.30093 | -0.20383 | -0.08728 | 0.05446 | 0.23041 |
| | 0.4 | -0.30230 | -0.20345 | -0.08502 | 0.05873 | 0.23679 |
| 0.8 | 0.0 | -0.29583 | -0.19791 | -0.08035 | 0.06254 | 0.23971 |
| | 0.4 | -0.29680 | -0.19366 | -0.07072 | 0.07765 | 0.26022 |
| | 0.8 | -0.28511 | -0.17449 | -0.04426 | 0.11087 | 0.29920 |
| 1.2 | 0.0 | -0.28736 | -0.18565 | -0.06401 | 0.08308 | 0.26432 |
| | 0.4 | -0.28197 | -0.17182 | -0.04203 | 0.11266 | 0.30052 |
| | 0.8 | -0.26068 | -0.14214 | -0.00505 | 0.15550 | 0.34719 |
| | 1.2 | -0.22893 | -0.10411 | 0.03766 | 0.20100 | 0.39319 |
| 1.6 | 0.0 | -0.27245 | -0.16378 | -0.03535 | 0.11801 | 0.30448 |
| | 0.4 | -0.25665 | -0.13872 | -0.00219 | 0.15778 | 0.34886 |
| | 0.8 | -0.22765 | -0.10301 | 0.03859 | 0.20174 | 0.39373 |
| | 1.2 | -0.19414 | -0.06564 | 0.07825 | 0.24202 | 0.43263 |
| | 1.6 | -0.16218 | -0.03177 | 0.11281 | 0.27591 | 0.46425 |
| 2.0 | 0.0 | -0.24855 | -0.13186 | 0.00351 | 0.16233 | 0.35218 |
| | 0.4 | -0.22380 | -0.09969 | 0.04137 | 0.20397 | 0.39534 |
| | 0.8 | -0.19244 | -0.06416 | 0.07951 | 0.24303 | 0.43336 |
| | 1.2 | -0.16159 | -0.03125 | 0.11326 | 0.27627 | 0.46450 |
| | 1.6 | -0.13418 | -0.00285 | 0.14168 | 0.30365 | 0.48955 |
| | 2.0 | -0.11068 | 0.02113 | 0.16536 | 0.32615 | 0.50978 |
| 2.4 | 0.0 | -0.21739 | -0.09419 | 0.04598 | 0.20766 | 0.39801 |
| | 0.4 | -0.18906 | -0.06121 | 0.08202 | 0.24505 | 0.43481 |
| | 0.8 | -0.15983 | -0.02966 | 0.11460 | 0.27735 | 0.46528 |
| | 1.2 | -0.13333 | -0.00208 | 0.14234 | 0.30418 | 0.48993 |
| | 1.6 | -0.11036 | 0.02142 | 0.16561 | 0.32635 | 0.50993 |
| | 2.0 | -0.09066 | 0.04138 | 0.18519 | 0.34481 | 0.52634 |
| | 2.4 | -0.07373 | 0.05843 | 0.20180 | 0.36033 | 0.53998 |
| 2.8 | 0.0 | -0.18402 | -0.05681 | 0.08575 | 0.24805 | 0.43698 |
| | 0.4 | -0.15691 | -0.02710 | 0.11681 | 0.27914 | 0.46657 |
| | 0.8 | -0.13162 | -0.00056 | 0.14365 | 0.30524 | 0.49069 |
| | 1.2 | -0.10940 | 0.02228 | 0.16636 | 0.32696 | 0.51036 |
| | 1.6 | -0.09017 | 0.04183 | 0.18558 | 0.34512 | 0.52656 |
| | 2.0 | -0.07354 | 0.05861 | 0.20195 | 0.36045 | 0.54007 |
| | 2.4 | -0.05908 | 0.07310 | 0.21601 | 0.37353 | 0.55147 |
| | 2.8 | -0.04645 | 0.08572 | 0.22819 | 0.38479 | 0.56120 |
| 3.2 | 0.0 | -0.15285 | -0.02351 | 0.11989 | 0.28163 | 0.46836 |
| | 0.4 | -0.12909 | 0.00170 | 0.14560 | 0.30682 | 0.49182 |
| | 0.8 | -0.10781 | 0.02371 | 0.16760 | 0.32796 | 0.51108 |
| | 1.2 | -0.08919 | 0.04271 | 0.18635 | 0.34574 | 0.52701 |
| | 1.6 | -0.07296 | 0.05913 | 0.20241 | 0.36082 | 0.54034 |
| | 2.0 | -0.05877 | 0.07338 | 0.21626 | 0.37373 | 0.55162 |
| | 2.4 | -0.04632 | 0.08584 | 0.22829 | 0.38487 | 0.56126 |
| | 2.8 | -0.03533 | 0.09678 | 0.23883 | 0.39456 | 0.56958 |
| | 3.2 | -0.02558 | 0.10647 | 0.24811 | 0.40307 | 0.57683 |

FRACTILES $W_\alpha$ RHO = 0.95

| DELTAS | | $\alpha$:0.5500 | 0.6000 | 0.6500 | 0.7000 | 0.7500 |
|---|---|---|---|---|---|---|
| 3.6 | 0.0 | -0.12575 | 0.00469 | 0.14818 | 0.30891 | 0.49333 |
| 3.6 | 0.4 | -0.10561 | 0.02570 | 0.16931 | 0.32936 | 0.51208 |
| 3.6 | 0.8 | -0.08773 | 0.04403 | 0.18749 | 0.34668 | 0.52767 |
| 3.6 | 1.2 | -0.07200 | 0.06000 | 0.20316 | 0.36144 | 0.54078 |
| 3.6 | 1.6 | -0.05816 | 0.07394 | 0.21675 | 0.37413 | 0.55190 |
| 3.6 | 2.0 | -0.04595 | 0.08618 | 0.22859 | 0.38511 | 0.56143 |
| 3.6 | 2.4 | -0.03513 | 0.09697 | 0.23899 | 0.39470 | 0.56968 |
| 3.6 | 2.8 | -0.02549 | 0.10654 | 0.24818 | 0.40312 | 0.57687 |
| 3.6 | 3.2 | -0.01688 | 0.11508 | 0.25634 | 0.41057 | 0.58318 |
| 3.6 | 3.6 | -0.00914 | 0.12273 | 0.26363 | 0.41720 | 0.58877 |
| 4.0 | 0.0 | -0.10280 | 0.02822 | 0.17150 | 0.33113 | 0.51336 |
| 4.0 | 0.4 | -0.08580 | 0.04578 | 0.18901 | 0.34791 | 0.52856 |
| 4.0 | 0.8 | -0.07066 | 0.06121 | 0.20422 | 0.36230 | 0.54139 |
| 4.0 | 1.2 | -0.05724 | 0.07478 | 0.21747 | 0.37472 | 0.55232 |
| 4.0 | 1.6 | -0.04533 | 0.08675 | 0.22909 | 0.38552 | 0.56172 |
| 4.0 | 2.0 | -0.03472 | 0.09735 | 0.23932 | 0.39497 | 0.56987 |
| 4.0 | 2.4 | -0.02524 | 0.10678 | 0.24838 | 0.40329 | 0.57699 |
| 4.0 | 2.8 | -0.01674 | 0.11521 | 0.25646 | 0.41067 | 0.58325 |
| 4.0 | 3.2 | -0.00908 | 0.12278 | 0.26368 | 0.41724 | 0.58880 |
| 4.0 | 3.6 | -0.00216 | 0.12961 | 0.27018 | 0.42313 | 0.59374 |
| 4.0 | 4.0 | 0.00412 | 0.13579 | 0.27605 | 0.42844 | 0.59817 |
| 6.0 | 0.0 | -0.02972 | 0.10192 | 0.24334 | 0.39826 | 0.57223 |
| 6.0 | 0.4 | -0.02147 | 0.11023 | 0.25142 | 0.40577 | 0.57877 |
| 6.0 | 0.8 | -0.01391 | 0.11780 | 0.25874 | 0.41254 | 0.58459 |
| 6.0 | 1.2 | -0.00698 | 0.12471 | 0.26538 | 0.41864 | 0.58980 |
| 6.0 | 1.6 | -0.00061 | 0.13103 | 0.27143 | 0.42416 | 0.59448 |
| 6.0 | 2.0 | 0.00523 | 0.13682 | 0.27695 | 0.42918 | 0.59871 |
| 6.0 | 2.4 | 0.01062 | 0.14213 | 0.28200 | 0.43376 | 0.60253 |
| 6.0 | 2.8 | 0.01558 | 0.14701 | 0.28664 | 0.43794 | 0.60601 |
| 6.0 | 3.2 | 0.02017 | 0.15151 | 0.29090 | 0.44178 | 0.60919 |
| 6.0 | 3.6 | 0.02441 | 0.15568 | 0.29483 | 0.44531 | 0.61210 |
| 6.0 | 4.0 | 0.02835 | 0.15953 | 0.29846 | 0.44856 | 0.61478 |
| 6.0 | 6.0 | 0.04438 | 0.17516 | 0.31314 | 0.46163 | 0.62542 |
| 12.0 | 0.0 | 0.04735 | 0.17793 | 0.31561 | 0.46368 | 0.62691 |
| 12.0 | 0.4 | 0.04952 | 0.18006 | 0.31762 | 0.46549 | 0.62840 |
| 12.0 | 0.8 | 0.05160 | 0.18210 | 0.31954 | 0.46720 | 0.62980 |
| 12.0 | 1.2 | 0.05358 | 0.18404 | 0.32136 | 0.46883 | 0.63113 |
| 12.0 | 1.6 | 0.05548 | 0.18589 | 0.32310 | 0.47038 | 0.63238 |
| 12.0 | 2.0 | 0.05730 | 0.18766 | 0.32476 | 0.47185 | 0.63358 |
| 12.0 | 2.4 | 0.05903 | 0.18935 | 0.32634 | 0.47325 | 0.63471 |
| 12.0 | 2.8 | 0.06069 | 0.19096 | 0.32785 | 0.47459 | 0.63579 |
| 12.0 | 3.2 | 0.06228 | 0.19250 | 0.32930 | 0.47587 | 0.63681 |
| 12.0 | 3.6 | 0.06380 | 0.19398 | 0.33067 | 0.47708 | 0.63779 |
| 12.0 | 4.0 | 0.06525 | 0.19539 | 0.33199 | 0.47824 | 0.63872 |
| 12.0 | 6.0 | 0.07169 | 0.20162 | 0.33779 | 0.48334 | 0.64278 |
| 12.0 | 12.0 | 0.08502 | 0.21448 | 0.34970 | 0.49374 | 0.65096 |
| INFINITY | | 0.12566 | 0.25335 | 0.38532 | 0.52440 | 0.67449 |

FRACTILES $W_\alpha$ RHO = 0.95

| DELTAS | | $\alpha$:0.8000 | 0.8500 | 0.9000 | 0.9250 | 0.9500 |
|---|---|---|---|---|---|---|
| 0.0 | 0.0 | 0.45444 | 0.75828 | 1.20592 | 1.53425 | 2.00884 |
| 0.4 | 0.0 | 0.45580 | 0.75945 | 1.20671 | 1.53470 | 2.00866 |
| | 0.4 | 0.46433 | 0.77001 | 1.21858 | 1.54636 | 2.01840 |
| 0.8 | 0.0 | 0.46626 | 0.77073 | 1.21767 | 1.54431 | 2.01476 |
| | 0.4 | 0.49175 | 0.79998 | 1.24735 | 1.57108 | 2.03337 |
| | 0.8 | 0.53465 | 0.84341 | 1.28423 | 1.59907 | 2.04404 |
| 1.2 | 0.0 | 0.49436 | 0.80081 | 1.24577 | 1.56784 | 2.02780 |
| | 0.4 | 0.53545 | 0.84358 | 1.28355 | 1.59781 | 2.04195 |
| | 0.8 | 0.58306 | 0.88757 | 1.31558 | 1.61774 | 2.04107 |
| | 1.2 | 0.62652 | 0.92402 | 1.33714 | 1.62622 | 2.02859 |
| 1.6 | 0.0 | 0.53786 | 0.84412 | 1.28156 | 1.59407 | 2.03578 |
| | 0.4 | 0.58402 | 0.88767 | 1.31449 | 1.61582 | 2.03802 |
| | 0.8 | 0.62682 | 0.92401 | 1.33671 | 1.62550 | 2.02746 |
| | 1.2 | 0.66177 | 0.95125 | 1.34964 | 1.62654 | 2.00998 |
| | 1.6 | 0.68901 | 0.97097 | 1.35627 | 1.62261 | 1.98987 |
| 2.0 | 0.0 | 0.58594 | 0.88787 | 1.31236 | 1.61207 | 2.03200 |
| | 0.4 | 0.62770 | 0.92399 | 1.33544 | 1.62336 | 2.02410 |
| | 0.8 | 0.66216 | 0.95120 | 1.34898 | 1.62546 | 2.00831 |
| | 1.2 | 0.68914 | 0.97094 | 1.35601 | 1.62220 | 1.98924 |
| | 1.6 | 0.71012 | 0.98526 | 1.35902 | 1.61619 | 1.96951 |
| | 2.0 | 0.72662 | 0.99581 | 1.35964 | 1.60897 | 1.95041 |
| 2.4 | 0.0 | 0.62919 | 0.92398 | 1.33338 | 1.61985 | 2.01859 |
| | 0.4 | 0.66293 | 0.95111 | 1.34769 | 1.62332 | 2.00498 |
| | 0.8 | 0.68954 | 0.97086 | 1.35526 | 1.62096 | 1.98733 |
| | 1.2 | 0.71031 | 0.98521 | 1.35862 | 1.61555 | 1.96852 |
| | 1.6 | 0.72669 | 0.99579 | 1.35948 | 1.60872 | 1.95002 |
| | 2.0 | 0.73982 | 1.00374 | 1.35887 | 1.60138 | 1.93251 |
| | 2.4 | 0.75051 | 1.00983 | 1.35741 | 1.59401 | 1.91624 |
| 2.8 | 0.0 | 0.66409 | 0.95100 | 1.34579 | 1.62017 | 2.00008 |
| | 0.4 | 0.69021 | 0.97073 | 1.35401 | 1.61893 | 1.98420 |
| | 0.8 | 0.71070 | 0.98511 | 1.35784 | 1.61429 | 1.96658 |
| | 1.2 | 0.72691 | 0.99572 | 1.35902 | 1.60797 | 1.94887 |
| | 1.6 | 0.73993 | 1.00370 | 1.35862 | 1.60098 | 1.93189 |
| | 2.0 | 0.75055 | 1.00981 | 1.35731 | 1.59385 | 1.91599 |
| | 2.4 | 0.75934 | 1.01457 | 1.35548 | 1.58687 | 1.90129 |
| | 2.8 | 0.76672 | 1.01833 | 1.35337 | 1.58020 | 1.88776 |
| 3.2 | 0.0 | 0.69115 | 0.97057 | 1.35231 | 1.61614 | 1.97988 |
| | 0.4 | 0.71128 | 0.98497 | 1.35668 | 1.61241 | 1.96369 |
| | 0.8 | 0.72727 | 0.99561 | 1.35825 | 1.60673 | 1.94697 |
| | 1.2 | 0.74015 | 1.00363 | 1.35813 | 1.60018 | 1.93068 |
| | 1.6 | 0.75068 | 1.00977 | 1.35701 | 1.59336 | 1.91525 |
| | 2.0 | 0.75941 | 1.01454 | 1.35531 | 1.58661 | 1.90088 |
| | 2.4 | 0.76674 | 1.01832 | 1.35330 | 1.58009 | 1.88760 |
| | 2.8 | 0.77297 | 1.02134 | 1.35114 | 1.57390 | 1.87537 |
| | 3.2 | 0.77831 | 1.02379 | 1.34892 | 1.56807 | 1.86413 |

FRACTILES $W_\alpha$ RHO = 0.95

| DELTAS | | $\alpha$:0.8000 | 0.8500 | 0.9000 | 0.9250 | 0.9500 |
|---|---|---|---|---|---|---|
| 3.6 | 0.0 | 0.71205 | 0.98479 | 1.35517 | 1.60994 | 1.95989 |
| | 0.4 | 0.72778 | 0.99547 | 1.35718 | 1.60501 | 1.94433 |
| | 0.8 | 0.74048 | 1.00352 | 1.35739 | 1.59900 | 1.92886 |
| | 1.2 | 0.75090 | 1.00969 | 1.35651 | 1.59256 | 1.91403 |
| | 1.6 | 0.75955 | 1.01449 | 1.35498 | 1.58608 | 1.90008 |
| | 2.0 | 0.76683 | 1.01829 | 1.35310 | 1.57976 | 1.88710 |
| | 2.4 | 0.77302 | 1.02132 | 1.35102 | 1.57371 | 1.87509 |
| | 2.8 | 0.77833 | 1.02378 | 1.34887 | 1.56799 | 1.86401 |
| | 3.2 | 0.78293 | 1.02580 | 1.34671 | 1.56260 | 1.85380 |
| | 3.6 | 0.78696 | 1.02746 | 1.34459 | 1.55755 | 1.84438 |
| 4.0 | 0.0 | 0.72843 | 0.99530 | 1.35584 | 1.60284 | 1.94100 |
| | 0.4 | 0.74093 | 1.00338 | 1.35642 | 1.59744 | 1.92647 |
| | 0.8 | 0.75120 | 1.00958 | 1.35581 | 1.59145 | 1.91232 |
| | 1.2 | 0.75976 | 1.01441 | 1.35449 | 1.58529 | 1.89887 |
| | 1.6 | 0.76697 | 1.01823 | 1.35276 | 1.57922 | 1.88626 |
| | 2.0 | 0.77311 | 1.02129 | 1.35079 | 1.57335 | 1.87453 |
| | 2.4 | 0.77839 | 1.02376 | 1.34873 | 1.56776 | 1.86365 |
| | 2.8 | 0.78297 | 1.02578 | 1.34663 | 1.56247 | 1.85359 |
| | 3.2 | 0.78697 | 1.02746 | 1.34455 | 1.55750 | 1.84429 |
| | 3.6 | 0.79050 | 1.02885 | 1.34253 | 1.55283 | 1.83569 |
| | 4.0 | 0.79362 | 1.03002 | 1.34057 | 1.54846 | 1.82772 |
| 6.0 | 0.0 | 0.77427 | 1.02084 | 1.34802 | 1.56891 | 1.86772 |
| | 0.4 | 0.77926 | 1.02341 | 1.34660 | 1.56436 | 1.85845 |
| | 0.8 | 0.78362 | 1.02551 | 1.34501 | 1.55988 | 1.84963 |
| | 1.2 | 0.78746 | 1.02725 | 1.34334 | 1.55555 | 1.84131 |
| | 1.6 | 0.79086 | 1.02869 | 1.34162 | 1.55138 | 1.83347 |
| | 2.0 | 0.79388 | 1.02990 | 1.33991 | 1.54741 | 1.82611 |
| | 2.4 | 0.79659 | 1.03093 | 1.33822 | 1.54363 | 1.81921 |
| | 2.8 | 0.79903 | 1.03179 | 1.33658 | 1.54005 | 1.81274 |
| | 3.2 | 0.80123 | 1.03253 | 1.33498 | 1.53666 | 1.80666 |
| | 3.6 | 0.80323 | 1.03317 | 1.33344 | 1.53345 | 1.80097 |
| | 4.0 | 0.80505 | 1.03371 | 1.33197 | 1.53041 | 1.79562 |
| | 6.0 | 0.81214 | 1.03555 | 1.32546 | 1.51752 | 1.77323 |
| 12.0 | 0.0 | 0.81288 | 1.03525 | 1.32362 | 1.51455 | 1.76863 |
| | 0.4 | 0.81389 | 1.03554 | 1.32276 | 1.51280 | 1.76556 |
| | 0.8 | 0.81483 | 1.03579 | 1.32191 | 1.51110 | 1.76259 |
| | 1.2 | 0.81571 | 1.03602 | 1.32108 | 1.50946 | 1.75974 |
| | 1.6 | 0.81655 | 1.03621 | 1.32027 | 1.50787 | 1.75699 |
| | 2.0 | 0.81733 | 1.03639 | 1.31947 | 1.50633 | 1.75434 |
| | 2.4 | 0.81807 | 1.03654 | 1.31870 | 1.50484 | 1.75180 |
| | 2.8 | 0.81877 | 1.03668 | 1.31795 | 1.50341 | 1.74936 |
| | 3.2 | 0.81943 | 1.03681 | 1.31722 | 1.50203 | 1.74700 |
| | 3.6 | 0.82006 | 1.03691 | 1.31651 | 1.50070 | 1.74474 |
| | 4.0 | 0.82065 | 1.03701 | 1.31583 | 1.49941 | 1.74256 |
| | 6.0 | 0.82321 | 1.03736 | 1.31270 | 1.49363 | 1.73284 |
| | 12.0 | 0.82822 | 1.03773 | 1.30577 | 1.48114 | 1.71213 |
| INFINITY | | 0.84162 | 1.03643 | 1.28155 | 1.43953 | 1.64485 |

FRACTILES $W_\alpha$ RHO = 0.95

| DELTAS | | $\alpha$:0.9625 | 0.9750 | 0.9825 | 0.9900 | 0.9925 |
|---|---|---|---|---|---|---|
| 0.0 | 0.0 | 2.35245 | 2.84466 | 3.28407 | 3.98343 | 4.34693 |
| 0.4 | 0.0 | 2.35172 | 2.84300 | 3.28145 | 3.97901 | 4.34145 |
| | 0.4 | 2.35897 | 2.84521 | 3.27778 | 3.96357 | 4.31881 |
| 0.8 | 0.0 | 2.35421 | 2.83886 | 3.27003 | 3.95362 | 4.30772 |
| | 0.4 | 2.36441 | 2.83398 | 3.24909 | 3.90289 | 4.23978 |
| | 0.8 | 2.35999 | 2.80505 | 3.19600 | 3.80795 | 4.12175 |
| 1.2 | 0.0 | 2.35721 | 2.82447 | 3.23756 | 3.88820 | 4.22346 |
| | 0.4 | 2.35733 | 2.80158 | 3.19182 | 3.80266 | 4.11589 |
| | 0.8 | 2.33959 | 2.75781 | 3.12336 | 3.69281 | 3.98375 |
| | 1.2 | 2.31084 | 2.70460 | 3.04746 | 3.57955 | 3.85059 |
| 1.6 | 0.0 | 2.34943 | 2.79128 | 3.17942 | 3.78698 | 4.09853 |
| | 0.4 | 2.33573 | 2.75282 | 3.11738 | 3.68530 | 3.97545 |
| | 0.8 | 2.30942 | 2.70278 | 3.04529 | 3.57683 | 3.84759 |
| | 1.2 | 2.27784 | 2.65024 | 2.97346 | 3.47347 | 3.72752 |
| | 1.6 | 2.24552 | 2.59990 | 2.90663 | 3.37980 | 3.61967 |
| 2.0 | 0.0 | 2.32813 | 2.74298 | 3.10560 | 3.67048 | 3.95907 |
| | 0.4 | 2.30520 | 2.69736 | 3.03882 | 3.56872 | 3.83864 |
| | 0.8 | 2.27574 | 2.64755 | 2.97026 | 3.46947 | 3.72312 |
| | 1.2 | 2.24472 | 2.59888 | 2.90542 | 3.37829 | 3.61801 |
| | 1.6 | 2.21468 | 2.55367 | 2.84635 | 3.29672 | 3.52457 |
| | 2.0 | 2.18669 | 2.51263 | 2.79343 | 3.22453 | 3.44222 |
| 2.4 | 0.0 | 2.29828 | 2.68846 | 3.02819 | 3.55540 | 3.82395 |
| | 0.4 | 2.27158 | 2.64223 | 2.96392 | 3.46154 | 3.71438 |
| | 0.8 | 2.24235 | 2.59585 | 2.90182 | 3.37379 | 3.61305 |
| | 1.2 | 2.21346 | 2.55211 | 2.84450 | 3.29441 | 3.52202 |
| | 1.6 | 2.18621 | 2.51202 | 2.79270 | 3.22362 | 3.44122 |
| | 2.0 | 2.16110 | 2.47577 | 2.74631 | 3.16082 | 3.36979 |
| | 2.4 | 2.13820 | 2.44317 | 2.70490 | 3.10515 | 3.30662 |
| 2.8 | 0.0 | 2.26544 | 2.63436 | 2.95454 | 3.44982 | 3.70145 |
| | 0.4 | 2.23844 | 2.59085 | 2.89586 | 3.36636 | 3.60487 |
| | 0.8 | 2.21103 | 2.54901 | 2.84081 | 3.28981 | 3.51696 |
| | 1.2 | 2.18478 | 2.51019 | 2.79053 | 3.22091 | 3.43824 |
| | 1.6 | 2.16034 | 2.47480 | 2.74516 | 3.15939 | 3.36821 |
| | 2.0 | 2.13789 | 2.44277 | 2.70443 | 3.10457 | 3.30598 |
| | 2.4 | 2.11743 | 2.41389 | 2.66790 | 3.05569 | 3.25061 |
| | 2.8 | 2.09881 | 2.38784 | 2.63511 | 3.01201 | 3.20121 |
| 3.2 | 0.0 | 2.23304 | 2.58395 | 2.88765 | 3.35610 | 3.59356 |
| | 0.4 | 2.20743 | 2.54441 | 2.83534 | 3.28299 | 3.50944 |
| | 0.8 | 2.18241 | 2.50717 | 2.78693 | 3.21643 | 3.43331 |
| | 1.2 | 2.15882 | 2.47286 | 2.74286 | 3.15652 | 3.36505 |
| | 1.6 | 2.13697 | 2.44160 | 2.70303 | 3.10283 | 3.30406 |
| | 2.0 | 2.11692 | 2.41325 | 2.66714 | 3.05474 | 3.24957 |
| | 2.4 | 2.09860 | 2.38758 | 2.63480 | 3.01162 | 3.20078 |
| | 2.8 | 2.08189 | 2.36432 | 2.60560 | 2.97285 | 3.15699 |
| | 3.2 | 2.06664 | 2.34321 | 2.57920 | 2.93789 | 3.11754 |

FRACTILES $W_\alpha$ RHO = 0.95

| DELTAS | | α:0.9625 | 0.9750 | 0.9825 | 0.9900 | 0.9925 |
|---|---|---|---|---|---|---|
| 3.6 | 0.0 | 2.20270 | 2.53836 | 2.82814 | 3.27400 | 3.49955 |
| | 0.4 | 2.17913 | 2.50298 | 2.78195 | 3.21022 | 3.42647 |
| | 0.8 | 2.15656 | 2.46998 | 2.73944 | 3.15225 | 3.36035 |
| | 1.2 | 2.13544 | 2.43965 | 2.70072 | 3.09994 | 3.30089 |
| | 1.6 | 2.11592 | 2.41197 | 2.66562 | 3.05284 | 3.24748 |
| | 2.0 | 2.09798 | 2.38678 | 2.63385 | 3.01044 | 3.19948 |
| | 2.4 | 2.08154 | 2.36387 | 2.60508 | 2.97219 | 3.15626 |
| | 2.8 | 2.06649 | 2.34303 | 2.57897 | 2.93761 | 3.11724 |
| | 3.2 | 2.05270 | 2.32402 | 2.55524 | 2.90627 | 3.08189 |
| | 3.6 | 2.04005 | 2.30665 | 2.53361 | 2.87776 | 3.04978 |
| 4.0 | 0.0 | 2.17497 | 2.49766 | 2.77563 | 3.20233 | 3.41778 |
| | 0.4 | 2.15358 | 2.46618 | 2.73492 | 3.14661 | 3.35415 |
| | 0.8 | 2.13332 | 2.43695 | 2.69750 | 3.09593 | 3.29647 |
| | 1.2 | 2.11442 | 2.41006 | 2.66335 | 3.05001 | 3.24436 |
| | 1.6 | 2.09695 | 2.38546 | 2.63228 | 3.00848 | 3.19732 |
| | 2.0 | 2.08085 | 2.36299 | 2.60402 | 2.97087 | 3.15481 |
| | 2.4 | 2.06605 | 2.34246 | 2.57830 | 2.93678 | 3.11631 |
| | 2.8 | 2.05245 | 2.32370 | 2.55486 | 2.90579 | 3.08137 |
| | 3.2 | 2.03994 | 2.30652 | 2.53345 | 2.87756 | 3.04956 |
| | 3.6 | 2.02842 | 2.29075 | 2.51384 | 2.85176 | 3.02052 |
| | 4.0 | 2.01779 | 2.27625 | 2.49584 | 2.82812 | 2.99393 |
| 6.0 | 0.0 | 2.07237 | 2.35215 | 2.59112 | 2.95477 | 3.13707 |
| | 0.4 | 2.05956 | 2.33417 | 2.56843 | 2.92445 | 3.10273 |
| | 0.8 | 2.04751 | 2.31739 | 2.54735 | 2.89640 | 3.07102 |
| | 1.2 | 2.03621 | 2.30175 | 2.52778 | 2.87047 | 3.04175 |
| | 1.6 | 2.02565 | 2.28721 | 2.50962 | 2.84649 | 3.01470 |
| | 2.0 | 2.01578 | 2.27368 | 2.49277 | 2.82429 | 2.98969 |
| | 2.4 | 2.00656 | 2.26109 | 2.47712 | 2.80371 | 2.96654 |
| | 2.8 | 1.99795 | 2.24935 | 2.46257 | 2.78462 | 2.94506 |
| | 3.2 | 1.98989 | 2.23841 | 2.44902 | 2.76686 | 2.92510 |
| | 3.6 | 1.98236 | 2.22820 | 2.43639 | 2.75033 | 2.90653 |
| | 4.0 | 1.97529 | 2.21865 | 2.42459 | 2.73491 | 2.88921 |
| | 6.0 | 1.94589 | 2.17907 | 2.37581 | 2.67135 | 2.81791 |
| 12.0 | 0.0 | 1.94014 | 2.17167 | 2.36696 | 2.66022 | 2.80562 |
| | 0.4 | 1.93608 | 2.16617 | 2.36016 | 2.65133 | 2.79562 |
| | 0.8 | 1.93217 | 2.16090 | 2.35365 | 2.64282 | 2.78606 |
| | 1.2 | 1.92842 | 2.15584 | 2.34741 | 2.63468 | 2.77692 |
| | 1.6 | 1.92482 | 2.15099 | 2.34143 | 2.62688 | 2.76817 |
| | 2.0 | 1.92135 | 2.14634 | 2.33570 | 2.61942 | 2.75980 |
| | 2.4 | 1.91803 | 2.14187 | 2.33021 | 2.61227 | 2.75179 |
| | 2.8 | 1.91484 | 2.13759 | 2.32495 | 2.60542 | 2.74411 |
| | 3.2 | 1.91177 | 2.13348 | 2.31990 | 2.59886 | 2.73676 |
| | 3.6 | 1.90882 | 2.12954 | 2.31505 | 2.59257 | 2.72971 |
| | 4.0 | 1.90599 | 2.12575 | 2.31040 | 2.58654 | 2.72295 |
| | 6.0 | 1.89335 | 2.10890 | 2.28975 | 2.55977 | 2.69298 |
| | 12.0 | 1.86659 | 2.07336 | 2.24630 | 2.50363 | 2.63021 |
| INFINITY | | 1.78046 | 1.95996 | 2.10836 | 2.32635 | 2.43238 |

FRACTILES $W_\alpha$ RHO = 0.95

| DELTAS | | $\alpha$:0.9950 | 0.9975 | 0.9990 | 0.9995 |
|---|---|---|---|---|---|
| 0.0 | 0.0 | 4.86316 | 5.75463 | 6.94710 | 7.85763 |
| 0.4 | 0.0 | 4.85602 | 5.74425 | 6.93164 | 7.83772 |
| | 0.4 | 4.82198 | 5.68753 | 6.83937 | 7.71479 |
| 0.8 | 0.0 | 4.80928 | 5.67209 | 6.82027 | 7.69292 |
| | 0.4 | 4.71515 | 5.52862 | 6.60428 | 7.41757 |
| | 0.8 | 4.56308 | 5.31494 | 6.30395 | 7.04863 |
| 1.2 | 0.0 | 4.69654 | 5.50610 | 6.57659 | 7.38598 |
| | 0.4 | 4.55642 | 5.30693 | 6.29416 | 7.03749 |
| | 0.8 | 4.39187 | 5.08477 | 5.99251 | 6.67373 |
| | 1.2 | 4.23001 | 4.87234 | 5.71095 | 6.33849 |
| 1.6 | 0.0 | 4.53670 | 5.28318 | 6.26511 | 7.00445 |
| | 0.4 | 4.38247 | 5.07349 | 5.97877 | 6.65814 |
| | 0.8 | 4.22661 | 4.86828 | 5.70601 | 6.33290 |
| | 1.2 | 4.08252 | 4.68201 | 5.46228 | 6.04469 |
| | 1.6 | 3.95431 | 4.51816 | 5.25001 | 5.79501 |
| 2.0 | 0.0 | 4.36391 | 5.05123 | 5.95165 | 6.62736 |
| | 0.4 | 4.21650 | 4.85618 | 5.69130 | 6.31624 |
| | 0.8 | 4.07754 | 4.67607 | 5.45507 | 6.03652 |
| | 1.2 | 3.95244 | 4.51593 | 5.24730 | 5.79194 |
| | 1.6 | 3.84197 | 4.37569 | 5.06668 | 5.58014 |
| | 2.0 | 3.74508 | 4.25341 | 4.90999 | 5.39694 |
| 2.4 | 0.0 | 4.19988 | 4.83629 | 5.66712 | 6.28885 |
| | 0.4 | 4.06767 | 4.66427 | 5.44075 | 6.02032 |
| | 0.8 | 3.94684 | 4.50925 | 5.23921 | 5.78279 |
| | 1.2 | 3.83910 | 4.37226 | 5.06253 | 5.57545 |
| | 1.6 | 3.74396 | 4.25207 | 4.90837 | 5.39510 |
| | 2.0 | 3.66016 | 4.14668 | 4.77377 | 5.23800 |
| | 2.4 | 3.58626 | 4.05408 | 4.65587 | 5.10062 |
| 2.8 | 0.0 | 4.05306 | 4.64681 | 5.41957 | 5.99635 |
| | 0.4 | 3.93760 | 4.49821 | 5.22583 | 5.76766 |
| | 0.8 | 3.83339 | 4.36545 | 5.05428 | 5.56613 |
| | 1.2 | 3.74060 | 4.24806 | 4.90352 | 5.38962 |
| | 1.6 | 3.65837 | 4.14455 | 4.77120 | 5.23509 |
| | 2.0 | 3.58554 | 4.05322 | 4.65483 | 5.09944 |
| | 2.4 | 3.52089 | 3.97238 | 4.55212 | 4.97990 |
| | 2.8 | 3.46331 | 3.90056 | 4.46107 | 4.87406 |
| 3.2 | 0.0 | 3.92483 | 4.48297 | 5.20735 | 5.74676 |
| | 0.4 | 3.82490 | 4.35532 | 5.04201 | 5.55226 |
| | 0.8 | 3.73503 | 4.24141 | 4.89547 | 5.38053 |
| | 1.2 | 3.65481 | 4.14031 | 4.76606 | 5.22928 |
| | 1.6 | 3.58338 | 4.05064 | 4.65171 | 5.09592 |
| | 2.0 | 3.51970 | 3.97097 | 4.55041 | 4.97798 |
| | 2.4 | 3.46282 | 3.89998 | 4.46036 | 4.87326 |
| | 2.8 | 3.41184 | 3.83648 | 4.37997 | 4.77988 |
| | 3.2 | 3.36599 | 3.77947 | 4.30790 | 4.69623 |

FRACTILES $W_\alpha$ RHO = 0.95

| DELTAS | | α:0.9950 | 0.9975 | 0.9990 | 0.9995 |
|---|---|---|---|---|---|
| 3.6 | 0.0 | 3.81373 | 4.34200 | 5.02586 | 5.53401 |
| | 0.4 | 3.72731 | 4.23220 | 4.88432 | 5.36792 |
| | 0.8 | 3.64951 | 4.13398 | 4.75840 | 5.22063 |
| | 1.2 | 3.57979 | 4.04636 | 4.64653 | 5.09007 |
| | 1.6 | 3.51735 | 3.96816 | 4.54701 | 4.97413 |
| | 2.0 | 3.46135 | 3.89823 | 4.45824 | 4.87086 |
| | 2.4 | 3.41102 | 3.83551 | 4.37878 | 4.77854 |
| | 2.8 | 3.36564 | 3.77906 | 4.30740 | 4.69567 |
| | 3.2 | 3.32459 | 3.72807 | 4.24302 | 4.62098 |
| | 3.6 | 3.28733 | 3.68186 | 4.18472 | 4.55341 |
| 4.0 | 0.0 | 3.71750 | 4.22050 | 4.87015 | 5.35191 |
| | 0.4 | 3.64251 | 4.12563 | 4.74828 | 5.20919 |
| | 0.8 | 3.57480 | 4.04041 | 4.63933 | 5.08193 |
| | 1.2 | 3.51383 | 3.96396 | 4.54193 | 4.96839 |
| | 1.6 | 3.45892 | 3.89532 | 4.45472 | 4.86688 |
| | 2.0 | 3.40938 | 3.83355 | 4.37642 | 4.77586 |
| | 2.4 | 3.36460 | 3.77782 | 4.30589 | 4.69397 |
| | 2.8 | 3.32400 | 3.72737 | 4.24216 | 4.62002 |
| | 3.2 | 3.28708 | 3.68156 | 4.18435 | 4.55300 |
| | 3.6 | 3.25340 | 3.63982 | 4.13175 | 4.49204 |
| | 4.0 | 3.22259 | 3.60167 | 4.08371 | 4.43642 |
| 6.0 | 0.0 | 3.38935 | 3.80963 | 4.34743 | 4.74309 |
| | 0.4 | 3.34926 | 3.75949 | 4.28369 | 4.66886 |
| | 0.8 | 3.31231 | 3.71341 | 4.22523 | 4.60087 |
| | 1.2 | 3.27825 | 3.67101 | 4.17156 | 4.53853 |
| | 1.6 | 3.24683 | 3.63197 | 4.12222 | 4.48127 |
| | 2.0 | 3.21781 | 3.59595 | 4.07677 | 4.42856 |
| | 2.4 | 3.19096 | 3.56268 | 4.03483 | 4.37996 |
| | 2.8 | 3.16607 | 3.53187 | 3.99604 | 4.33503 |
| | 3.2 | 3.14297 | 3.50330 | 3.96010 | 4.29342 |
| | 3.6 | 3.12149 | 3.47675 | 3.92672 | 4.25480 |
| | 4.0 | 3.10147 | 3.45203 | 3.89566 | 4.21888 |
| | 6.0 | 3.01912 | 3.35049 | 3.76830 | 4.07170 |
| 12.0 | 0.0 | 3.00518 | 3.33377 | 3.74791 | 4.04858 |
| | 0.4 | 2.99362 | 3.31946 | 3.72990 | 4.02772 |
| | 0.8 | 2.98256 | 3.30579 | 3.71271 | 4.00781 |
| | 1.2 | 2.97199 | 3.29273 | 3.69630 | 3.98883 |
| | 1.6 | 2.96188 | 3.28026 | 3.68063 | 3.97070 |
| | 2.0 | 2.95221 | 3.26833 | 3.66566 | 3.95339 |
| | 2.4 | 2.94296 | 3.25692 | 3.65135 | 3.93685 |
| | 2.8 | 2.93410 | 3.24601 | 3.63766 | 3.92104 |
| | 3.2 | 2.92561 | 3.23556 | 3.62456 | 3.90591 |
| | 3.6 | 2.91749 | 3.22555 | 3.61203 | 3.89143 |
| | 4.0 | 2.90969 | 3.21596 | 3.60001 | 3.87756 |
| | 6.0 | 2.87516 | 3.17351 | 3.54689 | 3.81625 |
| | 12.0 | 2.80292 | 3.08485 | 3.43617 | 3.68862 |
| INFINITY | | 2.57583 | 2.80703 | 3.09024 | 3.29051 |

FRACTILES $w_\alpha$ RHO = 1.00

| DELTAS | | $\alpha$:0.0005 | 0.0010 | 0.0025 | 0.0050 | 0.0075 |
|---|---|---|---|---|---|---|
| 0.0 | 0.0 | -0.70711 | -0.70711 | -0.70710 | -0.70708 | -0.70704 |
| 0.4 | 0.0 | -0.70763 | -0.70763 | -0.70762 | -0.70760 | -0.70757 |
| | 0.4 | -0.71393 | -0.71393 | -0.71392 | -0.71390 | -0.71387 |
| 0.8 | 0.0 | -0.71393 | -0.71393 | -0.71392 | -0.71390 | -0.71387 |
| | 0.4 | -0.73327 | -0.73326 | -0.73325 | -0.73323 | -0.73319 |
| | 0.8 | -0.76800 | -0.76800 | -0.76799 | -0.76797 | -0.76792 |
| 1.2 | 0.0 | -0.73327 | -0.73326 | -0.73325 | -0.73323 | -0.73319 |
| | 0.4 | -0.76800 | -0.76800 | -0.76799 | -0.76797 | -0.76792 |
| | 0.8 | -0.81650 | -0.81649 | -0.81649 | -0.81645 | -0.81640 |
| | 1.2 | -0.87591 | -0.87590 | -0.87589 | -0.87585 | -0.87578 |
| 1.6 | 0.0 | -0.76800 | -0.76800 | -0.76799 | -0.76797 | -0.76792 |
| | 0.4 | -0.81650 | -0.81649 | -0.81649 | -0.81645 | -0.81640 |
| | 0.8 | -0.87591 | -0.87590 | -0.87589 | -0.87585 | -0.87578 |
| | 1.2 | -0.94361 | -0.94361 | -0.94359 | -0.94352 | -0.94341 |
| | 1.6 | -1.01756 | -1.01755 | -1.01752 | -1.01741 | -1.01723 |
| 2.0 | 0.0 | -0.81650 | -0.81649 | -0.81649 | -0.81645 | -0.81640 |
| | 0.4 | -0.87591 | -0.87590 | -0.87589 | -0.87585 | -0.87578 |
| | 0.8 | -0.94361 | -0.94361 | -0.94359 | -0.94352 | -0.94341 |
| | 1.2 | -1.01756 | -1.01755 | -1.01752 | -1.01741 | -1.01723 |
| | 1.6 | -1.09622 | -1.09621 | -1.09616 | -1.09597 | -1.09564 |
| | 2.0 | -1.17851 | -1.17849 | -1.17838 | -1.17801 | -1.17738 |
| 2.4 | 0.0 | -0.87591 | -0.87590 | -0.87589 | -0.87585 | -0.87578 |
| | 0.4 | -0.94361 | -0.94361 | -0.94359 | -0.94352 | -0.94341 |
| | 0.8 | -1.01756 | -1.01755 | -1.01752 | -1.01741 | -1.01723 |
| | 1.2 | -1.09622 | -1.09621 | -1.09616 | -1.09597 | -1.09564 |
| | 1.6 | -1.17851 | -1.17849 | -1.17838 | -1.17801 | -1.17738 |
| | 2.0 | -1.26360 | -1.26356 | -1.26334 | -1.26254 | -1.26122 |
| | 2.4 | -1.35089 | -1.35082 | -1.35030 | -1.34848 | -1.34555 |
| 2.8 | 0.0 | -0.94361 | -0.94361 | -0.94359 | -0.94352 | -0.94341 |
| | 0.4 | -1.01756 | -1.01755 | -1.01752 | -1.01741 | -1.01723 |
| | 0.8 | -1.09622 | -1.09621 | -1.09616 | -1.09597 | -1.09564 |
| | 1.2 | -1.17851 | -1.17849 | -1.17838 | -1.17801 | -1.17738 |
| | 1.6 | -1.26360 | -1.26356 | -1.26334 | -1.26254 | -1.26122 |
| | 2.0 | -1.35089 | -1.35082 | -1.35030 | -1.34848 | -1.34555 |
| | 2.4 | -1.43994 | -1.43975 | -1.43846 | -1.43409 | -1.42753 |
| | 2.8 | -1.53035 | -1.52984 | -1.52643 | -1.51614 | -1.50288 |
| 3.2 | 0.0 | -1.01756 | -1.01755 | -1.01752 | -1.01741 | -1.01723 |
| | 0.4 | -1.09622 | -1.09621 | -1.09616 | -1.09597 | -1.09564 |
| | 0.8 | -1.17851 | -1.17849 | -1.17838 | -1.17801 | -1.17738 |
| | 1.2 | -1.26360 | -1.26356 | -1.26334 | -1.26254 | -1.26122 |
| | 1.6 | -1.35089 | -1.35082 | -1.35030 | -1.34848 | -1.34555 |
| | 2.0 | -1.43994 | -1.43975 | -1.43846 | -1.43409 | -1.42753 |
| | 2.4 | -1.53035 | -1.52984 | -1.52643 | -1.51614 | -1.50288 |
| | 2.8 | -1.62170 | -1.62022 | -1.61136 | -1.59030 | -1.56840 |
| | 3.2 | -1.71326 | -1.70892 | -1.68898 | -1.65460 | -1.62457 |

FRACTILES $W_\alpha$ RHO = 1.00

| DELTAS | | $\alpha$:0.0005 | 0.0010 | 0.0025 | 0.0050 | 0.0075 |
|---|---|---|---|---|---|---|
| 3.6 | 0.0 | -1.09622 | -1.09621 | -1.09616 | -1.09597 | -1.09564 |
| | 0.4 | -1.17851 | -1.17849 | -1.17838 | -1.17801 | -1.17738 |
| | 0.8 | -1.26360 | -1.26356 | -1.26334 | -1.26254 | -1.26122 |
| | 1.2 | -1.35089 | -1.35082 | -1.35030 | -1.34848 | -1.34555 |
| | 1.6 | -1.43994 | -1.43975 | -1.43846 | -1.43409 | -1.42753 |
| | 2.0 | -1.53035 | -1.52984 | -1.52643 | -1.51614 | -1.50288 |
| | 2.4 | -1.62170 | -1.62022 | -1.61136 | -1.59030 | -1.56840 |
| | 2.8 | -1.71326 | -1.70892 | -1.68898 | -1.65460 | -1.62457 |
| | 3.2 | -1.80314 | -1.79207 | -1.75704 | -1.71037 | -1.67347 |
| | 3.6 | -1.88749 | -1.86633 | -1.81669 | -1.75955 | -1.71672 |
| 4.0 | 0.0 | -1.17851 | -1.17849 | -1.17838 | -1.17801 | -1.17738 |
| | 0.4 | -1.26360 | -1.26356 | -1.26334 | -1.26254 | -1.26122 |
| | 0.8 | -1.35089 | -1.35082 | -1.35030 | -1.34848 | -1.34555 |
| | 1.2 | -1.43994 | -1.43975 | -1.43846 | -1.43409 | -1.42753 |
| | 1.6 | -1.53035 | -1.52984 | -1.52643 | -1.51614 | -1.50288 |
| | 2.0 | -1.62170 | -1.62022 | -1.61136 | -1.59030 | -1.56840 |
| | 2.4 | -1.71326 | -1.70892 | -1.68898 | -1.65460 | -1.62457 |
| | 2.8 | -1.80314 | -1.79207 | -1.75704 | -1.71037 | -1.67347 |
| | 3.2 | -1.88749 | -1.86633 | -1.81669 | -1.75955 | -1.71672 |
| | 3.6 | -1.96312 | -1.93195 | -1.86972 | -1.80343 | -1.75534 |
| | 4.0 | -2.03048 | -1.99065 | -1.91737 | -1.84289 | -1.79007 |
| 6.0 | 0.0 | -1.62170 | -1.62022 | -1.61136 | -1.59030 | -1.56840 |
| | 0.4 | -1.71326 | -1.70892 | -1.68898 | -1.65460 | -1.62457 |
| | 0.8 | -1.80314 | -1.79207 | -1.75704 | -1.71037 | -1.67347 |
| | 1.2 | -1.88749 | -1.86633 | -1.81669 | -1.75955 | -1.71672 |
| | 1.6 | -1.96312 | -1.93195 | -1.86972 | -1.80343 | -1.75534 |
| | 2.0 | -2.03048 | -1.99065 | -1.91737 | -1.84289 | -1.79007 |
| | 2.4 | -2.09113 | -2.04367 | -1.96046 | -1.87857 | -1.82145 |
| | 2.8 | -2.14622 | -2.09185 | -1.99962 | -1.91098 | -1.84995 |
| | 3.2 | -2.19651 | -2.13585 | -2.03536 | -1.94055 | -1.87595 |
| | 3.6 | -2.24262 | -2.17617 | -2.06811 | -1.96763 | -1.89974 |
| | 4.0 | -2.28503 | -2.21326 | -2.09821 | -1.99251 | -1.92161 |
| | 6.0 | -2.45458 | -2.36143 | -2.21839 | -2.09178 | -2.00877 |
| 12.0 | 0.0 | -2.45458 | -2.36143 | -2.21839 | -2.09178 | -2.00877 |
| | 0.4 | -2.48189 | -2.38529 | -2.23773 | -2.10774 | -2.02278 |
| | 0.8 | -2.50749 | -2.40765 | -2.25585 | -2.12269 | -2.03589 |
| | 1.2 | -2.53153 | -2.42864 | -2.27286 | -2.13672 | -2.04820 |
| | 1.6 | -2.55414 | -2.44839 | -2.28885 | -2.14991 | -2.05977 |
| | 2.0 | -2.57545 | -2.46700 | -2.30392 | -2.16233 | -2.07066 |
| | 2.4 | -2.59557 | -2.48456 | -2.31814 | -2.17406 | -2.08095 |
| | 2.8 | -2.61459 | -2.50117 | -2.33159 | -2.18514 | -2.09066 |
| | 3.2 | -2.63261 | -2.51689 | -2.34431 | -2.19563 | -2.09986 |
| | 3.6 | -2.64969 | -2.53180 | -2.35638 | -2.20557 | -2.10857 |
| | 4.0 | -2.66591 | -2.54596 | -2.36784 | -2.21501 | -2.11685 |
| | 6.0 | -2.73612 | -2.60722 | -2.41739 | -2.25583 | -2.15261 |
| | 12.0 | -2.87606 | -2.72927 | -2.51603 | -2.33699 | -2.22367 |
| INFINITY | | -3.29051 | -3.09024 | -2.80703 | -2.57583 | -2.43238 |

FRACTILES $W_\alpha$ RHO = 1.00

| DELTAS | | $\alpha$:0.0100 | 0.0175 | 0.0250 | 0.0375 | 0.0500 |
|---|---|---|---|---|---|---|
| 0.0 | 0.0 | -0.70700 | -0.70677 | -0.70641 | -0.70554 | -0.70433 |
| 0.4 | 0.0 | -0.70752 | -0.70729 | -0.70693 | -0.70606 | -0.70485 |
| | 0.4 | -0.71382 | -0.71358 | -0.71322 | -0.71233 | -0.71109 |
| 0.8 | 0.0 | -0.71382 | -0.71358 | -0.71322 | -0.71233 | -0.71109 |
| | 0.4 | -0.73314 | -0.73289 | -0.73250 | -0.73155 | -0.73022 |
| | 0.8 | -0.76786 | -0.76757 | -0.76713 | -0.76604 | -0.76451 |
| 1.2 | 0.0 | -0.73314 | -0.73289 | -0.73250 | -0.73155 | -0.73022 |
| | 0.4 | -0.76786 | -0.76757 | -0.76713 | -0.76604 | -0.76451 |
| | 0.8 | -0.81632 | -0.81596 | -0.81541 | -0.81405 | -0.81214 |
| | 1.2 | -0.87567 | -0.87518 | -0.87442 | -0.87257 | -0.86997 |
| 1.6 | 0.0 | -0.76786 | -0.76757 | -0.76713 | -0.76604 | -0.76451 |
| | 0.4 | -0.81632 | -0.81596 | -0.81541 | -0.81405 | -0.81214 |
| | 0.8 | -0.87567 | -0.87518 | -0.87442 | -0.87257 | -0.86997 |
| | 1.2 | -0.94326 | -0.94253 | -0.94140 | -0.93864 | -0.93480 |
| | 1.6 | -1.01698 | -1.01579 | -1.01395 | -1.00951 | -1.00340 |
| 2.0 | 0.0 | -0.81632 | -0.81596 | -0.81541 | -0.81405 | -0.81214 |
| | 0.4 | -0.87567 | -0.87518 | -0.87442 | -0.87257 | -0.86997 |
| | 0.8 | -0.94326 | -0.94253 | -0.94140 | -0.93864 | -0.93480 |
| | 1.2 | -1.01698 | -1.01579 | -1.01395 | -1.00951 | -1.00340 |
| | 1.6 | -1.09519 | -1.09308 | -1.08986 | -1.08221 | -1.07200 |
| | 2.0 | -1.17651 | -1.17248 | -1.16650 | -1.15291 | -1.13590 |
| 2.4 | 0.0 | -0.87567 | -0.87518 | -0.87442 | -0.87257 | -0.86997 |
| | 0.4 | -0.94326 | -0.94253 | -0.94140 | -0.93864 | -0.93480 |
| | 0.8 | -1.01698 | -1.01579 | -1.01395 | -1.00951 | -1.00340 |
| | 1.2 | -1.09519 | -1.09308 | -1.08986 | -1.08221 | -1.07200 |
| | 1.6 | -1.17651 | -1.17248 | -1.16650 | -1.15291 | -1.13590 |
| | 2.0 | -1.25941 | -1.25135 | -1.24011 | -1.21686 | -1.19060 |
| | 2.4 | -1.34165 | -1.32572 | -1.30609 | -1.27057 | -1.23468 |
| 2.8 | 0.0 | -0.94326 | -0.94253 | -0.94140 | -0.93864 | -0.93480 |
| | 0.4 | -1.01698 | -1.01579 | -1.01395 | -1.00951 | -1.00340 |
| | 0.8 | -1.09519 | -1.09308 | -1.08986 | -1.08221 | -1.07200 |
| | 1.2 | -1.17651 | -1.17248 | -1.16650 | -1.15291 | -1.13590 |
| | 1.6 | -1.25941 | -1.25135 | -1.24011 | -1.21686 | -1.19060 |
| | 2.0 | -1.34165 | -1.32572 | -1.30609 | -1.27057 | -1.23468 |
| | 2.4 | -1.41948 | -1.39124 | -1.36164 | -1.31422 | -1.27006 |
| | 2.8 | -1.48862 | -1.44640 | -1.40767 | -1.35027 | -1.29935 |
| 3.2 | 0.0 | -1.01698 | -1.01579 | -1.01395 | -1.00951 | -1.00340 |
| | 0.4 | -1.09519 | -1.09308 | -1.08986 | -1.08221 | -1.07200 |
| | 0.8 | -1.17651 | -1.17248 | -1.16650 | -1.15291 | -1.13590 |
| | 1.2 | -1.25941 | -1.25135 | -1.24011 | -1.21686 | -1.19060 |
| | 1.6 | -1.34165 | -1.32572 | -1.30609 | -1.27057 | -1.23468 |
| | 2.0 | -1.41948 | -1.39124 | -1.36164 | -1.31422 | -1.27006 |
| | 2.4 | -1.48862 | -1.44640 | -1.40767 | -1.35027 | -1.29935 |
| | 2.8 | -1.54764 | -1.49291 | -1.44659 | -1.38089 | -1.32428 |
| | 3.2 | -1.59820 | -1.53300 | -1.48025 | -1.40743 | -1.34589 |

FRACTILES $W_{\alpha}$ RHO = 1.00

| DELTAS | | α:0.0100 | 0.0175 | 0.0250 | 0.0375 | 0.0500 |
|---|---|---|---|---|---|---|
| 3.6 | 0.0 | -1.09519 | -1.09308 | -1.08986 | -1.08221 | -1.07200 |
| | 0.4 | -1.17651 | -1.17248 | -1.16650 | -1.15291 | -1.13590 |
| | 0.8 | -1.25941 | -1.25135 | -1.24011 | -1.21686 | -1.19060 |
| | 1.2 | -1.34165 | -1.32572 | -1.30609 | -1.27057 | -1.23468 |
| | 1.6 | -1.41948 | -1.39124 | -1.36164 | -1.31422 | -1.27006 |
| | 2.0 | -1.48862 | -1.44640 | -1.40767 | -1.35027 | -1.29935 |
| | 2.4 | -1.54764 | -1.49291 | -1.44659 | -1.38089 | -1.32428 |
| | 2.8 | -1.59820 | -1.53300 | -1.48025 | -1.40743 | -1.34589 |
| | 3.2 | -1.64235 | -1.56815 | -1.50979 | -1.43072 | -1.36483 |
| | 3.6 | -1.68145 | -1.59930 | -1.53597 | -1.45134 | -1.38157 |
| 4.0 | 0.0 | -1.17651 | -1.17248 | -1.16650 | -1.15291 | -1.13590 |
| | 0.4 | -1.25941 | -1.25135 | -1.24011 | -1.21686 | -1.19060 |
| | 0.8 | -1.34165 | -1.32572 | -1.30609 | -1.27057 | -1.23468 |
| | 1.2 | -1.41948 | -1.39124 | -1.36164 | -1.31422 | -1.27006 |
| | 1.6 | -1.48862 | -1.44640 | -1.40767 | -1.35027 | -1.29935 |
| | 2.0 | -1.54764 | -1.49291 | -1.44659 | -1.38089 | -1.32428 |
| | 2.4 | -1.59820 | -1.53300 | -1.48025 | -1.40743 | -1.34589 |
| | 2.8 | -1.64235 | -1.56815 | -1.50979 | -1.43072 | -1.36483 |
| | 3.2 | -1.68145 | -1.59930 | -1.53597 | -1.45134 | -1.38157 |
| | 3.6 | -1.71637 | -1.62712 | -1.55932 | -1.46970 | -1.39647 |
| | 4.0 | -1.74776 | -1.65210 | -1.58028 | -1.48616 | -1.40980 |
| 6.0 | 0.0 | -1.54764 | -1.49291 | -1.44659 | -1.38089 | -1.32428 |
| | 0.4 | -1.59820 | -1.53300 | -1.48025 | -1.40743 | -1.34589 |
| | 0.8 | -1.64235 | -1.56815 | -1.50979 | -1.43072 | -1.36483 |
| | 1.2 | -1.68145 | -1.59930 | -1.53597 | -1.45134 | -1.38157 |
| | 1.6 | -1.71637 | -1.62712 | -1.55932 | -1.46970 | -1.39647 |
| | 2.0 | -1.74776 | -1.65210 | -1.58028 | -1.48616 | -1.40980 |
| | 2.4 | -1.77613 | -1.67465 | -1.59919 | -1.50100 | -1.42180 |
| | 2.8 | -1.80188 | -1.69511 | -1.61633 | -1.51444 | -1.43266 |
| | 3.2 | -1.82535 | -1.71375 | -1.63194 | -1.52666 | -1.44252 |
| | 3.6 | -1.84684 | -1.73081 | -1.64621 | -1.53782 | -1.45153 |
| | 4.0 | -1.86658 | -1.74646 | -1.65931 | -1.54806 | -1.45977 |
| | 6.0 | -1.94523 | -1.80874 | -1.71133 | -1.58863 | -1.49240 |
| 12.0 | 0.0 | -1.94523 | -1.80874 | -1.71133 | -1.58863 | -1.49240 |
| | 0.4 | -1.95786 | -1.81873 | -1.71967 | -1.59512 | -1.49760 |
| | 0.8 | -1.96968 | -1.82808 | -1.72746 | -1.60119 | -1.50247 |
| | 1.2 | -1.98078 | -1.83685 | -1.73477 | -1.60687 | -1.50702 |
| | 1.6 | -1.99121 | -1.84509 | -1.74164 | -1.61221 | -1.51130 |
| | 2.0 | -2.00103 | -1.85285 | -1.74811 | -1.61723 | -1.51532 |
| | 2.4 | -2.01029 | -1.86016 | -1.75420 | -1.62196 | -1.51910 |
| | 2.8 | -2.01905 | -1.86707 | -1.75996 | -1.62643 | -1.52268 |
| | 3.2 | -2.02734 | -1.87361 | -1.76540 | -1.63065 | -1.52606 |
| | 3.6 | -2.03519 | -1.87981 | -1.77056 | -1.63466 | -1.52925 |
| | 4.0 | -2.04264 | -1.88568 | -1.77545 | -1.63845 | -1.53228 |
| | 6.0 | -2.07485 | -1.91107 | -1.79657 | -1.65481 | -1.54534 |
| | 12.0 | -2.13881 | -1.96141 | -1.83838 | -1.68712 | -1.57106 |
| INFINITY | | -2.32635 | -2.10836 | -1.95996 | -1.78046 | -1.64485 |

FRACTILES $W_{\alpha}$ RHO = 1.00

| DELTAS | | $\alpha$:0.0750 | 0.1000 | 0.1500 | 0.2000 | 0.2500 |
|---|---|---|---|---|---|---|
| 0.0 | 0.0 | -0.70084 | -0.69594 | -0.68182 | -0.66172 | -0.63531 |
| 0.4 | 0.0 | -0.70135 | -0.69645 | -0.68230 | -0.66218 | -0.63573 |
| | 0.4 | -0.70753 | -0.70253 | -0.68810 | -0.66759 | -0.64064 |
| 0.8 | 0.0 | -0.70753 | -0.70253 | -0.68810 | -0.66759 | -0.64064 |
| | 0.4 | -0.72642 | -0.72107 | -0.70565 | -0.68375 | -0.65502 |
| | 0.8 | -0.76014 | -0.75400 | -0.73635 | -0.71137 | -0.67876 |
| 1.2 | 0.0 | -0.72642 | -0.72107 | -0.70565 | -0.68375 | -0.65502 |
| | 0.4 | -0.76014 | -0.75400 | -0.73635 | -0.71137 | -0.67876 |
| | 0.8 | -0.80669 | -0.79906 | -0.77726 | -0.74670 | -0.70730 |
| | 1.2 | -0.86259 | -0.85233 | -0.82342 | -0.78385 | -0.73421 |
| 1.6 | 0.0 | -0.76014 | -0.75400 | -0.73635 | -0.71137 | -0.67876 |
| | 0.4 | -0.80669 | -0.79906 | -0.77726 | -0.74670 | -0.70730 |
| | 0.8 | -0.86259 | -0.85233 | -0.82342 | -0.78385 | -0.73421 |
| | 1.2 | -0.92400 | -0.90923 | -0.86898 | -0.81648 | -0.75372 |
| | 1.6 | -0.98662 | -0.96453 | -0.90811 | -0.83997 | -0.76354 |
| 2.0 | 0.0 | -0.80669 | -0.79906 | -0.77726 | -0.74670 | -0.70730 |
| | 0.4 | -0.86259 | -0.85233 | -0.82342 | -0.78385 | -0.73421 |
| | 0.8 | -0.92400 | -0.90923 | -0.86898 | -0.81648 | -0.75372 |
| | 1.2 | -0.98662 | -0.96453 | -0.90811 | -0.83997 | -0.76354 |
| | 1.6 | -1.04542 | -1.01286 | -0.93720 | -0.85377 | -0.76561 |
| | 2.0 | -1.09555 | -1.05084 | -0.95677 | -0.86070 | -0.76352 |
| 2.4 | 0.0 | -0.86259 | -0.85233 | -0.82342 | -0.78385 | -0.73421 |
| | 0.4 | -0.92400 | -0.90923 | -0.86898 | -0.81648 | -0.75372 |
| | 0.8 | -0.98662 | -0.96453 | -0.90811 | -0.83997 | -0.76354 |
| | 1.2 | -1.04542 | -1.01286 | -0.93720 | -0.85377 | -0.76561 |
| | 1.6 | -1.09555 | -1.05084 | -0.95677 | -0.86070 | -0.76352 |
| | 2.0 | -1.13503 | -1.07907 | -0.96984 | -0.86397 | -0.75987 |
| | 2.4 | -1.16552 | -1.10038 | -0.97916 | -0.86552 | -0.75588 |
| 2.8 | 0.0 | -0.92400 | -0.90923 | -0.86898 | -0.81648 | -0.75372 |
| | 0.4 | -0.98662 | -0.96453 | -0.90811 | -0.83997 | -0.76354 |
| | 0.8 | -1.04542 | -1.01286 | -0.93720 | -0.85377 | -0.76561 |
| | 1.2 | -1.09555 | -1.05084 | -0.95677 | -0.86070 | -0.76352 |
| | 1.6 | -1.13503 | -1.07907 | -0.96984 | -0.86397 | -0.75987 |
| | 2.0 | -1.16552 | -1.10038 | -0.97916 | -0.86552 | -0.75588 |
| | 2.4 | -1.18984 | -1.11731 | -0.98630 | -0.86623 | -0.75199 |
| | 2.8 | -1.21002 | -1.13130 | -0.99203 | -0.86650 | -0.74833 |
| 3.2 | 0.0 | -0.98662 | -0.96453 | -0.90811 | -0.83997 | -0.76354 |
| | 0.4 | -1.04542 | -1.01286 | -0.93720 | -0.85377 | -0.76561 |
| | 0.8 | -1.09555 | -1.05084 | -0.95677 | -0.86070 | -0.76352 |
| | 1.2 | -1.13503 | -1.07907 | -0.96984 | -0.86397 | -0.75987 |
| | 1.6 | -1.16552 | -1.10038 | -0.97916 | -0.86552 | -0.75588 |
| | 2.0 | -1.18984 | -1.11731 | -0.98630 | -0.86623 | -0.75199 |
| | 2.4 | -1.21002 | -1.13130 | -0.99203 | -0.86650 | -0.74833 |
| | 2.8 | -1.22719 | -1.14316 | -0.99675 | -0.86649 | -0.74492 |
| | 3.2 | -1.24203 | -1.15336 | -1.00070 | -0.86630 | -0.74176 |

FRACTILES $W_\alpha$ RHO = 1.00

| DELTAS | | α:0.0750 | 0.1000 | 0.1500 | 0.2000 | 0.2500 |
|---|---|---|---|---|---|---|
| 3.6 | 0.0 | -1.04542 | -1.01286 | -0.93720 | -0.85377 | -0.76561 |
| | 0.4 | -1.09555 | -1.05084 | -0.95677 | -0.86070 | -0.76352 |
| | 0.8 | -1.13503 | -1.07907 | -0.96984 | -0.86397 | -0.75987 |
| | 1.2 | -1.16552 | -1.10038 | -0.97916 | -0.86552 | -0.75588 |
| | 1.6 | -1.18984 | -1.11731 | -0.98630 | -0.86623 | -0.75199 |
| | 2.0 | -1.21002 | -1.13130 | -0.99203 | -0.86650 | -0.74833 |
| | 2.4 | -1.22719 | -1.14316 | -0.99675 | -0.86649 | -0.74492 |
| | 2.8 | -1.24203 | -1.15336 | -1.00070 | -0.86630 | -0.74176 |
| | 3.2 | -1.25499 | -1.16222 | -1.00404 | -0.86598 | -0.73884 |
| | 3.6 | -1.26641 | -1.16998 | -1.00689 | -0.86559 | -0.73613 |
| 4.0 | 0.0 | -1.09555 | -1.05084 | -0.95677 | -0.86070 | -0.76352 |
| | 0.4 | -1.13503 | -1.07907 | -0.96984 | -0.86397 | -0.75987 |
| | 0.8 | -1.16552 | -1.10038 | -0.97916 | -0.86552 | -0.75588 |
| | 1.2 | -1.18984 | -1.11731 | -0.98630 | -0.86623 | -0.75199 |
| | 1.6 | -1.21002 | -1.13130 | -0.99203 | -0.86650 | -0.74833 |
| | 2.0 | -1.22719 | -1.14316 | -0.99675 | -0.86649 | -0.74492 |
| | 2.4 | -1.24203 | -1.15336 | -1.00070 | -0.86630 | -0.74176 |
| | 2.8 | -1.25499 | -1.16222 | -1.00404 | -0.86598 | -0.73884 |
| | 3.2 | -1.26641 | -1.16998 | -1.00689 | -0.86559 | -0.73613 |
| | 3.6 | -1.27653 | -1.17683 | -1.00934 | -0.86515 | -0.73362 |
| | 4.0 | -1.28557 | -1.18291 | -1.01148 | -0.86467 | -0.73128 |
| 6.0 | 0.0 | -1.22719 | -1.14316 | -0.99675 | -0.86649 | -0.74492 |
| | 0.4 | -1.24203 | -1.15336 | -1.00070 | -0.86630 | -0.74176 |
| | 0.8 | -1.25499 | -1.16222 | -1.00404 | -0.86598 | -0.73884 |
| | 1.2 | -1.26641 | -1.16998 | -1.00689 | -0.86559 | -0.73613 |
| | 1.6 | -1.27653 | -1.17683 | -1.00934 | -0.86515 | -0.73362 |
| | 2.0 | -1.28557 | -1.18291 | -1.01148 | -0.86467 | -0.73128 |
| | 2.4 | -1.29368 | -1.18835 | -1.01334 | -0.86418 | -0.72912 |
| | 2.8 | -1.30099 | -1.19324 | -1.01498 | -0.86368 | -0.72710 |
| | 3.2 | -1.30762 | -1.19766 | -1.01643 | -0.86319 | -0.72522 |
| | 3.6 | -1.31366 | -1.20167 | -1.01772 | -0.86269 | -0.72346 |
| | 4.0 | -1.31918 | -1.20532 | -1.01888 | -0.86221 | -0.72181 |
| | 6.0 | -1.34090 | -1.21958 | -1.02317 | -0.85998 | -0.71496 |
| 12.0 | 0.0 | -1.34090 | -1.21958 | -1.02317 | -0.85998 | -0.71496 |
| | 0.4 | -1.34434 | -1.22183 | -1.02381 | -0.85957 | -0.71382 |
| | 0.8 | -1.34756 | -1.22392 | -1.02440 | -0.85918 | -0.71274 |
| | 1.2 | -1.35057 | -1.22587 | -1.02495 | -0.85880 | -0.71171 |
| | 1.6 | -1.35339 | -1.22770 | -1.02545 | -0.85844 | -0.71074 |
| | 2.0 | -1.35604 | -1.22941 | -1.02591 | -0.85809 | -0.70981 |
| | 2.4 | -1.35853 | -1.23102 | -1.02634 | -0.85775 | -0.70893 |
| | 2.8 | -1.36088 | -1.23253 | -1.02674 | -0.85742 | -0.70809 |
| | 3.2 | -1.36310 | -1.23396 | -1.02712 | -0.85711 | -0.70729 |
| | 3.6 | -1.36520 | -1.23531 | -1.02746 | -0.85680 | -0.70653 |
| | 4.0 | -1.36719 | -1.23658 | -1.02779 | -0.85651 | -0.70580 |
| | 6.0 | -1.37572 | -1.24204 | -1.02914 | -0.85519 | -0.70261 |
| | 12.0 | -1.39244 | -1.25261 | -1.03155 | -0.85230 | -0.69599 |
| INFINITY | | -1.43953 | -1.28155 | -1.03643 | -0.84162 | -0.67449 |

FRACTILES $w_\alpha$ RHO = 1.00

| DELTAS | | $\alpha$:0.3000 | 0.3500 | 0.4000 | 0.4500 | 0.5000 |
|---|---|---|---|---|---|---|
| 0.0 | 0.0 | -0.60212 | -0.56151 | -0.51266 | -0.45445 | -0.38542 |
| 0.4 | 0.0 | -0.60249 | -0.56182 | -0.51290 | -0.45461 | -0.38549 |
| | 0.4 | -0.60677 | -0.56536 | -0.51557 | -0.45629 | -0.38605 |
| 0.8 | 0.0 | -0.60677 | -0.56536 | -0.51557 | -0.45629 | -0.38605 |
| | 0.4 | -0.61898 | -0.57502 | -0.52229 | -0.45970 | -0.38580 |
| | 0.8 | -0.63810 | -0.58882 | -0.53018 | -0.46119 | -0.38049 |
| 1.2 | 0.0 | -0.61898 | -0.57502 | -0.52229 | -0.45970 | -0.38580 |
| | 0.4 | -0.63810 | -0.58882 | -0.53018 | -0.46119 | -0.38049 |
| | 0.8 | -0.65888 | -0.60115 | -0.53363 | -0.45558 | -0.36594 |
| | 1.2 | -0.67497 | -0.60636 | -0.52835 | -0.44052 | -0.34206 |
| 1.6 | 0.0 | -0.63810 | -0.58882 | -0.53018 | -0.46119 | -0.38049 |
| | 0.4 | -0.65888 | -0.60115 | -0.53363 | -0.45558 | -0.36594 |
| | 0.8 | -0.67497 | -0.60636 | -0.52835 | -0.44052 | -0.34206 |
| | 1.2 | -0.68209 | -0.60230 | -0.51453 | -0.41842 | -0.31317 |
| | 1.6 | -0.68047 | -0.59134 | -0.49605 | -0.39403 | -0.28426 |
| 2.0 | 0.0 | -0.65888 | -0.60115 | -0.53363 | -0.45558 | -0.36594 |
| | 0.4 | -0.67497 | -0.60636 | -0.52835 | -0.44052 | -0.34206 |
| | 0.8 | -0.68209 | -0.60230 | -0.51453 | -0.41842 | -0.31317 |
| | 1.2 | -0.68047 | -0.59134 | -0.49605 | -0.39403 | -0.28426 |
| | 1.6 | -0.67356 | -0.57752 | -0.47691 | -0.37086 | -0.25817 |
| | 2.0 | -0.66479 | -0.56368 | -0.45924 | -0.35034 | -0.23563 |
| 2.4 | 0.0 | -0.67497 | -0.60636 | -0.52835 | -0.44052 | -0.34206 |
| | 0.4 | -0.68209 | -0.60230 | -0.51453 | -0.41842 | -0.31317 |
| | 0.8 | -0.68047 | -0.59134 | -0.49605 | -0.39403 | -0.28426 |
| | 1.2 | -0.67356 | -0.57752 | -0.47691 | -0.37086 | -0.25817 |
| | 1.6 | -0.66479 | -0.56368 | -0.45924 | -0.35034 | -0.23563 |
| | 2.0 | -0.65601 | -0.55102 | -0.44364 | -0.33257 | -0.21636 |
| | 2.4 | -0.64789 | -0.53977 | -0.43003 | -0.31722 | -0.19984 |
| 2.8 | 0.0 | -0.68209 | -0.60230 | -0.51453 | -0.41842 | -0.31317 |
| | 0.4 | -0.68047 | -0.59134 | -0.49605 | -0.39403 | -0.28426 |
| | 0.8 | -0.67356 | -0.57752 | -0.47691 | -0.37086 | -0.25817 |
| | 1.2 | -0.66479 | -0.56368 | -0.45924 | -0.35034 | -0.23563 |
| | 1.6 | -0.65601 | -0.55102 | -0.44364 | -0.33257 | -0.21636 |
| | 2.0 | -0.64789 | -0.53977 | -0.43003 | -0.31722 | -0.19984 |
| | 2.4 | -0.64055 | -0.52983 | -0.41812 | -0.30389 | -0.18557 |
| | 2.8 | -0.63396 | -0.52102 | -0.40766 | -0.29224 | -0.17314 |
| 3.2 | 0.0 | -0.68047 | -0.59134 | -0.49605 | -0.39403 | -0.28426 |
| | 0.4 | -0.67356 | -0.57752 | -0.47691 | -0.37086 | -0.25817 |
| | 0.8 | -0.66479 | -0.56368 | -0.45924 | -0.35034 | -0.23563 |
| | 1.2 | -0.65601 | -0.55102 | -0.44364 | -0.33257 | -0.21636 |
| | 1.6 | -0.64789 | -0.53977 | -0.43003 | -0.31722 | -0.19984 |
| | 2.0 | -0.64055 | -0.52983 | -0.41812 | -0.30389 | -0.18557 |
| | 2.4 | -0.63396 | -0.52102 | -0.40766 | -0.29224 | -0.17314 |
| | 2.8 | -0.62803 | -0.51318 | -0.39840 | -0.28197 | -0.16222 |
| | 3.2 | -0.62266 | -0.50616 | -0.39016 | -0.27286 | -0.15257 |

FRACTILES $W_{\alpha}$ RHO = 1.00

| DELTAS | | α:0.3000 | 0.3500 | 0.4000 | 0.4500 | 0.5000 |
|---|---|---|---|---|---|---|
| 3.6 | 0.0 | -0.67356 | -0.57752 | -0.47691 | -0.37086 | -0.25817 |
| | 0.4 | -0.66479 | -0.56368 | -0.45924 | -0.35034 | -0.23563 |
| | 0.8 | -0.65601 | -0.55102 | -0.44364 | -0.33257 | -0.21636 |
| | 1.2 | -0.64789 | -0.53977 | -0.43003 | -0.31722 | -0.19984 |
| | 1.6 | -0.64055 | -0.52983 | -0.41812 | -0.30389 | -0.18557 |
| | 2.0 | -0.63396 | -0.52102 | -0.40766 | -0.29224 | -0.17314 |
| | 2.4 | -0.62803 | -0.51318 | -0.39840 | -0.28197 | -0.16222 |
| | 2.8 | -0.62266 | -0.50616 | -0.39016 | -0.27286 | -0.15257 |
| | 3.2 | -0.61780 | -0.49985 | -0.38278 | -0.26473 | -0.14398 |
| | 3.6 | -0.61338 | -0.49415 | -0.37613 | -0.25744 | -0.13628 |
| 4.0 | 0.0 | -0.66479 | -0.56368 | -0.45924 | -0.35034 | -0.23563 |
| | 0.4 | -0.65601 | -0.55102 | -0.44364 | -0.33257 | -0.21636 |
| | 0.8 | -0.64789 | -0.53977 | -0.43003 | -0.31722 | -0.19984 |
| | 1.2 | -0.64055 | -0.52983 | -0.41812 | -0.30389 | -0.18557 |
| | 1.6 | -0.63396 | -0.52102 | -0.40766 | -0.29224 | -0.17314 |
| | 2.0 | -0.62803 | -0.51318 | -0.39840 | -0.28197 | -0.16222 |
| | 2.4 | -0.62266 | -0.50616 | -0.39016 | -0.27286 | -0.15257 |
| | 2.8 | -0.61780 | -0.49985 | -0.38278 | -0.26473 | -0.14398 |
| | 3.2 | -0.61338 | -0.49415 | -0.37613 | -0.25744 | -0.13628 |
| | 3.6 | -0.60934 | -0.48897 | -0.37013 | -0.25086 | -0.12936 |
| | 4.0 | -0.60564 | -0.48425 | -0.36467 | -0.24489 | -0.12309 |
| 6.0 | 0.0 | -0.62803 | -0.51318 | -0.39840 | -0.28197 | -0.16222 |
| | 0.4 | -0.62266 | -0.50616 | -0.39016 | -0.27286 | -0.15257 |
| | 0.8 | -0.61780 | -0.49985 | -0.38278 | -0.26473 | -0.14398 |
| | 1.2 | -0.61338 | -0.49415 | -0.37613 | -0.25744 | -0.13628 |
| | 1.6 | -0.60934 | -0.48897 | -0.37013 | -0.25086 | -0.12936 |
| | 2.0 | -0.60564 | -0.48425 | -0.36467 | -0.24489 | -0.12309 |
| | 2.4 | -0.60224 | -0.47994 | -0.35969 | -0.23946 | -0.11740 |
| | 2.8 | -0.59910 | -0.47598 | -0.35513 | -0.23449 | -0.11220 |
| | 3.2 | -0.59620 | -0.47233 | -0.35094 | -0.22994 | -0.10743 |
| | 3.6 | -0.59352 | -0.46896 | -0.34708 | -0.22575 | -0.10305 |
| | 4.0 | -0.59102 | -0.46584 | -0.34351 | -0.22187 | -0.09901 |
| | 6.0 | -0.58080 | -0.45315 | -0.32905 | -0.20625 | -0.08276 |
| 12.0 | 0.0 | -0.58080 | -0.45315 | -0.32905 | -0.20625 | -0.08276 |
| | 0.4 | -0.57911 | -0.45107 | -0.32670 | -0.20371 | -0.08013 |
| | 0.8 | -0.57753 | -0.44911 | -0.32448 | -0.20133 | -0.07765 |
| | 1.2 | -0.57603 | -0.44727 | -0.32240 | -0.19908 | -0.07533 |
| | 1.6 | -0.57461 | -0.44553 | -0.32043 | -0.19697 | -0.07314 |
| | 2.0 | -0.57327 | -0.44388 | -0.31857 | -0.19497 | -0.07107 |
| | 2.4 | -0.57200 | -0.44233 | -0.31681 | -0.19308 | -0.06911 |
| | 2.8 | -0.57079 | -0.44085 | -0.31514 | -0.19129 | -0.06726 |
| | 3.2 | -0.56964 | -0.43944 | -0.31356 | -0.18959 | -0.06551 |
| | 3.6 | -0.56854 | -0.43811 | -0.31206 | -0.18798 | -0.06384 |
| | 4.0 | -0.56750 | -0.43684 | -0.31062 | -0.18645 | -0.06226 |
| | 6.0 | -0.56294 | -0.43130 | -0.30440 | -0.17979 | -0.05538 |
| | 12.0 | -0.55365 | -0.42007 | -0.29183 | -0.16638 | -0.04159 |
| INFINITY | | -0.52440 | -0.38532 | -0.25335 | -0.12566 | 0.0 |

FRACTILES $w_\alpha$ RHO = 1.00

| DELTAS | | $\alpha$:0.5500 | 0.6000 | 0.6500 | 0.7000 | 0.7500 |
|---|---|---|---|---|---|---|
| 0.0 | 0.0 | -0.30359 | -0.20624 | -0.08948 | 0.05246 | 0.22861 |
| 0.4 | 0.0 | -0.30357 | -0.20610 | -0.08921 | 0.05287 | 0.22917 |
| | 0.4 | -0.30286 | -0.20400 | -0.08558 | 0.05816 | 0.23621 |
| 0.8 | 0.0 | -0.30286 | -0.20400 | -0.08558 | 0.05816 | 0.23621 |
| | 0.4 | -0.29863 | -0.19550 | -0.07258 | 0.07577 | 0.25835 |
| | 0.8 | -0.28627 | -0.17602 | -0.04614 | 0.10870 | 0.29681 |
| 1.2 | 0.0 | -0.29863 | -0.19550 | -0.07258 | 0.07577 | 0.25835 |
| | 0.4 | -0.28627 | -0.17602 | -0.04614 | 0.10870 | 0.29681 |
| | 0.8 | -0.26315 | -0.14498 | -0.00817 | 0.15220 | 0.34387 |
| | 1.2 | -0.23162 | -0.10717 | 0.03436 | 0.19760 | 0.38987 |
| 1.6 | 0.0 | -0.28627 | -0.17602 | -0.04614 | 0.10870 | 0.29681 |
| | 0.4 | -0.26315 | -0.14498 | -0.00817 | 0.15220 | 0.34387 |
| | 0.8 | -0.23162 | -0.10717 | 0.03436 | 0.19760 | 0.38987 |
| | 1.2 | -0.19742 | -0.06913 | 0.07469 | 0.23852 | 0.42936 |
| | 1.6 | -0.16528 | -0.03500 | 0.10956 | 0.27276 | 0.46135 |
| 2.0 | 0.0 | -0.26315 | -0.14498 | -0.00817 | 0.15220 | 0.34387 |
| | 0.4 | -0.23162 | -0.10717 | 0.03436 | 0.19760 | 0.38987 |
| | 0.8 | -0.19742 | -0.06913 | 0.07469 | 0.23852 | 0.42936 |
| | 1.2 | -0.16528 | -0.03500 | 0.10956 | 0.27276 | 0.46135 |
| | 1.6 | -0.13726 | -0.00598 | 0.13859 | 0.30071 | 0.48690 |
| | 2.0 | -0.11346 | 0.01831 | 0.16259 | 0.32353 | 0.50745 |
| 2.4 | 0.0 | -0.23162 | -0.10717 | 0.03436 | 0.19760 | 0.38987 |
| | 0.4 | -0.19742 | -0.06913 | 0.07469 | 0.23852 | 0.42936 |
| | 0.8 | -0.16528 | -0.03500 | 0.10956 | 0.27276 | 0.46135 |
| | 1.2 | -0.13726 | -0.00598 | 0.13859 | 0.30071 | 0.48690 |
| | 1.6 | -0.11346 | 0.01831 | 0.16259 | 0.32353 | 0.50745 |
| | 2.0 | -0.09331 | 0.03872 | 0.18259 | 0.34238 | 0.52420 |
| | 2.4 | -0.07614 | 0.05601 | 0.19944 | 0.35814 | 0.53807 |
| 2.8 | 0.0 | -0.19742 | -0.06913 | 0.07469 | 0.23852 | 0.42936 |
| | 0.4 | -0.16528 | -0.03500 | 0.10956 | 0.27276 | 0.46135 |
| | 0.8 | -0.13726 | -0.00598 | 0.13859 | 0.30071 | 0.48690 |
| | 1.2 | -0.11346 | 0.01831 | 0.16259 | 0.32353 | 0.50745 |
| | 1.6 | -0.09331 | 0.03872 | 0.18259 | 0.34238 | 0.52420 |
| | 2.0 | -0.07614 | 0.05601 | 0.19944 | 0.35814 | 0.53807 |
| | 2.4 | -0.06138 | 0.07081 | 0.21380 | 0.37148 | 0.54970 |
| | 2.8 | -0.04857 | 0.08361 | 0.22616 | 0.38291 | 0.55959 |
| 3.2 | 0.0 | -0.16528 | -0.03500 | 0.10956 | 0.27276 | 0.46135 |
| | 0.4 | -0.13726 | -0.00598 | 0.13859 | 0.30071 | 0.48690 |
| | 0.8 | -0.11346 | 0.01831 | 0.16259 | 0.32353 | 0.50745 |
| | 1.2 | -0.09331 | 0.03872 | 0.18259 | 0.34238 | 0.52420 |
| | 1.6 | -0.07614 | 0.05601 | 0.19944 | 0.35814 | 0.53807 |
| | 2.0 | -0.06138 | 0.07081 | 0.21380 | 0.37148 | 0.54970 |
| | 2.4 | -0.04857 | 0.08361 | 0.22616 | 0.38291 | 0.55959 |
| | 2.8 | -0.03735 | 0.09478 | 0.23691 | 0.39280 | 0.56808 |
| | 3.2 | -0.02746 | 0.10460 | 0.24633 | 0.40143 | 0.57544 |

FRACTILES $W_\alpha$ RHO = 1.00

| DELTAS | | $\alpha$:0.5500 | 0.6000 | 0.6500 | 0.7000 | 0.7500 |
|---|---|---|---|---|---|---|
| 3.6 | 0.0 | -0.13726 | -0.00598 | 0.13859 | 0.30071 | 0.48690 |
| | 0.4 | -0.11346 | 0.01831 | 0.16259 | 0.32353 | 0.50745 |
| | 0.8 | -0.09331 | 0.03872 | 0.18259 | 0.34238 | 0.52420 |
| | 1.2 | -0.07614 | 0.05601 | 0.19944 | 0.35814 | 0.53807 |
| | 1.6 | -0.06138 | 0.07081 | 0.21380 | 0.37148 | 0.54970 |
| | 2.0 | -0.04857 | 0.08361 | 0.22616 | 0.38291 | 0.55959 |
| | 2.4 | -0.03735 | 0.09478 | 0.23691 | 0.39280 | 0.56808 |
| | 2.8 | -0.02746 | 0.10460 | 0.24633 | 0.40143 | 0.57544 |
| | 3.2 | -0.01868 | 0.11330 | 0.25465 | 0.40903 | 0.58188 |
| | 3.6 | -0.01083 | 0.12106 | 0.26205 | 0.41576 | 0.58756 |
| 4.0 | 0.0 | -0.11346 | 0.01831 | 0.16259 | 0.32353 | 0.50745 |
| | 0.4 | -0.09331 | 0.03872 | 0.18259 | 0.34238 | 0.52420 |
| | 0.8 | -0.07614 | 0.05601 | 0.19944 | 0.35814 | 0.53807 |
| | 1.2 | -0.06138 | 0.07081 | 0.21380 | 0.37148 | 0.54970 |
| | 1.6 | -0.04857 | 0.08361 | 0.22616 | 0.38291 | 0.55959 |
| | 2.0 | -0.03735 | 0.09478 | 0.23691 | 0.39280 | 0.56808 |
| | 2.4 | -0.02746 | 0.10460 | 0.24633 | 0.40143 | 0.57544 |
| | 2.8 | -0.01868 | 0.11330 | 0.25465 | 0.40903 | 0.58188 |
| | 3.2 | -0.01083 | 0.12106 | 0.26205 | 0.41576 | 0.58756 |
| | 3.6 | -0.00378 | 0.12802 | 0.26867 | 0.42177 | 0.59260 |
| | 4.0 | 0.00259 | 0.13429 | 0.27462 | 0.42715 | 0.59710 |
| 6.0 | 0.0 | -0.03735 | 0.09478 | 0.23691 | 0.39280 | 0.56808 |
| | 0.4 | -0.02746 | 0.10460 | 0.24633 | 0.40143 | 0.57544 |
| | 0.8 | -0.01868 | 0.11330 | 0.25465 | 0.40903 | 0.58188 |
| | 1.2 | -0.01083 | 0.12106 | 0.26205 | 0.41576 | 0.58756 |
| | 1.6 | -0.00378 | 0.12802 | 0.26867 | 0.42177 | 0.59260 |
| | 2.0 | 0.00259 | 0.13429 | 0.27462 | 0.42715 | 0.59710 |
| | 2.4 | 0.00838 | 0.13997 | 0.28001 | 0.43201 | 0.60114 |
| | 2.8 | 0.01364 | 0.14514 | 0.28490 | 0.43641 | 0.60479 |
| | 3.2 | 0.01847 | 0.14987 | 0.28936 | 0.44042 | 0.60810 |
| | 3.6 | 0.02289 | 0.15420 | 0.29345 | 0.44409 | 0.61112 |
| | 4.0 | 0.02697 | 0.15819 | 0.29721 | 0.44745 | 0.61388 |
| | 6.0 | 0.04334 | 0.17416 | 0.31220 | 0.46079 | 0.62474 |
| 12.0 | 0.0 | 0.04334 | 0.17416 | 0.31220 | 0.46079 | 0.62474 |
| | 0.4 | 0.04599 | 0.17673 | 0.31461 | 0.46293 | 0.62647 |
| | 0.8 | 0.04848 | 0.17915 | 0.31687 | 0.46493 | 0.62808 |
| | 1.2 | 0.05081 | 0.18141 | 0.31899 | 0.46680 | 0.62959 |
| | 1.6 | 0.05301 | 0.18355 | 0.32098 | 0.46856 | 0.63101 |
| | 2.0 | 0.05508 | 0.18556 | 0.32285 | 0.47022 | 0.63234 |
| | 2.4 | 0.05704 | 0.18746 | 0.32462 | 0.47178 | 0.63359 |
| | 2.8 | 0.05889 | 0.18925 | 0.32630 | 0.47326 | 0.63477 |
| | 3.2 | 0.06065 | 0.19096 | 0.32788 | 0.47465 | 0.63588 |
| | 3.6 | 0.06232 | 0.19257 | 0.32938 | 0.47597 | 0.63694 |
| | 4.0 | 0.06390 | 0.19410 | 0.33081 | 0.47723 | 0.63794 |
| | 6.0 | 0.07076 | 0.20074 | 0.33697 | 0.48264 | 0.64223 |
| | 12.0 | 0.08451 | 0.21398 | 0.34923 | 0.49334 | 0.65065 |
| INFINITY | | 0.12566 | 0.25335 | 0.38532 | 0.52440 | 0.67449 |

FRACTILES $W_\alpha$ RHO = 1.00

| DELTAS | | $\alpha$:0.8000 | 0.8500 | 0.9000 | 0.9250 | 0.9500 |
|---|---|---|---|---|---|---|
| 0.0 | 0.0 | 0.45423 | 0.75820 | 1.20600 | 1.53446 | 2.00921 |
| 0.4 | 0.0 | 0.45496 | 0.75909 | 1.20701 | 1.53546 | 2.01008 |
| | 0.4 | 0.46375 | 0.76946 | 1.21814 | 1.54604 | 2.01834 |
| 0.8 | 0.0 | 0.46375 | 0.76946 | 1.21814 | 1.54604 | 2.01834 |
| | 0.4 | 0.48995 | 0.79839 | 1.24628 | 1.57053 | 2.03372 |
| | 0.8 | 0.53218 | 0.84108 | 1.28248 | 1.59794 | 2.04400 |
| 1.2 | 0.0 | 0.48995 | 0.79839 | 1.24628 | 1.57053 | 2.03372 |
| | 0.4 | 0.53218 | 0.84108 | 1.28248 | 1.59794 | 2.04400 |
| | 0.8 | 0.57993 | 0.88496 | 1.31407 | 1.61719 | 2.04209 |
| | 1.2 | 0.62349 | 0.92160 | 1.33589 | 1.62597 | 2.02990 |
| 1.6 | 0.0 | 0.53218 | 0.84108 | 1.28248 | 1.59794 | 2.04400 |
| | 0.4 | 0.57993 | 0.88496 | 1.31407 | 1.61719 | 2.04209 |
| | 0.8 | 0.62349 | 0.92160 | 1.33589 | 1.62597 | 2.02990 |
| | 1.2 | 0.65895 | 0.94920 | 1.34892 | 1.62690 | 2.01200 |
| | 1.6 | 0.68656 | 0.96926 | 1.35583 | 1.62318 | 1.99199 |
| 2.0 | 0.0 | 0.57993 | 0.88496 | 1.31407 | 1.61719 | 2.04209 |
| | 0.4 | 0.62349 | 0.92160 | 1.33589 | 1.62597 | 2.02990 |
| | 0.8 | 0.65895 | 0.94920 | 1.34892 | 1.62690 | 2.01200 |
| | 1.2 | 0.68656 | 0.96926 | 1.35583 | 1.62318 | 1.99199 |
| | 1.6 | 0.70796 | 0.98386 | 1.35890 | 1.61708 | 1.97192 |
| | 2.0 | 0.72474 | 0.99465 | 1.35967 | 1.60994 | 1.95278 |
| 2.4 | 0.0 | 0.62349 | 0.92160 | 1.33589 | 1.62597 | 2.02990 |
| | 0.4 | 0.65895 | 0.94920 | 1.34892 | 1.62690 | 2.01200 |
| | 0.8 | 0.68656 | 0.96926 | 1.35583 | 1.62318 | 1.99199 |
| | 1.2 | 0.70796 | 0.98386 | 1.35890 | 1.61708 | 1.97192 |
| | 1.6 | 0.72474 | 0.99465 | 1.35967 | 1.60994 | 1.95278 |
| | 2.0 | 0.73814 | 1.00277 | 1.35907 | 1.60249 | 1.93499 |
| | 2.4 | 0.74902 | 1.00901 | 1.35768 | 1.59512 | 1.91863 |
| 2.8 | 0.0 | 0.65895 | 0.94920 | 1.34892 | 1.62690 | 2.01200 |
| | 0.4 | 0.68656 | 0.96926 | 1.35583 | 1.62318 | 1.99199 |
| | 0.8 | 0.70796 | 0.98386 | 1.35890 | 1.61708 | 1.97192 |
| | 1.2 | 0.72474 | 0.99465 | 1.35967 | 1.60994 | 1.95278 |
| | 1.6 | 0.73814 | 1.00277 | 1.35907 | 1.60249 | 1.93499 |
| | 2.0 | 0.74902 | 1.00901 | 1.35768 | 1.59512 | 1.91863 |
| | 2.4 | 0.75800 | 1.01387 | 1.35584 | 1.58805 | 1.90370 |
| | 2.8 | 0.76550 | 1.01773 | 1.35376 | 1.58136 | 1.89008 |
| 3.2 | 0.0 | 0.68656 | 0.96926 | 1.35583 | 1.62318 | 1.99199 |
| | 0.4 | 0.70796 | 0.98386 | 1.35890 | 1.61708 | 1.97192 |
| | 0.8 | 0.72474 | 0.99465 | 1.35967 | 1.60994 | 1.95278 |
| | 1.2 | 0.73814 | 1.00277 | 1.35907 | 1.60249 | 1.93499 |
| | 1.6 | 0.74902 | 1.00901 | 1.35768 | 1.59512 | 1.91863 |
| | 2.0 | 0.75800 | 1.01387 | 1.35584 | 1.58805 | 1.90370 |
| | 2.4 | 0.76550 | 1.01773 | 1.35376 | 1.58136 | 1.89008 |
| | 2.8 | 0.77186 | 1.02083 | 1.35158 | 1.57508 | 1.87766 |
| | 3.2 | 0.77730 | 1.02334 | 1.34937 | 1.56922 | 1.86632 |

FRACTILES $W_\alpha$ RHO = 1.00

| DELTAS | | $\alpha$:0.8000 | 0.8500 | 0.9000 | 0.9250 | 0.9500 |
|---|---|---|---|---|---|---|
| 3.6 | 0.0 | 0.70796 | 0.98386 | 1.35890 | 1.61708 | 1.97192 |
| | 0.4 | 0.72474 | 0.99465 | 1.35967 | 1.60994 | 1.95278 |
| | 0.8 | 0.73814 | 1.00277 | 1.35907 | 1.60249 | 1.93499 |
| | 1.2 | 0.74902 | 1.00901 | 1.35768 | 1.59512 | 1.91863 |
| | 1.6 | 0.75800 | 1.01387 | 1.35584 | 1.58805 | 1.90370 |
| | 2.0 | 0.76550 | 1.01773 | 1.35376 | 1.58136 | 1.89008 |
| | 2.4 | 0.77186 | 1.02083 | 1.35158 | 1.57508 | 1.87766 |
| | 2.8 | 0.77730 | 1.02334 | 1.34937 | 1.56922 | 1.86632 |
| | 3.2 | 0.78200 | 1.02540 | 1.34719 | 1.56376 | 1.85596 |
| | 3.6 | 0.78609 | 1.02711 | 1.34507 | 1.55867 | 1.84645 |
| 4.0 | 0.0 | 0.72474 | 0.99465 | 1.35967 | 1.60994 | 1.95278 |
| | 0.4 | 0.73814 | 1.00277 | 1.35907 | 1.60249 | 1.93499 |
| | 0.8 | 0.74902 | 1.00901 | 1.35768 | 1.59512 | 1.91863 |
| | 1.2 | 0.75800 | 1.01387 | 1.35584 | 1.58805 | 1.90370 |
| | 1.6 | 0.76550 | 1.01773 | 1.35376 | 1.58136 | 1.89008 |
| | 2.0 | 0.77186 | 1.02083 | 1.35158 | 1.57508 | 1.87766 |
| | 2.4 | 0.77730 | 1.02334 | 1.34937 | 1.56922 | 1.86632 |
| | 2.8 | 0.78200 | 1.02540 | 1.34719 | 1.56376 | 1.85596 |
| | 3.2 | 0.78609 | 1.02711 | 1.34507 | 1.55867 | 1.84645 |
| | 3.6 | 0.78969 | 1.02854 | 1.34302 | 1.55394 | 1.83772 |
| | 4.0 | 0.79287 | 1.02974 | 1.34106 | 1.54954 | 1.82968 |
| 6.0 | 0.0 | 0.77186 | 1.02083 | 1.35158 | 1.57508 | 1.87766 |
| | 0.4 | 0.77730 | 1.02334 | 1.34937 | 1.56922 | 1.86632 |
| | 0.8 | 0.78200 | 1.02540 | 1.34719 | 1.56376 | 1.85596 |
| | 1.2 | 0.78609 | 1.02711 | 1.34507 | 1.55867 | 1.84645 |
| | 1.6 | 0.78969 | 1.02854 | 1.34302 | 1.55394 | 1.83772 |
| | 2.0 | 0.79287 | 1.02974 | 1.34106 | 1.54954 | 1.82968 |
| | 2.4 | 0.79570 | 1.03076 | 1.33918 | 1.54543 | 1.82225 |
| | 2.8 | 0.79823 | 1.03163 | 1.33739 | 1.54160 | 1.81537 |
| | 3.2 | 0.80051 | 1.03237 | 1.33569 | 1.53802 | 1.80899 |
| | 3.6 | 0.80258 | 1.03301 | 1.33407 | 1.53466 | 1.80306 |
| | 4.0 | 0.80445 | 1.03357 | 1.33253 | 1.53152 | 1.79752 |
| | 6.0 | 0.81170 | 1.03545 | 1.32591 | 1.51838 | 1.77470 |
| 12.0 | 0.0 | 0.81170 | 1.03545 | 1.32591 | 1.51838 | 1.77470 |
| | 0.4 | 0.81283 | 1.03570 | 1.32477 | 1.51617 | 1.77092 |
| | 0.8 | 0.81388 | 1.03593 | 1.32368 | 1.51409 | 1.76734 |
| | 1.2 | 0.81486 | 1.03612 | 1.32265 | 1.51211 | 1.76397 |
| | 1.6 | 0.81578 | 1.03630 | 1.32166 | 1.51023 | 1.76077 |
| | 2.0 | 0.81663 | 1.03646 | 1.32071 | 1.50844 | 1.75773 |
| | 2.4 | 0.81743 | 1.03660 | 1.31981 | 1.50674 | 1.75485 |
| | 2.8 | 0.81819 | 1.03672 | 1.31895 | 1.50513 | 1.75211 |
| | 3.2 | 0.81890 | 1.03684 | 1.31812 | 1.50358 | 1.74950 |
| | 3.6 | 0.81956 | 1.03694 | 1.31733 | 1.50211 | 1.74702 |
| | 4.0 | 0.82019 | 1.03703 | 1.31657 | 1.50070 | 1.74465 |
| | 6.0 | 0.82288 | 1.03736 | 1.31319 | 1.49450 | 1.73426 |
| | 12.0 | 0.82803 | 1.03772 | 1.30605 | 1.48164 | 1.71295 |
| INFINITY | | 0.84162 | 1.03643 | 1.28155 | 1.43953 | 1.64485 |

FRACTILES $W_\alpha$ RHO = 1.00

| DELTAS | | $\alpha$:0.9625 | 0.9750 | 0.9825 | 0.9900 | 0.9925 |
|---|---|---|---|---|---|---|
| 0.0 | 0.0 | 2.35294 | 2.84532 | 3.28488 | 3.98447 | 4.34810 |
| 0.4 | 0.0 | 2.35360 | 2.84556 | 3.28461 | 3.98314 | 4.34608 |
| | 0.4 | 2.35914 | 2.84576 | 3.27874 | 3.96526 | 4.32091 |
| 0.8 | 0.0 | 2.35914 | 2.84576 | 3.27874 | 3.96526 | 4.32091 |
| | 0.4 | 2.36552 | 2.83628 | 3.25255 | 3.90835 | 4.24633 |
| | 0.8 | 2.36084 | 2.80731 | 3.19961 | 3.81383 | 4.12887 |
| 1.2 | 0.0 | 2.36552 | 2.83628 | 3.25255 | 3.90835 | 4.24633 |
| | 0.4 | 2.36084 | 2.80731 | 3.19961 | 3.81383 | 4.12887 |
| | 0.8 | 2.34183 | 2.76187 | 3.12912 | 3.70138 | 3.99380 |
| | 1.2 | 2.31334 | 2.70888 | 3.05338 | 3.58815 | 3.86061 |
| 1.6 | 0.0 | 2.36084 | 2.80731 | 3.19961 | 3.81383 | 4.12887 |
| | 0.4 | 2.34183 | 2.76187 | 3.12912 | 3.70138 | 3.99380 |
| | 0.8 | 2.31334 | 2.70888 | 3.05338 | 3.58815 | 3.86061 |
| | 1.2 | 2.28110 | 2.65534 | 2.98024 | 3.48299 | 3.73848 |
| | 1.6 | 2.24879 | 2.60488 | 2.91317 | 3.38887 | 3.63008 |
| 2.0 | 0.0 | 2.34183 | 2.76187 | 3.12912 | 3.70138 | 3.99380 |
| | 0.4 | 2.31334 | 2.70888 | 3.05338 | 3.58815 | 3.86061 |
| | 0.8 | 2.28110 | 2.65534 | 2.98024 | 3.48299 | 3.73848 |
| | 1.2 | 2.24879 | 2.60488 | 2.91317 | 3.38887 | 3.63008 |
| | 1.6 | 2.21824 | 2.55889 | 2.85310 | 3.30593 | 3.53508 |
| | 2.0 | 2.19012 | 2.51761 | 2.79981 | 3.23320 | 3.45209 |
| 2.4 | 0.0 | 2.31334 | 2.70888 | 3.05338 | 3.58815 | 3.86061 |
| | 0.4 | 2.28110 | 2.65534 | 2.98024 | 3.48299 | 3.73848 |
| | 0.8 | 2.24879 | 2.60488 | 2.91317 | 3.38887 | 3.63008 |
| | 1.2 | 2.21824 | 2.55889 | 2.85310 | 3.30593 | 3.53508 |
| | 1.6 | 2.19012 | 2.51761 | 2.79981 | 3.23320 | 3.45209 |
| | 2.0 | 2.16460 | 2.48076 | 2.75266 | 3.16937 | 3.37949 |
| | 2.4 | 2.14154 | 2.44790 | 2.71089 | 3.11318 | 3.31573 |
| 2.8 | 0.0 | 2.28110 | 2.65534 | 2.98024 | 3.48299 | 3.73848 |
| | 0.4 | 2.24879 | 2.60488 | 2.91317 | 3.38887 | 3.63008 |
| | 0.8 | 2.21824 | 2.55889 | 2.85310 | 3.30593 | 3.53508 |
| | 1.2 | 2.19012 | 2.51761 | 2.79981 | 3.23320 | 3.45209 |
| | 1.6 | 2.16460 | 2.48076 | 2.75266 | 3.16937 | 3.37949 |
| | 2.0 | 2.14154 | 2.44790 | 2.71089 | 3.11318 | 3.31573 |
| | 2.4 | 2.12075 | 2.41855 | 2.67377 | 3.06351 | 3.25946 |
| | 2.8 | 2.10198 | 2.39226 | 2.64066 | 3.01939 | 3.20956 |
| 3.2 | 0.0 | 2.24879 | 2.60488 | 2.91317 | 3.38887 | 3.63008 |
| | 0.4 | 2.21824 | 2.55889 | 2.85310 | 3.30593 | 3.53508 |
| | 0.8 | 2.19012 | 2.51761 | 2.79981 | 3.23320 | 3.45209 |
| | 1.2 | 2.16460 | 2.48076 | 2.75266 | 3.16937 | 3.37949 |
| | 1.6 | 2.14154 | 2.44790 | 2.71089 | 3.11318 | 3.31573 |
| | 2.0 | 2.12075 | 2.41855 | 2.67377 | 3.06351 | 3.25946 |
| | 2.4 | 2.10198 | 2.39226 | 2.64066 | 3.01939 | 3.20956 |
| | 2.8 | 2.08501 | 2.36863 | 2.61101 | 2.98000 | 3.16506 |
| | 3.2 | 2.06961 | 2.34732 | 2.58433 | 2.94467 | 3.12519 |

FRACTILES $W_{\alpha}$ RHO = 1.00

| DELTAS | | α:0.9625 | 0.9750 | 0.9825 | 0.9900 | 0.9925 |
|---|---|---|---|---|---|---|
| 3.6 | 0.0 | 2.21824 | 2.55889 | 2.85310 | 3.30593 | 3.53508 |
| | 0.4 | 2.19012 | 2.51761 | 2.79981 | 3.23320 | 3.45209 |
| | 0.8 | 2.16460 | 2.48076 | 2.75266 | 3.16937 | 3.37949 |
| | 1.2 | 2.14154 | 2.44790 | 2.71089 | 3.11318 | 3.31573 |
| | 1.6 | 2.12075 | 2.41855 | 2.67377 | 3.06351 | 3.25946 |
| | 2.0 | 2.10198 | 2.39226 | 2.64066 | 3.01939 | 3.20956 |
| | 2.4 | 2.08501 | 2.36863 | 2.61101 | 2.98000 | 3.16506 |
| | 2.8 | 2.06961 | 2.34732 | 2.58433 | 2.94467 | 3.12519 |
| | 3.2 | 2.05561 | 2.32801 | 2.56022 | 2.91283 | 3.08929 |
| | 3.6 | 2.04283 | 2.31046 | 2.53835 | 2.88400 | 3.05681 |
| 4.0 | 0.0 | 2.19012 | 2.51761 | 2.79981 | 3.23320 | 3.45209 |
| | 0.4 | 2.16460 | 2.48076 | 2.75266 | 3.16937 | 3.37949 |
| | 0.8 | 2.14154 | 2.44790 | 2.71089 | 3.11318 | 3.31573 |
| | 1.2 | 2.12075 | 2.41855 | 2.67377 | 3.06351 | 3.25946 |
| | 1.6 | 2.10198 | 2.39226 | 2.64066 | 3.01939 | 3.20956 |
| | 2.0 | 2.08501 | 2.36863 | 2.61101 | 2.98000 | 3.16506 |
| | 2.4 | 2.06961 | 2.34732 | 2.58433 | 2.94467 | 3.12519 |
| | 2.8 | 2.05561 | 2.32801 | 2.56022 | 2.91283 | 3.08929 |
| | 3.2 | 2.04283 | 2.31046 | 2.53835 | 2.88400 | 3.05681 |
| | 3.6 | 2.03113 | 2.29445 | 2.51844 | 2.85780 | 3.02732 |
| | 4.0 | 2.02040 | 2.27980 | 2.50024 | 2.83390 | 3.00042 |
| 6.0 | 0.0 | 2.08501 | 2.36863 | 2.61101 | 2.98000 | 3.16506 |
| | 0.4 | 2.06961 | 2.34732 | 2.58433 | 2.94467 | 3.12519 |
| | 0.8 | 2.05561 | 2.32801 | 2.56022 | 2.91283 | 3.08929 |
| | 1.2 | 2.04283 | 2.31046 | 2.53835 | 2.88400 | 3.05681 |
| | 1.6 | 2.03113 | 2.29445 | 2.51844 | 2.85780 | 3.02732 |
| | 2.0 | 2.02040 | 2.27980 | 2.50024 | 2.83390 | 3.00042 |
| | 2.4 | 2.01051 | 2.26634 | 2.48355 | 2.81200 | 2.97579 |
| | 2.8 | 2.00138 | 2.25394 | 2.46819 | 2.79188 | 2.95318 |
| | 3.2 | 1.99293 | 2.24248 | 2.45401 | 2.77333 | 2.93234 |
| | 3.6 | 1.98509 | 2.23186 | 2.44089 | 2.75617 | 2.91307 |
| | 4.0 | 1.97779 | 2.22200 | 2.42871 | 2.74027 | 2.89522 |
| | 6.0 | 1.94782 | 2.18165 | 2.37899 | 2.67549 | 2.82255 |
| 12.0 | 0.0 | 1.94782 | 2.18165 | 2.37899 | 2.67549 | 2.82255 |
| | 0.4 | 1.94287 | 2.17501 | 2.37083 | 2.66487 | 2.81065 |
| | 0.8 | 1.93821 | 2.16876 | 2.36313 | 2.65487 | 2.79944 |
| | 1.2 | 1.93380 | 2.16285 | 2.35587 | 2.64544 | 2.78888 |
| | 1.6 | 1.92962 | 2.15726 | 2.34901 | 2.63654 | 2.77890 |
| | 2.0 | 1.92567 | 2.15197 | 2.34252 | 2.62811 | 2.76946 |
| | 2.4 | 1.92192 | 2.14696 | 2.33637 | 2.62012 | 2.76052 |
| | 2.8 | 1.91835 | 2.14220 | 2.33052 | 2.61255 | 2.75204 |
| | 3.2 | 1.91496 | 2.13767 | 2.32497 | 2.60535 | 2.74398 |
| | 3.6 | 1.91173 | 2.13336 | 2.31969 | 2.59851 | 2.73632 |
| | 4.0 | 1.90865 | 2.12925 | 2.31466 | 2.59199 | 2.72902 |
| | 6.0 | 1.89518 | 2.11132 | 2.29269 | 2.56355 | 2.69720 |
| | 12.0 | 1.86764 | 2.07476 | 2.24801 | 2.50583 | 2.63267 |
| INFINITY | | 1.78046 | 1.95996 | 2.10836 | 2.32635 | 2.43238 |

FRACTILES $W_\alpha$ RHO = 1.00

| DELTAS | | $\alpha$:0.9950 | 0.9975 | 0.9990 | 0.9995 |
|---|---|---|---|---|---|
| 0.0 | 0.0 | 4.86450 | 5.75627 | 6.94914 | 7.85996 |
| 0.4 | 0.0 | 4.86138 | 5.75087 | 6.93996 | 7.84737 |
| | 0.4 | 4.82472 | 5.69148 | 6.84509 | 7.72198 |
| 0.8 | 0.0 | 4.82472 | 5.69148 | 6.84509 | 7.72198 |
| | 0.4 | 4.72331 | 5.53969 | 6.61944 | 7.43597 |
| | 0.8 | 4.57201 | 5.32709 | 6.32058 | 7.06875 |
| 1.2 | 0.0 | 4.72331 | 5.53969 | 6.61944 | 7.43597 |
| | 0.4 | 4.57201 | 5.32709 | 6.32058 | 7.06875 |
| | 0.8 | 4.40407 | 5.10076 | 6.01367 | 6.69891 |
| | 1.2 | 4.24208 | 4.88801 | 5.73151 | 6.36285 |
| 1.6 | 0.0 | 4.57201 | 5.32709 | 6.32058 | 7.06875 |
| | 0.4 | 4.40407 | 5.10076 | 6.01367 | 6.69891 |
| | 0.8 | 4.24208 | 4.88801 | 5.73151 | 6.36285 |
| | 1.2 | 4.09555 | 4.69867 | 5.48386 | 6.07006 |
| | 1.6 | 3.96663 | 4.53384 | 5.27023 | 5.81873 |
| 2.0 | 0.0 | 4.40407 | 5.10076 | 6.01367 | 6.69891 |
| | 0.4 | 4.24208 | 4.88801 | 5.73151 | 6.36285 |
| | 0.8 | 4.09555 | 4.69867 | 5.48386 | 6.07006 |
| | 1.2 | 3.96663 | 4.53384 | 5.27023 | 5.81873 |
| | 1.6 | 3.85434 | 4.39131 | 5.08669 | 5.60354 |
| | 2.0 | 3.75667 | 4.26800 | 4.92864 | 5.41872 |
| 2.4 | 0.0 | 4.24208 | 4.88801 | 5.73151 | 6.36285 |
| | 0.4 | 4.09555 | 4.69867 | 5.48386 | 6.07006 |
| | 0.8 | 3.96663 | 4.53384 | 5.27023 | 5.81873 |
| | 1.2 | 3.85434 | 4.39131 | 5.08669 | 5.60354 |
| | 1.6 | 3.75667 | 4.26800 | 4.92864 | 5.41872 |
| | 2.0 | 3.67150 | 4.16090 | 4.79188 | 5.25909 |
| | 2.4 | 3.59690 | 4.06739 | 4.67279 | 5.12032 |
| 2.8 | 0.0 | 4.09555 | 4.69867 | 5.48386 | 6.07006 |
| | 0.4 | 3.96663 | 4.53384 | 5.27023 | 5.81873 |
| | 0.8 | 3.85434 | 4.39131 | 5.08669 | 5.60354 |
| | 1.2 | 3.75667 | 4.26800 | 4.92864 | 5.41872 |
| | 1.6 | 3.67150 | 4.16090 | 4.79188 | 5.25909 |
| | 2.0 | 3.59690 | 4.06739 | 4.67279 | 5.12032 |
| | 2.4 | 3.53120 | 3.98525 | 4.56844 | 4.99887 |
| | 2.8 | 3.47303 | 3.91267 | 4.47640 | 4.89187 |
| 3.2 | 0.0 | 3.96663 | 4.53384 | 5.27023 | 5.81873 |
| | 0.4 | 3.85434 | 4.39131 | 5.08669 | 5.60354 |
| | 0.8 | 3.75667 | 4.26800 | 4.92864 | 5.41872 |
| | 1.2 | 3.67150 | 4.16090 | 4.79188 | 5.25909 |
| | 1.6 | 3.59690 | 4.06739 | 4.67279 | 5.12032 |
| | 2.0 | 3.53120 | 3.98525 | 4.56844 | 4.99887 |
| | 2.4 | 3.47303 | 3.91267 | 4.47640 | 4.89187 |
| | 2.8 | 3.42123 | 3.84816 | 4.39473 | 4.79701 |
| | 3.2 | 3.37487 | 3.79051 | 4.32184 | 4.71241 |

FRACTILES $W_\alpha$ RHO = 1.00

| DELTAS | | $\alpha$:0.9950 | 0.9975 | 0.9990 | 0.9995 |
|---|---|---|---|---|---|
| 3.6 | 0.0 | 3.85434 | 4.39131 | 5.08669 | 5.60354 |
| | 0.4 | 3.75667 | 4.26800 | 4.92864 | 5.41872 |
| | 0.8 | 3.67150 | 4.16090 | 4.79188 | 5.25909 |
| | 1.2 | 3.59690 | 4.06739 | 4.67279 | 5.12032 |
| | 1.6 | 3.53120 | 3.98525 | 4.56844 | 4.99887 |
| | 2.0 | 3.47303 | 3.91267 | 4.47640 | 4.89187 |
| | 2.4 | 3.42123 | 3.84816 | 4.39473 | 4.79701 |
| | 2.8 | 3.37487 | 3.79051 | 4.32184 | 4.71241 |
| | 3.2 | 3.33317 | 3.73872 | 4.25644 | 4.63655 |
| | 3.6 | 3.29549 | 3.69197 | 4.19746 | 4.56818 |
| 4.0 | 0.0 | 3.75667 | 4.26800 | 4.92864 | 5.41872 |
| | 0.4 | 3.67150 | 4.16090 | 4.79188 | 5.25909 |
| | 0.8 | 3.59690 | 4.06739 | 4.67279 | 5.12032 |
| | 1.2 | 3.53120 | 3.98525 | 4.56844 | 4.99887 |
| | 1.6 | 3.47303 | 3.91267 | 4.47640 | 4.89187 |
| | 2.0 | 3.42123 | 3.84816 | 4.39473 | 4.79701 |
| | 2.4 | 3.37487 | 3.79051 | 4.32184 | 4.71241 |
| | 2.8 | 3.33317 | 3.73872 | 4.25644 | 4.63655 |
| | 3.2 | 3.29549 | 3.69197 | 4.19746 | 4.56818 |
| | 3.6 | 3.26128 | 3.64958 | 4.14404 | 4.50627 |
| | 4.0 | 3.23011 | 3.61098 | 4.09543 | 4.44998 |
| 6.0 | 0.0 | 3.42123 | 3.84816 | 4.39473 | 4.79701 |
| | 0.4 | 3.37487 | 3.79051 | 4.32184 | 4.71241 |
| | 0.8 | 3.33317 | 3.73872 | 4.25644 | 4.63655 |
| | 1.2 | 3.29549 | 3.69197 | 4.19746 | 4.56818 |
| | 1.6 | 3.26128 | 3.64958 | 4.14404 | 4.50627 |
| | 2.0 | 3.23011 | 3.61098 | 4.09543 | 4.44998 |
| | 2.4 | 3.20159 | 3.57569 | 4.05102 | 4.39857 |
| | 2.8 | 3.17542 | 3.54332 | 4.01031 | 4.35146 |
| | 3.2 | 3.15130 | 3.51353 | 3.97287 | 4.30814 |
| | 3.6 | 3.12903 | 3.48602 | 3.93830 | 4.26817 |
| | 4.0 | 3.10839 | 3.46054 | 3.90632 | 4.23118 |
| | 6.0 | 3.02447 | 3.35709 | 3.77656 | 4.08125 |
| 12.0 | 0.0 | 3.02447 | 3.35709 | 3.77656 | 4.08125 |
| | 0.4 | 3.01074 | 3.34017 | 3.75537 | 4.05677 |
| | 0.8 | 2.99781 | 3.32426 | 3.73543 | 4.03376 |
| | 1.2 | 2.98563 | 3.30926 | 3.71665 | 4.01208 |
| | 1.6 | 2.97412 | 3.29511 | 3.69893 | 3.99162 |
| | 2.0 | 2.96324 | 3.28172 | 3.68217 | 3.97228 |
| | 2.4 | 2.95294 | 3.26905 | 3.66631 | 3.95397 |
| | 2.8 | 2.94316 | 3.25702 | 3.65127 | 3.93662 |
| | 3.2 | 2.93387 | 3.24561 | 3.63699 | 3.92015 |
| | 3.6 | 2.92505 | 3.23476 | 3.62342 | 3.90449 |
| | 4.0 | 2.91664 | 3.22443 | 3.61050 | 3.88959 |
| | 6.0 | 2.88000 | 3.17943 | 3.55425 | 3.82472 |
| | 12.0 | 2.80575 | 3.08832 | 3.44050 | 3.69360 |
| INFINITY | | 2.57583 | 2.80703 | 3.09024 | 3.29051 |

ABCDEFGHIJ–CM–8987654321